T0319288

Laminar Composites

To Ellen, Dan, Ben, and Jen

Laminar Composites

Second Edition

George H. Staab

AMSTERDAM • BOSTON • HEIDELBERG • LONDON
NEW YORK • OXFORD • PARIS • SAN DIEGO
SAN FRANCISCO • SINGAPORE • SYDNEY • TOKYO
Butterworth-Heinemann is an imprint of Elsevier

Butterworth-Heinemann is an imprint of Elsevier
The Boulevard, Langford Lane, Kidlington, Oxford OX5 1GB, UK
225 Wyman Street, Waltham, MA 02451, USA

Notices

Knowledge and best practice in this field are constantly changing. As new research and experience broaden our understanding, changes in research methods, professional practices, or medical treatment may become necessary.

Practitioners and researchers may always rely on their own experience and knowledge in evaluating and using any information, methods, compounds, or experiments described herein. In using such information or methods they should be mindful of their own safety and the safety of others, including parties for whom they have a professional responsibility.

To the fullest extent of the law, neither the Publisher nor the authors, contributors, or editors, assume any liability for any injury and/or damage to persons or property as a matter of products liability, negligence or otherwise, or from any use or operation of any methods, products, instructions, or ideas contained in the material herein.

Library of Congress Cataloging-in-Publication Data
A catalog record for this book is available from the Library of Congress

British Library Cataloguing-in-Publication Data
A catalogue record for this book is available from the British Library

ISBN: 978-0-12-802400-3

For information on all Butterworth-Heinemann publications
visit our website at http://store.elsevier.com/

Publisher: Matthew Deans
Acquisition Editor: David Jackson
Editorial Project Manager: Peter Gane
Production Project Manager: Melissa Read
Designer: Victoria Pearson

Printed and bound in the USA

Contents

Preface

As an introduction to composite materials, the texts that have been published present topics from either a materials science or applied mechanics viewpoint. This text presents the subject from an applied mechanics point of view and limits discussions to continuous fiber composites. Topics are developed at a level suitable for terminal undergraduate students and beginning graduate students. As a prerequisite, students should have completed a course in strength of materials. Additionally, they should be familiar with stress–strain relations for isotropic materials and load–stress relationships. The philosophy behind this text is that it should be fundamentally simple enough for a senior undergraduate to understand and apply the concepts forwarded, while at the same time not too trivial for a beginning graduate student.

The scope of this text is limited to topics associated with the analysis and design of continuous fiber-laminated composite materials. Lamina and laminate analysis is presented with a blend of theoretical developments and examples. The analysis of laminated composites relies heavily on concepts developed in undergraduate statics and mechanics of materials courses. Examples presented in this text require an understanding of free-body diagrams and analysis techniques introduced in undergraduate mechanics courses. Experimental techniques applicable to defining the constitutive relationships for orthotropic lamina are presented, as are failure theories for orthotropic materials.

After establishing the stress–strain relationships, discussing special testing considerations, and covering failure criteria for orthotropic lamina, classical lamination theory is developed. An attempt has been made to present material in an easy-to-follow, logical manner. Loading conditions involving mechanical, thermal, and hygral loads are considered after the effect of each is discussed and developed independently.

Chapters on beams, plates, and shells have been added to the original text. The chapter on beams should prove useful to undergraduates. Beams are a fundamental structural element and this chapter is an extension of what undergraduates learned in their introductory strength of materials courses. Although the plates and shells chapters may be too difficult for undergraduates, they have been added for completeness, and to serve as a starting point for students interested in these topics. These two chapters are necessarily brief since the solutions to many types plate and shell problems require numerical techniques beyond the scope of this text.

Many of the topics covered in this text are a compilation of the topics covered in preceding books, such as *Primer on Composite Materials: Analysis* by Ashton, Halpin, and Petit; *Mechanics of Composite Materials* by Jones; *Introduction to Composite Materials* by Tsai and Hahn; *Experimental Mechanics of Fiber Reinforced*

Composite Materials by Whitney, Daniel, and Pipes; and *The Behavior of Structures Composed of Composite Materials* by Vinson and Sierakowski; *Mechanics of Laminated Composite Plates Theory and Analysis* and *Mechanics of Laminated Composite Plates and Shells* by J.N. Reddy. These texts served as the foundation upon which this text was developed. The present text incorporates many of the standard equations and formulations found in the preceding texts and builds upon them.

The original edition of this text contained an appendix on matrix arithmetic and a section containing additional references. Due to the advances in personal computing, it was felt that a section on matrix arithmetic is no longer needed. Along similar lines, the advancement of web search engines makes a section containing additional references somewhat obsolete. Therefore, this section was similarly deleted.

I am deeply thankful to my longtime friend and colleague Dr. H.R. Busby, emeritus professor of Mechanical and Aerospace Engineering at The Ohio State University. His friendship, helpful comments, suggestions, and notes that we used to develop the composite materials courses at OSU formed the basis of this manuscript. Finally, I wish to thank my wife, Ellen, for her long-term patience and eventual understanding of how engineers are.

Answers to the Problems throughout this book are available on the book's companion website. Go online to access it at: http://booksite.elsevier.com/9780128024003

Introduction to composite materials

1

1.1 Historic and introductory comments

In the most general of terms, a composite is a material which consists of two or more constituent materials or phases. Traditional engineering materials (steel, aluminum, etc.) contain impurities which can represent different phases of the same material and fit the broad definition of a composite, but are not considered composites because the elastic modulus or strength of each phase are nearly identical. The definition of a composite material is flexible and can be augmented to fit specific requirements. In this text, a composite material is considered to be the one which contains two or more distinct constituents with significantly different macroscopic behavior and a distinct interface between each constituent (on the microscopic level). This includes the continuous fiber-laminated composites of primary concern herein, as well as a variety of composites not specifically addressed.

Composite materials have been in existence for many centuries. No record exists as to when people first started using composites. Some of the earliest records of their use date back to the Egyptians, who are credited with the introduction of plywood, paper mache, and the use of straw in mud for strengthening bricks. Similarly, the ancient Inca and Mayan civilizations used plant fibers to strengthen bricks and pottery. Swords and armor were plated to add strength in medieval times. An example is the Samurai sword, which was produced by repeated folding and reshaping to form a multilayered composite (it is estimated that several million layers could have been used). Eskimos use moss to strengthen ice in forming igloos. Similarly, it is not uncommon to find horse hair in plaster for enhanced strength. The automotive industry introduced large scale use of composites with the 1953 Chevrolet Corvette. All of these are examples of man-made composite materials. Bamboo, bone, and celery are examples of cellular composites which exist in nature. Muscle tissue is a multidirectional fibrous laminate. There are numerous other examples of both natural and man-made composite materials.

The structural materials most commonly used in design can be categorized into four primary groups: metals, polymers, composites, and ceramics. These materials have been used to various degrees since the beginning of time. Their relative importance to various societies throughout history has fluctuated. Ashby [1] presents a chronological variation of the relative importance of each group from 10,000 BC, and extrapolates their importance through the year 2020. The information contained in Ref. [1] has been partially reproduced in Figure 1.1. The importance of composites experienced steady growth since about 1960, and is projected to increase in importance through the next several decades. The relative importance of each group of materials is not associated with any specific unit of measure (net tonnage, etc.).

Laminar Composites. http://dx.doi.org/10.1016/B978-0-12-802400-3.00001-5

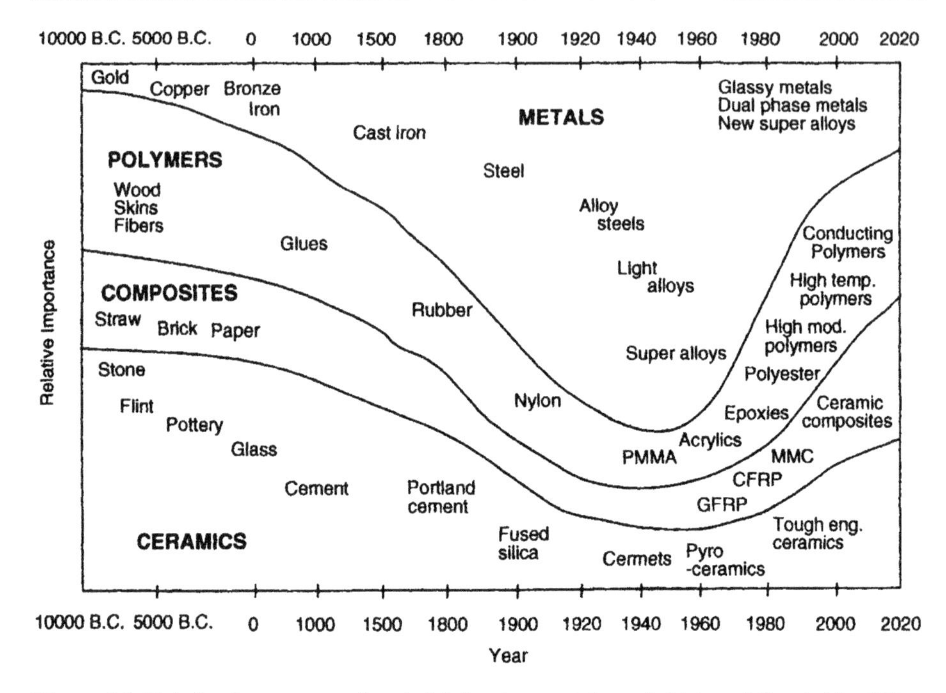

Figure 1.1 Relative importance of material development through history (after Ashby [1]).

As with many advances throughout history, advances in material technology (from both a manufacturing and analysis viewpoint) typically have its origins in military applications. Subsequently this technology filters into the general population and alters many aspects of society. This is most recently seen in the marked increase in relative importance of such structural materials such as composites starting around 1960, when the race for space dominated many aspects of research and development. Similarly, the Strategic Defense Initiative (SDI) program in the 1980s prompted increased research activities in the development of new material systems. Advances in material systems research, manufacturing techniques, and the reduced cost of raw materials have made the use of composite materials a common practice in most aspects of everyday life. The use of composites has grown so much that Roberts [2] estimates the global demand for carbon fibers alone in 2015 will exceed 67,000 metric tons (147,400,000 lbs).

The composites generally used in structural applications are best classified as high performance. They are typically made from synthetic materials, have high strength to weight ratios, and require controlled manufacturing environments for optimum performance. The aircraft industry uses composites to meet performance requirements beyond the capabilities of metals. The Boeing 757, for example, uses approximately 760 ft^3 of composites in its body and wing components, with an additional 361 ft^3 used in rudder, elevator, edge panels, and tip fairings. An accurate breakdown of specific components and materials can be found in Ref. [3]. The B-2 bomber contains carbon and glass fibers, epoxy resin matrices, high-temperature polyimides as well as other materials in more than 10,000 composite components. It is considered to be one of the first major steps in making aircraft structures primary from composites. Composites are also used in race cars, tennis rackets, golf clubs, and other sports and leisure products. Although composite materials technology has grown rapidly, it is not fully developed. New combinations of fiber/resin systems, and even new materials are constantly being developed. The best one can hope to do is identify the types of composites that exist through broad characterizations and classifications.

1.2 Characteristics of a composite material

The constituents of a composite are generally arranged so that one or more discontinuous phase is embedded in a continuous phase. The discontinuous phase is termed the *reinforcement* and the continuous phase is the *matrix*. An exception to this is rubber particles suspended in a rigid rubber matrix, which produces a class of materials known as rubber-modified polymers. In general, the reinforcements are much stronger and stiffer than the matrix. Both constituents are required, and each must accomplish specific tasks if the composite is to perform as intended.

A material is generally stronger and stiffer in fiber form than in bulk form. The number of microscopic flaws which act as fracture initiation sites in bulk materials are reduced when the material is drawn into a thinner section. In fiber form, the material will typically contain very few microscopic flaws from which cracks may initiate to produce catastrophic failure. Therefore, the strength of the fiber is greater than that of the bulk material. Individual fibers are hard to control and form into useable components. Without a binder material to separate them, they can become knotted, twisted, and hard to separate. The binder (matrix) material must be continuous and surround each fiber so that they are kept distinctly separate from adjacent fibers and the entire material system is easier to handle and work with.

The physical and mechanical properties of composites are dependent on the properties, geometry, and concentration of the constituents. Increasing the volume content of reinforcements can increase the strength and stiffness of a composite to a point. If the volume content of reinforcements is too high there will not be enough matrix to keep them separate and they can become tangled. Similarly, the geometry of individual reinforcements and their arrangement within the matrix can affect the performance of a composite. There are many factors to be considered when designing with composite materials. The type of reinforcement and matrix, the geometric arrangement and volume fraction of each constituent, the anticipated mechanical loads, the operating environment for the composite, etc., must all be taken into account.

Analysis of composites subjected to various mechanical, thermal, and hygral conditions is the main thrust of this text. Discussions are limited to continuous fiber-laminated composites. In introductory strength of materials, the constitutive relationship between stress and strain was established for homogeneous isotropic materials as Hooke's law. A composite material is analyzed in a similar manner, by establishing a constitutive relationship between stress and strain.

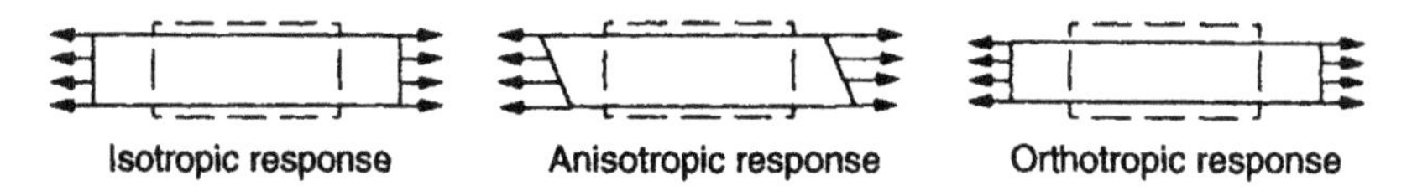

Figure 1.2 Typical material responses for isotropic, anisotropic, and orthotropic materials subjected to axial tension.

Isotropic, homogeneous materials (steel, aluminum, etc.) are assumed to be uniform throughout, and have the same elastic properties in all directions. Upon application of a uniaxial tensile load an isotropic material deforms in a manner similar to that indicated in Figure 1.2 (the dashed lines represent the undeformed specimen). Assuming a unit width and thickness for the specimen, the transverse in-plane and out-of-plane displacements are the same. Unlike conventional engineering materials, a composite material is generally nonhomogeneous and does not behave as an isotropic material. Most composites behave as either an *anisotropic* or *orthotropic* material.

The material properties of an anisotropic material are different in all directions. There is typically a coupling of extension and shear deformation under conditions of uniaxial tension. The response of an anisotropic material subjected to uniaxial tension is also illustrated in Figure 1.2. There are varying degrees of anisotropic material behavior, and the actual deformation resulting from applied loads depends on the material.

The material properties of an orthotropic material are different in three mutually perpendicular planes, but there is generally no shear-extension coupling as with an anisotropic material. The transverse in-plane and out-of-plane displacements are not typically the same since Poisson's ratio is different in these two directions. Figure 1.2 also illustrates orthotropic material response. Although it appears similar to that of an isotropic material, the magnitude of the in-plane and out-of-plane displacements are different.

1.3 Composite materials classifications

Composite materials are usually classified according to the type of reinforcement used. Two broad classes of composites are fibrous and particulate. Each has unique properties and application potential, and can be subdivided into specific categories as discussed below.

Fibrous: A fibrous composite consists of either continuous (long) or chopped (whiskers) fibers suspended in a matrix material. Both continuous fibers and whiskers can be identified from a geometric viewpoint:

Continuous fibers. A continuous fiber is geometrically characterized as having a very high length to diameter ratio. They are generally stronger and stiffer than bulk material. Fiber diameters generally range between 0.00012 and 0.0074-in. (3–200 μm), depending upon the fiber [4].

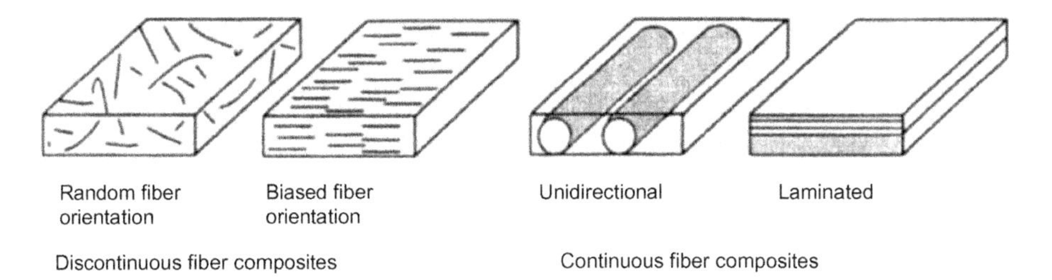

Figure 1.3 Schematic representations of fibrous composites.

Whiskers. A whisker is generally considered to be a short, stubby fiber. It can be broadly defined as having a length to diameter ratio of $5 < l/d << 1000$ and beyond [5]. Whisker diameters generally range between 0.787 and 3937 μin. (0.02 and 100 μm).

Composites in which the reinforcements are discontinuous fibers or whiskers can be produced so that the reinforcements have either a random or biased orientation. Material systems composed of discontinuous reinforcements are considered single-layer composites. The discontinuities can produce a material response which is anisotropic, but in many instances the random reinforcements produce nearly isotropic composites. Continuous fiber composites can be either single layer or multilayered. The single-layer continuous fiber composites can be either unidirectional or woven, and multilayered composites are generally referred to as laminates. The material response of a continuous fiber composite is generally orthotropic. Schematics of both types of fibrous composites are shown in Figure 1.3.

Particulate: A particulate composite is characterized as being composed of particles suspended in a matrix. Particles can have virtually any shape, size, or configuration. Examples of well-known particulate composites are concrete and particle board. There are two subclasses of particulates; flake and filled/skeletal:

Flake. A flake composite is generally composed of flakes with large ratios of planform area to thickness, suspended in a matrix material (e.g., particle board).
Filled/skeletal. A filled/skeletal composite is composed of a continuous skeletal matrix filled by a second material. For example, a honeycomb core filled with an insulating material.

The response of a particulate composite can be either anisotropic or orthotropic. They are used for many applications in which strength is not a significant component of the design. A schematic of several types of particulate composites is shown in Figure 1.4.

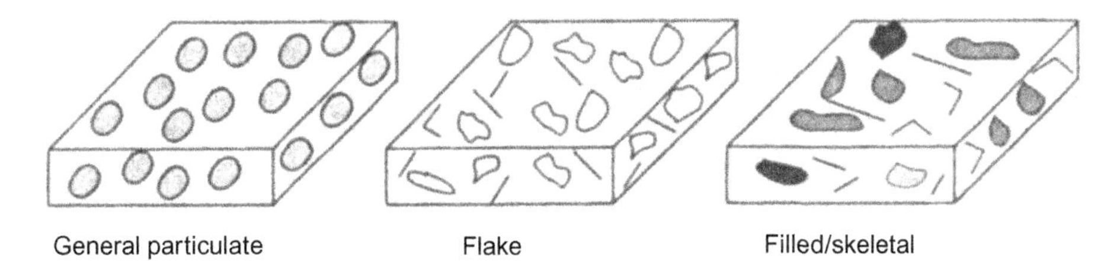

Figure 1.4 Schematic representations of particulate composites.

1.4 Fundamental composite material terminology

Some of the more prominent terms used with composite materials are defined below. A more detailed list can be found in the Glossary as well as Ref. [6].

Lamina. A lamina is a flat (or sometimes curved) arrangement of unidirectional (or woven) fibers suspended in a matrix material. A lamina is generally assumed to be orthotropic, and its thickness depends on the material from which it is made. For example, a graphite/epoxy (graphite fibers suspended in an epoxy matrix) lamina may be on the order of 0.005 in. (0.127 mm) thick. For the purpose of analysis, a lamina is typically modeled as having one layer of fibers through the thickness. This is only a model and not a true representation of fiber arrangement. Both unidirectional and woven lamina are schematically shown in Figure 1.5.

Reinforcements. Reinforcements are used to make the composite structure or component stronger. The most commonly used reinforcements are boron, glass, graphite (often referred to as simply carbon), and Kevlar, but there are other types of reinforcements such as alumina, aluminum, silicon carbide, silicon nitride, and titanium.

Fibers. Fibers are a special case of reinforcements. They are generally continuous and have diameters ranging from 120 to 7400 μin. (3–200 μm). Fibers are typically linear elastic or elastic-perfectly plastic, and are generally stronger and stiffer than the same material in bulk form. The most commonly used fibers are boron, glass, carbon, and Kevlar. Fiber and whisker technology is continuously changing [4,5,7].

Matrix. The matrix is the binder material which supports, separates, and protects the fibers. It provides a path by which load is both transferred to the fibers and redistributed among the fibers in the event of fiber breakage. The matrix typically has a lower density, stiffness, and strength than the fibers. Matrices can be brittle, ductile, elastic, or plastic. They can have either linear or nonlinear stress–strain behavior. In addition, the matrix material must be capable of being forced around the reinforcement during some stage in the manufacture of the composite. Fibers must often be chemically treated in order to assure proper adhesion to the matrix. The most commonly used matrices are polymeric (PMC), ceramic (CMC), metal (MMC), carbon, and

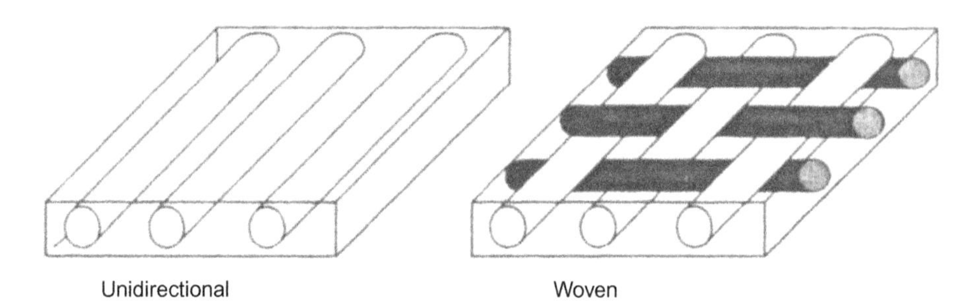

Figure 1.5 Schematic representation of unidirectional and woven composite lamina.

ceramic. Each has special appeal and usefulness, as well as limitations. Richardson [8] presents a comprehensive discussion of matrices, which guided the presentation below.

1. *Carbon matrix*. High heat capacity per unit weight. They have been used as rocket nozzles, ablative shields for reentry vehicles, clutch, and brake pads for aircraft.
2. *Ceramic matrix*. A ceramic matrix is usually brittle. Carbon, ceramic, metal, and glass fibers are typically used with ceramic matrices in areas where extreme environments (high temperatures, etc.) are anticipated.
3. *Glass matrix*. Glass and glass-ceramic composites usually have an elastic modulus much lower than the reinforcement. Carbon and metal oxide fibers are the most common reinforcements with glass matrix composites. The best characteristics of glass or ceramic matrix composites are their strength at high service temperatures. The primary applications of glass matrix composites are for heat-resistant parts in engines, exhaust systems, and electrical components.
4. *Metal matrix*. A metal matrix is especially good for high temperature use in oxidizing environments. The most commonly used metals are iron, nickel, tungsten, titanium, magnesium, and aluminum. There are three classes of metal matrix composites:

   Class I. The reinforcement and matrix are insoluble (there is little chance that degradation will affect service life of the part). Reinforcement/matrix combinations in this class include tungsten or alumina/copper, BN-coated B or boron/aluminum and boron/magnesium.

   Class II. The reinforcement/matrix exhibits some solubility (generally over a period of time and during processing) and the interaction will alter the physical properties of the composite. Reinforcement/matrix combinations included in this class are carbon or tungsten/nickel, tungsten/columbium, and tungsten/copper (chromium).

   Class III. The most critical situations in terms of matrix and reinforcement are in this class. The problems encountered here are generally of a manufacturing nature and can be solved through processing controls. Within this class the reinforcement/matrix combinations include alumina or boron or silicon carbide/titanium, carbon or silica/aluminum, and tungsten/copper (titanium).
5. *Polymer matrix*. Polymeric matrices are the most common and least expensive. They are found in nature as amber, pitch, and resin. Some of the earliest composites were layers of fiber, cloth, and pitch. Polymers are easy to process, offer good mechanical properties, generally wet reinforcements well, and provide good adhesion. They are a low-density material. Due to low processing temperatures many organic reinforcements can be used. A typical polymeric matrix is either viscoelastic or viscoplastic, meaning it is affected by time, temperature, and moisture. The terms thermoset and thermoplastic are often used to identify a special property of many polymeric matrices.

   Thermoplastic. A thermoplastic matrix has polymer chains which are not cross-linked. Although the chains can be in contact, they are not linked to each other. A thermoplastic can be remolded to a new shape when it is heated to approximately the same temperature at which it was formed.

   Thermoset. A thermoset matrix has highly cross-linked polymer chains. A thermoset cannot be remolded after it has been processed. Thermoset matrices are sometimes used at higher temperatures for composite applications.

Laminate. A laminate is a stack of lamina, as illustrated in Figure 1.6, oriented in a specific manner to achieve a desired result. Individual lamina is bonded together by a curing procedure which depends on the material system used. The mechanical

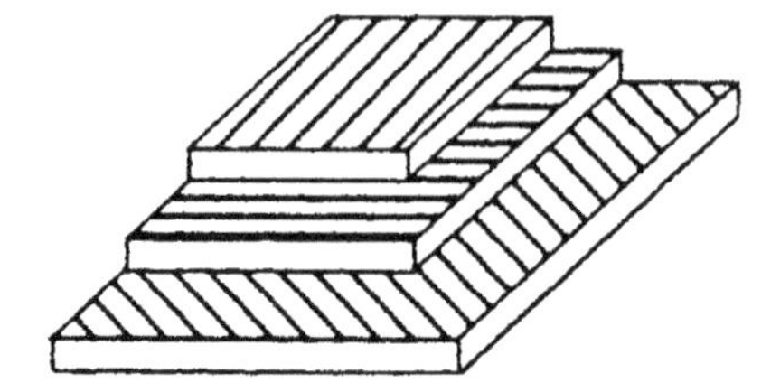

Figure 1.6 Schematic of a laminated composite.

response of a laminate is different than that of the individual lamina which forms it. The laminate's response depends on the properties of each lamina, as well as the order in which the lamina are stacked.

Hybrid. A hybrid composite typically consists of multiple constituents such as glass and carbon fibers suspended in an appropriate matrix. There are several types of hybrid composites which can be characterized as:

1. Interply or tow by tow, in which tows of the two or more constituent types of fiber are mixed in a regular or random manner.
2. Sandwich hybrids, also known as core–shell, in which one material is sandwiched between two layers of another.
3. Interply or laminated, where alternate layers of the two (or more) materials are stacked in a regular manner.
4. Intimately mixed hybrids, where the constituent fibers are made to mix as randomly as possible so that no overconcentration of any one type is present in the material.
5. Other kinds, such as those reinforced with ribs, pultruded wires, thin veils of fiber, or combinations of the above.

Micromechanics. A specialized area of composites involving a study of the interaction of constituent materials on the microscopic level. This study is generally conducted by use of a mathematical model describing the response of each constituent material.

Macromechanics. A study of the overall response of a lamina (or laminate) in which the effects of constituent materials are averaged to achieve an apparent response on the macroscopic level.

1.5 Advantages afforded by composite materials

Materials are often expected to perform multiple tasks. An example is a smart material, in which sensors embedded in the material are used to determine conditions within the material. The use of an embedded sensor to define real time conditions in a structure is beneficial in predicting critical component life, or identifying when reassigned parameters reach a critical stage and require specific action. One approach to developing *smart structures* is to use fiber optics embedded in a composite. They can be directly embedded into the structure during manufacture, and are somewhat protected from damage. The general nature of a smart structure raises issues not generally considered with conventional laminates. Journal articles, conference proceedings, and books dedicated to smart structures are available. Several pertinent references to smart materials and structures are [9–11].

Composites offer a wide range of characteristics suitable for many design requirements. The apparent elastic modulus and tensile strength for several types of fibers shown in Table 1.1 indicates a wide range of possible material responses, which can be altered by changing the procedure used to develop each fiber. Of the fiber properties shown, carbon offers the most variety. Agarwal and Broutman [12] identified a total of 38 carbon fibers with elastic moduli and strength ranges of $4 \times 10^6 - 88 \times 10^6$ psi (28–607 GPa) and 140–450 ksi (966–3105 MPa), respectively, in 1980. By 1986, there were 17 worldwide manufacturers producing 74 different grades of high-modulus carbon fibers according to Ref. [7]. Between 1985 and 1998, the net tonnage of carbon fiber used in the aircraft industry alone grew at a rate of approximately 3800 tons/year [13]. The three major areas of carbon fiber use are typically considered to be aerospace, consumer, and industrial. Using on current trends in the usage of carbon fibers (woven, prepreg, raw, etc.) Roberts [2] projected the net worldwide usage of carbon fibers through 2020 to be as depicted in Figure 1.7.

The high elastic modulus and strength in Table 1.1 do not reveal the actual behavior of a composite once the fibers have been suspended in a matrix. The properties of the matrix also contribute to the strength and stiffness of the material system. Since the matrix is generally much weaker and less stiff than the fiber, the composite will not be as strong or stiff as the fibers themselves. In addition, the properties cited above refer only to the fiber direction. In a composite there are three directions to consider; one parallel to the fibers (longitudinal direction) and two perpendicular to the fibers (transverse directions). The properties in the longitudinal direction are superior to those in the transverse directions, in which the matrix is the dominant constituent.

Table 1.1 Apparent elastic modulus and strength of selected fibers (from Refs. [5,12])

Fiber	Tensile modulus Msi (GPa)	Tensile strength ksi (MPa)
Beryllium	35 (240)	189 (1300)
Boron	56 (385)	405 (2800)
Carbon: high tenacity, high modulus	10.2–87.5 (70–600)	254–509 (1750–3500)
PAN (polyacrylonitrite)	29.2–56.9 (200–390)	305–494 (2100–3400)
Pitch (mesophase)	24.8–100.6 (170–690)	189–349 (1300–2400)
Glass:		
E-glass	10.5 (72.4)	508 (3500)
S-glass	12.4 (85.5)	68 (4600)
M-glass	15.9 (110)	508 (3500)
Kevlar-29	8.6 (59)	384 (2640)
Kevlar-49	18.9 (128)	406 (2800)
Silica	16.5 (72.4)	482 (5800)
Tungsten	60 (414)	610 (4200)

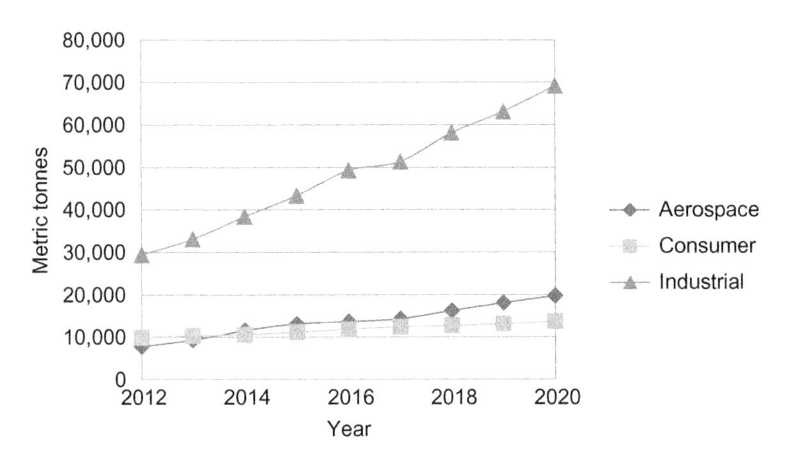

Figure 1.7 Worldwide usage of carbon fibers from 2012 to 2020 (after Roberts [2]).

Composites are attractive for many design considerations. Since the fiber direction can be altered throughout the thickness of a laminate, a components response can be tailored to fit specific requirements. Designing a component with a zero coefficient of thermal expansion can be an important consideration for space applications where one side of a structure can be exposed to extremely high temperatures, while the other side experiences extremely low temperatures.

A composite offers strength to weight and stiffness to weight ratios superior to those of conventional materials. Figure 1.8 illustrates this for several composite material systems in terms of strength and stiffness, respectively. As seen in these figures a wide range of specific strength (σ_u/ρ) and specific modulus (E/ρ) are available. In some instances

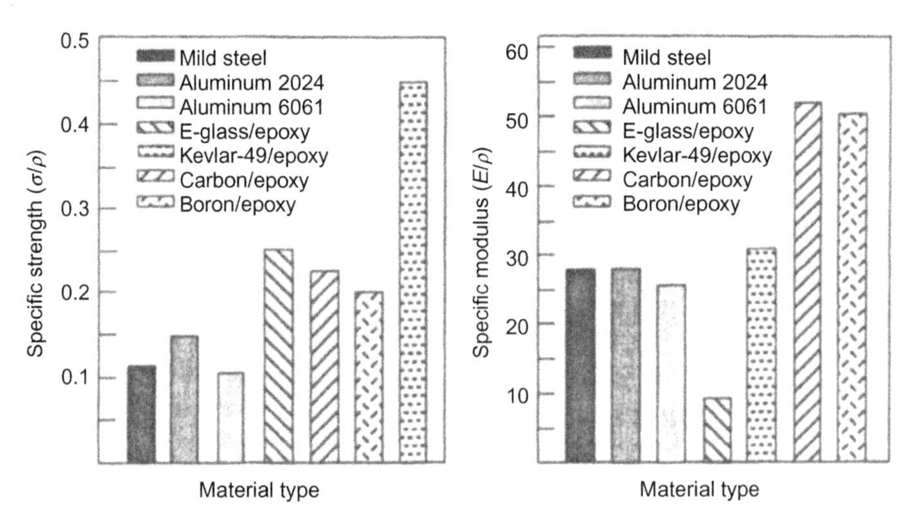

Figure 1.8 Specific strength and stiffness comparison for selected composite and conventional bulk materials.

strength may be a primary consideration, while in others the stiffness is more important. In all cases shown the specific strength for the composite material systems is better than for the conventional materials, while the specific modulus is not always superior.

Composites offer better structural response characteristic, but they are expensive. Both the raw material and many manufacturing techniques used with composites are more expensive than for conventional materials. There are many possible uses for composites not discussed herein. A sampling of applications and techniques ranging from automotive parts to orthopedic applications are found in Drozda [13].

1.6 Selected manufacturing techniques for composites

There are many manufacturing, fabricating, processing, and forming processes for composites. In 1987, Richardson [8] identified seven major processes by which polymer matrix composites are formed: (1) molding, (2) casting, (3) thermoforming, (4) expansion, (5) coating, (6) fabrication, and (7) radiation. Within each of these processes different techniques can be used. For example, in the category of molding, there are nine subprocesses containing sixteen subsets, including injection; coinjection; reaction injection; compression; caldenering; blow (extrusion-blow molding, injection-blow molding, stretch-blow molding, and multilayer-blow molding); extrusion; laminating; and reinforcing (match-die molding, hand layup, spray-up molding, vacuum bag molding, filament winding, continuous reinforcing, cold molding, cold forming/stamping, sintering, liquid-resin molding, vacuum-injection molding, and thermal expansion resin transfer). The other major processes for forming composites also contain subprocesses.

The process selected for the production of a component depends on many variables, and can influence the final product. Continuous fiber composites can be processed in various manners. The technique used depends to a large extent on the type of matrix used. Some techniques for polymer matrix composites are not appropriate for metal, ceramic, or thermoplastic matrix composites. Vinson and Sierakowski [14] present some of the primary techniques for processing metal and nonmetal matrix as well as short-fiber composites. Table 1.2 provides a synopsis of some procedures applicable to continuous fiber composites, arranged according to the type of matrix. The diverse nature of composite material systems dictates that no single procedure can be applied to all composites. The discussions in this section are limited to selected procedures for continuous fiber polymer matrix composites. Discussions of fabrication processes for metal matrix, ceramic matrix, and thermoplastic matrix composites can be found in other references, including [3,15–20]. Similarly, sheet-molding compounds (SMC) and bulk-molding compounds (BMC), which are composed of chopped fibers, are not addressed herein.

The matrix of a polymeric composite can be either a thermoplastic or thermoset. Both types are generally available as a prepreg tape, which means the fibers have been precoated with the resin and arranged on a backing sheet in either unidirectional or woven configurations. A thermoset must be stored in a refrigeration system since

Table 1.2 **Common fabrication and curing processes for continuous fiber-laminated composite materials (from Ref. [14])**

Processing technique	Nonmetal matrix composites	Metal matrix composites
Hand layup	X	
Vacuum bag/autoclave	X	
Matched die molding	X	X
Filament winding	X	X
Pressure and roll bonding		X
Plasma spraying	X	X
Powder metallurgy		X
Liquid infiltration	X	X
Coextrusion		X
Controlled solidification		X
Rotational molding		X
Pultrusion		X
Injection molding		X
Centrifugal casting		X
Pneumatic impaction		X
Thermoplastic molding	X	
Resin transfer molding (RTM)	X	

the resin is partially cured, and exposure to room temperature for extended periods can complete the curing process. A thermoplastic may be stored at room temperature until the matrix is melted during the final stages of processing. Prepreg tape allows fabricators flexibility since it eliminates the concern of mixing resin components in the correct proportions and subsequently combining resin and fiber.

Continuous fiber polymer matrix composites are most effectively cured at elevated temperatures and pressures. Prior to the final cure procedure, the fiber arrangement through the thickness of the composite must be defined using either prepreg tape or individual fibers coated with resin. Two approaches for doing this are hand layup and filament winding. Hand layup is generally used for sample preparation in laboratory applications involving prepreg tape. It is also used in areas where a tailored laminate is required. In the aircraft industry, it is not uncommon for several plies to be built-up or dropped-off in specific areas of a large structural component where strength or stiffness requirements vary. The hand layup procedure consists of using a prepreg tape or woven mat to individually position each ply. In some applications, the woven mat form of composite is placed directly into a mold and coated with resin prior to curing. In using this procedure, a *bleeder* cloth should be used to absorb any extra resin which may be squeezed from the component when pressure is applied during the cure procedure. In the case of a neat resin system, a breather (without a bleeder) is used. In addition, a peel-ply or mold release agent which allows for easy removal of the specimen after curing is needed. After the laminate is layedup, it is cured (generally using an autoclave or lamination press).

Filament winding is perhaps the oldest fabrication procedure used for continuous fiber composites. Either individual fibers or tapes may be used with this process. A schematic of a generalized winding system for individual fibers is shown in Figure 1.9. The fiber is initially placed on a spool, and fed through a resin bath. The resin-impregnated fiber is then passed through a feeder arm which is free to move at various speeds in the transverse direction. The fiber is then wound onto a mandrel (or a form, which in general is removed after forming is complete). The mandrel is turned by a lathe at a specified rate. The lathe may also be rotated as indicated in Figure 1.9. By controlling both the feeding and rotational rates, various ply orientations can be achieved. Specific names are given to the type of winding associated with different operating speeds (v_1, v_2, ω_1, ω_2) of the systems components. Table 1.3 identifies each type of winding and indicates which combinations of angular and linear velocity produce them (note that an entry of 0 implies no motion). By proper control of the motion parameters, a filament wound vessel can contain helical, circumferential, and polar windings, each illustrated in Figure 1.10. Braid-wrap and loop-wrap windings are also possible, but cannot be achieved using the system shown in Figure 1.9. Winding procedures involving prepreg tapes are also possible. Examples of continuous helical, normal-axial, and rotating mandrel wraps are presented in

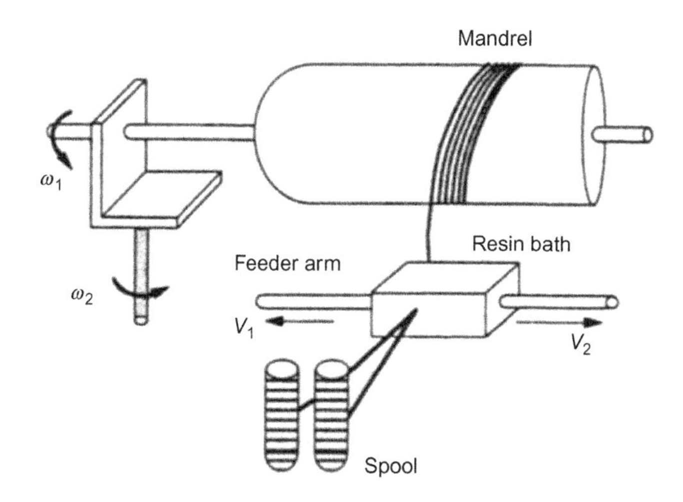

Figure 1.9 Schematic of a generalized filament winding operation.

Table 1.3 Basic filament winding patterns as a function of winding system motions defined in Figure 1.9

Pattern type	v_1	v_2	ω_1	ω_2
Helical wind	X	X	X	0
Circumferential wind	X	0	X	0
Polar wind	0	0	X	X

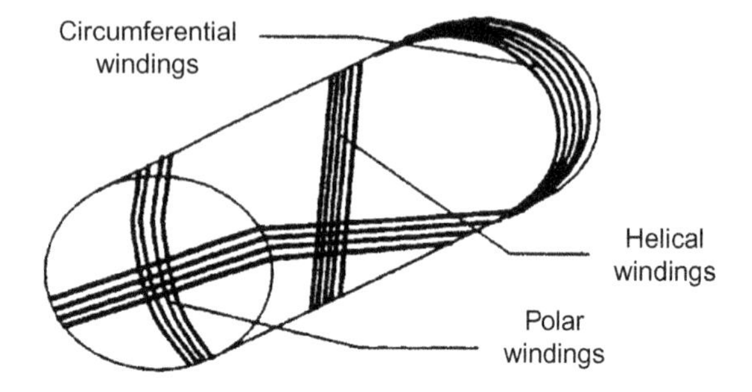

Figure 1.10 Polar, circumferential, and helical windings.

Ref. [7]. After winding is complete, the part is cured. Pressure vessels, rocket motor cases, power transmission shafts, and chemical tanks are well suited for filament winding.

After layup, polymer matrix composites must be cured using specified combinations of temperature and pressure. The most commonly recommended curing procedures are a vacuum bag, autoclave, or lamination press. In the vacuum bag procedure, the specimen is first layedup, then peel and bleeder (or breather) plies are placed around the specimen, and the entire unit is placed on a tool plate. A boundary support (or dam) is placed around the periphery of the specimen, and a pressure plate is placed over the specimen. The pressure bag is then sealed around the tool plate. A vacuum is drawn and the specimen is cured. A schematic of the pressure bag assembly is shown in Figure 1.11. An autoclave curing procedure is somewhat different, since the autoclave can serve both functions of temperature control as well as creating a vacuum. An autoclave cure often uses a vacuum bag too.

During the final cure process, temperature and pressure must be controlled for specified time periods, and at specified rates. A schematic of a cure cycle is presented in Figure 1.12. The specifications of temperature pressure are typically defined by the material manufacturers. Failure to follow recommended procedures can result in composites which are not structurally adequate. They may contain an unacceptable number of voids, or regions in which intraply adhesion is weak.

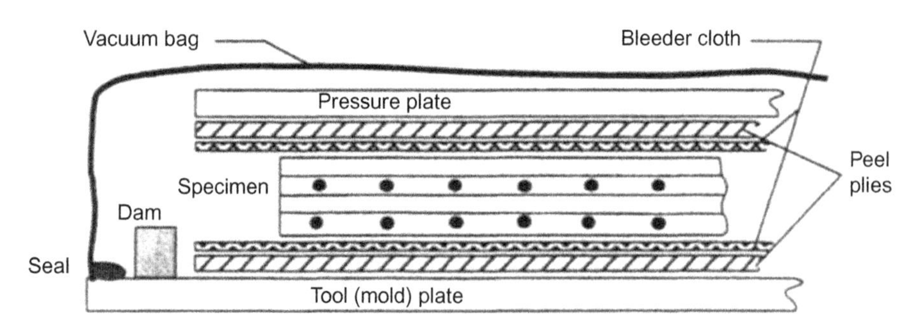

Figure 1.11 Schematic of vacuum bag method.

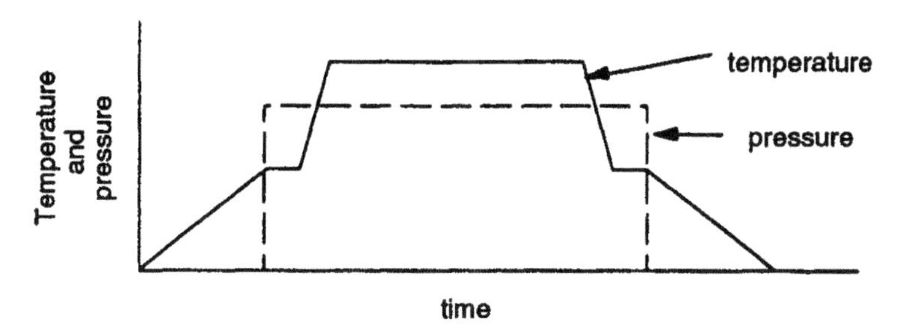

Figure 1.12 Pressure–temperature–time relations for a possible cure cycle.

Many other processing methods for composite materials are available, including reaction injection molding (RIM), vacuum-assisted resin transfer molding (VARTM), resin film infusion (RFI), automated fiber placement (AFP), and automated tape layup (ATL). The ones presented herein illustrate some of the procedures applicable to continuous fiber-laminated composites. Other composite material systems require different processing methods. The selection of a specific process is highly dependent on the intended application of the component being made. It is best to consult experts in manufacturing and processing before selecting a process.

References

[1] Ashby MF. Technology in the 1990s: advanced materials and predictive design. Philos Trans R Soc Lond A 1987;322:393–407.
[2] Roberts T. Carbon fiber industry worldwide 2011 to 2020. UK: Materials Technology Publications; 2011.
[3] Reinhart TJ, editor. Engineered materials handbook. Composites, vol. 1. USA: ASM International; 1987.
[4] Watt W, Perov BV. In: Handbook of composites. Strong fibers, vol. 1. Amsterdam: North-Holland; 1985.
[5] Evans CC. Whiskers. UK: Mills & Boon Limited; 1972.
[6] Tsai SW. Composites design 1986. USA: Think Composites; 1986.
[7] Plastics and Rubber Institute, editor. Carbon fibers technology, uses and prospects. London, England: Noyes Publications; 1986.
[8] Richardson T. Composites: a design guide. USA: Industrial Press Inc.; 1987.
[9] Rogers CA, editor. Smart materials, structures, and mathematical issues. USA: Technomic Press; 1985.
[10] Leo DJ. Engineering analysis of smart material systems. USA: John Wiley and Sons; 2007.
[11] Schwartz M, editor. Smart materials. USA: CRC Press; 2008.
[12] Agarwal BD, Broutman LJ. Analysis and performance of fiber composites. USA: John Wiley and Sons; 1980.
[13] Drozda T, editor. Composites applications the future is now. USA: Society of Manufacturing Engineers; 1989.
[14] Vinson JR, Sierakowski RL. The behavior of structures composed of composite materials. The Netherlands: Martinus Nijhoff; 1986.

[15] Katz HS, Milewski JV, editors. Handbook of fibers and reinforcements for plastics. USA: Van Nostrand Reinhold Co.; 1978.
[16] Lubin G, editor. Handbook of composites. USA: Van Nostrand Reinhold Co.; 1982.
[17] Schwartz MM. Composite materials handbook. USA: McGraw-Hill; 1984.
[18] Strong AB. Fundamentals of composite manufacturing. USA: Society of Manufacturing Engineering; 1989.
[19] Beland S. High performance thermoplastic resins and their composites. USA: Noyes Data Corporation; 1990.
[20] Advani S, Hsiao K-T, editors. Manufacturing techniques for polymer matrix composites (PMCs). The Netherlands: Elsevier; 2012.

A review of stress–strain and material behavior

2

2.1 Introduction

In developing methodologies for the analysis and design of laminated composite materials, a consistent nomenclature is required. Stress and strain are presented in terms of *Cartesian* coordinates for two reason: (1) continuity with developments in undergraduate strength of materials courses and (2) simplification of the analysis procedures involving thermal and hygral effects as well as the general form of the load-strain relationships. A shorthand notation (termed *contracted*) is used to identify stresses and strains. The coordinate axes are an x–y–z system or a numerical system of 1–2–3. The 1–2–3 system is termed the *material* or *on-axis* system. Figure 2.1 shows the relationship between the x–y–z and 1–2–3 coordinate systems. All rotations of coordinate axes are assumed to be about the z-axis, so z is coincident with the three directions, which is consistent with the assumption that individual lamina are modeled as orthotropic materials. The notational relationship between the Cartesian, tensor, material, and contracted stresses and strains is presented below for the special case when the x-, y-, and z-axes coincide with the 1, 2, and 3 axes.

Cartesian	Tensor	Material	Contracted
$\sigma_x - \varepsilon_x$	$\sigma_{xx} - \varepsilon_{xx}$	$\sigma_1 - \varepsilon_1$	$\sigma_1 - \varepsilon_1$
$\sigma_y - \varepsilon_y$	$\sigma_{yy} - \varepsilon_{yy}$	$\sigma_2 - \varepsilon_2$	$\sigma_2 - \varepsilon_2$
$\sigma_z - \varepsilon_z$	$\sigma_{zz} - \varepsilon_{zz}$	$\sigma_3 - \varepsilon_3$	$\sigma_3 - \varepsilon_3$
$\tau_{yz} - \gamma_{yz}$	$\sigma_{yz} - 2\varepsilon_{yz}$	$\tau_{23} - \gamma_{23}$	$\sigma_4 - \varepsilon_4$
$\tau_{xz} - \gamma_{xz}$	$\sigma_{xz} - 2\varepsilon_{xz}$	$\tau_{13} - \gamma_{13}$	$\sigma_5 - \varepsilon_5$
$\tau_{xy} - \gamma_{xy}$	$\sigma_{xy} - 2\varepsilon_{xy}$	$\tau_{12} - \gamma_{12}$	$\sigma_6 - \varepsilon_6$

2.2 Strain–displacement relations

When external forces are applied to an elastic body, material points within the body are displaced. If this results in a change in distance between two or more of these points, a deformation exists. A displacement which does not result in distance changes between any two material points is termed a rigid body translation or rotation. The displacement fields for an elastic body can be denoted by $U(x,y,z,t)$, $V(x,y,z,t)$, and $W(x,y,z,t)$, where U, V, and W represent the displacements in the x-, y-, and z-directions, respectively, and t represents time. For the time being, our discussions are limited to static analysis and time is eliminated from the displacement fields.

Laminar Composites. http://dx.doi.org/10.1016/B978-0-12-802400-3.00002-7

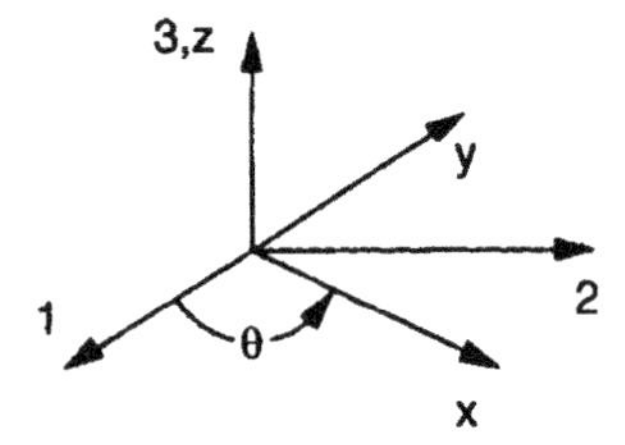

Figure 2.1 Cartesian and material coordinate axes.

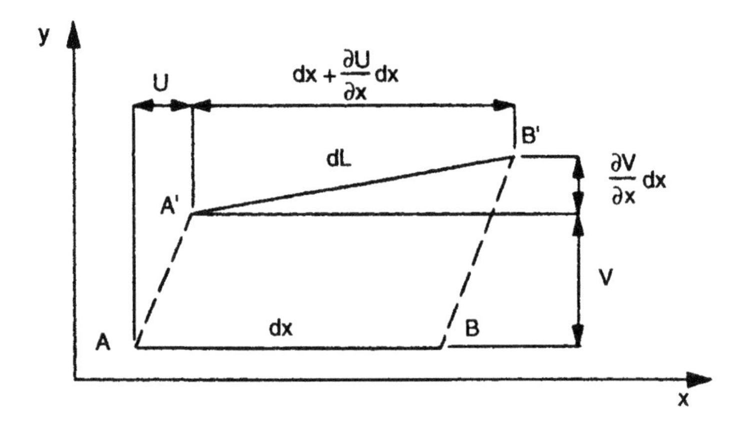

Figure 2.2 Displacement of material points A and B.

The displacement fields are denoted simply as U, V, and W. For many cases of practical interest, these reduce to planar (two-dimensional) fields.

Assume two adjacent material points A and B in Figure 2.2 are initially a distance $\mathrm{d}x$ apart. Assume line AB is parallel to the x-axis and displacements take place in the x–y plane. Upon application of a load, the two points are displaced and occupy new positions denoted as A' and B'. The change in length of $\mathrm{d}x$ is denoted as $\mathrm{d}L$, and is expressed as:

$$\mathrm{d}L = \sqrt{\left(\mathrm{d}x + \frac{\partial U}{\partial x}\mathrm{d}x\right)^2 + \left(\frac{\partial V}{\partial x}\mathrm{d}x\right)^2} = \sqrt{\left(1 + 2\frac{\partial U}{\partial x}\right) + \left(\frac{\partial U}{\partial x}\right)^2 + \left(\frac{\partial V}{\partial x}\right)^2}\,\mathrm{d}x$$

Assuming $\partial U/\partial x \ll 1$ and $\partial V/\partial x \ll 1$, the terms $(\partial U/\partial x)^2$ and $(\partial V/\partial x)^2$ are considered to be zero and the expression for $\mathrm{d}L$ becomes:

$$\mathrm{d}L = \sqrt{1 + 2\frac{\partial U}{\partial x}}\mathrm{d}x$$

Expanding this in a binomial series, $\mathrm{d}L = [1 + \partial U/\partial x + \text{h.o.t}]\mathrm{d}x$. The higher order terms (h.o.t.) are neglected since they are small. The normal strain in the x-direction is defined as $\varepsilon_x = (\mathrm{d}L - \mathrm{d}x)/\mathrm{d}x$. Substituting $\mathrm{d}L = [1 + \partial U/\partial x + \text{h.o.t}]\mathrm{d}x$, the strain in the x-direction is defined. This approach can be extended to include the y- and z-directions. The resulting relationships are:

$$\varepsilon_x = \frac{\partial U}{\partial x} \quad \varepsilon_y = \frac{\partial V}{\partial y} \quad \varepsilon_z = \frac{\partial W}{\partial z}$$

Shear strain is associated with a net change in right angles of a representative volume element (RVE). The deformation associated with a positive shear is shown in Figure 2.3 for pure shear in the x–y plane. Material points O, A, B, and C deform to O', A', B', and C' as shown. Since a condition of pure shear is assumed, the original lengths dx and dy are unchanged. Therefore, $\partial U/\partial x = 0$ and $\partial V/\partial y = 0$, and the angles θ_{yx} and θ_{xy} can be defined from the trigonometric relationships

$$\sin\theta_{xy} = \frac{(\partial V/\partial x)dx}{dx} \quad \sin\theta_{yx} = \frac{(\partial U/\partial y)dy}{dy}$$

Small deformations are assumed so approximations of $\sin\theta_{xy} \approx \theta_{xy}$ and $\sin\theta_{yx} \approx \theta_{yx}$ are used. The shear strain in the x–y plane is defined as $\gamma_{xy} = \theta_{xy} + \theta_{yx} = \pi/2 - \phi = \partial U/\partial y + \partial V/\partial x$, where ϕ represents the angle between two originally orthogonal sides after deformation. For a negative shear strain, the angle ϕ increases. Similar expressions can be established for the x–z and y–z planes. The relationships between shear strain and displacement are:

$$\gamma_{xy} = \frac{\partial U}{\partial y} + \frac{\partial V}{\partial x} \quad \gamma_{xz} = \frac{\partial U}{\partial z} + \frac{\partial W}{\partial x} \quad \gamma_{yz} = \frac{\partial V}{\partial z} + \frac{\partial W}{\partial y}$$

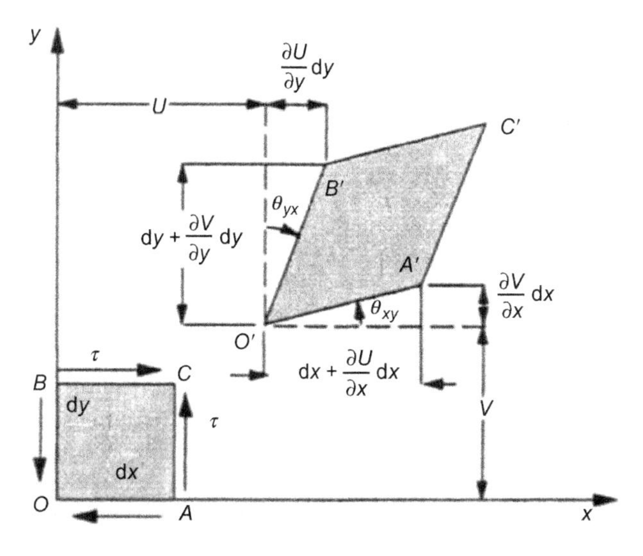

Figure 2.3 Deformation under conditions of pure shear.

These are the Cartesian representations of shear strains and are related to the tensor form by: $\gamma_{xy} = 2\varepsilon_{xy}$, $\gamma_{xz} = 2\varepsilon_{xz}$, and $\gamma_{yz} = 2\varepsilon_{yz}$. The 2 in this relationship can make the tensor form of laminate analysis complicated, especially when thermal and hygral effects are considered.

Since strain is directly related to displacement, it is possible to establish the displacement fields U, V, and W from a strain field. For a displacement field to be valid, it must satisfy a set of equations known as the compatibility equations. These equations are generally expressed either in terms of strain or stress components. The compatibility equations ensure that the displacement fields will be single-valued functions of the coordinates when evaluated by integrating displacement gradients along any path in the region. The equations of compatibility can be found in numerous texts on elasticity, such as [1]. The strain component form of the constitutive equations is:

$$\frac{\partial^2 \gamma_{xy}}{\partial x \partial y} = \frac{\partial^2 \varepsilon_x}{\partial y^2} + \frac{\partial^2 \varepsilon_y}{\partial x^2} \quad 2\frac{\partial^2 \varepsilon_x}{\partial y \partial z} = \frac{\partial}{\partial x}\left(-\frac{\partial \gamma_{yz}}{\partial x} + \frac{\partial \gamma_{xz}}{\partial y} + \frac{\partial \gamma_{xy}}{\partial z}\right)$$

$$\frac{\partial^2 \gamma_{yz}}{\partial y \partial z} = \frac{\partial^2 \varepsilon_y}{\partial z^2} + \frac{\partial^2 \varepsilon_z}{\partial y^2} \quad 2\frac{\partial^2 \varepsilon_y}{\partial x \partial z} = \frac{\partial}{\partial y}\left(\frac{\partial \gamma_{yz}}{\partial x} - \frac{\partial \gamma_{xz}}{\partial y} + \frac{\partial \gamma_{xy}}{\partial z}\right)$$

$$\frac{\partial^2 \gamma_{xz}}{\partial x \partial z} = \frac{\partial^2 \varepsilon_z}{\partial x^2} + \frac{\partial^2 \varepsilon_x}{\partial z^2} \quad 2\frac{\partial^2 \varepsilon_z}{\partial x \partial y} = \frac{\partial}{\partial z}\left(\frac{\partial \gamma_{yz}}{\partial x} + \frac{\partial \gamma_{xz}}{\partial y} - \frac{\partial \gamma_{xy}}{\partial z}\right)$$

Each strain component is defined in terms of a displacement; therefore, these equations can be expressed in terms of displacements. In addition, the constitutive relationship (relating stress to strain) can be used to express these equations in terms of stress components.

Example 2.1 Assume the only nonzero strains are $\varepsilon_x = 10\,\mu\text{in./in.}$, $\varepsilon_y = -2\,\mu\text{in./in.}$, and $\gamma_{xy} = 5\,\mu\text{in./in.}$ The displacement fields U and V will be functions of x and y only. In the z-direction, the displacement field will be $W = 0$. From the definitions of axial strain:

$$U = \int \varepsilon_x \mathrm{d}x = 10x + A \quad U = \int \varepsilon_y \mathrm{d}y = -2y + B$$

where A and B are constants of integration. From the definition of shear strain:

$$\gamma_{xy} = 5 = \frac{\partial U}{\partial y} + \frac{\partial V}{\partial x} = \frac{\partial A}{\partial y} + \frac{\partial B}{\partial x}$$

Therefore, A and B must be functions of y and x, respectively. Otherwise a shear strain would not exist. These functions can be arbitrarily expressed as $A(y) = C_1 + C_2 y$ and $B(x) = C_3 + C_4 x$. After differentiation, $\gamma_{xy} = 5 = C_2 + C_4$. Therefore, $U = 10x + C_1 + C_2 y$ and $V = -2y + C_3 + C_4 x$. Constants C_1 and C_2 are rigid body translations. For conditions of equilibrium rigid body translations vanish, therefore, $C_1 = C_2 = 0$. Similarly, rigid body rotations are not allowed, therefore:

$$\frac{\partial U}{\partial y} = \frac{\partial V}{\partial x} \Rightarrow C_2 = C_4 = \frac{\gamma_{xy}}{2}$$

The displacement field can thus be established as:

$$U = (10x + 2.5y)\ \mu\text{in.} \quad V = (-2y + 2.5x)\ \mu\text{in.}$$

These displacement fields can easily be checked to see that the compatibility requirements are satisfied. Since $W = 0$ and U and V are only functions of x and y, it is easy to establish, through the definitions of strain that $\varepsilon_z = \gamma_{yz} = \gamma_{xz} = 0$. Since the in-plane strains $(\varepsilon_x, \varepsilon_y, \gamma_{xy})$ are constant, the compatibility conditions are satisfied.

Sometimes it is convenient to use the strain–displacement definition from undergraduate strength, $\varepsilon = \Delta L/L$, instead of $\varepsilon_x = \partial U/\partial x$. Similarly, shear strains can be established using the small angle approximations to define changes with respect to both the x- and y-axes.

Example 2.2 Assume the two-dimensional solid shape shown is deformed as illustrated by the dashed lines in Figure E2.2.

Since the x-displacement of point A is 0.020-in., and the y-displacement of point B is 0.036-in., the normal strains in x and y are:

$$\varepsilon_x = \frac{\Delta x}{x} = \frac{0.02}{2.0} = 0.010 \quad \varepsilon_y = \frac{\Delta y}{y} = \frac{0.036}{3.0} = 0.012$$

The shear strain is defined from identifying θ_{yx} and θ_{xy} using small angle approximations as shown in Figure 2.3. Therefore,

$$\theta_{yx} = \frac{0.012}{3.0} = 0.004 \quad \theta_{xy} = \frac{0.010}{2.0} = 0.005$$

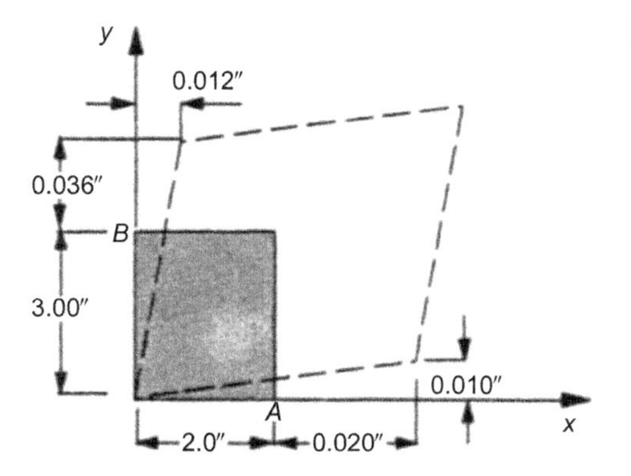

Figure E2.2 Deformed material element.

The shear strain is $\gamma_{xy}=\theta_{yx}+\theta_{xy}=0.004+0.005=0.009$. Had the original right angle between lines OA and OB been increased instead of decreased, the shear strain would be negative as per the definitions of positive and negative shear strain.

2.2.1 Strain transformations

Consider the undeformed two-dimensional triangle ABC shown in Figure 2.4. Assume that the triangle is subjected to a pure shear deformation, so points A, B, and C occupy positions A', B', and C' in the deformed state. The deformed length of $\mathrm{d}L$, denoted as $\mathrm{d}L'$ is:

$$\mathrm{d}L'=\sqrt{\left(\mathrm{d}x+\frac{\partial U}{\partial x}\mathrm{d}x-\frac{\partial U}{\partial y}\mathrm{d}y\right)^2+\left(\mathrm{d}y+\frac{\partial V}{\partial y}\mathrm{d}y-\frac{\partial V}{\partial x}\mathrm{d}x\right)^2}$$

Recall that $\varepsilon_x=\frac{\partial U}{\partial x}$, $\varepsilon_y=\frac{\partial V}{\partial y}$, $\theta_{yx}=\frac{\partial V}{\partial x}$, $\theta_{xy}=\frac{\partial U}{\partial y}$, and note $\sin^2\theta=\frac{(\mathrm{d}x)^2}{(\mathrm{d}x)^2+(\mathrm{d}y)^2}$, $\cos^2\theta=\frac{(\mathrm{d}y)^2}{(\mathrm{d}x)^2+(\mathrm{d}y)^2}$, and $\sin\theta\cos\theta=\frac{(\mathrm{d}x)(\mathrm{d}y)}{(\mathrm{d}x)^2+(\mathrm{d}y)^2}$.

Using these terms in the expression for $\mathrm{d}L'$, expanding the equation and neglecting higher order terms such as ε_x^2 and $\varepsilon_x\gamma_{xy}$ yields:

$$\mathrm{d}L'=\sqrt{1+2\varepsilon_x\sin^2\theta+2\varepsilon_y\cos^2\theta-2\gamma_{xy}\sin\theta\cos\theta\mathrm{d}L}$$

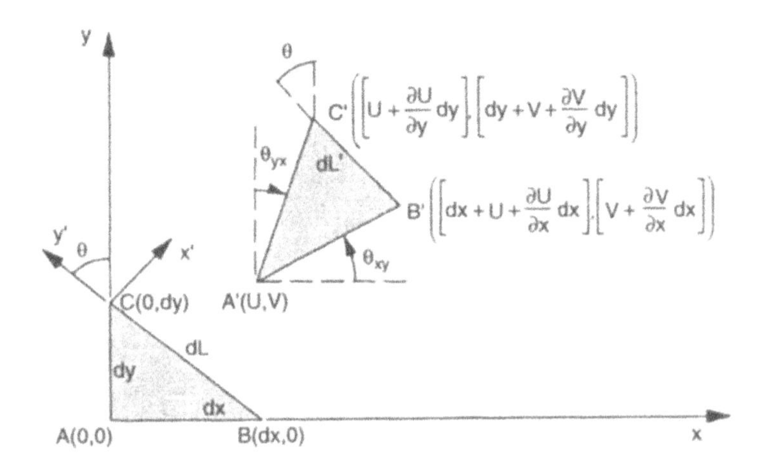

Figure 2.4 Geometry and coordinate changes for pure shear deformation.

This expression shows that the change in length of the original line dL (for a pure shear deformation) includes extensional terms ε_x and ε_y. The normal strain in the y'-direction (indicated in Figure 2.4) is:

$$\varepsilon_{y'} = \frac{\mathrm{d}L' - \mathrm{d}L}{\mathrm{d}L}$$

Following the same procedure as for the normal strain (expanding in a binomial series), yields $\varepsilon_{y'} = \varepsilon_x \sin^2\theta + \varepsilon_y \cos^2\theta - \gamma_{xy}\sin\theta\cos\theta$. Similar expressions can be established for $\varepsilon_{x'}$ and $\gamma_{xy'}$. Rotations in the x- and y-directions cause deformations in other directions. The strain transformation relating strains in the primed and unprimed system are:

$$\begin{Bmatrix} \varepsilon_{x'} \\ \varepsilon_{y'} \\ \gamma_{xy'} \end{Bmatrix} = [T_\varepsilon] \begin{Bmatrix} \varepsilon_x \\ \varepsilon_y \\ \gamma_{xy} \end{Bmatrix}$$

where

$$[T_\varepsilon] = \begin{bmatrix} m^2 & n^2 & mn \\ n^2 & m^2 & -mn \\ -2mn & 2mn & m^2 - n^2 \end{bmatrix} \tag{2.1}$$

and

$$m = \cos\theta, \quad n = \sin\theta$$

This transformation can be extended to three dimensions to include ε_z, γ_{xz}, and γ_{yz}. Assuming a rotation about the z-axis only:

$$\begin{Bmatrix} \varepsilon_{x'} \\ \varepsilon_{y'} \\ \varepsilon_{z'} \\ \gamma_{yz'} \\ \gamma_{xz'} \\ \gamma_{xy'} \end{Bmatrix} = \begin{bmatrix} m^2 & n^2 & 0 & 0 & 0 & mn \\ n^2 & m^2 & 0 & 0 & 0 & -mn \\ 0 & 0 & 1 & 0 & 0 & 0 \\ 0 & 0 & 0 & m & -n & 0 \\ 0 & 0 & 0 & n & m & 0 \\ -2mn & 2mn & 0 & 0 & 0 & m^2 - n^2 \end{bmatrix} \begin{Bmatrix} \varepsilon_x \\ \varepsilon_y \\ \varepsilon_z \\ \gamma_{yz} \\ \gamma_{xz} \\ \gamma_{xy} \end{Bmatrix} \tag{2.2}$$

Assuming arbitrary rotations about any axis are allowed, a different transformation results. This representation is described by a more general set of transformation equations as presented in Appendix A, and in various texts, including Boresi and Lynn [2] and Dally and Riley [3]. If the tensor form of shear strain is used mn becomes $2mn$, and $2mn$ becomes mn, since $\gamma_{xy} = 2\varepsilon_{xy}$.

2.3 Stress and stress transformations

The positive sign convention used for stresses is shown on the three-dimensional representative volume element (RVE) in Figure 2.5. In this element, it is assumed that the

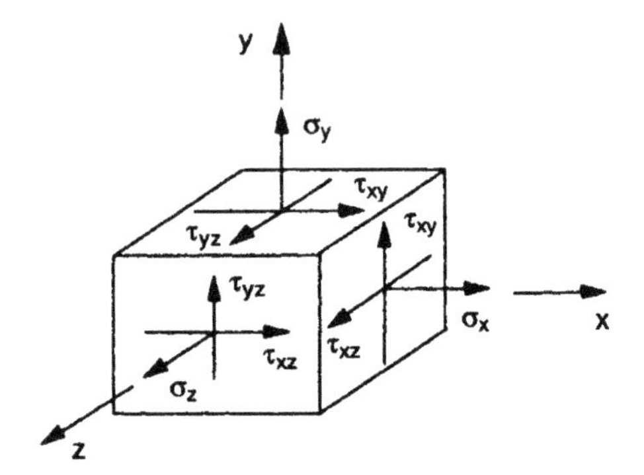

Figure 2.5 Positive sign convention for a 3D state of stress.

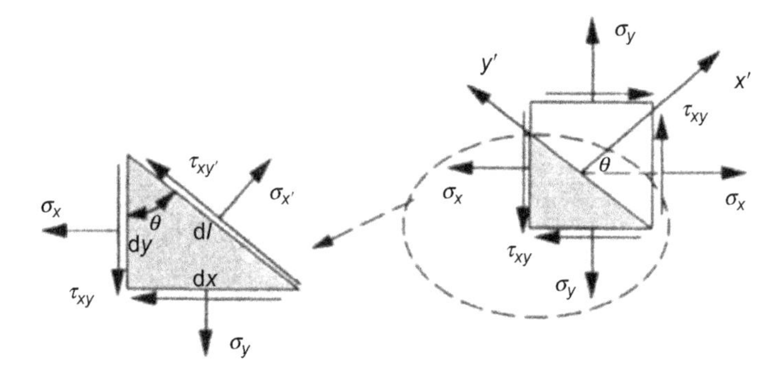

Figure 2.6 Transformation of stresses from an x–y to an x'–y' coordinate system.

conditions of equilibrium have been satisfied (with $\tau_{xy} = -\tau_{yx}$, etc.). Stress components on the hidden faces exist, but are not shown.

For simplicity, consider the transformation of the state of stress shown in Figure 2.6 from the x–y system to the x'–y' system. The transformations are established by assuming the element has a unit thickness t, and summing forces on the free-body diagram representing the exposed x' plane. The summation of forces in x' yields:

$$\sum F_{x'} = \sigma_{x'}(t\mathrm{d}l) - \sigma_x(t\mathrm{d}y)\cos\theta - \sigma_y(t\mathrm{d}x)\sin\theta - \tau_{xy}(t\mathrm{d}x)\sin\theta - \tau_{xy}(t\mathrm{d}x)\cos\theta = 0$$

Using the notation $m = \cos\theta = \mathrm{d}y/\mathrm{d}l$ and $n = \sin\theta = \mathrm{d}x/\mathrm{d}l$, it is a simple matter to show:

$$\sigma_{x'} = \sigma_x m^2 + \sigma_y n^2 + 2\tau_{xy} mn$$

Similar procedures can be followed by drawing a free-body diagram exposing the normal stress in the y'-direction in order to establish the complete set of transformation equations of stress from an x–y coordinate system to an x'–y' system. The general form of this transformation is:

$$\begin{Bmatrix} \sigma_{x'} \\ \sigma_{y'} \\ \tau_{xy'} \end{Bmatrix} = [T_\sigma] \begin{Bmatrix} \sigma_x \\ \sigma_y \\ \tau_{xy} \end{Bmatrix}$$

where

$$[T_\sigma] = \begin{bmatrix} m^2 & n^2 & 2mn \\ n^2 & m^2 & -2mn \\ -mn & mn & m^2 - n^2 \end{bmatrix} \tag{2.3}$$

In three dimensions, the transformation from an x–y system to an x'–y' system (assuming all rotations to be about the z-axis only) is:

$$\begin{Bmatrix} \sigma_{x'} \\ \sigma_{y'} \\ \sigma_{z'} \\ \tau_{yz'} \\ \tau_{xz'} \\ \tau_{xy'} \end{Bmatrix} = \begin{bmatrix} m^2 & n^2 & 0 & 0 & 0 & 2mn \\ n^2 & m^2 & 0 & 0 & 0 & -2mn \\ 0 & 0 & 1 & 0 & 0 & 0 \\ 0 & 0 & 0 & m & -n & 0 \\ 0 & 0 & 0 & n & m & 0 \\ -mn & mn & 0 & 0 & 0 & m^2 - n^2 \end{bmatrix} \begin{Bmatrix} \sigma_x \\ \sigma_y \\ \sigma_z \\ \tau_{yz} \\ \tau_{xz} \\ \tau_{xy} \end{Bmatrix} \tag{2.4}$$

A more populated transformation matrix exists if arbitrary rotations are allowed. The fundamental assumptions from which simple lamination theory is developed depend on rotations about the z-axis only. It should be noted that if the tensor notation were being used for strains then the strain and stress transformations are related by $[T_\varepsilon] = [T_\sigma]$.

2.4 Stress–strain relationships

The generalized form of Hooke's law relating stress to strain is $\{\sigma\}[C] = \{\varepsilon\}$, where $[C]$ is a 6×6 stiffness matrix. The coefficients of $[C]$ are generally not constants. They depend on location within an elastic body, as well as time and temperature. The relationship between stress and strain through $[C]$ is an approximation which is valid for small strains. For a homogeneous linearly elastic material, the material properties are assumed to be the same at every point within the material, and the strain energy density (U_0) is equal to the complementary internal energy density, or complementary strain energy density (C_0). Through the use of energy methods (formulated from considerations based on the first law of thermodynamics), the strain energy density can be related to the stress and a subsequent stress–strain relationship [2]. The complementary strain energy concept which relates U_0 to C_0 is generally used to relate stress to strain through the stiffness matrix $[C]$. In addition, using energy methods, it can be shown that $[C]$ is symmetric, and the terms within it are related by $C_{ij} = C_{ji}$. Therefore,

21 independent elastic constants must be determined. The entire stiffness matrix, however, contains 36 nonzero terms. Using contracted notation, the generalized form of the stress–strain relationship for an anisotropic material is:

$$\begin{Bmatrix} \sigma_1 \\ \sigma_2 \\ \sigma_3 \\ \sigma_4 \\ \sigma_5 \\ \sigma_6 \end{Bmatrix} = \begin{bmatrix} C_{11} & C_{12} & C_{13} & C_{14} & C_{15} & C_{16} \\ C_{12} & C_{22} & C_{23} & C_{24} & C_{25} & C_{26} \\ C_{13} & C_{23} & C_{33} & C_{34} & C_{35} & C_{36} \\ C_{14} & C_{24} & C_{34} & C_{44} & C_{45} & C_{46} \\ C_{15} & C_{25} & C_{35} & C_{45} & C_{55} & C_{56} \\ C_{16} & C_{26} & C_{36} & C_{46} & C_{56} & C_{66} \end{bmatrix} \begin{Bmatrix} \varepsilon_1 \\ \varepsilon_2 \\ \varepsilon_3 \\ \varepsilon_4 \\ \varepsilon_5 \\ \varepsilon_6 \end{Bmatrix}$$

The stiffness matrix $[C]$ can be shown to be invariant. In addition to being termed *anisotropic*, this type of material behavior is also termed *triclinic*. This relationship does not distinguish between tensile and compressive behavior. The material is assumed to have the same stiffness in tension and compression. The response characteristics of this material, as defined by $[C]$, show that shear and extension are coupled. This means that even under conditions of uniaxial tension, a shear deformation will develop. In a similar manner, a pure shear load will create normal deformations. Characterization of an anisotropic material is difficult from an experimental viewpoint, since 21 independent elastic constants must be determined.

2.4.1 Monoclinic materials

If any material symmetry exists, the number of terms in $[C]$ reduces. Assume, for example, that the x–y (or $z=0$) plane is a plane of material symmetry. The effect of this symmetry on the stresses and strains is seen by allowing a rotation of 180° from the x–y coordinate system to the x'–y' coordinate system as depicted in Figure 2.7. Using the strain and stress transformation equations given by (2.2) and (2.4), the primed and unprimed stresses and strains are related by:

$$\begin{Bmatrix} \sigma_{x'} \\ \sigma_{y'} \\ \sigma_{z'} \\ \tau_{yz'} \\ \tau_{xz'} \\ \tau_{xy'} \end{Bmatrix} = \begin{Bmatrix} \sigma_x \\ \sigma_y \\ \sigma_z \\ -\tau_{yz} \\ -\tau_{xz} \\ \tau_{xy} \end{Bmatrix} \quad \begin{Bmatrix} \varepsilon_{x'} \\ \varepsilon_{y'} \\ \varepsilon_{z'} \\ \gamma_{yz'} \\ \gamma_{xz'} \\ \gamma_{xy'} \end{Bmatrix} = \begin{Bmatrix} \varepsilon_x \\ \varepsilon_y \\ \varepsilon_z \\ -\gamma_{yz} \\ -\gamma_{xz} \\ \gamma_{xy} \end{Bmatrix}$$

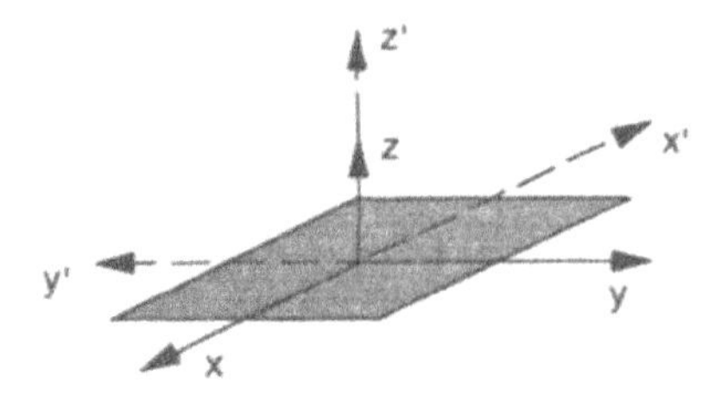

Figure 2.7 Plane of material symmetry for a monoclinic material.

Consider the σ_x and $\sigma_{x'}$ stress components. Using the general constitutive relationship for an anisotropic material, the stress in the primed and unprimed systems are:

$$\sigma_{x'} = C_{11}\varepsilon_{x'} + C_{12}\varepsilon_{y'} + C_{13}\varepsilon_{z'} + C_{14}\gamma_{yz'} + C_{15}\gamma_{xz'} + C_{16}\gamma_{xy'}$$

$$\sigma_x = C_{11}\varepsilon_x + C_{12}\varepsilon_y + C_{13}\varepsilon_z - C_{14}\gamma_{yz} - C_{15}\gamma_{xz} + C_{16}\gamma_{xy}$$

In order for $[C]$ to be invariant, these two equations must be equal, which can happen only if $C_{14} = C_{15} = 0$. Similar procedures can be followed for the remaining stress components in both the primed and unprimed systems, which results in $C_{24} = C_{25} = C_{34} = C_{35} = C_{46} = C_{56} = 0$. The stiffness matrix reduces to 13 independent elastic constants, and is:

$$[C] = \begin{bmatrix} C_{11} & C_{12} & C_{13} & 0 & 0 & C_{16} \\ C_{12} & C_{22} & C_{23} & 0 & 0 & C_{26} \\ C_{13} & C_{23} & C_{33} & 0 & 0 & C_{36} \\ 0 & 0 & 0 & C_{44} & C_{45} & 0 \\ 0 & 0 & 0 & C_{45} & C_{55} & 0 \\ C_{16} & C_{26} & C_{36} & 0 & 0 & C_{66} \end{bmatrix}$$

Note that *extension-shear coupling* exists due to terms such as C_{16}, etc. Although 13 independent elastic constants are present in the stiffness matrix, there are 20 nonzero terms. As the number of conditions of material symmetry increase, the number of elastic constants required to describe the material decreases.

2.4.2 Orthotropic materials

The relationship between stresses and strains in the primed and unprimed coordinate systems for an orthotropic material can be established by allowing rotations of 180° from the original reference frame about the z- and x-axes. The rotations about the z-axis follow those of the previous section using Equations (2.2) and (2.4). These two equations cannot be directly applied to rotations about the x-axis, since they were defined only for rotations about the z-axis. A more general set of transformations (as presented in Appendix A) is required to establish the primed and unprimed relations for rotations about the x-axis. It should also be noted that a third rotation about the remaining axis does not reduce the stiffness matrix further. Proceeding as in the previous case, relating the stresses and strains in the primed and unprimed coordinate systems, coupled with the invariance of $[C]$ establishes the orthotropic stiffness matrix, expressed as:

$$[C] = \begin{bmatrix} C_{11} & C_{12} & C_{13} & 0 & 0 & 0 \\ C_{12} & C_{22} & C_{23} & 0 & 0 & 0 \\ C_{13} & C_{23} & C_{33} & 0 & 0 & 0 \\ 0 & 0 & 0 & C_{44} & 0 & 0 \\ 0 & 0 & 0 & 0 & C_{55} & 0 \\ 0 & 0 & 0 & 0 & 0 & C_{66} \end{bmatrix}$$

There are 9 independent elastic constants associated with an orthotropic material and a total of 12 nonzero terms. In addition, there is no shear-extension coupling.

2.4.3 Transversely isotropic materials

For a transversely isotropic material, there is an axis of material symmetry (defined as a direction with respect to which the material has identical properties) in addition to three planes of symmetry. Therefore, any two material fibers having symmetrical positions with respect to the axis of symmetry have the same stiffness. If the z-axis is assumed to coincide with the axis of symmetry, the x- and y-axes can be in any direction (provided they remain perpendicular to each other) without altering the value of $[C]$. In this case, the x–y plane is referred to as an isotropic plane. Under the assumption that the z-axis coincides with the axis of symmetry, the stiffness matrix is:

$$[C]=\begin{bmatrix} C_{11} & C_{12} & C_{13} & 0 & 0 & 0 \\ C_{12} & C_{11} & C_{13} & 0 & 0 & 0 \\ C_{13} & C_{13} & C_{33} & 0 & 0 & 0 \\ 0 & 0 & 0 & C_{44} & 0 & 0 \\ 0 & 0 & 0 & 0 & C_{44} & 0 \\ 0 & 0 & 0 & 0 & 0 & C_{66} \end{bmatrix}$$

where

$$C_{66}=\frac{C_{11}-C_{12}}{2}$$

A transversely isotropic material has 5 independent elastic constants and 12 nonzero terms. The form of the stiffness matrix would be different if another axis were chosen to represent the axis of symmetry as shown in Tsai [4].

2.4.4 Isotropic materials

For an isotropic material, all planes are planes of material symmetry and are isotropic. There are 2 independent elastic constants associated with an isotropic material and 12 nonzero terms in the stiffness matrix. The resulting stiffness matrix for an isotropic material is:

$$[C]=\begin{bmatrix} C_{11} & C_{12} & C_{12} & 0 & 0 & 0 \\ C_{12} & C_{11} & C_{12} & 0 & 0 & 0 \\ C_{12} & C_{12} & C_{11} & 0 & 0 & 0 \\ 0 & 0 & 0 & C_{44} & 0 & 0 \\ 0 & 0 & 0 & 0 & C_{44} & 0 \\ 0 & 0 & 0 & 0 & 0 & C_{44} \end{bmatrix}$$

where

$$C_{44}=\frac{C_{11}-C_{12}}{2}$$

2.4.5 Summary of material responses

To summarize stress–strain relationships for anisotropic, monoclinic, orthotropic, transversely isotropic, and isotropic materials, we note that the form of each stiffness matrix above is valid only for rotations about those axes specified. The total number of independent elastic constants in each case remains the same, but the number of nonzero terms for each stiffness matrix changes. Stress–strain relationships for a variety of material property symmetry conditions are found in Refs. [4,5]. A summary of the number of independent elastic constants and nonzero terms for each material considered is presented in Table 2.1. In this table the term *on-axis* is used to indicate rotations about an axis of symmetry, while *off-axis* refers to a rotation about one of the reference axes, and *general* refers to rotations about any axis. An orthotropic material in an on-axis configuration has 12 nonzero terms in its stiffness matrix. If a general rotation were used for the same material system there would be 36 nonzero terms, as for an anisotropic material. Similar observations can be made for other types of materials. Even though one generally begins by assuming an orthotropic material response for individual lamina, the final laminate may behave as an anisotropic material in which extension and shear are coupled.

2.5 Strain–stress relationships

The strain–stress relation is obtained by inverting the stiffness matrix in the stress–strain relation $\{\sigma\} = [C]\{\varepsilon\}$, resulting in:

$$\{\varepsilon\} = [C]^{-1}\{\sigma\} = [S]\{\sigma\}$$

where $[S]$ is the *elastic compliance matrix*, and is symmetric. The general form of the strain–stress relation is $\{\varepsilon\} = [S]\{\sigma\}$. For an anisotropic material:

$$[S] = \begin{bmatrix} S_{11} & S_{12} & S_{13} & S_{14} & S_{15} & S_{16} \\ S_{12} & S_{22} & S_{23} & S_{24} & S_{25} & S_{26} \\ S_{13} & S_{23} & S_{33} & S_{34} & S_{35} & S_{36} \\ S_{14} & S_{24} & S_{34} & S_{44} & S_{45} & S_{46} \\ S_{15} & S_{25} & S_{35} & S_{45} & S_{55} & S_{56} \\ S_{16} & S_{26} & S_{36} & S_{46} & S_{56} & S_{66} \end{bmatrix}$$

Table 2.1 Summary of material symmetries (after [4])

Type of material symmetry	Number of independent constants	Number of nonzero terms (on-axis)	Number of nonzero terms (off-axis)	Number of nonzero terms (general)
Anisotropic	21	36	36	36
Monoclinic	13	20	36	36
Orthotropic	9	12	20	36
Transversely isotropic	5	12	20	36
Isotropic	2	12	12	12

As various forms of material symmetry are considered, this matrix reduces in the same manner as the stiffness matrix. For the monoclinic and orthotropic materials, the elastic constants above change to:

Monoclinic: $S_{14}=S_{15}=S_{24}=S_{25}=S_{34}=S_{35}=S_{46}=S_{56}=0$
Orthotropic: (all monoclinic constants) and $S_{16}=S_{26}=S_{36}=S_{45}=0$

The compliance matrices for transversely isotropic and isotropic materials are:

Transversely isotropic:

$$[S]=\begin{bmatrix} S_{11} & S_{12} & S_{13} & 0 & 0 & 0 \\ S_{12} & S_{11} & S_{13} & 0 & 0 & 0 \\ S_{13} & S_{13} & S_{33} & 0 & 0 & 0 \\ 0 & 0 & 0 & S_{44} & 0 & 0 \\ 0 & 0 & 0 & 0 & S_{44} & 0 \\ 0 & 0 & 0 & 0 & 0 & S_{66} \end{bmatrix}$$

Isotropic:

$$[S]=\begin{bmatrix} S_{11} & S_{12} & S_{12} & 0 & 0 & 0 \\ S_{12} & S_{11} & S_{12} & 0 & 0 & 0 \\ S_{12} & S_{12} & S_{11} & 0 & 0 & 0 \\ 0 & 0 & 0 & S_{44} & 0 & 0 \\ 0 & 0 & 0 & 0 & S_{44} & 0 \\ 0 & 0 & 0 & 0 & 0 & S_{44} \end{bmatrix}$$

where $S_{66}=2(S_{11}-S_{12})$ $S_{44}=2(S_{11}-S_{12})$.

2.6 Thermal and hygral effects

The previously defined relationships between stress and strain are valid as long as temperature (thermal) and moisture (hygral) effects are not present. In many structural applications involving traditional engineering materials, the above relationships would be sufficient for most stress analysis. In the case of laminated composites, however, this is not true. The effects of both temperature and moisture (relative humidity) on the stiffness and strength of polymeric composites are schematically illustrated in Figure 2.8, where ΔM represents the change in moisture content (measured as a percentage). In space applications, thermal effects can be severe. One side of a structure may be subjected to direct sunlight while the other is subjected to freezing conditions. This results in a thermal gradient within the structure and a complex state of stress. This thermal gradient may even be cyclic, further complicating the analysis.

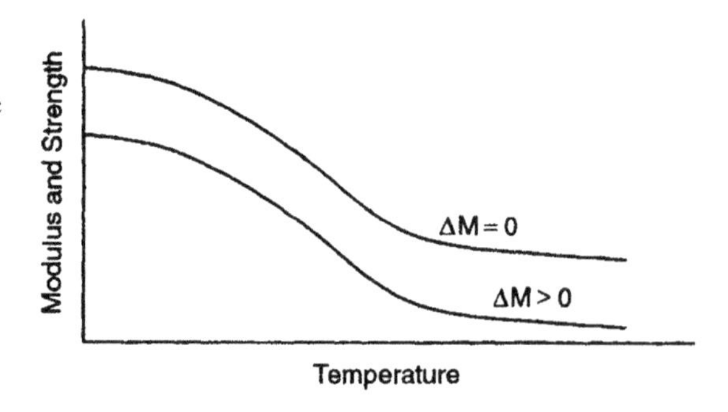

Figure 2.8 Schematic representations of the effects of temperature and moisture on elastic modulus and strength.

In addition to the effects of temperature and moisture on the modulus and strength, it affects the strains. Consider a simple one-dimensional case where the total strain for an elastic body is given by:

$$\{\varepsilon\} = \{\varepsilon\}_{\mathrm{m}} + \{\varepsilon\}^{\mathrm{T}} + \{\varepsilon\}^{\mathrm{H}} + \{\varepsilon\}_{\mathrm{misc}}$$

where $\{\varepsilon\}_{\mathrm{m}}$ is the elastic or mechanical strain; $\{\varepsilon\}^{\mathrm{T}}$ is the thermal strain; $\{\varepsilon\}^{\mathrm{H}}$ is the hygral strain; $\{\varepsilon\}_{\mathrm{misc}}$ is the miscellaneous strains, such as plastic strain, etc.

The thermally induced strains can be expressed as:

$$\{\varepsilon\}^{\mathrm{T}} = \alpha(\Delta T)$$

where ΔT is the change in temperature from an initial reference temperature to a final operating temperature. In general, the initial temperature (T_0) is regarded as the *stress-free* temperature, or the temperature at which no thermal stress exists. The final, or operating temperature (T) is the temperature at which the structural component is required to perform, and $\Delta T = T - T_0$. The term α is the coefficient of thermal expansion, and in general has different values in each direction. A simple means of describing α is with subscripts, such as α_i, where $i = 1,2,3$.

In a similar manner, the hygral strains are expressed as:

$$\{\varepsilon\}^{\mathrm{H}} = \beta(\Delta M)$$

where ΔM is the percentage change in moisture and β is the coefficient of hygral expansion. As with thermal effects, the β terms are also subscripted as β_i, where $i = 1,2,3$.

The α_i and β_i terms used to describe thermal and hygral coefficients are dilatational, meaning that they affect only expansion and contraction, not shear.

For a case in which there are no miscellaneous strains ($\{\varepsilon\}_{\mathrm{misc}} = 0$) we write $\{\varepsilon\} = \{\varepsilon\}_{\mathrm{m}} + \{\varepsilon\}^{\mathrm{T}} + \{\varepsilon\}^{\mathrm{H}} = \{\varepsilon\}_{\mathrm{m}} + \alpha\Delta T + \beta\Delta M$, and the mechanical strain is:

$$\{\varepsilon\}_{\mathrm{m}} = \{\varepsilon\} - \alpha\Delta T - \beta\Delta M$$

2.7 Complete anisotropic response

The complete set of governing equations for anisotropic polymeric composites can expressed in shorthand notation as:

$$\varepsilon_i = S_{ij}\sigma_i + \alpha_i\Delta T + \beta_i\Delta M \quad (i = 1,2,3),\ (j = 1-6) \tag{2.5}$$

$$\varepsilon_i = S_{ij}\sigma_j \quad (i = 4,5,6),\ (j = 1-6)$$

where ε_i is the total strain and $S_{ij}\sigma_j$ is the mechanical strain. In expanded matrix notation, this equation becomes:

$$\begin{Bmatrix} \varepsilon_1 \\ \varepsilon_2 \\ \varepsilon_3 \\ \varepsilon_4 \\ \varepsilon_5 \\ \varepsilon_6 \end{Bmatrix} = \begin{bmatrix} S_{11} & S_{12} & S_{13} & S_{14} & S_{15} & S_{16} \\ S_{12} & S_{22} & S_{23} & S_{24} & S_{25} & S_{26} \\ S_{13} & S_{23} & S_{33} & S_{34} & S_{35} & S_{36} \\ S_{14} & S_{24} & S_{34} & S_{44} & S_{45} & S_{46} \\ S_{15} & S_{25} & S_{35} & S_{45} & S_{55} & S_{56} \\ S_{16} & S_{26} & S_{36} & S_{46} & S_{56} & S_{66} \end{bmatrix} \begin{Bmatrix} \sigma_1 \\ \sigma_2 \\ \sigma_3 \\ \sigma_4 \\ \sigma_5 \\ \sigma_6 \end{Bmatrix} + \begin{Bmatrix} \alpha_1 \\ \alpha_2 \\ \alpha_3 \\ 0 \\ 0 \\ 0 \end{Bmatrix} \Delta T + \begin{Bmatrix} \beta_1 \\ \beta_2 \\ \beta_3 \\ 0 \\ 0 \\ 0 \end{Bmatrix} \Delta M$$

The relationship above can also be expressed in terms of the stiffness matrix $[C]$, obtained by multiplying the entire expression by $[S_{ij}]^{-1}$. This results in:

$$\sigma_i = C_{ij}(\varepsilon_j - \alpha_j \Delta T - \beta_j \Delta M) \quad (i=1,2,3),\ (j=1-6)$$

$$\sigma_i = C_{ij}(\varepsilon_j - \alpha_j \Delta T - \beta_j \Delta M) \quad (i=4,5,6),\ (j=1-6)$$

In these expressions, we note from the previous section we can write $\{\varepsilon_j\}_\mathrm{m} = \{\varepsilon_j\} - \{\alpha_j\}\Delta T - \{\beta_j\}\Delta M$, therefore $\{\sigma_i\} = [C_{ij}]\{\varepsilon_j\}_\mathrm{m}$, where $\{\varepsilon_j\}_\mathrm{m}$ is the mechanical strain.

Example 2.3 Assume the compliance matrix for a particular material is:

$$[S] = \begin{bmatrix} 7 & 11 & -8 & 12 & 5 & -8 \\ 11 & 25 & -15 & 25 & 8 & -15 \\ -8 & -15 & 2 & -30 & 5 & 12 \\ 12 & 25 & -30 & 15 & 19 & 11 \\ 5 & 8 & 5 & 19 & -1 & -12 \\ -8 & -15 & 12 & 11 & -12 & 2 \end{bmatrix} \times 10^{-9}$$

Assume the thermal and hygral coefficients of expansion are constant over the ranges of ΔT and ΔM of interest, and are:

$$\begin{Bmatrix} \alpha_1 \\ \alpha_2 \\ \alpha_3 \end{Bmatrix} = \begin{Bmatrix} 10 \\ 5 \\ 2 \end{Bmatrix} \mu\text{in./in./}^\circ\text{F} \qquad \begin{Bmatrix} \beta_1 \\ \beta_2 \\ \beta_3 \end{Bmatrix} = \begin{Bmatrix} 0.10 \\ 0.40 \\ 0.20 \end{Bmatrix}$$

Assume a state of plane stress in which the only nonzero stresses are $\sigma_1 = 20\,\text{ksi}$, $\sigma_2 = 10\,\text{ksi}$ and $\sigma_6 (= \tau_{12}) = 5\,\text{ksi}$. Using the expanded form of Equation (2.5) the strains are represented as:

$$\begin{Bmatrix} \varepsilon_1 \\ \varepsilon_2 \\ \varepsilon_3 \\ \varepsilon_4 \\ \varepsilon_5 \\ \varepsilon_6 \end{Bmatrix} = \begin{bmatrix} 7 & 11 & -8 & 12 & 5 & -8 \\ 11 & 25 & -15 & 25 & 8 & -15 \\ -8 & -15 & 2 & -30 & 5 & 12 \\ 12 & 25 & -30 & 15 & 19 & 11 \\ 5 & 8 & 5 & 19 & -1 & -12 \\ -8 & -15 & 12 & 11 & -12 & 2 \end{bmatrix} \times 10^{-9} \begin{Bmatrix} 20 \\ 10 \\ 0 \\ 0 \\ 0 \\ 5 \end{Bmatrix} \times 10^3 + \begin{Bmatrix} 10 \\ 5 \\ 2 \\ 0 \\ 0 \\ 0 \end{Bmatrix} \times 10^{-6} \Delta T + \begin{Bmatrix} 0.10 \\ 0.40 \\ 0.20 \\ 0 \\ 0 \\ 0 \end{Bmatrix} \Delta M$$

$$\begin{Bmatrix} \varepsilon_1 \\ \varepsilon_2 \\ \varepsilon_3 \\ \varepsilon_4 \\ \varepsilon_5 \\ \varepsilon_6 \end{Bmatrix} = \begin{Bmatrix} 210 \\ 395 \\ -250 \\ 545 \\ 120 \\ -300 \end{Bmatrix} \times 10^{-6} + \begin{Bmatrix} 10 \\ 5 \\ 2 \\ 0 \\ 0 \\ 0 \end{Bmatrix} \times 10^{-6} \Delta T + \begin{Bmatrix} 0.10 \\ 0.40 \\ 0.20 \\ 0 \\ 0 \\ 0 \end{Bmatrix} \Delta M$$

From this we observe that the shear strains ($\varepsilon_4, \varepsilon_5, \varepsilon_6$) remain constant for any ΔT and ΔM values considered. The normal strains are affected by both ΔT and ΔM, and vary linearly. It is also observed that even though a state of plane stress exists the out-of-plane shear strains (ε_4 and ε_5) are present.

2.8 Problems

2.1 Determine the displacement field for the following states of strain. If a strain component is not specified, assume it to be 0.
(A) $\varepsilon_x = -1000\,\mu\text{in./in.}$ $\varepsilon_y = 1000\,\mu\text{in./in.}$ $\gamma_{xy} = 400\,\mu$
(B) $\varepsilon_x = 600\,\mu\text{in./in.}$ $\varepsilon_y = -1000\,\mu\text{in./in.}$ $\gamma_{xy} = 400\,\mu$
(C) $\varepsilon_x = -500\,\mu\text{in./in.}$ $\varepsilon_y = -1000\,\mu\text{in./in.}$ $\gamma_{xy} = -200\,\mu$

2.2 Determine the deformed lengths of lines AB and CD, and the angle between lines after deformation. Note that lines AB and CD are perpendicular before deformation. Assume that the only nonzero strains on the undeformed cubic solid are:
(A) $\varepsilon_x = 10{,}000\,\mu\text{in./in.}$, $\varepsilon_y = 20{,}000\,\mu\text{in./in.}$, $\gamma_{xy} = 30{,}000\,\mu$
(B) $\varepsilon_x = 3000\,\mu\text{in./in.}$, $\varepsilon_y = -1200\,\mu\text{in./in.}$, $\gamma_{xy} = 900\,\mu$

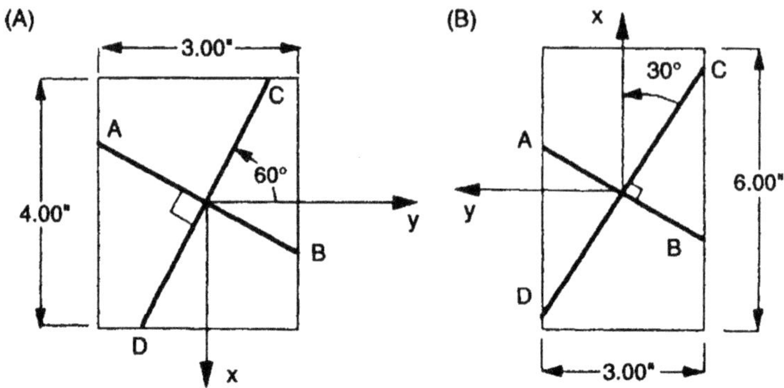

2.3 Determine the state of stress that exists if the nonzero strains at a point in a material are $\varepsilon_x = 300\,\mu\text{in./in.}$, $\varepsilon_y = 150\,\mu\text{in./in.}$, $\gamma_{xy} = -200\,\mu$, and the stiffness of the material is:

$$[C] = \begin{bmatrix} 200 & -50 & 0 & 0 & 0 & 50 \\ -50 & 150 & 0 & 0 & 0 & 50 \\ 0 & 0 & 150 & 0 & 0 & 0 \\ 0 & 0 & 0 & 100 & 0 & 0 \\ 0 & 0 & 0 & 0 & 100 & 0 \\ 50 & 50 & 0 & 0 & 0 & 100 \end{bmatrix} \times 10^5\,\text{psi}$$

2.4 Work problem 2.3 for the following stiffness matrix:

$$[C] = \begin{bmatrix} 500 & -50 & 50 & -100 & -50 & 50 \\ -50 & 250 & 100 & -100 & -50 & 25 \\ 50 & 100 & 125 & -50 & 100 & 50 \\ -100 & -100 & -50 & 100 & 100 & 25 \\ -50 & -50 & 100 & 100 & 200 & 50 \\ 50 & 25 & 50 & 25 & 50 & 100 \end{bmatrix} \times 10^5\,\text{psi}$$

2.5 Use the stiffness matrix of problem 2.3 to determine the state of strain for the following states of stress. The stresses for each part correspond to: $(\sigma_x, \sigma_y, \sigma_z, \tau_{xz}, \tau_{yz}, \tau_{xy})$

(A) $\{50,25,10,20,-10,10\}$ ksi

(B) $\{0,0,0,10,10,10\}$ ksi

(C) $\{-10,0,40,0,20,-20\}$ ksi

2.6 A cubic solid is deformed as shown. Determine the state of strain which exists. The solid lines represent the undeformed state.

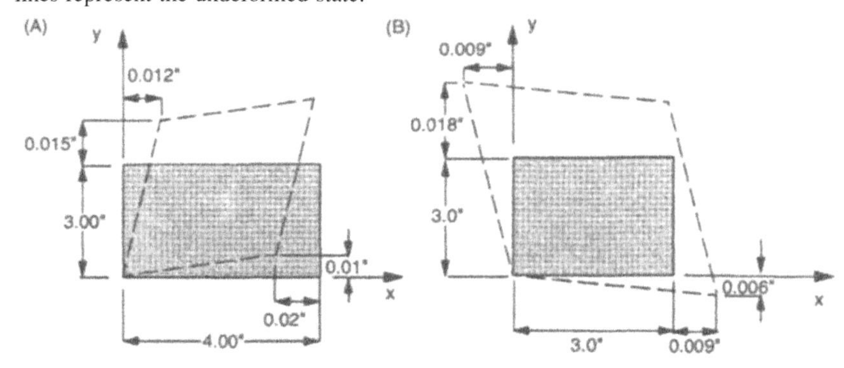

2.7 Use the stiffness matrix of problem 2.3 to determine the stresses required to produce the displacements defined in problem 2.6 (A) under the assumption that *x*,*y*,*z* are the 1,2,3 directions.

2.8 Use the stiffness matrix of problem 2.4 to determine the stresses required to produce the displacements defined in problem 2.6 (B) under the assumption that *x*,*y*,*z* are the 1,2,3 directions.

2.9 The stiffness matrix of problem 2.3 can be inverted to produce:

$$[S]=\begin{bmatrix} 7.69 & 4.62 & 0 & 0 & 0 & -6.15 \\ 4.62 & 10.77 & 0 & 0 & 0 & -7.69 \\ 0 & 0 & 6.67 & 0 & 0 & 0 \\ 0 & 0 & 0 & 10.0 & 0 & 0 \\ 0 & 0 & 0 & 0 & 10.0 & 0 \\ -6.15 & -7.69 & 0 & 0 & 0 & 16.92 \end{bmatrix}\times 10^{-8}$$

The coefficients of thermal and hygral expansion are:

$$\begin{Bmatrix} \alpha_1 \\ \alpha_2 \\ \alpha_3 \end{Bmatrix}=\begin{Bmatrix} 12 \\ 4 \\ 4 \end{Bmatrix}\mu\text{in./in./}^\circ\text{F} \quad \begin{Bmatrix} \beta_1 \\ \beta_2 \\ \beta_3 \end{Bmatrix}=\begin{Bmatrix} 0.0 \\ 0.40 \\ 0.40 \end{Bmatrix}$$

Assume $\Delta T=-300°\text{F}$ and $\Delta M=0.005$. Determine the strains in the principal material directions for a state of stress described by the notation $(\sigma_1,\sigma_2,\sigma_3,\sigma_4,\sigma_5,\sigma_6)$.

(A) {10, 5, 0, 0, 0, 5} ksi

(B) {−10, 10, 5, 0, 5, 10} ksi

(C) {0, 10, 5, 0, 5, 0} ksi

2.10 Assume that the material described by the compliance matrix of problem 2.9 is subjected to the state of stress shown. Furthermore, assume that the coefficients of thermal and hygral expansion given in problem 2.9 remain applicable, and that $\Delta T=-300\,°\text{F}$. On the same graph, plot the $(\varepsilon_1,\varepsilon_2,\varepsilon_3,\varepsilon_4,\varepsilon_5,\varepsilon_6)$ as a function of ΔM for $0.0\le\Delta M\le 0.05$.

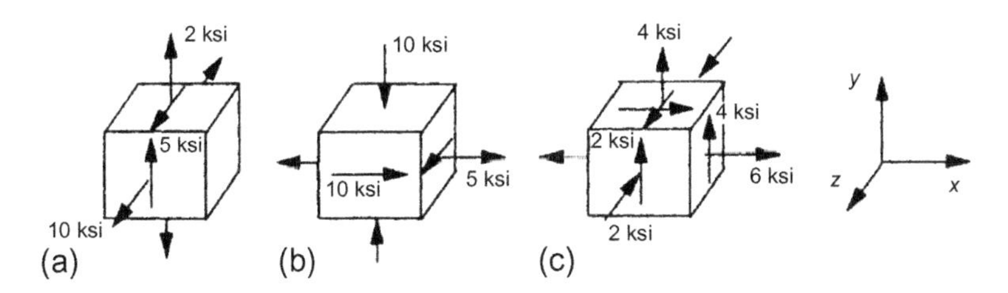

2.11 Work problem 2.10 (A), (B), or (C) assuming ΔM remains constant ($\Delta M=0.005$), and ΔT varies in the range $-300°\text{F}\le\Delta T\le 300°\text{F}$. The resulting plot for this problem should be in terms of ΔT instead of ΔM.

2.12 Assume a material which has thermal and hygral properties defined as:

$$\begin{Bmatrix} \alpha_1 \\ \alpha_2 \\ \alpha_3 \end{Bmatrix}=\begin{Bmatrix} 10 \\ 6 \\ 3 \end{Bmatrix}\mu\text{in./in./}^\circ\text{F} \quad \begin{Bmatrix} \beta_1 \\ \beta_2 \\ \beta_3 \end{Bmatrix}=\begin{Bmatrix} 0.05 \\ 0.20 \\ 0.10 \end{Bmatrix}$$

Furthermore, assume that these properties remain constant for all ranges of ΔT and ΔM considered. For a constant $\Delta T\le-300°\text{F}$, and $0.0\le\Delta M\le 0.020$, plot, on the same graph, the stress (as a function of ΔM) required to produce displacements corresponding to problem 2.6 (A), if:

$$[C] = \begin{bmatrix} 500 & -50 & 50 & -100 & -50 & 50 \\ -50 & 250 & 100 & -100 & -50 & 25 \\ 50 & 100 & 125 & -50 & 100 & 50 \\ -100 & -100 & -50 & 100 & 100 & 25 \\ -50 & -50 & 100 & 100 & 200 & 50 \\ 50 & 25 & 50 & 25 & 50 & 100 \end{bmatrix} \times 10^4\,\text{psi}$$

References

[1] Timoshenko SP, Goodier JN. Theory of elasticity. USA: McGraw-Hill; 1970.
[2] Boresi AP, Lynn PP. Elasticity in engineering mechanics. USA: Prentice-Hall; 1974.
[3] Dally JW, Riley WF. Experimental stress analysis. USA: McGraw-Hill; 1978.
[4] Tasi SW. Composites design. USA: Think Composites; 1987.
[5] Tsai SW. Mechanics of composite materials, part II, theoretical aspects, AFML-TR-66-149; 1966.

Lamina analysis

3

3.1 Introduction

One difference between laminated composites and traditional engineering materials is that a composite's response to loads is direction dependent. In order to analyze the response of a composite, we must be able to predict the behavior of individual lamina. A lamina (considered a unidirectional composite) is characterized by having all fibers (either a single ply or multiple plies) oriented in the same direction. This model allows one to treat the lamina as an orthotropic material. In reality, fibers are not perfectly straight or uniformly oriented within the lamina. There are generally several layers of fibers nested within a single lamina. The model used to represent a lamina consists of a single fiber per layer. In developing relations between material response and applied loads, the simplified model is an accepted representation. A schematic of an actual and modeled lamina is presented in Figure 3.1. The 1, 2, and 3 axes are the principal directions of orthotropic material behavior, defined as:

1. principal fiber direction
2. in-plane direction perpendicular to fibers
3. out-of-plane direction perpendicular to fibers

3.2 Mechanical response of lamina

In order to evaluate the response of a lamina, each component of the stiffness matrix $[C]$ must be determined. The stress–strain relationships needed to define $[C]$ are obtained by experimental procedures as discussed in Chapter 4. A uniform stress is easier to approximate than uniform strain; therefore, the stiffness matrix is established by first developing the compliance matrix $[S]$ and inverting it to obtain $[C]$. The lamina is orthotropic, so extension and shear are uncoupled in the principal material directions. The extension and shear components of the compliance are determined independent of each other, with uniaxial tension used to determine the extension components. Figure 3.2 shows the directions of normal load application required to establish the normal stress components of $[S]$ in each direction.

Applying a normal stress in the 1-direction (with all other stresses being zero) results in normal strains in the 2 and 3 directions. There is no shear–extension coupling, so the relationship between each normal strain and the applied stress is

$$\varepsilon_1 = \frac{\sigma_1}{E_1}, \quad \varepsilon_2 = -\frac{\nu_{12}\sigma_1}{E_1}, \quad \varepsilon_3 = -\frac{\nu_{13}\sigma_1}{E_1}$$

where E_1 is the elastic modulus in the 1-direction (parallel to the fibers), ν_{12} the Poisson's ratio in the 2-direction when the lamina is loaded in the 1-direction, and ν_{13} the Poisson's ratio in the 3-direction when the lamina is loaded in the 1-direction.

Laminar Composites. http://dx.doi.org/10.1016/B978-0-12-802400-3.00003-9

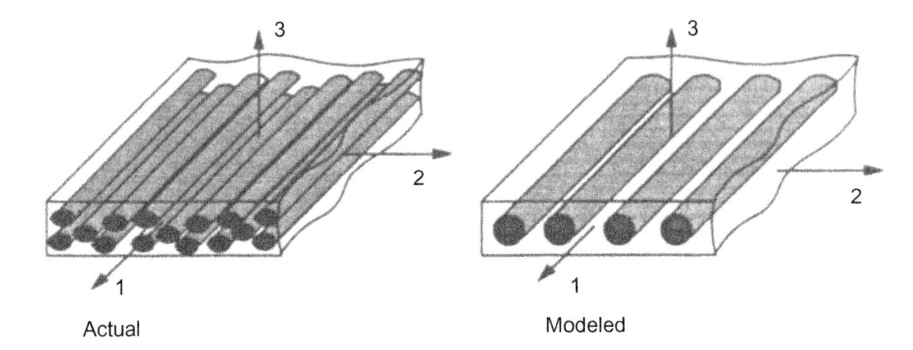

Figure 3.1 Schematic of actual and modeled lamina.

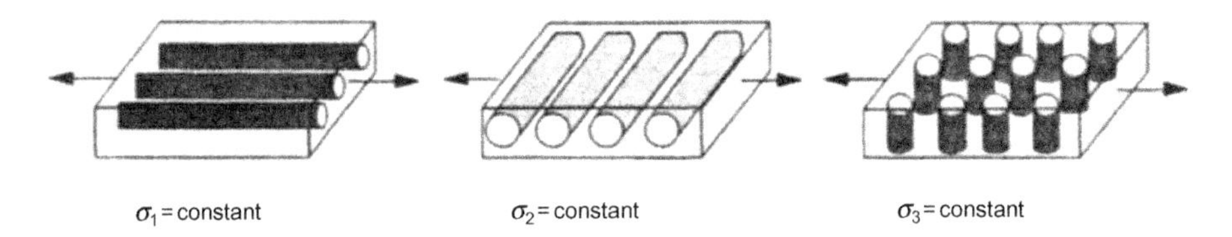

Figure 3.2 Schematic of applied stress for determining components of an orthotropic compliance matrix.

Similarly, by application of a normal stress in the 2-direction (with all other stresses zero), the relationship between strains in the 1-, 2-, and 3-directions and the only nonzero applied stress component σ_2 is

$$\varepsilon_1 = -\frac{\nu_{21}\sigma_2}{E_2}, \quad \varepsilon_2 = \frac{\sigma_2}{E_2}, \quad \varepsilon_3 = -\frac{\nu_{23}\sigma_2}{E_2}$$

The strains developed with σ_3 as the only nonzero stress component are

$$\varepsilon_1 = -\frac{\nu_{31}\sigma_3}{E_3}, \quad \varepsilon_2 = -\frac{\nu_{32}\sigma_3}{E_3}, \quad \varepsilon_3 = \frac{\sigma_3}{E_3}$$

Combining these results, and recalling $\{\varepsilon_i\} = [S_{ij}]\{\sigma_j\}$, the extension terms are

$$\begin{aligned} S_{11} &= \frac{1}{E_1}, \quad S_{12} = -\frac{\nu_{21}}{E_2}, \quad S_{13} = -\frac{\nu_{31}}{E_3} \\ S_{21} &= -\frac{\nu_{12}}{E_1}, \quad S_{22} = \frac{1}{E_2}, \quad S_{23} = -\frac{\nu_{32}}{E_3} \\ S_{31} &= -\frac{\nu_{13}}{E_1}, \quad S_{32} = -\frac{\nu_{23}}{E_2}, \quad S_{33} = \frac{1}{E_3} \end{aligned} \tag{3.1}$$

The elastic modulus and Poisson's ratio can be expressed as:

E_i is the elastic modulus in the i-direction, with a normal stress applied in the i-direction.
ν_{ij} is the Poisson's ratio for transverse strain in the j-direction, with a normal stress applied in the i-direction.

Since the compliance matrix is symmetric, a simplifying relationship exists between Poisson's ratio and the elastic moduli.

$$\frac{\nu_{ij}}{E_i}=\frac{\nu_{ji}}{E_j} \quad (i,j=1,2,3) \tag{3.2}$$

Alternately, $\nu_{12}/E_1=\nu_{21}/E_2$, $\nu_{13}/E_1=\nu_{31}/E_3$, $\nu_{23}/E_2=\nu_{32}/E_3$. Fibers are generally stronger and stiffer than the matrix. Therefore, E_1 (associated with the fiber direction) is typically greater than either E_2 or E_3 (associated with the matrix direction).
The relationship defined by (3.2) can be used to express (3.1) as

$$\begin{aligned}
S_{11}&=\frac{1}{E_1}, \quad S_{12}=-\frac{\nu_{12}}{E_1}, \quad S_{13}=-\frac{\nu_{13}}{E_1}\\
S_{21}&=-\frac{\nu_{12}}{E_1}, \quad S_{22}=\frac{1}{E_2}, \quad S_{23}=-\frac{\nu_{23}}{E_2}\\
S_{31}&=-\frac{\nu_{13}}{E_1}, \quad S_{32}=-\frac{\nu_{23}}{E_2}, \quad S_{33}=\frac{1}{E_3}
\end{aligned} \tag{3.3}$$

In addition to the normal components of the compliance matrix, the shear terms S_{44}, S_{55}, and S_{66} must be determined. In principle this is a simple matter, since there is no shear–extension coupling. By application of a pure shear on the 2–3, 1–3, and 1–2 planes, the relationship between shear stress and strain are

$$S_{44}=\frac{1}{G_{23}}, \quad S_{55}=\frac{1}{G_{13}}, \quad S_{66}=\frac{1}{G_{12}} \tag{3.4}$$

where G_{ij} is the shear modulus corresponding to a shear stress applied to the ij-plane. In order to completely characterize an orthotropic lamina, nine elastic constants are required. Not all nine of these constants are generally determined. In many cases the 1–2 direction is given most of the attention since properties in the 1–3 and 2–3 directions arc difficult to establish.

3.2.1 Stiffness matrix

The stiffness matrix is obtained by inverting the compliance matrix. The stiffness matrix is, by convention, expressed as [Q] instead of [C]. The form of the stiffness matrix presented in Chapter 2 for an orthotropic material is more accurately referred to as *specially orthotropic*. The stress–strain relationship for a specially orthotropic lamina is

$$\begin{Bmatrix} \sigma_1 \\ \sigma_2 \\ \sigma_3 \\ \sigma_4 \\ \sigma_5 \\ \sigma_6 \end{Bmatrix} = \begin{bmatrix} Q_{11} & Q_{12} & Q_{13} & 0 & 0 & 0 \\ Q_{12} & Q_{22} & Q_{23} & 0 & 0 & 0 \\ Q_{13} & Q_{23} & Q_{33} & 0 & 0 & 0 \\ 0 & 0 & 0 & Q_{44} & 0 & 0 \\ 0 & 0 & 0 & 0 & Q_{55} & 0 \\ 0 & 0 & 0 & 0 & 0 & Q_{66} \end{bmatrix} \begin{Bmatrix} \varepsilon_1 \\ \varepsilon_2 \\ \varepsilon_3 \\ \varepsilon_4 \\ \varepsilon_5 \\ \varepsilon_6 \end{Bmatrix} \tag{3.5}$$

The individual components of the stiffness matrix $[Q]$ are expressed in terms of the elastic constants as

$$\begin{aligned} Q_{11} &= E_1(1-\nu_{23}\nu_{32})/\Delta \\ Q_{22} &= E_2(1-\nu_{31}\nu_{13})/\Delta \\ Q_{33} &= E_3(1-\nu_{12}\nu_{21})/\Delta \\ Q_{12} &= E_1(\nu_{21}+\nu_{31}\nu_{23})/\Delta = E_2(\nu_{12}+\nu_{32}\nu_{13})/\Delta \\ Q_{13} &= E_1(\nu_{31}+\nu_{21}\nu_{32})/\Delta = E_3(\nu_{13}+\nu_{12}\nu_{23})/\Delta \\ Q_{23} &= E_2(\nu_{32}+\nu_{12}\nu_{31})/\Delta = E_3(\nu_{23}+\nu_{21}\nu_{13})/\Delta \\ Q_{44} &= G_{23} \\ Q_{55} &= G_{13} \\ Q_{66} &= G_{12} \end{aligned} \tag{3.6}$$

where $\Delta = 1-\nu_{12}\nu_{21}-\nu_{23}\nu_{32}-\nu_{31}\nu_{13}-2\nu_{13}\nu_{21}\nu_{32}$.

Under appropriate conditions these expressions can be simplified. For example, the elastic moduli in the 2- and 3-directions are generally assumed to be the same ($E_2 = E_3$), which implies $\nu_{23} = \nu_{32}$ and $\nu_{21}\nu_{13} = \nu_{12}\nu_{31}$. Simplifications to these equations can be made by assumptions such as plane stress.

The relation between elastic constants (shear, bulk, elastic moduli, and Poisson's ratio), which must be satisfied for an isotropic material, places restrictions on the possible range of values for Poisson's ratio of $-1 < \nu < 1/2$. In a similar manner, there are restrictions on the relationships between ν_{ij} and E_i in orthotropic materials. These constraints, first generalized by Lempriere [1], are based on first law of thermodynamics considerations. In formulating these constraints, both the stiffness and compliance matrices must be positive-definite. Therefore, each major diagonal term of both matrices must be >0. This results in two restrictions on elastic moduli and Poisson's ratio. They are deduced from (3.3), (3.4), and (3.6) to be $E_1,\ E_2,\ E_3,\ G_{23},\ G_{13},\ G_{12} > 0$ and $(1-\nu_{23}\nu_{32}),\ (1-\nu_{31}\nu_{13}),\ (1-\nu_{12}\nu_{21}) > 0$. From the second of these relationships and (3.2), it can be shown that $|\nu_{ij}| < \sqrt{E_i/E_j}$, which results in

$$|\nu_{21}| < \sqrt{E_2/E_1},\quad |\nu_{12}| < \sqrt{E_1/E_2},\quad |\nu_{32}| < \sqrt{E_3/E_2},\quad |\nu_{23}| < \sqrt{E_2/E_3},$$
$$|\nu_{31}| < \sqrt{E_3/E_1},\quad |\nu_{13}| < \sqrt{E_1/E_3}$$

In conjunction with these, it can also be shown that $\Delta = 1 - \nu_{12}\nu_{21} - \nu_{23}\nu_{32} - \nu_{31}\nu_{13} - 2\nu_{13}\nu_{21}\nu_{32} > 0$. This expression can be rearranged to show that $2\nu_{13}\nu_{21}\nu_{32} < \left[1 - \nu_{21}^2(E_1/E_2) - \nu_{32}^2(E_2/E_3) - \nu_{13}^2(E_3/E_1)\right] < 1$.

In turn, this relationship can be manipulated to show that

$$\left[1 - \nu_{32}^2(E_2/E_3)\right]\left[1 - \nu_{13}^2(E_3/E_1)\right] - \left\{\nu_{21}\sqrt{E_1/E_2} + \nu_{13}\nu_{32}\sqrt{E_2/E_1}\right\}^2 > 0$$

A relationship between ν_{12} and the other terms in this expression can be obtained by further rearrangement of the above expression. The form of this relationship is, from Jones [2],

$$-\left\{\nu_{32}\nu_{13}\left(\frac{E_2}{E_1}\right) + \sqrt{1 - \nu_{32}^2\left(\frac{E_2}{E_3}\right)}\sqrt{1 - \nu_{13}^2\left(\frac{E_3}{E_1}\right)}\sqrt{\frac{E_2}{E_1}}\right\} < \nu_{21}$$
$$< -\left\{\nu_{32}\nu_{13}\left(\frac{E_2}{E_1}\right) - \sqrt{1 - \nu_{32}^2\left(\frac{E_2}{E_3}\right)}\sqrt{1 - \nu_{13}^2\left(\frac{E_3}{E_1}\right)}\sqrt{\frac{E_2}{E_1}}\right\}$$

Additional expressions involving ν_{23} and ν_{13} can be formulated but are not presented. These restrictions on the engineering constants for orthotropic materials can be used to evaluate experimental data and assess whether or not it is physically consistent within the framework of the elasticity model developed. Numerical values of Poisson's ratio determined from experimental techniques may appear to be unrealistically high but fit within the constraints presented [2].

3.2.2 Transformation of stresses

Equation (3.5) is based on elastic constants for the special case of the x-axis coinciding with the 1-axis of the material (an *on-axis* configuration). In this arrangement, the $x(1)$-direction is associated with the maximum lamina stiffness, while the $y(2)$-direction corresponds to the direction of minimum lamina stiffness. It is not always practical for the x-axis of a lamina to correspond to the 1-axis of the material. An orientation in which x–y and 1–2 do not coincide is an *off-axis* configuration. Both configurations are illustrated in Figure 3.3.

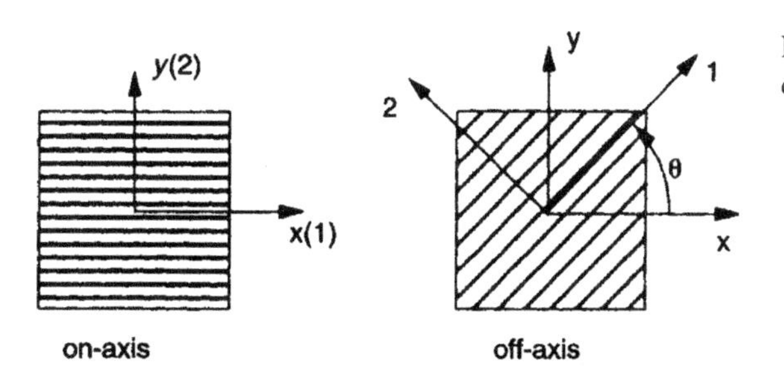

Figure 3.3 On- and off-axis configurations.

The on-axis stress–strain relationship of (3.5) is not adequate for the analysis of an off-axis configuration. Relating stresses and strains in the x–y system to the constitutive relations developed for the 1–2 system requires the use of transformation Equations (2.2) and (2.4) (repeated here in abbreviated form)

$$\begin{Bmatrix} \varepsilon_1 \\ \cdot \\ \varepsilon_6 \end{Bmatrix} = [T_\varepsilon] \begin{Bmatrix} \varepsilon_x \\ \cdot \\ \gamma_{xy} \end{Bmatrix} \quad \begin{Bmatrix} \sigma_1 \\ \cdot \\ \sigma_6 \end{Bmatrix} = [T_\sigma] \begin{Bmatrix} \sigma_x \\ \cdot \\ \tau_{xy} \end{Bmatrix}$$

From Equation (3.5) and the stress and strain transformations above, the principal material direction stress components in terms of Cartesian strain components are

$$\begin{Bmatrix} \sigma_1 \\ \cdot \\ \sigma_6 \end{Bmatrix} = [Q] \begin{Bmatrix} \varepsilon_1 \\ \cdot \\ \varepsilon_6 \end{Bmatrix} = [Q][T_\varepsilon] \begin{Bmatrix} \varepsilon_x \\ \cdot \\ \gamma_{xy} \end{Bmatrix}$$

Cartesian and principal material direction stresses are related by

$$\begin{Bmatrix} \sigma_x \\ \cdot \\ \tau_{xy} \end{Bmatrix} = [T_\sigma]^{-1} \begin{Bmatrix} \sigma_1 \\ \cdot \\ \sigma_6 \end{Bmatrix}$$

The Cartesian stress–strain relationship can be written as

$$\begin{Bmatrix} \sigma_x \\ \cdot \\ \tau_{xy} \end{Bmatrix} = [T_\sigma]^{-1}[Q][T_\varepsilon] \begin{Bmatrix} \varepsilon_x \\ \cdot \\ \gamma_{xy} \end{Bmatrix}$$

This expression can be simplified by defining a new matrix $[\bar{Q}] = [T_\sigma]^{-1}[Q][T_\varepsilon]$. The expanded form of the Cartesian stress–strain relationship is

$$\begin{Bmatrix} \sigma_x \\ \sigma_y \\ \sigma_z \\ \tau_{yz} \\ \tau_{xz} \\ \tau_{xy} \end{Bmatrix} = \begin{bmatrix} \bar{Q}_{11} & \bar{Q}_{12} & \bar{Q}_{13} & 0 & 0 & \bar{Q}_{16} \\ \bar{Q}_{12} & \bar{Q}_{22} & \bar{Q}_{23} & 0 & 0 & \bar{Q}_{26} \\ \bar{Q}_{13} & \bar{Q}_{23} & \bar{Q}_{33} & 0 & 0 & \bar{Q}_{36} \\ 0 & 0 & 0 & \bar{Q}_{44} & \bar{Q}_{45} & 0 \\ 0 & 0 & 0 & \bar{Q}_{45} & \bar{Q}_{55} & 0 \\ \bar{Q}_{16} & \bar{Q}_{26} & \bar{Q}_{36} & 0 & 0 & \bar{Q}_{66} \end{bmatrix} \begin{Bmatrix} \varepsilon_x \\ \varepsilon_y \\ \varepsilon_z \\ \gamma_{yz} \\ \gamma_{xz} \\ \gamma_{xy} \end{Bmatrix} \tag{3.7}$$

Each element of $[\bar{Q}]$ is defined as

$$\begin{aligned}
\bar{Q}_{11} &= Q_{11}m^4 + 2(Q_{12} + 2Q_{66})m^2n^2 + Q_{22}n^4 \\
\bar{Q}_{12} &= (Q_{11} + Q_{22} - 4Q_{66})m^2n^2 + Q_{12}(m^4 + n^4) \\
\bar{Q}_{13} &= Q_{13}m^2 + Q_{23}n^2 \\
\bar{Q}_{16} &= -Q_{22}mn^3 + Q_{11}m^3n - (Q_{12} + 2Q_{66})mn(m^2 - n^2) \\
\bar{Q}_{22} &= Q_{11}n^4 + 2(Q_{12} + 2Q_{66})m^2n^2 + Q_{22}m^4 \\
\bar{Q}_{23} &= Q_{13}n^2 + Q_{23}m^2 \\
\bar{Q}_{26} &= -Q_{22}m^3n + Q_{11}mn^3 - (Q_{12} + 2Q_{66})mn(m^2 - n^2) \\
\bar{Q}_{33} &= Q_{33} \\
\bar{Q}_{36} &= (Q_{13} - Q_{23})mn \\
\bar{Q}_{44} &= Q_{44}m^2 + Q_{55}n^2 \\
\bar{Q}_{45} &= (Q_{55} - Q_{44})mn \\
\bar{Q}_{55} &= Q_{55}m^2 + Q_{44}n^2 \\
\bar{Q}_{66} &= (Q_{11} + Q_{22} - 2Q_{12})m^2n^2 + Q_{66}(m^2 - n^2)^2
\end{aligned} \tag{3.8}$$

where $m = \cos\theta$ and $n = \sin\theta$. Equations (3.7) and (3.8) allow for the analysis of an off-axis lamina provided an on-axis constitutive relationship exists. Note that the off-axis relations (3.7) indicate extension–shear coupling.

The orthotropic material response of (3.7) appears to be different than the specially orthotropic response of (3.5) because of the reference system chosen to define the material behavior. In Chapter 2 (Table 2.1), the number of independent elastic constants and nonzero stiffness matrix terms for an orthotropic material were presented. Comparing (3.5) and (3.7), it should now be easier to see the relationship between a material response and a selected reference axis. The same behavior is predicted by both (3.5) and (3.7). The difference between them is the reference system used to define the response. Although there are still only 9 independent elastic constants represented by Equation (3.7), there are 20 nonzero terms in the stiffness matrix. The material response defined by (3.7) is termed *generally orthotropic*. The material response previously termed *specially orthotropic* is typically reserved for on-axis cases in which there is no shear–extension coupling.

The development of (3.8) follows directly from the transformation equations in Chapter 2. In examining the terms associated with each Q_{ij}, it is evident that coupling between $\sin\theta$ and $\cos\theta$ terms exists. Figure 3.4 shows the sign convention for which Equation (3.8) is valid. Although it will generally not affect most terms of the $[\bar{Q}]$ matrix, a mistake in the + or − sense of θ can influence the shear terms ($\bar{Q}_{16}$, $\bar{Q}_{26}$, $\bar{Q}_{36}$, and $\bar{Q}_{45}$). In some texts a positive angle θ is measured from the principal material axis (1) to the x-axis. For our development, this would cause the n ($\sin\theta$) term in (3.8) to be negative, thus causing a sign change. Although the sign of the shear stress generally does not affect the shear failure strength of a lamina, it does have an effect on other stress components.

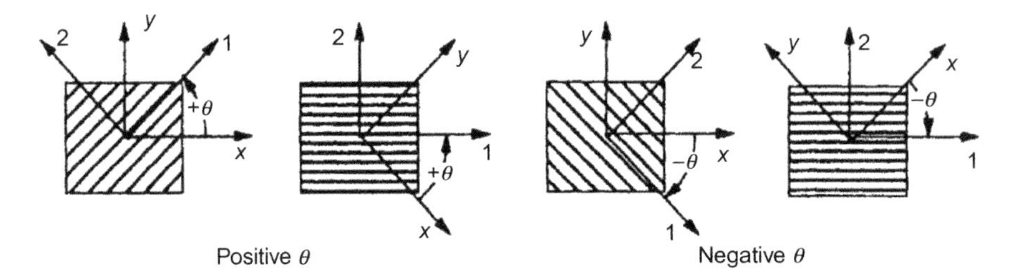

Figure 3.4 Sign convention of positive and negative fiber orientations.

3.2.3 Plane stress analysis

For plane stress, Equations (3.5) and (3.7) reduce since one normal and two shear components of stress are zero. As with the case of an isotropic material, the elimination of stress components does not imply that strain components become zero. Either the stiffness or compliance matrix can be used for plane stress analysis.

3.2.3.1 Stiffness matrix

In the case of plane stress we assume that in the material coordinate system $\sigma_3 = \sigma_4 = \sigma_5 = 0$. The stiffness matrix for plane stress is termed the *reduced stiffness matrix*. The on-axis form of the reduced stiffness matrix is similar to the $[Q]$ of (3.5) and is

$$[Q] = \begin{bmatrix} Q_{11} & Q_{12} & 0 \\ Q_{12} & Q_{22} & 0 \\ 0 & 0 & Q_{66} \end{bmatrix} \tag{3.9}$$

where the individual terms are

$$Q_{11} = \frac{E_1}{1 - \nu_{12}\nu_{21}}, \quad Q_{22} = \frac{E_2}{1 - \nu_{12}\nu_{21}}, \quad Q_{12} = \frac{\nu_{12}E_2}{1 - \nu_{12}\nu_{21}} = \frac{\nu_{21}E_1}{1 - \nu_{12}\nu_{21}}, \quad Q_{66} = G_{12}$$

Although both out-of-plane shear strains are zero for this case, the normal strain (ε_3) exists and is easily derived from (3.5).

The off-axis form of the reduced stiffness matrix contains out-of-plane strains. Both normal (ε_z) and shear strains (γ_{yz} and γ_{xz}) may be present under the plane stress assumptions of $\sigma_z = \tau_{xz} = \tau_{yz} = 0$. These shear strains are not generally included for in-plane analysis. In some cases, these terms are included but are separated from the in-plane portion of the analysis. Beam, plate, and shell problems formulated from a displacement field approach generally include these terms. Such

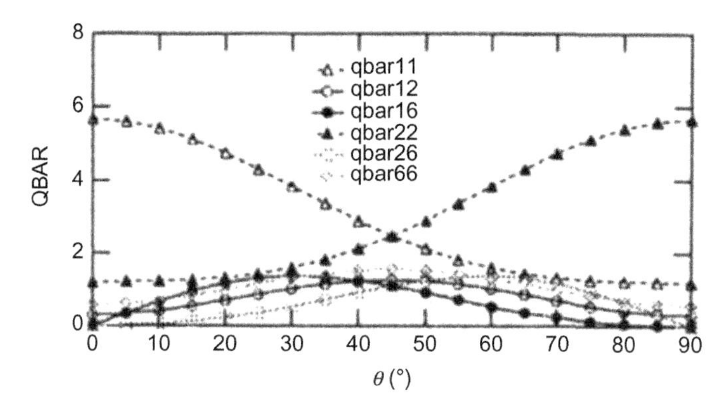

Figure 3.5 Variation of $[\bar{Q}]$ with θ for a carbon/epoxy lamina.

formulations are considered advanced topics and are not addressed in this text. The off-axis form of the reduced stiffness is formulated using the stress and strain transformations of Chapter 2 along with (3.9). Following the same procedures as before results in

$$\begin{Bmatrix} \sigma_x \\ \sigma_y \\ \tau_{xy} \end{Bmatrix} = \begin{bmatrix} \bar{Q}_{11} & \bar{Q}_{12} & \bar{Q}_{16} \\ \bar{Q}_{12} & \bar{Q}_{22} & \bar{Q}_{26} \\ \bar{Q}_{16} & \bar{Q}_{26} & \bar{Q}_{66} \end{bmatrix} \begin{Bmatrix} \varepsilon_x \\ \varepsilon_y \\ \gamma_{xy} \end{Bmatrix} \tag{3.10}$$

where the corresponding terms from Equation (3.8) remain applicable for the plane stress case given by (3.10).

The variation of each component of $[\bar{Q}]$ with θ is illustrated in Figure 3.5 for a carbon/epoxy lamina. Although $\bar{Q}_{11}$ and $\bar{Q}_{22}$ are the dominant terms for all fiber orientations, all components of $[\bar{Q}]$ contribute to defining the overall material response characteristics.

3.2.3.2 *Invariant form of* $[\bar{Q}]$

The components of $[\bar{Q}]$ in (3.10) can be expressed in a form different than (3.8). This alternate representation is known as the invariant form of $[\bar{Q}]$ and was first introduced by Tsai and Pagano [3]. In order to establish the invariant forms of $[\bar{Q}]$, recall the trig identity $\cos^2\theta = (1+\cos 2\theta)/2$ and the similar one for $\sin^2\theta$. Using these yields

$$\cos^4\theta = (3+4\cos 2\theta+\cos 4\theta)/8, \quad \sin^4\theta = (3-4\cos 2\theta+\cos 4\theta)/8$$

$$\sin\theta\cos^3\theta = (2\sin 2\theta+\sin 4\theta)/8, \quad \cos\theta\sin^3\theta = (2\sin 2\theta-\sin 4\theta)/8$$

$$\cos^2\theta\sin^2\theta = (1-\cos 4\theta)/8$$

Evaluating $\bar{Q}_{11}$ from (3.8) and using the trig functions above results in

$$\bar{Q}_{11}=(3+4\cos 2\theta+\cos 4\theta)\frac{Q_{11}}{8}+2(1-\cos 4\theta)\frac{Q_{12}+2Q_{66}}{8}+(3-4\cos 2\theta+\cos 4\theta)\frac{Q_{22}}{8}$$

$$\bar{Q}_{11}=\frac{3Q_{11}+2Q_{12}+4Q_{66}+3Q_{22}}{8}+\left(\frac{4Q_{11}-4Q_{22}}{8}\right)\cos 2\theta$$
$$+\left(\frac{Q_{11}-2Q_{12}-4Q_{66}+Q_{22}}{8}\right)\sin 4\theta$$

This expression, as well as similar expressions for the remainder of the $[\bar{Q}]$ matrix, can be simplified. The following definitions are introduced:

$$\begin{aligned}U_1&=\frac{3Q_{11}+2Q_{12}+4Q_{66}+3Q_{22}}{8}\\U_2&=\frac{Q_{11}-Q_{22}}{2}\\U_3&=\frac{Q_{11}+Q_{22}-2Q_{12}-4Q_{66}}{8}\\U_4&=\frac{Q_{11}+Q_{22}+6Q_{12}-4Q_{66}}{8}\\U_5&=\frac{Q_{11}+Q_{22}-2Q_{12}+4Q_{66}}{8}\end{aligned} \tag{3.11}$$

The explicit form of $[\bar{Q}]$ can now be expressed as

$$\begin{Bmatrix}\bar{Q}_{11}\\\bar{Q}_{22}\\\bar{Q}_{12}\\\bar{Q}_{66}\\\bar{Q}_{16}\\\bar{Q}_{26}\end{Bmatrix}=\begin{bmatrix}U_1 & \cos 2\theta & \cos 4\theta\\U_1 & -\cos 2\theta & \cos 4\theta\\U_4 & 0 & -\cos 4\theta\\U_5 & 0 & -\cos 4\theta\\0 & \sin 2\theta/2 & \sin 4\theta\\0 & \sin 2\theta/2 & -\sin 4\theta\end{bmatrix}\begin{Bmatrix}1\\U_2\\U_3\end{Bmatrix} \tag{3.12}$$

This expression for $[\bar{Q}]$ provides an alternate representation for several terms in Equation (3.8) but does not prove the invariance of any parameter in (3.11) or (3.12). In order to prove that some of these quantities are invariant (meaning they do not depend on fiber orientation), we examine the U_1 term. Initially, it is assumed that an off-axis configuration is being investigated. In the off-axis configuration U_1 is expressed as U_1', and $[Q]$ is replaced by $[\bar{Q}]$. Therefore, $U_1'=(3\bar{Q}_{11}+3\bar{Q}_{22}+2\bar{Q}_{12}+4\bar{Q}_{66})/8$. Substituting the appropriate expressions for each $[\bar{Q}]$ term from (3.12) and simplifying leads to $U_1'=(6U_1+2U_4+4U_5)/8$. Substituting the expressions for U_1, U_4, and U_5 from (3.11) into this equation yields $U_1'=(3Q_{11}+3Q_{22}+2Q_{12}+4Q_{66})/8$. This is the expression for U_1 given in (3.11). Therefore, U_1 is invariant, which means that it does not change with the orientation of the lamina. In addition, it can be shown that

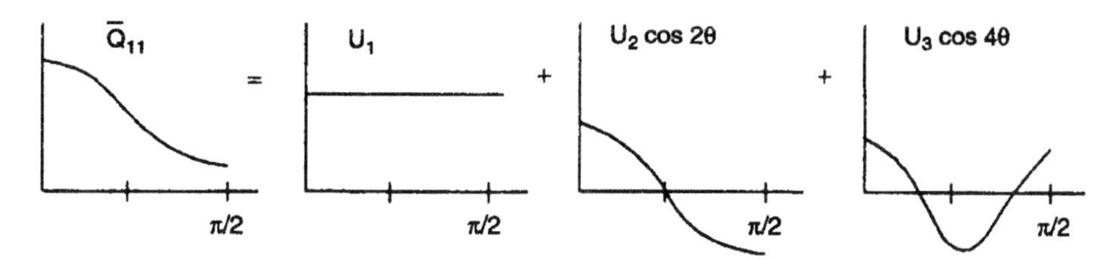

Figure 3.6 Components of $\bar{Q}_{11}$.

U_4 and U_5 are also invariant. Similarly, it can be shown that $U_5 = (U_1 - U_4)/2$. The primary advantage of expressing the components of $[\bar{Q}]$ in invariant form is that it can lead to simplifications in the design process, since several terms which do not vary with orientation are involved. Consider, for example, the $\bar{Q}_{11}$ term, which can be expressed as

$$\bar{Q}_{11} = U_1 + U_2 \cos 2\theta + U_3 \sin 4\theta$$

This term can be decomposed into its components, with each plotted as a function of θ. As seen in Figure 3.6, the total response of $[\bar{Q}]$ is linked to two components which vary with θ and one which is invariant. The usefulness of the invariant form of $[\bar{Q}]$ becomes more evident when laminate analysis is considered.

3.2.3.3 Compliance matrix

As with the stiffness matrix, the compliance matrix reduces for cases of plane stress. The strain–stress relationship for an on-axis configuration is

$$\begin{Bmatrix} \varepsilon_1 \\ \varepsilon_2 \\ \gamma_{12} \end{Bmatrix} = \begin{bmatrix} S_{11} & S_{12} & 0 \\ S_{12} & S_{22} & 0 \\ 0 & 0 & S_{66} \end{bmatrix} \begin{Bmatrix} \sigma_1 \\ \sigma_2 \\ \tau_{12} \end{Bmatrix} \quad (3.13)$$

where each component is expressed in terms of the elastic constants as

$$S_{11} = \frac{1}{E_1}, \quad S_{12} = -\frac{\nu_{12}}{E_1} = -\frac{\nu_{21}}{E_2}, \quad S_{22} = \frac{1}{E_2}, \quad S_{66} = \frac{1}{G_{12}} \quad (3.14)$$

Following transformation procedures similar to those for establishing $[\bar{Q}]$, it can be shown that the off-axis strain–stress relationship for a case of plane stress is

$$\begin{Bmatrix} \varepsilon_x \\ \varepsilon_y \\ \gamma_{xy} \end{Bmatrix} = \begin{bmatrix} \bar{S}_{11} & \bar{S}_{12} & \bar{S}_{16} \\ \bar{S}_{12} & \bar{S}_{22} & \bar{S}_{26} \\ \bar{S}_{16} & \bar{S}_{26} & \bar{S}_{66} \end{bmatrix} \begin{Bmatrix} \sigma_x \\ \sigma_y \\ \tau_{xy} \end{Bmatrix} \quad (3.15)$$

where

$$
\begin{aligned}
\bar{S}_{11} &= S_{11}m^4 + (2S_{12} + S_{66})m^2n^2 + S_{22}n^4 \\
\bar{S}_{12} &= (S_{11} + S_{22} - S_{66})m^2n^2 + S_{12}(m^4 + n^4) \\
\bar{S}_{16} &= (2S_{11} - 2S_{12} - S_{66})m^3n - (2S_{22} - 2S_{12} - S_{66})mn^3 \\
\bar{S}_{22} &= S_{11}n^4 + (2S_{12} + S_{66})m^2n^2 + S_{22}m^4 \\
\bar{S}_{26} &= (2S_{11} - 2S_{12} - S_{66})mn^3 - (2S_{22} - 2S_{12} - S_{66})m^3n \\
\bar{S}_{66} &= 2(2S_{11} + 2S_{22} - 4S_{12} - S_{66})m^2n^2 + S_{66}(m^4 + n^4)
\end{aligned}
\tag{3.16}
$$

The sign convention used for θ must be the same as that used for Equation (3.8), as defined in Figure 3.4.

The compliance matrix can also be expressed in terms of invariants. The derivation is analogous to that for the stiffness matrix. Using the same notation as before, we define

$$
\begin{aligned}
U_1 &= \frac{3S_{11} + 3S_{22} + 2S_{12} + S_{66}}{8} \\
U_2 &= \frac{S_{11} - S_{22}}{2} \\
U_4 &= \frac{S_{11} + S_{22} + 6S_{12} - S_{66}}{8} \\
U_5 &= \frac{S_{11} + S_{22} - 2S_{12} + S_{66}}{2}
\end{aligned}
\tag{3.17}
$$

The expressions for $[\bar{S}]$ in terms of U_1 through U_5 are

$$
\begin{Bmatrix} \bar{S}_{11} \\ \bar{S}_{22} \\ \bar{S}_{12} \\ \bar{S}_{66} \\ \bar{S}_{16} \\ \bar{S}_{26} \end{Bmatrix} = \begin{bmatrix} U_1 & \cos 2\theta & \cos 4\theta \\ U_1 & -\cos 2\theta & \cos 4\theta \\ U_4 & 0 & -\cos 4\theta \\ U_5 & 0 & -4\cos 4\theta \\ 0 & \sin 2\theta & 2\sin 4\theta \\ 0 & \sin 2\theta & -4\sin 4\theta \end{bmatrix} \begin{Bmatrix} 1 \\ U_2 \\ U_3 \end{Bmatrix}
\tag{3.18}
$$

The compliance matrix can be used to obtain useful information regarding off-axis lamina response. Consider, for example, the off-axis lamina shown in Figure 3.7.

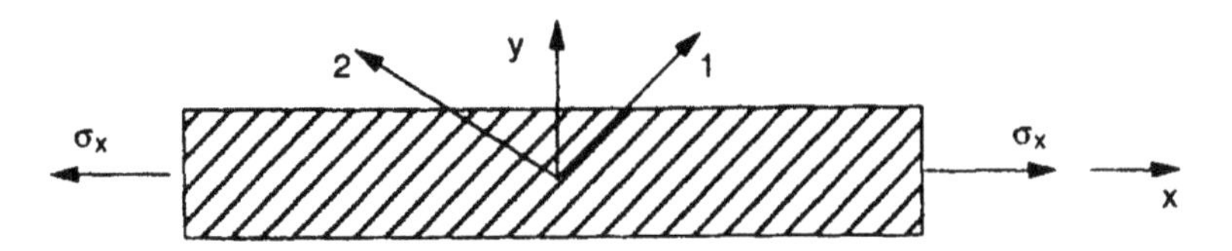

Figure 3.7 Off-axis lamina subjected to uniaxial tension.

This lamina is subjected to an applied stress σ_x. The strain in the direction of loading (ε_x) is to be monitored during loading. The strain and stress for this case are related by

$$\begin{Bmatrix} \varepsilon_x \\ \varepsilon_y \\ \gamma_{xy} \end{Bmatrix} = [\bar{S}]\begin{Bmatrix} \sigma_x \\ \sigma_y \\ \tau_{xy} \end{Bmatrix} = [\bar{S}]\begin{Bmatrix} \sigma_x \\ 0 \\ 0 \end{Bmatrix}$$

From this it is apparent that $\varepsilon_x = \bar{S}_{11}\sigma_x$. From the definition of elastic modulus, we write $\sigma_x = E_x\varepsilon_x$. For the apparent engineering modulus in the x-direction, the above expression can be written as

$$\frac{\varepsilon_x}{\sigma_x} = \frac{1}{E_x} = \bar{S}_{11} = S_{11}m^4 + (2S_{12} + S_{66})m^2n^2 + S_{22}n^4$$

Using the definitions of S_{11}, S_{12}, etc., this expression is written as

$$\frac{1}{E_x} = \frac{m^4}{E_1} + \left(\frac{1}{G_{12}} - \frac{2\nu_{12}}{E_1}\right)m^2n^2 + S_{22}\frac{n^4}{E_2}$$

Following similar procedures, one can relate the apparent engineering constants in an off-axis configuration (E_x, E_y, etc.) to on-axis material properties by loading a lamina in its non-principal directions. The following expressions result

$$\begin{aligned}
&\frac{1}{E_x} = \frac{m^4}{E_1} + \left(\frac{1}{G_{12}} - \frac{2\nu_{12}}{E_1}\right)m^2n^2 + S_{22}\frac{n^4}{E_2} \\
&\frac{1}{E_y} = \frac{n^4}{E_1} + \left(\frac{1}{G_{12}} - \frac{2\nu_{12}}{E_1}\right)m^2n^2 + S_{22}\frac{m^4}{E_2} \\
&\frac{1}{G_{xy}} = 2m^2n^2\left(\frac{2}{E_1} + \frac{2}{E_2} + \frac{4\nu_{12}}{E_1} - \frac{1}{G_{12}}\right) + \frac{m^4 + n^4}{G_{12}} \\
&\nu_{xy} = E_x\left[\frac{\nu_{12}(m^4 + n^4)}{E_1} - m^2n^2\left(\frac{1}{E_1} + \frac{1}{E_2} - \frac{1}{G_{12}}\right)\right] \\
&\eta_{xy,x} = E_x[C_1m^3n - C_2mn^3] \\
&\eta_{xy,y} = E_y[C_1mn^3 - C_2m^3n]
\end{aligned} \tag{3.19}$$

where

$$C_1 = \frac{2}{E_1} + \frac{2\nu_{12}}{E_1} - \frac{1}{G_{12}}, \quad C_2 = \frac{2}{E_2} + \frac{2\nu_{12}}{E_1} - \frac{1}{G_{12}}$$

The ηs are called *coefficients of mutual influence*, credited to Lekhniski [4], and are defined as

$\eta_{i,ij}$ is the coefficient of mutual influence of the first kind, which characterizes stretching in the i-direction caused by shear in the ij-plane, and can be written as

$\eta_{i,ij} = \frac{\varepsilon_i}{\gamma_{ij}}$ for $\tau_{ij} = \tau$

$\eta_{ij,i}$ is the coefficient of mutual influence of the second kind, which characterizes shearing in the ij-plane caused by a normal stress in the i-direction, and can be written as

$$\eta_{ij,i} = \frac{\gamma_{ij}}{\varepsilon_i} \text{ for } \sigma_i = \sigma$$

The coefficients of mutual influence are not frequently used in classical lamination theory but are useful in relating out-of-plane shear strains to in-plane shear and normal stresses. These relationships are generally presented in terms of coefficients of mutual influence and Chentsov coefficients. The Chentsov coefficients are expressed as $\mu_{ij,kl}$ and characterize shearing strain in the ij-plane due to shearing stress in the kl-plane. Mathematically it is defined as

$$\mu_{ij,kl} = \frac{\gamma_{ij}}{\gamma_{kl}}$$

For $\tau_{kl} = \tau$, with all other stress components being zero. The Chentsov coefficients are subject to the reciprocal relationship $\mu_{ij,kl}/G_{kl} = \mu_{kl,ij}/G_{ij}$. The relationship between out-of-plane shear strains and in-plane stress components for the 1–3 and 2–3 planes is

$$\gamma_{13} = \frac{\eta_{1,13}\sigma_1 + \eta_{2,13}\sigma_2 + \eta_{12,13}\tau_{12}}{G_{13}} \quad \text{and} \quad \gamma_{23} = \frac{\eta_{1,23}\sigma_1 + \eta_{2,23}\sigma_2 + \eta_{12,23}\tau_{12}}{G_{23}}$$

Similar representations can also be developed by solving (3.7) for strains under conditions of plane stress. The most widely used relationship in (3.19) for apparent engineering constants is generally the one for E_x, which is valuable in establishing material constants.

Example 3.1 Consider the case of uniaxial tension shown in Figure E3.1-1. For this problem, the strain–stress relationship is

$$\begin{Bmatrix} \varepsilon_x \\ \varepsilon_y \\ \gamma_{xy} \end{Bmatrix} = \begin{bmatrix} \bar{S}_{11} & \bar{S}_{12} & \bar{S}_{16} \\ \bar{S}_{12} & \bar{S}_{22} & \bar{S}_{26} \\ \bar{S}_{16} & \bar{S}_{26} & \bar{S}_{66} \end{bmatrix} \begin{Bmatrix} 0 \\ \sigma_0 \\ 0 \end{Bmatrix} = \begin{Bmatrix} \bar{S}_{12} \\ \bar{S}_{22} \\ \bar{S}_{26} \end{Bmatrix} \sigma_0$$

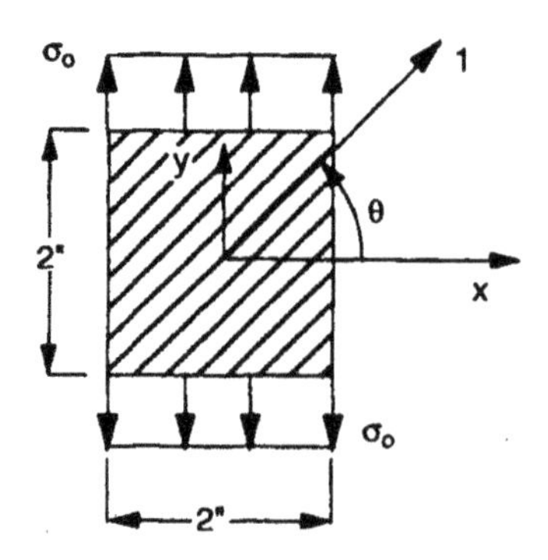

Figure E3.1-1 Off-axis lamina.

Therefore

$$\begin{Bmatrix} \varepsilon_x \\ \varepsilon_y \\ \gamma_{xy} \end{Bmatrix} = \begin{Bmatrix} (S_{11}+S_{22}-S_{66})m^2n^2+S_{12}(m^4+n^4) \\ S_{11}n^4+(2S_{12}+S_{66})m^2n^2+S_{22}m^4 \\ (2S_{11}-2S_{12}-S_{66})mn^3-(2S_{22}-2S_{12}-S_{66})m^3n \end{Bmatrix}\sigma_0$$

From these equations, it is obvious that there is shear–extension coupling for any angle of θ other than 0° or 90°. Even under a simple uniaxial load, the deformation will be similar to that of a state of stress including shear. Assume specimen dimensions as shown in Figure E3.1-1, $\sigma_0 = 50\,\text{ksi}$, $\theta = 45^\circ$ and material properties of $E_1 = 25 \times 10^6\,\text{psi}$, $E_2 = 1 \times 10^6\,\text{psi}$, $G_{12} = 0.5 \times 10^6\,\text{psi}$, and $\nu_{12} = 0.25$. This results in

$$S_{11} = 1/E_1 = 4 \times 10^{-8}, \quad S_{22} = 1/E_2 = 1 \times 10^{-6}, \quad S_{12} = -\nu_{12}S_{11}$$
$$= -1 \times 10^{-8}, \quad S_{66} = 1/G_{12} = 2 \times 10^{-6}$$

At $\theta = 45^\circ$ and with $\sigma_0 = 50\,\text{ksi}$ we get

$$\begin{Bmatrix} \varepsilon_x \\ \varepsilon_y \\ \gamma_{xy} \end{Bmatrix} = \begin{Bmatrix} 0.25(S_{11}+S_{22}-S_{66})+0.5S_{12} \\ 0.25(S_{11}+(2S_{12}+S_{66})+S_{22}m^4) \\ 0.25((2S_{11}-2S_{12}-S_{66})-(2S_{22}-2S_{12}-S_{66})) \end{Bmatrix}\sigma_0$$

$$= \begin{Bmatrix} -2.45 \\ 7.55 \\ -4.8 \end{Bmatrix}(50) \times 10^{-4} = \begin{Bmatrix} -0.01225 \\ 0.03775 \\ -0.025 \end{Bmatrix}\text{in/in}$$

The displacement field is obtained from the definitions of axial and shear strain as follows:

$$\varepsilon_x = \frac{\partial U}{\partial x} = -0.01225 \Rightarrow U = -0.01225x + f(y)$$

$$\varepsilon_y = \frac{\partial V}{\partial y} = 0.03775 \Rightarrow V = 0.03775y + g(x)$$

$$\gamma_{xy} = \frac{\partial U}{\partial y} + \frac{\partial V}{\partial x} = -0.024 \Rightarrow f'(y) + g'(x) = -0.024$$

The functions $f(y)$ and $g(x)$ can be arbitrarily assumed in order to fit the anticipated displacement field. Since the theory developed herein is for small deformations, linear functions are assumed. Therefore, we assume $f(y) = C_1 + C_2y$ and $g(x) = C_3 + C_4x$. Taking the partial derivative of each function with respect to its variable yields $f'(y) = C_2$ and $g'(x) = C_4$. Thus, according to the definition of shear strain $C_2 + C_4 = -0.024$. This results in

$$U = -0.01225x + C_1 + C_2y, \quad V = 0.03775y + C_3 + C_4x$$

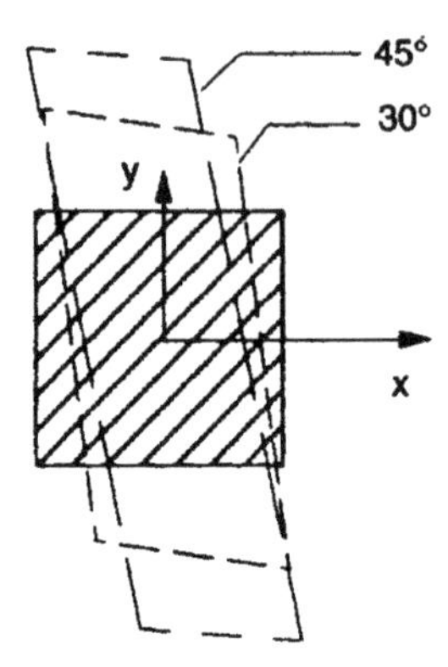

Figure E3.1-2 Deformed shape for $\theta=30°$ and $\theta=45°$.

At the center of the plate ($x=y=0$) the displacements must vanish, so that $U=V=0$. Using this condition, we arrive at $C_1=C_3=0$. The rigid body rotations are eliminated by the requirement that

$$\frac{\partial U}{\partial y}-\frac{\partial V}{\partial x}=0 \;\Rightarrow\; C_2=C_4$$

Thus, $C_2=C_4=-0.012$, and the displacement fields become

$$U=-0.01225x-0.012y, \quad V=0.03775y-0.012x$$

An exaggerated plot of the deformed shape for fiber orientations of both $\theta=45°$ and $\theta=30°$ is shown in Figure E3.1-2. These plots illustrate the effects of shear–extension coupling on the deformation. The actual deformation field which results depends on applied loads, elastic properties of the material, and fiber orientation. This dependency of load, material, and orientation extends from deformations to stress analysis. In the case of laminates in which each ply can have a different fiber orientation (or even be a different material) the coupling between load, material, and orientation is even more pronounced.

Example 3.2 Assume the clamp shown in Figure E3.2-1 is constructed from a unidirectional composite with elastic properties $E_1=30.3\times10^6$ psi, $E_2=2.8\times10^6$ psi, $G_{12}=0.93\times10^6$ psi, and $\nu_{12}=0.21$. The compressive force at points B and C of the clamp is 1.5 kip, and screw ED can only experience a tensile force. For this problem, we will establish the displacement field corresponding to point F, located on plane a–a of the clamp. Thermal and hygral effects are neglected. The fiber orientation is defined in Figure E3.2-1.

The loads acting on section a–a are established by defining an appropriate free-body diagram (FBD). Two possible FBDs can be used, as shown in Figure E3.2-2; one for portion AC of the clamp, or one for portion AB. In either case the unknown tensile force in the screw must be determined.

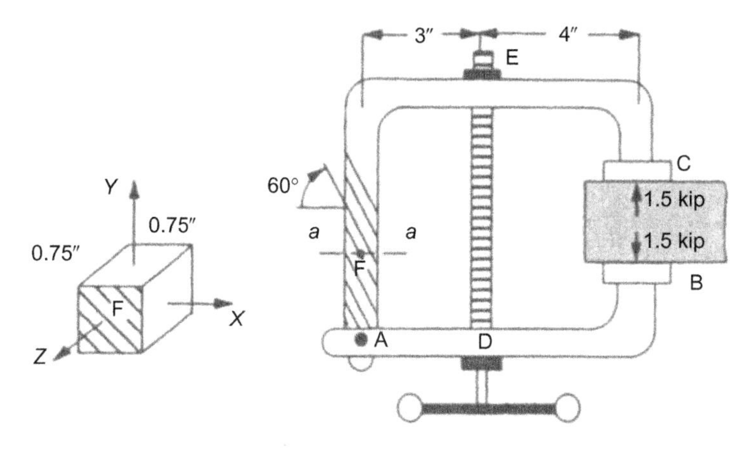

Figure E3.2-1 Composite clamp assembly.

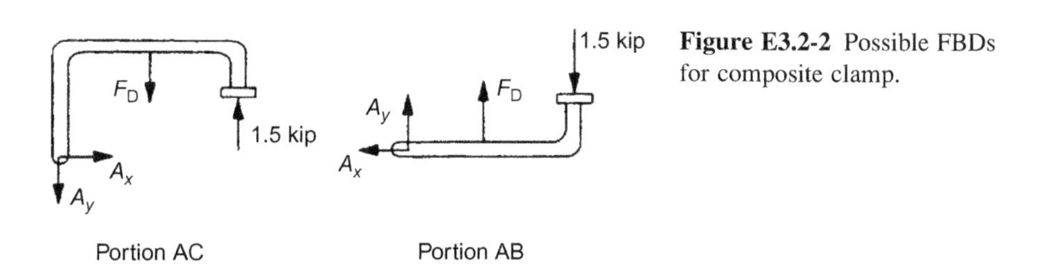

Figure E3.2-2 Possible FBDs for composite clamp.

Either FBD will work for defining the screw tension F_D, which is required to define the loads at section *a–a*. Arbitrarily using the FBD for portion *AB* from Figure E3.2-2, the tensile force F_D can be defined by taking moments about point *A*

$$\sum M_A = 0 = 7(1.5) - 3F_D \Rightarrow F_D = 3.5\,\text{kips}$$

From the FBD of Figure E3.2-3, the internal forces and moments at section *a–a* can be established from

$$\sum F_x = 0 = -V$$

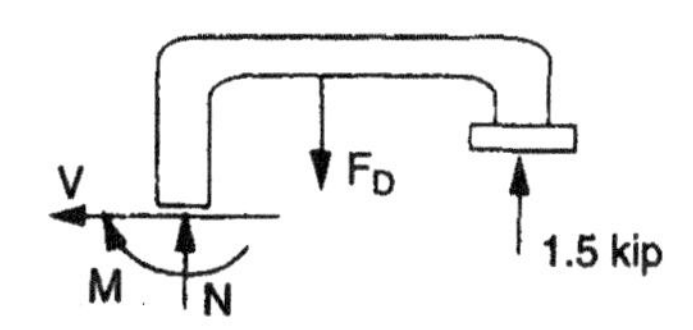

Figure E3.2-3 FBD for internal reactions at section *a–a*.

$$\sum F_y = 0 = N + 1.5 - 3.5 \Rightarrow N = 2.0\,\text{kips}$$

$$\sum M_F = 0 = M + 7(1.5) - 3(1.5) \Rightarrow M = 0$$

Since the internal moment and shear on section *a–a* are both zero, the only stress component to be considered is σ_y, which is easily defined by the compressive force (N) divided by the cross-sectional area of the clamp

$$\sigma_y = \frac{-2.0}{(0.75)(0.75)} = -3.55\,\text{ksi}$$

The state of stress at point F of the cross-section is shown in Figure E3.2-4. Due to the loads on section *a–a*, this state of stress would be identical for all points along plane *a–a*. Furthermore, due to the geometry of the clamp, this state of stress would exist at any point within the vertical section of the clamp, provided stress concentrations due to fillets are neglected.

Having established the state of stress at point F, the resulting displacement field can be defined. Since displacements are required, the compliance matrix relating strain to stress must be defined. The relationship between strain and stress is

$$\begin{Bmatrix} \varepsilon_x \\ \varepsilon_y \\ \gamma_{xy} \end{Bmatrix} = \begin{bmatrix} \bar{S}_{11} & \bar{S}_{12} & \bar{S}_{16} \\ \bar{S}_{12} & \bar{S}_{22} & \bar{S}_{26} \\ \bar{S}_{16} & \bar{S}_{26} & \bar{S}_{66} \end{bmatrix} \begin{Bmatrix} 0 \\ \sigma_y \\ 0 \end{Bmatrix} = \begin{Bmatrix} \bar{S}_{12} \\ \bar{S}_{22} \\ \bar{S}_{26} \end{Bmatrix} (-3.55 \times 10^3)$$

where the values of $[\bar{S}]$ are determined from Equation (3.16) to be $\bar{S}_{12} = -0.1328 \times 10^{-6}$, $\bar{S}_{22} = 0.2399 \times 10^{-6}$, and $\bar{S}_{26} = 0.2875 \times 10^{-6}$. Therefore,

$$\begin{Bmatrix} \varepsilon_x \\ \varepsilon_y \\ \gamma_{xy} \end{Bmatrix} = \begin{Bmatrix} -0.1328 \\ 0.2399 \\ 0.2875 \end{Bmatrix} \times 10^{-6}(-3.55 \times 10^3) = \begin{Bmatrix} 471.4 \\ -857.6 \\ -1014.2 \end{Bmatrix} \mu\text{in./in.}$$

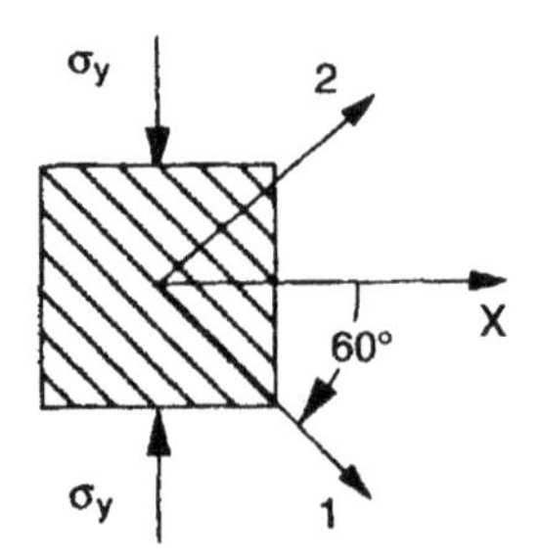

Figure E3.2-4 Stress at point F of the composite clamp.

Using the procedures of the previous example, the displacement fields are found to be

$$U = (471.4x - 507.1y) \times 10^{-6}, \quad V = (-507.1x - 851.6y) \times 10^{-6}$$

These numerical results would be different had another material and/or a different fiber orientation been used. For example, a fiber orientation of $\theta = +60°$, instead of $\theta = -60°$, results in

$$\begin{Bmatrix} \varepsilon_x \\ \varepsilon_y \\ \gamma_{xy} \end{Bmatrix}_{+60} = \begin{Bmatrix} 471.4 \\ -857.6 \\ 1014.2 \end{Bmatrix} \mu\text{in./in.}$$

Since only the shear term ($\bar{S}_{26}$) changes sign in going from $-60°$ to $+60°$, γ_{xy} is the only affected strain. This sign change is reflected in the displacement field, which would be

$$U_{+60} = (471.4x + 507.1y) \times 10^{-6}, \quad V_{+60} = (-507.1x + 851.6y) \times 10^{-6}$$

Example 3.3 Assume a 72-in. diameter closed end pressure vessel is designed to operate under an applied pressure of 100 psi. A unidirectional composite reinforcement is to be circumferentially wound around the vessel at selected intervals along the span. Due to space limitations, the reinforcement has a cross-sectional area of 0.50 in.^{2} The vessel is shown in Figure E3.3-1. Two materials are considered for the reinforcement. We wish to define the reinforcement spacing (s) as a function of the arbitrary fiber orientation angle θ, assuming that the reinforcements sustain all forces typically expressed as the circumferential stress in the vessel. In addition, the normal strain in the circumferential direction is not allowed to exceed 6000 μin./in. The two materials selected have the following elastic properties:

Property	**Material 1**	**Material 2**
E_1 ($\times 10^6$ psi)	30.3	8.29
E_2 ($\times 10^6$ psi)	2.10	2.92
G_{12} ($\times 10^6$ psi)	0.93	0.86
ν_{12}	0.21	0.26

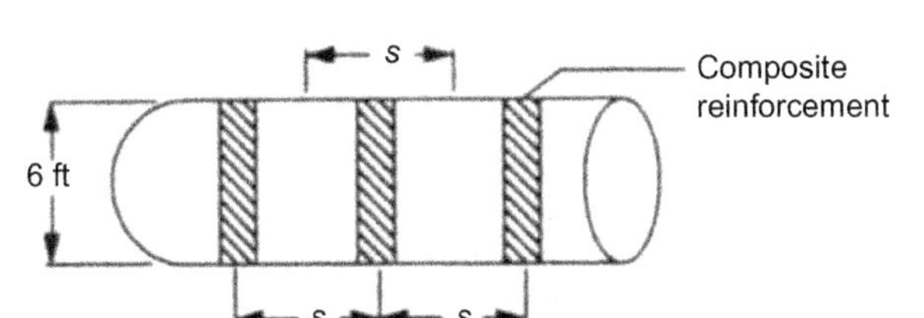

Figure E3.3-1 Schematic of composite reinforced pressure vessel.

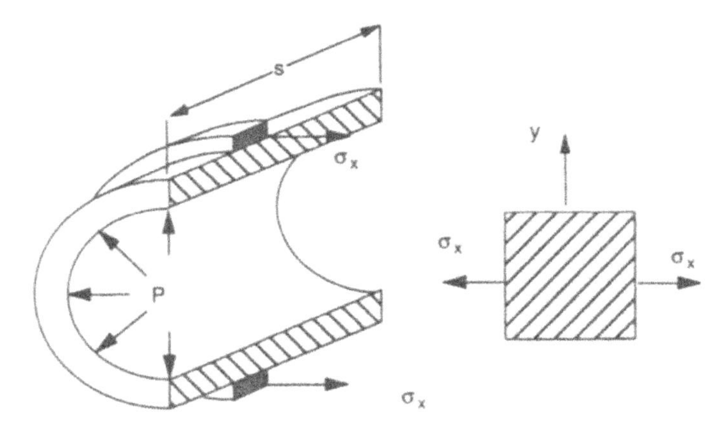

Figure E3.3-2 FBD for reinforced pressure vessel.

A relationship between internal pressure, stress in the reinforcement, and its spacing can be established using the FBD shown in Figure E3.3-2.

The stress in the reinforcement is denoted as σ_x. The reinforcement is subjected to the uniaxial state of stress depicted in Figure E3.3-2. The fiber orientation is assumed to be arbitrary and may be either positive or negative. Summation of forces in the x-direction results in

$$\sum F_x = 0 = 2A\sigma_x - P(6)(12)(s) = 2(0.5)\sigma_x - 100(72)$$

From this, we establish the relationship between reinforcement stress and spacing as $\sigma_x = 7200s$. There are two possible approaches to solving this problem. Either by using the stiffness matrix or the compliance matrix and solving one of the following sets of equations:

$$\begin{Bmatrix} \sigma_x \\ 0 \\ 0 \end{Bmatrix} = [\bar{Q}] \begin{Bmatrix} 6000 \\ \varepsilon_y \\ \gamma_{xy} \end{Bmatrix} \times 10^{-6} \quad \text{or} \quad \begin{Bmatrix} 6000 \\ \varepsilon_y \\ \gamma_{xy} \end{Bmatrix} \times 10^{-6} = [\bar{S}] \begin{Bmatrix} \sigma_x \\ 0 \\ 0 \end{Bmatrix}$$

A solution involving the stiffness matrix requires evaluation of ε_y and γ_{xy}. If the compliance matrix is used, the solution is forthcoming without an intermediate determination of strains. This is a result of the constraint $\varepsilon_x = 6000\,\mu\text{in./in.}$ Had a more rigorous constraint been involved, such as specific limits on ε_y and γ_{xy}, the solution involving stiffness may have been more appropriate. Using the compliance matrix, we establish $6000 \times 10^{-6} = \bar{S}_{11}\sigma_x = \bar{S}_{11}(7200s)$. Solving this equation for s,

$$s = \frac{8.33 \times 10^{-7}}{\bar{S}_{11}}$$

Since only $\bar{S}_{11}$ is involved in the solution, the sign of the fiber orientation in the reinforcement does not influence the solution. This would not have been the case if

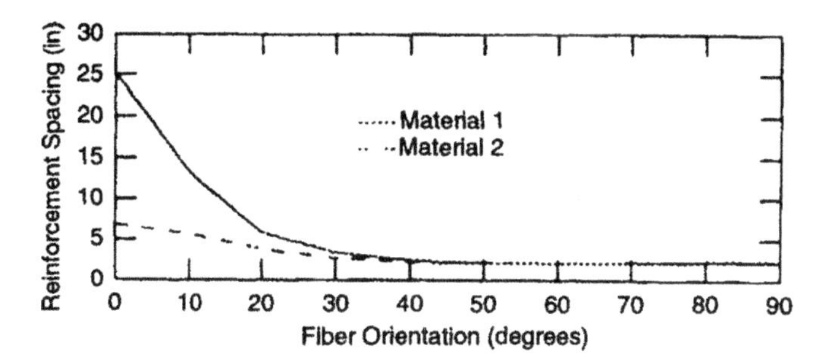

Figure E3.3-3 Reinforcement spacing as a function of fiber orientation.

additional constraints on ε_y and γ_{xy} had been imposed, since shear related terms of both the compliance and stiffness matrices depend on the sign of θ. These results are presented in graphical form in Figure E3.3-3.

It is easy to see that material 1 is better at smaller angles, but at fiber angles of approximately 40° either material can be used with approximately the same reinforcement spacing.

3.3 Thermal and hygral behavior of lamina

A lamina subjected to temperature and moisture changes will deform. The matrix is generally more susceptible to thermal (temperature) and hygral (moisture) deformations than the fiber. Neither constituent (fiber or matrix) is allowed to undergo free thermal and/or hygral expansion, so their responses are coupled and the entire lamina behaves orthotropically. Composites are often exposed to environments in which their temperature and moisture content vary with time. The variation of temperature and moisture within an orthotropic lamina is direction dependent. Assuming an arbitrary direction (x) within the lamina, basic thermodynamic considerations define the flux as

$$q_x = -K_x \frac{\partial G}{\partial x}$$

where q_x is the flux (either thermal or moisture) per unit area per unit time in the x-direction, K_x is the thermal conductivity or moisture diffusion coefficient in the x-direction, and $\frac{\partial G}{\partial x}$ the gradient of temperature or moisture in the x-direction.

The quantities q_x and K_x are often given superscripts of T or H to identify them as thermal or hygral parameters, respectively. The gradient G is generally replaced by either T or H, as appropriate. The thermal conductivity and moisture diffusion coefficients in the matrix directions (2- and 3-directions) are generally equal ($K_2^{\mathrm{T}} = K_3^{\mathrm{T}}$ and $K_2^{\mathrm{H}} = K_3^{\mathrm{H}}$). The governing equation for heat flow is developed by an energy balance using the one-dimensional model in Figure 3.8.

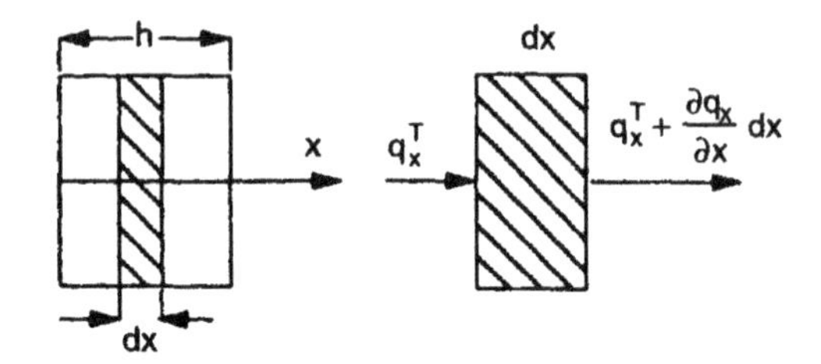

Figure 3.8 One-dimensional heat flux model.

The increase in energy stored per unit time within the representative volume element (RVE) of Figure 3.8 is $\rho c(\partial T/\partial t)\mathrm{d}x$, where c is the specific heat, ρ the mass density of the material, and t is time. A simple energy balance requires

$$q_x^{\mathrm{T}} - \left[q_x^{\mathrm{T}} + \left(\frac{\partial q_x^{\mathrm{T}}}{\partial x}\right)\mathrm{d}x\right] = \rho c\left(\frac{\partial T}{\partial t}\right)\mathrm{d}x \Rightarrow -\left(\frac{\partial q_x^{\mathrm{T}}}{\partial x}\right) = \rho c\left(\frac{\partial T}{\partial t}\right)$$

Using the definition of q_x^{T} above yields $\partial q_x^{\mathrm{T}}/\partial x = (\partial/\partial x)\left[-K_x^{\mathrm{T}}(\partial T/\partial x)\right]$. If K_x^{T}, ρ, and c are constant

$$\frac{\partial T}{\partial x} = \left[\frac{K_x^{\mathrm{T}}}{\rho c}\right]\left(\frac{\partial^2 T}{\partial x^2}\right) \tag{3.20}$$

The $K_x^{\mathrm{T}}/\rho c$ term is the thermal diffusivity and is a measure of the rate of temperature change within the material.

3.3.1 Thermal stress–strain relationships

In considering the effect of temperature on the stresses and strains in a lamina, the thermal conductivity and/or heat flux are not considered once the temperature has reached an equilibrium state (i.e., it has stabilized and is no longer increasing or decreasing). Since a lamina (or laminate) is cured at a temperature which is generally above the operating temperature, thermal stresses may be present in the lamina prior to application of loads.

The general form of this stress–strain relationship for a lamina in which temperature effects are considered was presented in Section 2.7 as $\sigma_i = Q_{ij}\left(\varepsilon_j - \alpha_j \Delta T\right)$. In this expression α_j is the coefficient of thermal expansion in the j-direction and ΔT is the temperature difference $(T - T_{\mathrm{o}})$ from the operating (T) to the stress free (or curing) temperature (T_{o}).

The coefficients of thermal expansion in a lamina are direction dependent, and in the principal material directions are α_1, α_2, and α_3. The subscripts denote the material direction in which each coefficient is applicable. There is no shear coefficient in the principal material direction due to thermal expansion since it is a dilatational quantity associated with volume change. The stress–strain relation given by Equation (3.5) must be appended to account for thermal expansion and is expressed as

$$\begin{Bmatrix}\sigma_1\\\sigma_2\\\sigma_3\\\sigma_4\\\sigma_5\\\sigma_6\end{Bmatrix}=\begin{bmatrix}Q_{11}&Q_{12}&Q_{13}&0&0&0\\Q_{12}&Q_{22}&Q_{23}&0&0&0\\Q_{13}&Q_{23}&Q_{33}&0&0&0\\0&0&0&Q_{44}&0&0\\0&0&0&0&Q_{55}&0\\0&0&0&0&0&Q_{66}\end{bmatrix}\begin{Bmatrix}\varepsilon_1-\alpha_1\Delta T\\\varepsilon_2-\alpha_2\Delta T\\\varepsilon_3-\alpha_3\Delta T\\\varepsilon_4\\\varepsilon_5\\\varepsilon_6\end{Bmatrix}\tag{3.21}$$

This relationship is valid only for on-axis configurations. In an off-axis configuration, the coefficients of thermal expansion are expressed in a different form. Consider a lamina subjected to thermal strains in the principal material directions, which are $\varepsilon_i^{\mathrm{T}}=\alpha_i\Delta T$. Using the strain transformation matrix (2.2) the relationship between thermal strains in the material and x–y coordinate systems is

$$\begin{Bmatrix}\varepsilon_1\\\varepsilon_2\\\varepsilon_3\\0\\0\\0\end{Bmatrix}^{\mathrm{T}}=\begin{Bmatrix}\alpha_1\\\alpha_2\\\alpha_3\\0\\0\\0\end{Bmatrix}\Delta T=[T_\varepsilon]\begin{Bmatrix}\varepsilon_x\\\varepsilon_y\\\varepsilon_y\\\gamma_{yz}\\\gamma_{xz}\\\gamma_{xy}\end{Bmatrix}^{\mathrm{T}}$$

The thermal strains in the x–y coordinate system are expressed as

$$\begin{Bmatrix}\varepsilon_x\\\varepsilon_y\\\varepsilon_y\\\gamma_{yz}\\\gamma_{xz}\\\gamma_{xy}\end{Bmatrix}^{\mathrm{T}}=[T_\varepsilon]^{-1}\begin{Bmatrix}\alpha_1\\\alpha_2\\\alpha_3\\0\\0\\0\end{Bmatrix}\Delta T=\begin{Bmatrix}\alpha_x\\\alpha_y\\\alpha_z\\0\\0\\\alpha_{xy}\end{Bmatrix}\Delta T$$

where the coefficients of thermal expansion in the x–y system are

$$\begin{aligned}\alpha_x&=m^2\alpha_1+n^2\alpha_2\\\alpha_y&=n^2\alpha_1+m^2\alpha_2\\\alpha_z&=\alpha_3\\\alpha_{xy}&=2mn(\alpha_1-\alpha_2)\end{aligned}\tag{3.22}$$

The term α_{xy} is called the *apparent* coefficient of thermal expansion.

The stress–strain relationship for Cartesian components of stress and strain given by Equation (3.7) must be appended to account for thermal strains and is

$$\begin{Bmatrix} \sigma_x \\ \sigma_y \\ \sigma_z \\ \tau_{yz} \\ \tau_{xz} \\ \tau_{xy} \end{Bmatrix} = \begin{bmatrix} \bar{Q}_{11} & \bar{Q}_{12} & \bar{Q}_{13} & 0 & 0 & \bar{Q}_{16} \\ \bar{Q}_{12} & \bar{Q}_{22} & \bar{Q}_{23} & 0 & 0 & \bar{Q}_{26} \\ \bar{Q}_{13} & \bar{Q}_{23} & \bar{Q}_{33} & 0 & 0 & \bar{Q}_{36} \\ 0 & 0 & 0 & \bar{Q}_{44} & \bar{Q}_{45} & 0 \\ 0 & 0 & 0 & \bar{Q}_{45} & \bar{Q}_{55} & 0 \\ \bar{Q}_{16} & \bar{Q}_{26} & \bar{Q}_{36} & 0 & 0 & \bar{Q}_{66} \end{bmatrix} \begin{Bmatrix} \varepsilon_x - \alpha_x \Delta T \\ \varepsilon_y - \alpha_y \Delta T \\ \varepsilon_z - \alpha_z \Delta T \\ \gamma_{yz} \\ \gamma_{xz} \\ \gamma_{xy} - \alpha_{xy} \Delta T \end{Bmatrix} \quad (3.23)$$

For the case of plane stress, the on-axis (3.21) and off-axis (3.23) stress–strain relationships reduce to

$$\begin{Bmatrix} \sigma_1 \\ \sigma_2 \\ \sigma_6 \end{Bmatrix} = \begin{bmatrix} Q_{11} & Q_{12} & 0 \\ Q_{12} & Q_{22} & 0 \\ 0 & 0 & Q_{66} \end{bmatrix} \begin{Bmatrix} \varepsilon_1 - \alpha_1 \Delta T \\ \varepsilon_2 - \alpha_2 \Delta T \\ \gamma_{12} \end{Bmatrix} \quad (3.24)$$

for the on-axis representation, and for the off-axis representation

$$\begin{Bmatrix} \sigma_x \\ \sigma_y \\ \tau_{xy} \end{Bmatrix} = \begin{bmatrix} \bar{Q}_{11} & \bar{Q}_{12} & \bar{Q}_{16} \\ \bar{Q}_{12} & \bar{Q}_{22} & \bar{Q}_{26} \\ \bar{Q}_{16} & \bar{Q}_{26} & \bar{Q}_{66} \end{bmatrix} \begin{Bmatrix} \varepsilon_x - \alpha_x \Delta T \\ \varepsilon_y - \alpha_y \Delta T \\ \gamma_{xy} - \alpha_{xy} \Delta T \end{Bmatrix} \quad (3.25)$$

3.3.2 *Hygral effects*

The equation for moisture diffusion is derived from a mass balance similar to the energy balance for temperature. The moisture diffusion process applicable to a variety of composite material systems is termed Fickian diffusion and follows Fick's law. In general, low temperatures and humid air promote Fickian diffusion, while high temperatures and immersion in liquids cause deviations from it. The Fickian diffusion process is a reasonable approximation for many composites. The moisture diffusion equation for a one-dimensional problem is

$$\frac{\partial}{\partial x}\left\{K_x^{\mathrm{H}}\left(\frac{\partial H}{\partial x}\right)\right\} = \frac{\partial H}{\partial t}$$

If K_x^{H} is constant this relationship reduces to Fick's law, given as

$$K_x^{\mathrm{H}}\left(\frac{\partial^2 H}{\partial x^2}\right) = \frac{\partial H}{\partial t}$$

Comparing thermal and hygral diffusion parameters (K_x^H for hygral and $K_x^T/\rho c$ for thermal) shows the rates at which moisture and temperature change within a material. Over a large range of temperatures and moisture concentrations encountered in many composite materials applications, the two parameters can be approximately related by $K_x^T/\rho c \approx 10^6 K_x^H$. In problems coupling temperature and moisture, the lamina reaches thermal equilibrium before hygral equilibrium, and although no temperature gradient exists, a moisture gradient may.

The moisture concentration is generally replaced by the specific moisture concentration, defined as $M = H/\rho$. Using this definition of M to represent the hygral term (H) in Fick's law, it can be rewritten as

$$K^H \frac{\partial^2 M}{\partial x^2} = \frac{\partial M}{\partial t} \tag{3.26}$$

The subscript x has been dropped from K, with the understanding that diffusion occurs in the x-direction. The boundary conditions required to solve this equation are

for $0 < x < h$: $M = M_0$ at $t = 0$

for $x = 0, x = h$: $M = M_\infty$ at $t > 0$

where h is the lamina thickness and t is time. The solution to Equation (3.26) using the boundary conditions above is expressed as a series

$$\frac{M - M_0}{M_\infty - M_0} = 1 - \frac{4}{\pi}\sum_{j=0}^{\infty} \frac{1}{2j+1} \sin\frac{(2j+1)\pi x}{h} e^{-\gamma} \tag{3.27}$$

where

$$\gamma = (2j+1)^2 \pi^2 K^H t / h^2$$

M_0 is the initial moisture content and M_∞ the equilibrium moisture content.

The equilibrium moisture content is generally higher than the initial moisture content for a moisture absorption test. The converse is true for moisture desorption, and Equation (3.27) is also valid for that case. In a typical moisture absorption test, specimens are subjected to specified relative humidities and temperatures and frequently weighed to determine moisture content. The average moisture content $\bar{M}$ is defined in terms of the moisture content M as

$$\bar{M} = \frac{1}{h}\int_0^h M \, dx$$

Substituting M from (3.27) into this equation and noting that $\bar{M} = M_0$ at time $t = 0$, and $\bar{M} = M_\infty$ at time $t = \infty$, yields

$$\frac{\bar{M} - M_0}{M_\infty - M_0} = 1 - \frac{8}{\pi^2}\sum_{j=0}^{\infty} \frac{e^{-\gamma}}{(2j+1)^2} \tag{3.28}$$

where γ is defined as before. For large times t, this expression can be approximated by the first term of the series as

$$\frac{\bar{M}-M_0}{M_\infty-M_0}=1-\frac{8}{\pi^2}e^{-\left(\pi^2K^{\mathrm{H}}t/h^2\right)} \tag{3.29}$$

For short times the approximation is given by an alternate expression from Ref. [5] as

$$\frac{\bar{M}-M_0}{M_\infty-M_0}=4\left(\frac{K^{\mathrm{H}}t}{\pi h^2}\right)^{1/2} \tag{3.30}$$

In the initial phase of moisture absorption the moisture content changes as a simple function of $\sqrt{t/h^2}$ (where t is time of exposure), while it becomes an exponential function at later times. The diffusion coefficient K^{H} must be determined in order to use either relationship. This coefficient is determined through moisture absorption tests conducted over long periods of time. A schematic of a complete moisture absorption test is shown in Figure 3.9, where moisture content is plotted against $\sqrt{t}$. The value of $\bar{M}$ is obtained by weighing specimens at various times during the test. Using Equation (3.30) and two different values of $\bar{M}$ in the linear region, K^{H} can be determined to be

$$K^{\mathrm{H}}=\frac{\pi}{16}\left(\frac{M_2-M_1}{M_\infty-M_0}\right)^2\left(\frac{h}{t_2-t_1}\right)^2$$

The diffusion coefficient can also be determined from the long time approximation. It is assumed that the long time absorption equation applies when the moisture content has reached 50% of M_∞. By setting $(\bar{M}-M_0)/(M_\infty-M_0)=1/2$ in the long time

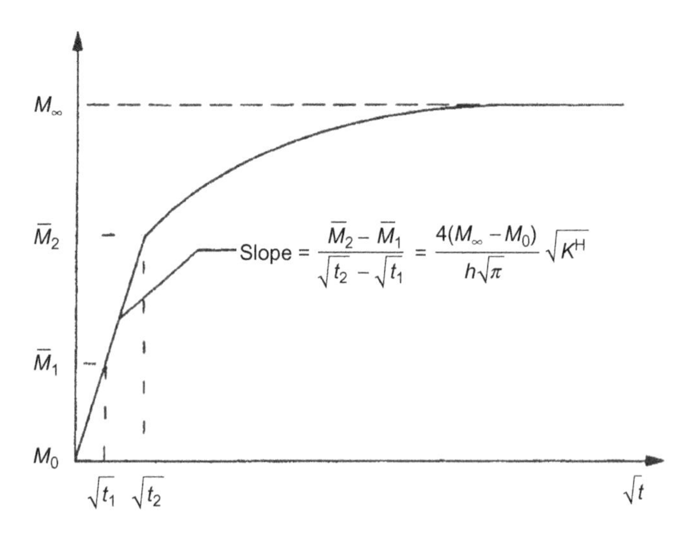

Figure 3.9 Determination of K^{H} from M vs. $\sqrt{t}$ data at temperature T_1.

equation, the time to reach 50% of M_∞ is determined, from which we get $K^H = 0.04895h^2/t_{1/2}$.

The equilibrium moisture content (M_∞) depends on environmental conditions and can be expressed in terms of the relative humidity as

$$M_\infty = a\left(\frac{\phi}{100}\right)^b$$

where ϕ is the percentage of relative humidity, and a and b are experimentally determined material constants. The amount of data available regarding these constants is limited, but approximate values of a and b for selected composite material systems are found in Refs. [5,6]. The equilibrium moisture content and the rate at which it is reached vary from material to material and to a large extent is controlled by the matrix. This is illustrated in Figure 3.10 which shows a moisture absorption profile for graphite fibers in both epoxy and PEEK (polyetheretherketone) matrices at 95% RH and 160 °F [7]. As seen here the graphite/epoxy system reaches an equilibrium moisture content of 2.23%, while the graphite/PEEK system has an equilibrium moisture content of only 0.15%. Since the fiber is the same in both materials, one concludes that the matrix is the dominant in moisture absorption.

The previously established relationships for K^H do not reflect the actual behavior of the diffusion coefficient which is highly dependent on temperature and can be represented by $K^H = K_o^H e^{-E_d/RT}$, where K_o^H is the pre-exponential factor, E_d the activation energy, R the gas constant [1.987 cal/(mole K)], and T the temperature.

Both K_o^H and E_d are material properties. Most of the parameters considered in the preceding discussion are obtained through experimental techniques. Springer [8–10] is a good reference for further information regarding moisture diffusion and the general effects of environment on composites. Tsai and Hahn [5] present the moisture properties for a limited number of materials.

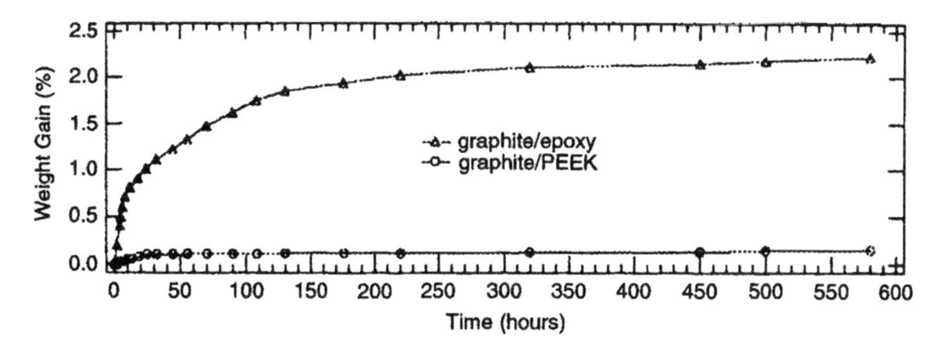

Figure 3.10 Moisture absorption of graphite/epoxy and graphite/PEEK composites at 95% RH and 160 °F.
After Ref. [7].

3.3.2.1 Hygral stress–strain relationships

Stresses resulting from hygral effects are analogous to those due to thermal effects. In order to assess hygral stresses, the hygral strains (sometimes termed swelling strains) must be considered. Prolonged exposure to moisture results in a weight gain and volume change for many composite materials. The weight gain is due to moisture absorption. The corresponding volume change results in strains (swelling strains) expressed in terms of the moisture content as $\varepsilon^{\mathrm{H}} = \beta M$, where M is used in place of ΔM (moisture change) from Equation (2.5). The swelling coefficient β is direction dependent in the same manner as the thermal expansion coefficient and is determined by a moisture absorption test of a unidirectional composite of thickness h subjected to the same relative humidity ϕ on both sides. Frequent measurements are made to determine the amount of swelling as a function of moisture content. The stress–strain relationship applicable when hygral effects are considered was given in Section 2.7 as

$$\sigma_i = Q_{ij}\left(\varepsilon_j - B_j M\right)$$

The hygral strains in both on-axis and off-axis configurations are

$$\begin{Bmatrix} \varepsilon_1 \\ \varepsilon_2 \\ \varepsilon_3 \\ \varepsilon_4 \\ \varepsilon_5 \\ \varepsilon_6 \end{Bmatrix} = \begin{Bmatrix} \beta_1 \\ \beta_2 \\ \beta_3 \\ 0 \\ 0 \\ 0 \end{Bmatrix} M, \quad \begin{Bmatrix} \varepsilon_x \\ \varepsilon_y \\ \varepsilon_z \\ \gamma_{yz} \\ \gamma_{xz} \\ \gamma_{xy} \end{Bmatrix} = \begin{Bmatrix} \beta_x \\ \beta_y \\ \beta_z \\ 0 \\ 0 \\ \beta_{xy} \end{Bmatrix} M$$

Only principal material directions are affected by moisture absorption in an on-axis configuration. On-axis swelling strains can be transformed into off-axis strains in the same manner as the thermal strains were. In the off-axis configuration the swelling coefficients are

$$\begin{aligned} \beta_x &= m^2\beta_1 + n^2\beta_2 \\ \beta_y &= n^2\beta_1 + m^2\beta_2 \\ \beta_z &= \beta_3 \;\; \beta_z = \beta_3 \\ \beta_{xy} &= 2mn(\beta_1 - \beta_2) \end{aligned} \tag{3.31}$$

where β_{xy} is termed the apparent coefficient.

Hygral strains are given in terms of M, which is a gradient. The time required to reach equilibrium for moisture is slower than for temperature, so the moisture induced strains are more variable through the thickness than thermal strains. In order to eliminate the problem of defining a moisture gradient, we assume that M can be replaced with the average moisture content $\bar{M}$, obtainable from either (3.29) or (3.30), as appropriate. Including moisture in the stress–strain relationship for an on-axis configuration and appending (3.21) yields

$$\begin{Bmatrix} \sigma_1 \\ \sigma_2 \\ \sigma_3 \\ \sigma_4 \\ \sigma_5 \\ \sigma_6 \end{Bmatrix} = \begin{bmatrix} Q_{11} & Q_{12} & Q_{13} & 0 & 0 & 0 \\ Q_{12} & Q_{22} & Q_{23} & 0 & 0 & 0 \\ Q_{13} & Q_{23} & Q_{33} & 0 & 0 & 0 \\ 0 & 0 & 0 & Q_{44} & 0 & 0 \\ 0 & 0 & 0 & 0 & Q_{55} & 0 \\ 0 & 0 & 0 & 0 & 0 & Q_{66} \end{bmatrix} \begin{Bmatrix} \varepsilon_1 - \alpha_1 \Delta T - \beta_1 \bar{M} \\ \varepsilon_2 - \alpha_2 \Delta T - \beta_2 \bar{M} \\ \varepsilon_3 - \alpha_3 \Delta T - \beta_3 \bar{M} \\ \gamma_{23} \\ \gamma_{13} \\ \gamma_{12} \end{Bmatrix} \quad (3.32)$$

For the off-axis configuration the appended form of (3.23) becomes

$$\begin{Bmatrix} \sigma_x \\ \sigma_y \\ \sigma_z \\ \tau_{yz} \\ \tau_{xz} \\ \tau_{xy} \end{Bmatrix} = \begin{bmatrix} \bar{Q}_{11} & \bar{Q}_{12} & \bar{Q}_{13} & 0 & 0 & \bar{Q}_{16} \\ \bar{Q}_{12} & \bar{Q}_{22} & \bar{Q}_{23} & 0 & 0 & \bar{Q}_{26} \\ \bar{Q}_{13} & \bar{Q}_{23} & \bar{Q}_{33} & 0 & 0 & \bar{Q}_{36} \\ 0 & 0 & 0 & \bar{Q}_{44} & \bar{Q}_{45} & 0 \\ 0 & 0 & 0 & \bar{Q}_{45} & \bar{Q}_{55} & 0 \\ \bar{Q}_{16} & \bar{Q}_{26} & \bar{Q}_{36} & 0 & 0 & \bar{Q}_{66} \end{bmatrix} \begin{Bmatrix} \varepsilon_x - \alpha_x \Delta T - \beta_x \bar{M} \\ \varepsilon_y - \alpha_y \Delta T - \beta_y \bar{M} \\ \varepsilon_3 - \alpha_z \Delta T - \beta_z \bar{M} \\ \gamma_{yz} \\ \gamma_{xz} \\ \gamma_{yy} - \alpha_{xy} \Delta T - \beta_{xy} \bar{M} \end{Bmatrix} \quad (3.33)$$

For the special case of plane stress, the on-axis form of the stress–strain relationship reduces to

$$\begin{Bmatrix} \sigma_1 \\ \sigma_2 \\ \sigma_6 \end{Bmatrix} = \begin{bmatrix} Q_{11} & Q_{12} & 0 \\ Q_{12} & Q_{22} & 0 \\ 0 & 0 & Q_{66} \end{bmatrix} \begin{Bmatrix} \varepsilon_1 - \alpha_1 \Delta T - \beta_1 \bar{M} \\ \varepsilon_2 - \alpha_2 \Delta T - \beta_2 \bar{M} \\ \gamma_{12} \end{Bmatrix} \quad (3.34)$$

The off-axis form of the stress–strain relationship for plane stress is

$$\begin{Bmatrix} \sigma_x \\ \sigma_y \\ \tau_{xy} \end{Bmatrix} = \begin{bmatrix} \bar{Q}_{11} & \bar{Q}_{12} & \bar{Q}_{16} \\ \bar{Q}_{12} & \bar{Q}_{22} & \bar{Q}_{26} \\ \bar{Q}_{16} & \bar{Q}_{26} & \bar{Q}_{66} \end{bmatrix} \begin{Bmatrix} \varepsilon_x - \alpha_x \Delta T - \beta_x \bar{M} \\ \varepsilon_y - \alpha_y \Delta T - \beta_y \bar{M} \\ \gamma_{xy} - \alpha_{xy} \Delta T - \beta_{xy} \bar{M} \end{Bmatrix} \quad (3.35)$$

In a state of free thermal and/or hygral expansion (or contraction) where $\{\sigma\} = 0$, the lamina strains (either on-axis or off-axis) can be represented as $\{\varepsilon\} = \{\alpha\}\Delta T + \{\beta\}\bar{M}$. There are no stresses associated with this state of strain since a stress is not applied. Now consider the one-dimensional case shown in Figure 3.11. The rigid walls impose a constraint on free expansion such that the overall deformation in the y-direction is zero. This causes $\varepsilon_y = 0$, while ε_x and γ_{xy} may exist. In order for the constraint of $\varepsilon_y = 0$ to be valid, a stress σ_y is imposed on the lamina by the wall. This stress is the thermal (or hygral) stress and can be expressed in terms of $\alpha\Delta T$ or $\beta\bar{M}$ from Equation (3.35) by setting the normal strain $\varepsilon_y = 0$. The stress components σ_x and τ_{xy} would be zero, since there are no constraints restricting deformations associated with these stresses.

Residual stresses from curing do not exist for flat lamina where only free thermal expansion or contraction is possible. Individual lamina does not generally have residual stresses from curing, but a laminate (composed of several lamina) does. This results from the varying expansion coefficients through the thickness. Since each

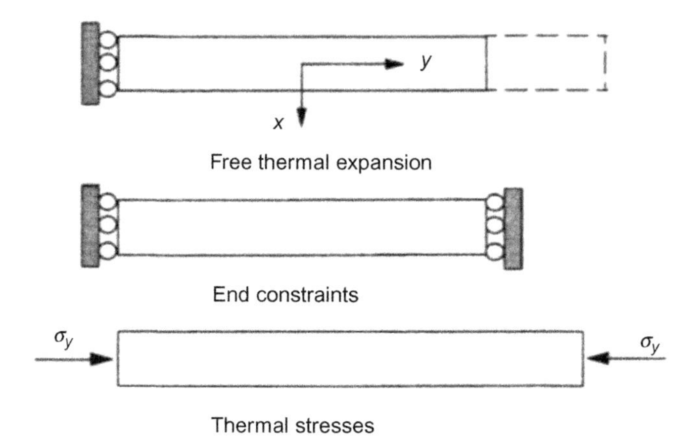

Figure 3.11 Schematic of constraints producing one-dimensional thermal stresses.

lamina may have a different expansion coefficient, it will not deform exactly like an adjacent lamina. Due to compatibility between lamina, a deformation constraint is placed on each lamina resulting in a stress.

Example 3.4 The composite reinforced pressure vessel considered in Example 3.3 is used to illustrate the effects of thermal and hygral strains on analysis. The relation between the normal stress in the composite reinforcement and spacing ($\sigma_x = 7200s$) defined in Example 3.3 is used again, as is the constraint $\varepsilon_x \leq 6000\,\mu$ in./in. As before, two possible equations can be solved

$$\begin{Bmatrix} \sigma_x \\ 0 \\ 0 \end{Bmatrix} = \begin{bmatrix} \bar{Q}_{11} & \bar{Q}_{12} & \bar{Q}_{16} \\ \bar{Q}_{12} & \bar{Q}_{22} & \bar{Q}_{26} \\ \bar{Q}_{16} & \bar{Q}_{26} & \bar{Q}_{66} \end{bmatrix} \begin{Bmatrix} 6000 \times 10^{-6} - \alpha_x \Delta T - \beta_x \bar{M} \\ \varepsilon_y - \alpha_y \Delta T - \beta_y \bar{M} \\ \gamma_{xy} - \alpha_{xy} \Delta T - \beta_{xy} \bar{M} \end{Bmatrix}$$

or

$$\begin{Bmatrix} 6000 \times 10^{-6} - \alpha_x \Delta T - \beta_x \bar{M} \\ \varepsilon_y - \alpha_y \Delta T - \beta_y \bar{M} \\ \gamma_{xy} - \alpha_{xy} \Delta T - \beta_{xy} \bar{M} \end{Bmatrix} = \begin{bmatrix} \bar{S}_{11} & \bar{S}_{12} & \bar{S}_{16} \\ \bar{S}_{12} & \bar{S}_{22} & \bar{S}_{26} \\ \bar{S}_{16} & \bar{S}_{26} & \bar{S}_{66} \end{bmatrix} \begin{Bmatrix} \sigma_x \\ 0 \\ 0 \end{Bmatrix}$$

As with Example 3.3, the second equation is selected since the state of stress is explicitly defined and establishing ε_y and γ_{xy} is not required. Therefore, the problem reduces to solving

$$6000 \times 10^{-6} - \alpha_x \Delta T - \beta_x \bar{M} = \bar{S}_{11}(7200s)$$

α_x and β_x are functions of fiber orientation and are established from $\alpha_x = m^2\alpha_1 + n^2\alpha_2$ and $\beta_x = m^2\beta_1 + n^2\beta_2$. For the two materials considered in Example 3.3, the thermal and hygral coefficients of expansion are assumed to be

	Material 1	Material 2
α_1	3.4×10^{-6} in./in./°F	12.0×10^{-6} in./in./°F
α_2	5.0×10^{-6} in./in./°F	8.0×10^{-6} in./in./°F
β_1	0.0	0.0
β_2	0.20	0.40

For this problem, it is assumed that $\Delta T = -280$°F and the average moisture content is $\bar{M} = 0.05$. Manipulation of the governing equation for reinforcement spacing results in

$$s = \frac{8.33 \times 10^{-7} - 1.389 \times 10^{4}(\alpha_x \Delta T + \beta_x \bar{M})}{\bar{S}_{11}}$$

In this expressions for α_x, β_x, and $\bar{S}_{11}$ are functions of θ.

Individual contributions of thermal and hygral effects on the reinforcement spacing are illustrated by separating them and examining one at a time. Figure E3.4-1 shows the effect of temperature compared to the solution for Example 3.3 for both materials. The effects of $\Delta T = -280$°F (and $\bar{M} = 0$) increase the required reinforcement spacing for each material considered.

The effects of moisture alone are presented in Figure E3.4-2 for $\bar{M} = 0.05$ and $\Delta T = 0$ and are compared to the results of Example 3.3. The negative reinforcement spacing indicates that the constraint on ε_x has been violated. This does not imply that no reinforcement is required, since the constraint on ε_x is an artificially imposed failure criteria. The swelling strains produced from inclusion of hygral effects will reduce the strain in the reinforcement.

The combination of thermal and hygral effects on predicted reinforcement spacing is shown in Figure E3.4-3. The originally predicted spacing from Example 3.3 is not presented in this figure. Thermal and hygral effects can influence the state of stress in a composite and have an effect on lamina failure.

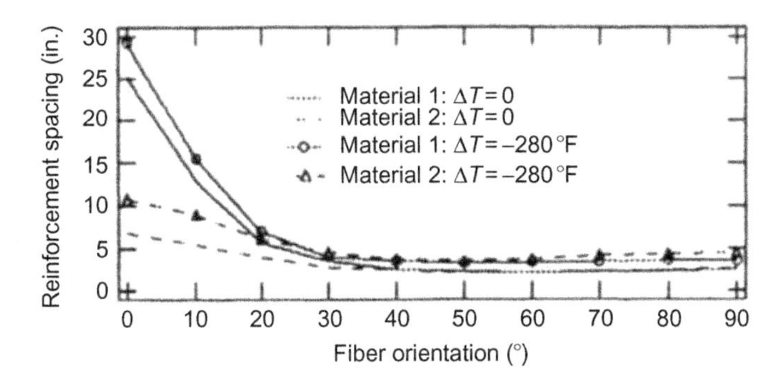

Figure E3.4-1 Effects of temperature on reinforcement spacing, $\bar{M} = 0$.

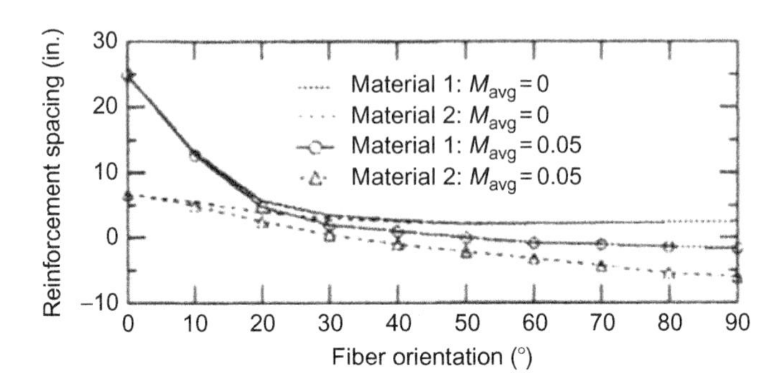

Figure E3.4-2 Effects of moisture on reinforcement spacing with $\Delta T = 0$.

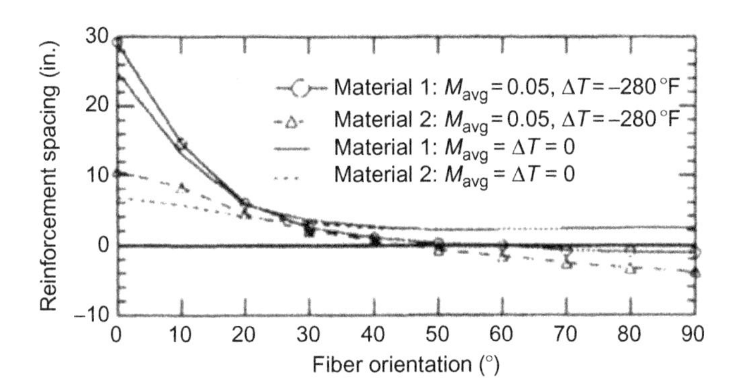

Figure E3.4-3 Effects of ΔT and $\bar{M}$ on reinforcement spacing.

3.4 Prediction of lamina properties (micromechanics)

Micromechanics considers the behavior of each constituent (fiber and matrix) as it relates to the prediction of lamina properties. The properties of lamina determined from experimental procedures are macroscopic properties. They do not reflect the interactions between fibers and matrix, nor do they offer insight into possible improvements of material response. Early developments of micromechanics consisted of three types of formulations: strength of materials, elasticity, and empirical, as summarized by Hashin [11,12] and Chamis and Sendeckyj [13]. Strength of materials approximations are easy to formulate, but the assumptions used often violate strict elasticity formulations. The empirical approach is generally a curve-fitting procedure which incorporates experimental data and either elasticity or strength of materials solutions to provide a set of lamina design equations.

The most widely used relationships between constituent properties and the macroscopic behavior of continuous fiber composites were developed prior to the 1980s. These relationships, although adequate for predicting elastic moduli, are not sufficient for the analysis of damage mechanics. In 1958, Kachanov [14] modeled the creep characteristics of metals by introducing the effects of microcrack growth and dislocations through the use of internal state variables. This provided the impetus for the development of what has become known as continuum damage mechanics [15–17].

The discussions of micromechanics presented herein pertain to strength of materials models, discussions of elasticity solutions, and empirical relationships. The simplest approach to determining lamina properties is based on assuming that each constituent material is homogenous and isotropic. Consider an RVE of a lamina as shown in Figure 3.12. In this RVE three distinct regions exist: fiber (given a subscript "f"), matrix (given a subscript "m"), and voids (given a subscript "v").

The total mass (M) and volume (V) of the RVE are $M = M_f + M_m$ and $V = V_f + V_m + V_v$, respectively. Dividing the mass by total mass and volume by total volume yields the mass and volume fractions, defined by $m_f + m_m = 1$ and $v_f + v_m = 1$, where $m_f = M_f/M$, $m_m = M_m/M$, $v_f = V_f/V$, $v_m = V_m/V$, and $v_v = V_v/V$. These represent the mass and volume fractions of matrix, fiber, and voids. The density of the lamina can be expressed as

$$\rho = \frac{M}{V} = \frac{\rho_f V_f + \rho_m V_m}{V} = \rho_f v_f + \rho_m v_m$$

Introducing mass fractions results in v_f expressed as $v_f = V_f/V = (M_f/\rho_f)/(M/\rho) = (\rho_f/\rho)m_f$ and a similar expression for v_m, we express the density as

$$\rho = \frac{1}{\frac{m_f}{\rho_f} + \frac{m_m}{\rho_m} + \frac{v_v}{\rho}}$$

This expression can be used to determine the volume fraction of voids in a lamina, provided the mass fractions and densities of each constituent are known. Solving for v_v,

$$v_v = 1 - \rho\left[\frac{m_f}{\rho_f} + \frac{m_m}{\rho_m}\right]$$

The volume fraction of fibers and matrix depends, to a large extent, on fiber geometry and packing arrangement within a lamina. Assume, for example, circular fibers of diameter d, contained in a matrix such that three possible RVEs can be defined, as shown in Figure 3.13. Each packing arrangement, triangular, square, and hexagonal,

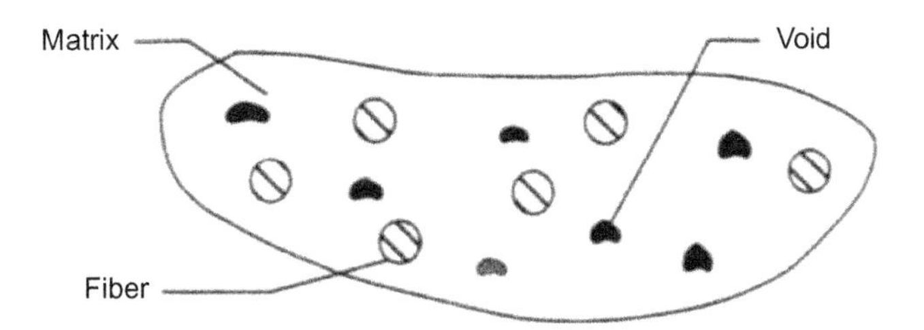

Figure 3.12 Representative volume element of a lamina.

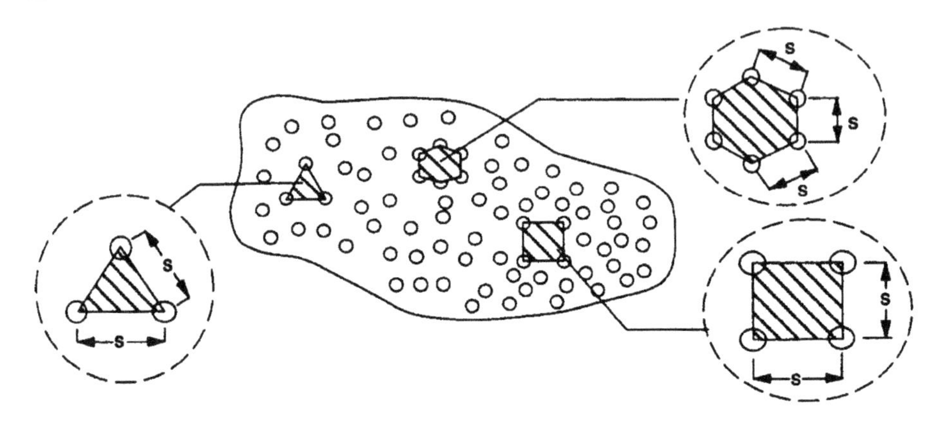

Figure 3.13 Triangular, square, and hexagonal fiber arrays.

produces a different volume fraction of fibers. The fiber volume fraction of the square array, for example, can be obtained by first assuming a unit width into the plane, and dividing the total area of fiber contained within the RVE by the area of the square. The area enclosed by the square RVE is $A_{\text{square}} = s^2$. The area within this square which contains fibers is $A_{\text{fiber}} = \pi d^2/4$. The volume fraction of fiber for this array as given in Ref. [18] is $v_f = A_{\text{fiber}}/A_{\text{square}} = \pi d^2/4s^2$. For triangular and hexagonal arrays, a similar procedure results in

$$\text{Triangular: } v_f = \frac{\pi d^2}{(2\sqrt{3})s^2}$$

$$\text{Hexagonal: } v_f = \frac{\pi d^2}{(3\sqrt{3})s^2}$$

For each of these fiber arrays, the maximum volume fraction of fibers occurs when $d/s = 1$. For this condition, the v_f for triangular, square, and hexagonal arrays are 0.907, 0.785, and 0.605, respectively. These limits should not be expected in practice, since in most continuous fiber composites the packing is random and processing lowers the actual v_f.

3.4.1 Mechanical properties of lamina

Various approaches of estimating the mechanical response of lamina from constitutive material behavior have evolved. They range in complexity from a simple rule of mixtures approach to the more sophisticated concentric cylinders approach. Between these extremes are several estimation schemes based on experimental observations and interpretations.

3.4.1.1 Strength of materials approach

In developing stress–strain relationships involving $[Q]$ and $[\bar{Q}]$, the material properties used are often termed *apparent* and are generally established through mechanical testing. It is useful to establish procedures for estimating apparent properties by

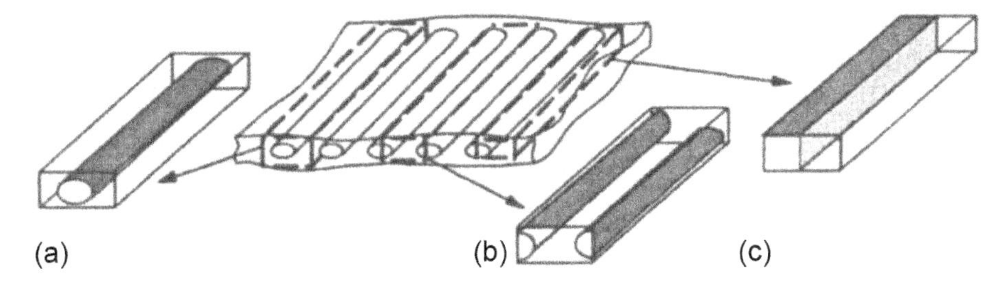

Figure 3.14 Possible lamina RVE configurations.

knowing the behavior of each constituent. The strength of materials approach is straight forward and simple to formulate. Consider a section of lamina as shown in Figure 3.14. Several possible RVEs are suitable for model development. In order to simplify geometric interactions, the model used closely resembles that of Figure 3.14c.

Two conditions can be applied to determine mechanical response in the fiber direction: constant stress or constant strain. Constant strain requires uniform displacements (Figure 3.15). This results in the fiber and matrix experiencing the same strain, with the stress distribution in the lamina as shown. A uniform displacement requires $\varepsilon_f = \varepsilon_m = \varepsilon$, and since fibers and matrix are assumed to be isotropic and homogenous, $\sigma_f = E_f\varepsilon$ and $\sigma_m = E_m\varepsilon$. Conversely, if a constant (uniform) stress test is conducted as shown in Figure 3.15, the resulting strain distribution is as indicated in the figure, since $\varepsilon_f = \sigma_o/E_f$ and $\varepsilon_m = \sigma_o/E_m$.

Since the interface between fiber and matrix is assumed to be a perfect bond, $\varepsilon_f = \varepsilon_m$. Therefore, the constant (uniform) strain approximation is closer to the actual physical conditions than the constant (uniform) stress condition for establishing material behavior in the fiber direction. There are cases when constant stress conditions are more appropriate. The constant strain model is sometimes termed the Voigt model, while the constant stress model is the Reuss model.

Assume the lamina is modeled as shown in Figure 3.16, with length L, width W, and a unit thickness.

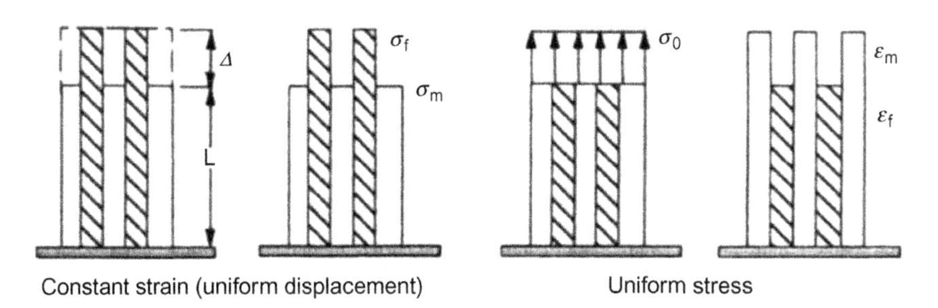

Figure 3.15 Constant strain and stress models.

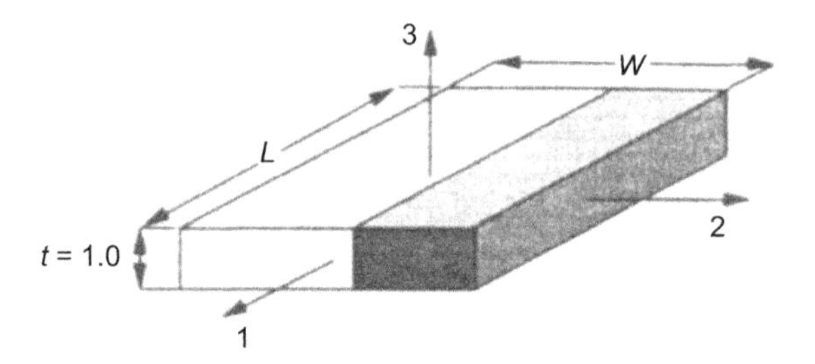

Figure 3.16 Lamina model for determining elastic moduli.

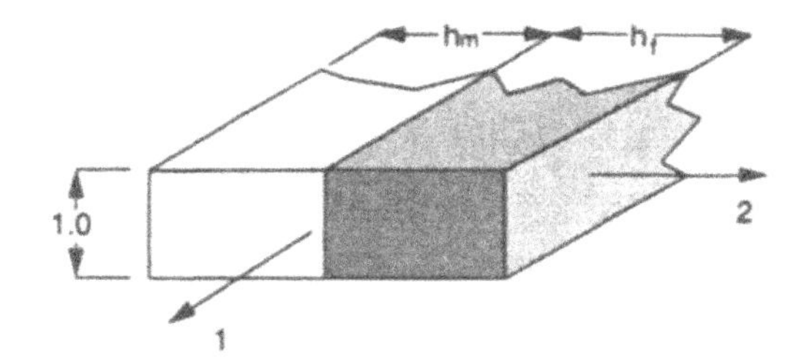

Figure 3.17 RVE for determination of E_1.

In establishing the material properties in each direction, both constituent materials are assumed to be isotropic, homogenous, and linear elastic, with elastic constants E_f, G_f, ν_f and E_m, G_m, ν_m.

Determination of E_1: The model shown in Figure 3.17 is used to determine E_1. Based on previous discussions, the uniform strain model is appropriate one for approximating E_1. The strains in the fiber and matrix are related by $\varepsilon_1^f = \varepsilon_1^m = \varepsilon$. The stress in the fiber direction is approximated as

$$\sigma_1 = \frac{\text{Resultant force}}{\text{Area}}$$

Considering the stress in both the fiber and matrix as contributors to the total stress yields

$$\sigma_1 = \frac{\sigma_1^f(h_f)(1) + \sigma_1^m(h_m)(1)}{(h_f + h_m)(1)} = \sigma_1^f \frac{h_f}{h_f + h_m} + \sigma_1^m \frac{h_m}{h_f + h_m}$$

Since the lamina thickness is unity, and both fiber and matrix are the same length, the volume fractions of fiber and matrix are expressed in terms of h_f and h_m as

$$v_f = \frac{h_f}{h_f + h_m} \quad \text{and} \quad v_m = \frac{h_m}{h_f + h_m}$$

These expressions for volume fractions are applicable only to this model. Another model may result in different forms of v_f and v_m. Using volume fractions, the stress is expressed as $\sigma_1 = \sigma_1^f v_f + \sigma_1^m v_m$. Since $\sigma_1^f = E_f \varepsilon_f = E_f \varepsilon$, and $\sigma_1^m = E_m \varepsilon_m = E_m \varepsilon$, the stress can be written as $\sigma_1 = (E_f v_f + E_m v_m)\varepsilon$, which can be expressed as $E_1 = \sigma_1/\varepsilon$, where

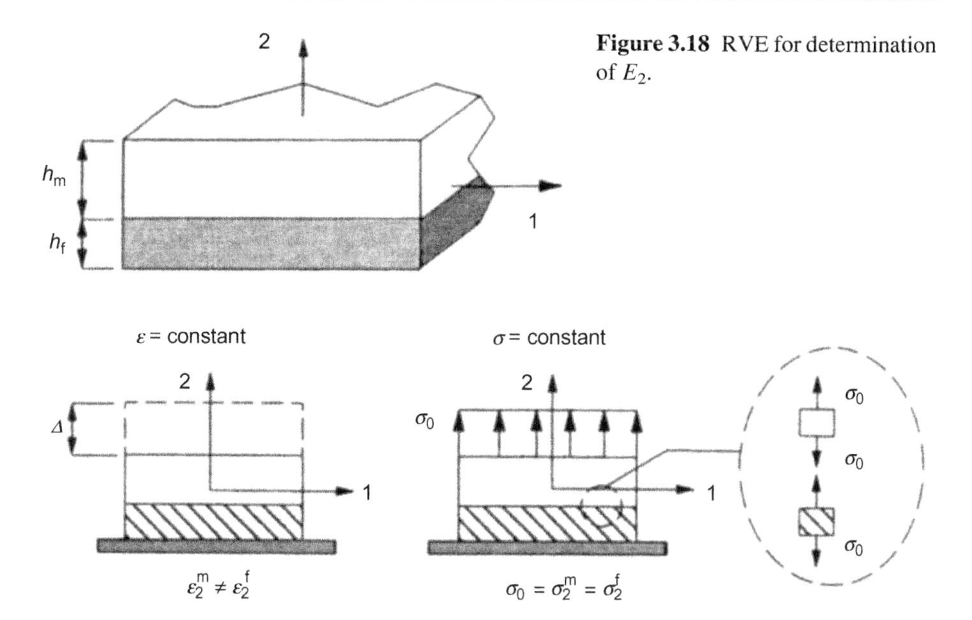

Figure 3.18 RVE for determination of E_2.

Figure 3.19 Constant stress and strain models for determining E_2.

$$E_1 = E_f v_f + E_m v_m \tag{3.36}$$

Equation (3.36) is known as the rule of mixtures and is a fairly accurate approximation of E_1.

Determination of E_2: To determine E_2, the RVE shown in Figure 3.18 is used. An appropriate first approximation to model the mechanical response in the 2-direction is the constant stress (Reuss) model, as shown in Figure 3.19.

The stress in both fiber and matrix is the same in the constant (uniform) stress model. Since the modulus of each constituent is different, the strain will not be equal in the fiber and matrix. The stress and strain in the 2-direction for both the fiber and matrix are related by

$$\varepsilon_2^f = \frac{\sigma_2^f}{E_f}, \quad \varepsilon_2^m = \frac{\sigma_2^m}{E_m}$$

The deformation under conditions of constant stress is expected to resemble that shown in Figure 3.20. The strain in the 2-direction is simply expressed as

$$\varepsilon_2 = \frac{\Delta h}{h} = \frac{\Delta h}{h_f + h_m}$$

where $\Delta h = \varepsilon_2^f h_f + \varepsilon_2^m h_m = \left(\frac{h_f}{E_f} + \frac{h_m}{E_m}\right)\sigma_o$

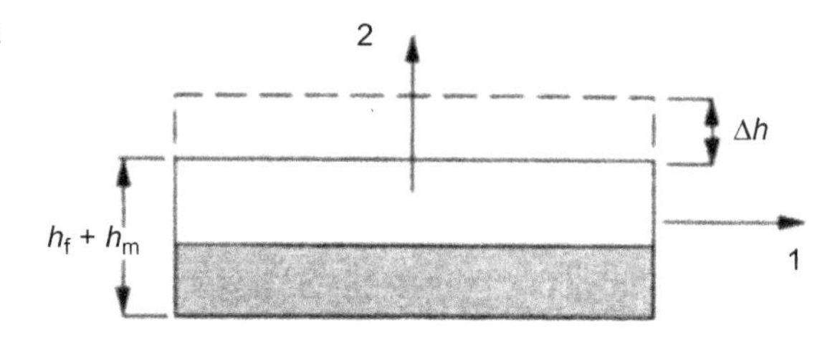

Figure 3.20 Deformations of RVE for E_2 under constant stress conditions.

The strain in the 2-direction is therefore

$$\varepsilon_2 = \frac{\Delta h}{h} = \left(\frac{h_f}{E_f} + \frac{h_m}{E_m}\right)\frac{\sigma_o}{h_f + h_m} = \left(\frac{v_f}{E_f} + \frac{v_m}{E_m}\right)\sigma_o$$

Since $E_2 = \sigma_o / \varepsilon_2$

$$E_2 = \frac{E_f E_m}{E_m v_f + E_f v_m} \tag{3.37}$$

Determination of G_{12}: In a manner analogous to the previous two derivations, the shear modulus can be established by considering a free-body-diagram as shown in Figure 3.21. From this, it is obvious that the stresses are related by $\tau_{12}^f = \tau_{12}^m = \tau_{12}$. The shear strains for both fiber and matrix are a function of the shear modulus for each constituent, with $\gamma_{12}^f = \tau_{12}^f / G_f$ and $\gamma_{12}^m = \tau_{12}^m / G_m$. The shear strain in the matrix does not equal that in the fiber. The shear deformation is

$$\gamma_{12} = \frac{\Delta}{h_f + h_m}$$

where $\Delta = h_f \gamma_{12}^f + h_m \gamma_{12}^m$. Using the definitions of shear strain and Δ presented above, the shear stress–strain relationship is

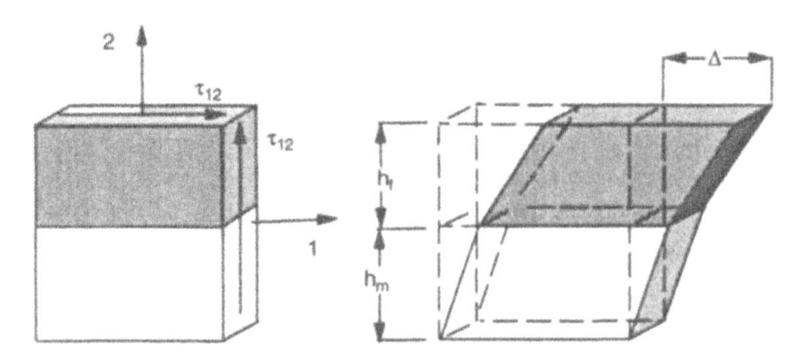

Figure 3.21 FBD and deformations for determining G_{12}.

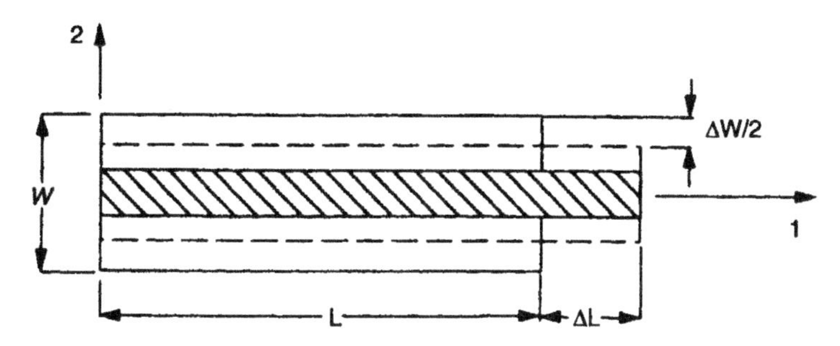

Figure 3.22 RVE for determining ν_{12}.

$$\gamma_{12} = \left(\frac{v_f}{G_f} + \frac{v_m}{G_m}\right)\tau_{12}$$

Since $\tau_{12} = G_{12}\gamma_{12}$, the expression for G_{12} is

$$G_{12} = \frac{G_f G_m}{G_m v_f + G_f v_m} \tag{3.38}$$

Determination of ν_{12}: A similar procedure is used in determining Poisson's ratio. Instead of considering the applicability of either a uniform stress or strain, displacements must be considered. Using the RVE in Figure 3.22, and the definition $\nu_{12} = -\varepsilon_2/\varepsilon_1$.

The lateral displacement of the model and each constituent are related by $\Delta W = -W\varepsilon_2 = W\nu_{12}\varepsilon_1 = \Delta_{mw} + \Delta_{fw}$. The lateral displacement of the matrix is $\Delta_{mw} = h_m\varepsilon_{mw} = -h_m\nu_m\varepsilon_1$. Additionally, $h_m = \nu_m(h_f + h_m) = v_m W$. Therefore $\Delta_{mw} = -Wv_m\nu_m\varepsilon_1$ and $\Delta_{fw} = -Wv_f\nu_f\varepsilon_1$, which leads to $\Delta W = -W\varepsilon_1(v_f\nu_f + v_m\nu_m)$. Since $\nu_{12} = (\Delta W/W)\varepsilon_1$, the expression for Poisson's ratio is

$$\nu_1 = \nu_f v_f + \nu_m v_m \tag{3.39}$$

Example 3.5 Assume a chopped fiber reinforced lamina can be modeled as shown in Figure E3.5-1. Further, assume both the fiber and matrix are isotropic and homogenous, with elastic constants E_f, G_f, ν_f and E_m, G_m, ν_m. We wish to estimate the elastic constants E_1 and E_2.

In order to determine E_1 and E_2 an appropriate RVE must be selected. For the purpose of illustration, the one shown in Figure E3.5-2 is used.

In order to determine the elastic moduli, two different materials are considered. Material A is orthotropic and material B is isotropic, since it consists of matrix only. Each material is considered separately, and then combined into one material.

Material B. Since material B is isotropic, its elastic moduli in the 1- and 2-directions are the modulus of the matrix, expressed as $E_1^B = E_2^B / E_m$.

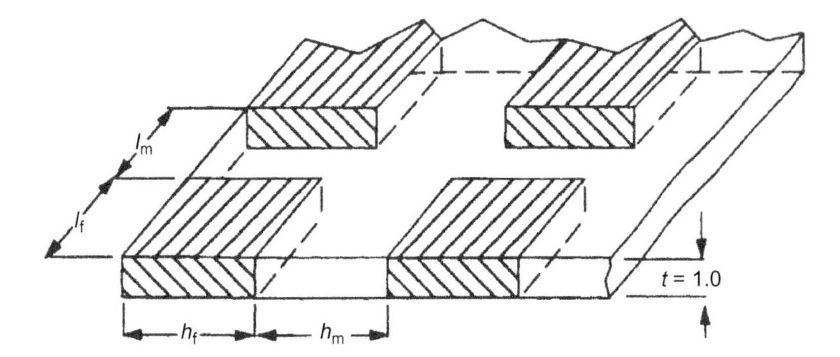

Figure E3.5-1 Rule of mixture model of a chopped fiber lamina.

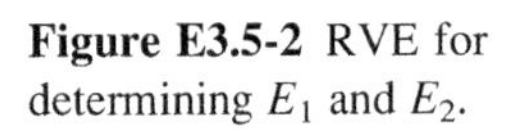
Figure E3.5-2 RVE for determining E_1 and E_2.

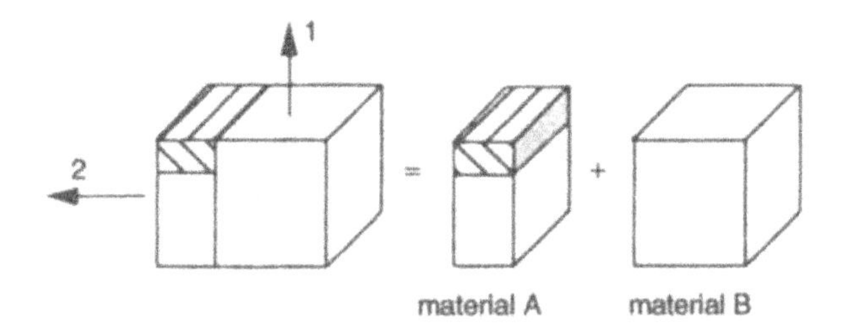

Figure E3.5-3 Model for material A.

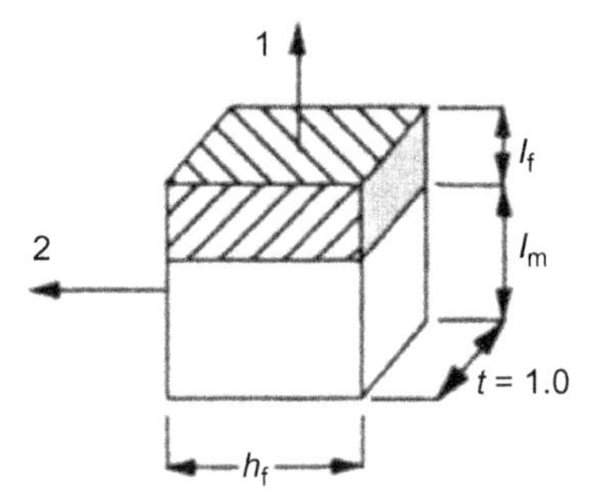

Material A. Two directions are considered for material A. From Figure E3.5-3, each direction is modeled differently.

x_1-direction: The constant stress model is appropriate for this direction. Therefore $\varepsilon_1^{\mathrm{f}} = \sigma_{\mathrm{o}}/E_{\mathrm{f}}$ and $\varepsilon_1^{\mathrm{m}} = \sigma_{\mathrm{o}}/E_{\mathrm{m}}$. The strain in the 1-direction is expressed as $\varepsilon_1 = \Delta L/(l_{\mathrm{f}} + l_{\mathrm{m}})$, where $\Delta L = \varepsilon_1^{\mathrm{f}}(l_{\mathrm{f}}) + \varepsilon_1^{\mathrm{m}}(l_{\mathrm{m}})$. By defining the terms $L_{\mathrm{f}} = l_{\mathrm{f}}/(l_{\mathrm{f}} + l_{\mathrm{m}})$ and $L_{\mathrm{m}} = l_{\mathrm{m}}/(l_{\mathrm{f}} + l_{\mathrm{m}})$, the strain can be written as

$$\varepsilon_1 = \left(\frac{L_{\mathrm{f}}}{E_{\mathrm{f}}} + \frac{L_{\mathrm{m}}}{E_{\mathrm{m}}}\right)\sigma_{\mathrm{o}}$$

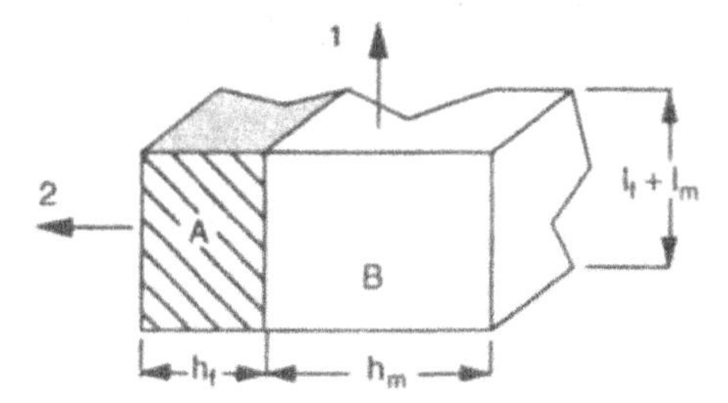

Figure E3.5-4 Model for combined material.

From this it follows directly that for material A

$$E_1^A = \frac{E_f E_m}{L_f E_m + L_m E_f}$$

x_2-direction: The constant strain model is most applicable in this direction. Therefore, the strains and stresses in each constituent material is expressed as $\varepsilon_2^f = \varepsilon_2^m = \varepsilon_2$, $\sigma_2^f = E_f \varepsilon_2^f$, and $\sigma_2^m = E_m \varepsilon_2^m$. The stress in the 2-direction is expressed as

$$\sigma_2 = \frac{\sigma_2^f(l_f) + \sigma_2^m(l_m)}{l_f + l_m} = \sigma_2^f(L_f) + \sigma_2^m(L_m) = (E_f L_f + E_m L_m)\varepsilon_2$$

From this it follows directly that $E_2^A = E_f L_f + E_m L_m$.

Combined material: After determining the material properties in each direction for both models, we combine them into one model. If both materials are combined as shown in Figure E3.5-4, one can identify the model (constant stress or strain) most applicable for each direction.

x_1-direction: The constant strain model is appropriate in this direction. The procedure followed is identical to that previously described, and explicit details are eliminated. A new notation is introduced and is applied in conjunction with the definitions for L_f and L_m. The new notation is $H_f = h_f/(h_f + h_m)$ and $H_m = h_m/(h_f + h_m)$. The stress in the 1-direction is $\sigma_1 = \sigma_1^A(H_f) + \sigma_1^B(H_m) = (E_1^A H_f + E_m H_m)\varepsilon_1$. Using this definition of stress and the previously determined expression for E_1^A it is obvious that

$$E_1 = E_1^A H_f + E_m H_m = \frac{E_f E_m H_f}{E_f L_m + E_m L_f} + E_m H_m$$

x_2-direction: The constant stress model is used for this direction. Following the same procedures as before, the strain is

$$\varepsilon_2 = \left(\frac{H_f}{E_2^A} + \frac{H_m}{E_m}\right)\sigma_2$$

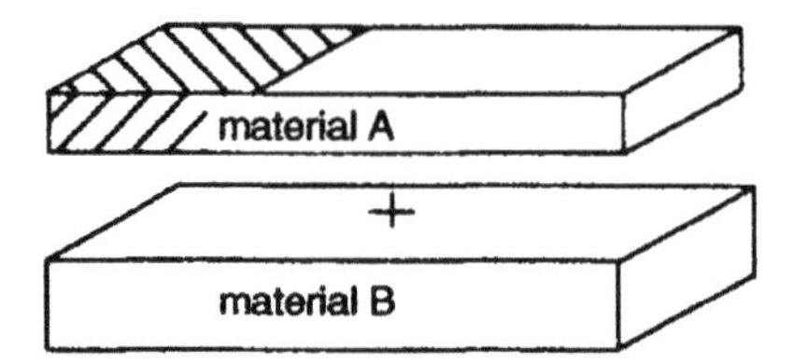

Figure E3.5-5 Alternate model for chopped fiber lamina.

From this expression it is easy to show that

$$E_2 = \frac{E_2^A E_m}{E_2^A H_m + E_m H_f} = \frac{(E_f L_f + E_m L_m) E_m}{(E_f L_f + E_m L_m) H_m + E_m H_f}$$

Neither of these expressions contains the classical form of the volume fractions as introduced in Section 3.4.1.1. The volume fractions for the fiber and matrix can be expressed using the definition of volume fraction. For the chopped fiber composite of this example, they are

Fiber	Matrix
$v_f = \frac{l_f h_f}{(l_f + l_m)(h_f + h_m)} = L_f H_f$	$v_m = \frac{(l_f + l_m) l_m + l_m h_f}{(l_f + l_m)(h_f + h_m)} = H_m + L_m H_f = 1 - v_f$

The model selected to represent the chopped fiber composite in this example was not the only possibility. An alternative is shown in Figure E3.5-5. The procedure for finding E_1 and E_2 does not change.

3.4.1.2 Modifications of E_2 approximations

The approximations for E_2 are less accurate than for E_1. It is fairly well established [19,20] that by considering strain energy, upper and lower bounds of E_i can be determined and expressed by

$$E_f v_f + E_m v_m \geq E_i \geq \frac{E_f E_m}{E_m v_f + E_f v_m}$$

This shows that the constant strain model represents an upper bound, and the constant stress model a lower bound on the actual modulus in either the 1- or 2-directions, as illustrated in Figure 3.23. The actual value of E is between the two solutions. Equations (3.36) and (3.39) generally tend to overestimate E_1 and ν_{12}, while underestimates of E_2 and G_{12} are given by (3.37) and (3.38).

A reason for the underestimation of E_2 is seen by considering the actual displacements which occur in developing the expression for E_2. Following the procedures of Ref. [5], consider the deformed shape of the uniformly stressed RVE in Figure 3.24,

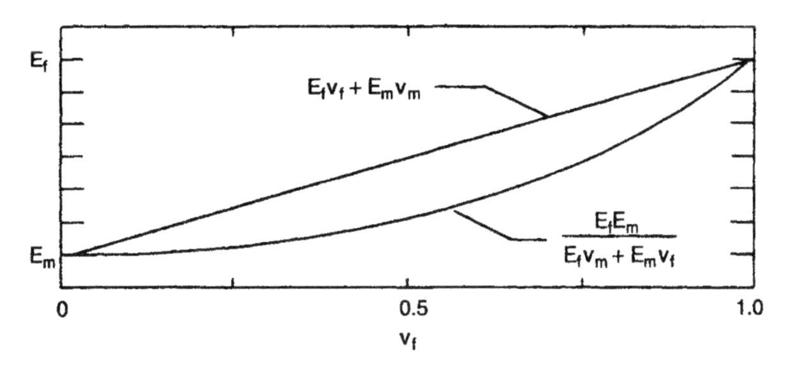

Figure 3.23 Upper and lower bounds on elastic moduli.

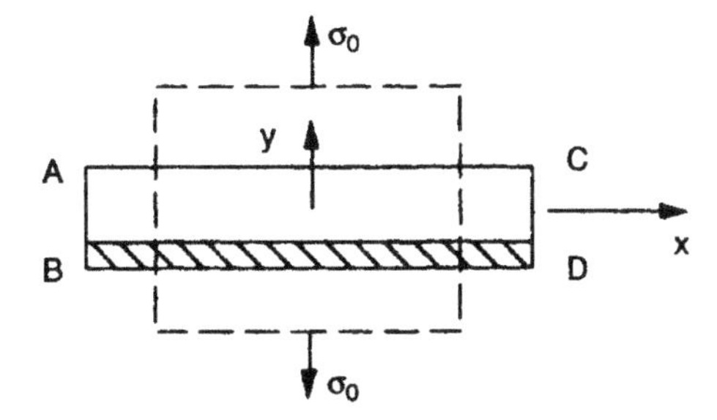

Figure 3.24 RVE for determining E_2 including Poisson's ratio.

which includes the effect of Poisson's ratio along edges *AB* and *CD*. The strains and stresses in the fiber and matrix are more accurately expressed as

$$\varepsilon_x = \varepsilon_{fx} = \varepsilon_{mx} = c = \text{constant}, \quad \sigma_y = \sigma_{fy} = \sigma_{my} = \sigma_o$$

Using conventional stress–strain relations for isotropic materials, the constituent materials experience stresses and strains which are

$$\varepsilon_{fx} = \frac{1}{E_f}\left(\sigma_{fx} - \nu_f \sigma_{fy}\right) = \frac{1}{E_f}\left(\sigma_{fx} - \nu_f \sigma_o\right) = c, \quad \varepsilon_{mx} = \frac{1}{E_m}\left(\sigma_{mx} - \nu_m \sigma_o\right) = c$$

$$\varepsilon_{fy} = \frac{1}{E_f}\left(\sigma_{fy} - \nu_f \sigma_{fx}\right) = \frac{1}{E_f}\left(\sigma_o - \nu_f \sigma_{fx}\right) = c, \quad \varepsilon_{my} = \frac{1}{E_m}\left(\sigma_o - \nu_m \sigma_{mx}\right) = c$$

The strain in the x-direction can be written as $\varepsilon_x = \left(\sigma_x - \nu_x \sigma_y\right)/E_x$. Since $\sigma_x = 0$, $\varepsilon_x = -\nu_x \sigma_y / E_x$, a solution for c can be found and the stresses in the x-direction for both fibers and matrix expressed in terms of the applied stress σ_o as

$$\sigma_{fx} = \frac{E_x \nu_f - E_f \nu_x}{E_x}\sigma_o, \quad \sigma_{mx} = \frac{E_x \nu_m - E_m \nu_x}{E_x}\sigma_o$$

The modulus E_2 is $E_2 = \sigma_o/\varepsilon_y$, where $\varepsilon_y = v_f \varepsilon_{fy} + v_m \varepsilon_{my}$. Substitution of the relationships for E_x and ν_x from Equations (3.36) and (3.39) yields

$$\frac{1}{E_2} = \frac{v_f}{E_f} + \frac{v_m}{E_m} - v_f v_m \frac{\nu_f^2 E_m/E_f + \nu_m^2 E_f/E_m - 2\nu_f \nu_m}{v_f E_f + v_m E_m} \tag{3.40}$$

This approximation works well for E_2 but not for G_{12}.

3.4.1.3 Semiemperical estimates of E_2 *and* G_{12}

Tsai and Hahn [5] present a semiemperical approach to estimating E_2 and G_{12} which requires experimental data. They argue that, since the matrix is softer than the fiber, it is assumed that the stress carried by the matrix in both tension and shear is a ratio of that carried by the fiber. For tension $\sigma_{my} = \eta_y \sigma_{fy}$ $(0 < \eta_y < 1)$ and for shear $\sigma_{ms} = \eta_s \sigma_{fs}$ $(0 < \eta_s < 1)$. Note that the subscript s refers to shear. The normal stress in the y-direction is

$$\sigma_o = \sigma_y = v_f \sigma_{fy} + v_m \sigma_{my} = \left(v_f + v_m \eta_y\right) \sigma_{fy}$$

Therefore,

$$\sigma_{fy} = \frac{\sigma_o}{v_f + v_m \eta_y}$$

The strain in the y-direction of the composite is

$$\varepsilon_y = v_f \varepsilon_{fy} + v_m \varepsilon_{my} = \frac{v_f}{E_f} \sigma_{fy} + \frac{v_m}{E_m} \sigma_{my} = \left(\frac{v_f}{E_f} + \frac{v_m \eta_y}{E_m}\right) \sigma_{fy} = \frac{\dfrac{v_f}{E_f} + \dfrac{v_m \eta_y}{E_m}}{v_f + v_m \eta_y} \sigma_o$$

Since $\varepsilon_y = \sigma_o/E_y$

$$\frac{1}{E_2} = \frac{1}{E_y} = \frac{v_f/E_f + v_m \eta_y/E_m}{v_f + v_m \eta_y} \tag{3.41}$$

In a similar manner

$$\frac{1}{G_{12}} = \frac{v_f/G_f + v_m \eta_s/G_m}{v_f + v_m \eta_s} \tag{3.42}$$

These equations provide better estimates of elastic moduli than the simple rule-of-mixtures expressions (3.37) and (3.38). When η_y and η_s are set to unity, (3.37) and (3.38) are recovered. The η parameters are useful in correlating experimental data. Data from Tsai [21] is used to show the relationship between Equations (3.40) and

(3.41) in Figure 3.25 for several η values. The material properties used in these plots are $E_f = 73.1\text{GPa}\ (10.6 \times 10^6\text{psi})$, $E_m = 3.45\text{GPa}\ (0.5 \times 10^6\text{psi})$, $\nu_f = 0.22$, and $\nu_m = 0.35$.

Figure 3.25 shows how η can be used to model a specific modulus from experimental data. The value of η which produced the closest correlation for this material may not be the appropriate value for a different material. Similarly, the η which provides the best correlation for E_2 may not be appropriate for G_{12}. The usefulness of η in predicting G_{12} can be seen in Figure 3.26, where data from Noyes and Jones [22] is shown with predictions from Equation (3.42). The material properties for the example of Figure 3.26 are $G_f = 30.2\text{GPa}\ (4.38 \times 10^6\text{psi})$ and $G_m = 1.8\text{GPa}\ (0.26 \times 10^6\text{psi})$.

The equations presented here were put in a general form by Tsai and Hahn [5] and are summarized in Table 3.1.

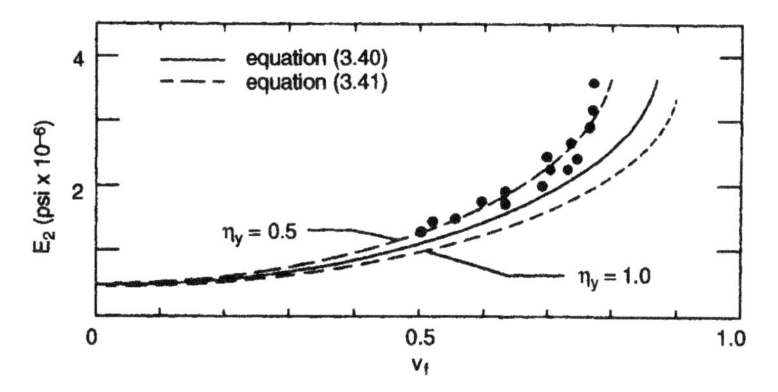

Figure 3.25 Correlation of Equations (3.40) and (3.41) with data from Tsai [21].

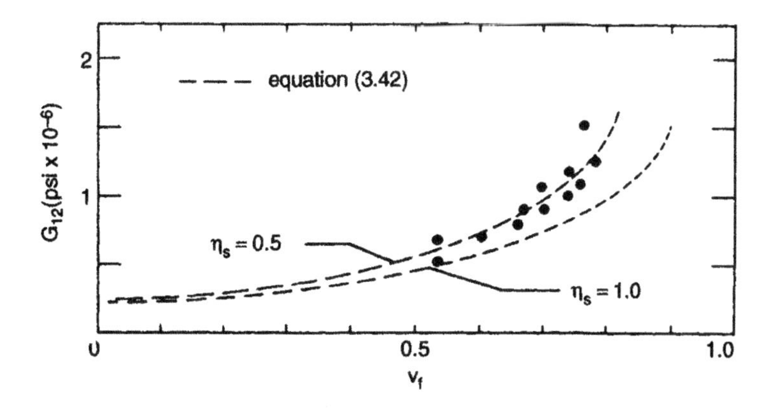

Figure 3.26 Correlation of G_{12} from Equation (3.42) with data from Noyes and Jones [22].

Table 3.1 Summary of formulas for predicting composite moduli $P = \frac{v_f P_f + \eta v_m P_m}{v_f + \eta v_m}$

Engineering constant	P	P_f	P_m	η
E_1	E_1	E_f	E_m	1
ν_{12}	ν_{12}	v_f	ν_m	1
E_2	$1/E_2$	$1/E_f$	$1/E_m$	η_s
k_y	$1/k_y$	$1/k_f$	$1/k_m$	η_k
G_{12}	$1/G_{12}$	$1/G_f$	$1/G_m$	η_G

After Ref. [5].

3.4.1.4 Elasticity solutions with contiguity

Contiguity was introduced by Tsai [21] as a method of making sense out of experimental data in comparison to theoretical predictions and is based on fiber spacing and arrangement. The contiguity factor, C, has a range of $0 < C < 1$ corresponding to the cases illustrated in Figure 3.27. Either none of the fibers contact adjacent fibers ($C = 0$), or all fibers contact adjacent fibers ($C = 1$). Additional elastic constants are based on assuming each constituent is elastic and isotropic and are

$$K_f = \frac{E_f}{2(1-\nu_f)}, \quad G_f = \frac{E_f}{2(1+\nu_f)}, \quad K_m = \frac{E_m}{2(1-\nu_m)}, \quad G_m = \frac{E_m}{2(1+\nu_m)}$$

In general the contiguity factor does not affect E_1. It is generally assumed that fibers are both continuous and straight. During processing, the fibers within a lamina may become somewhat curved. They may also become nested within fibers from an adjacent ply. In order to account for this possibility, and to provide a better correlation between theory and experiment, the misalignment factor k was introduced. The resulting expression for E_1 is

$$E_1 = k(E_f v_f + E_m v_m) \tag{3.43}$$

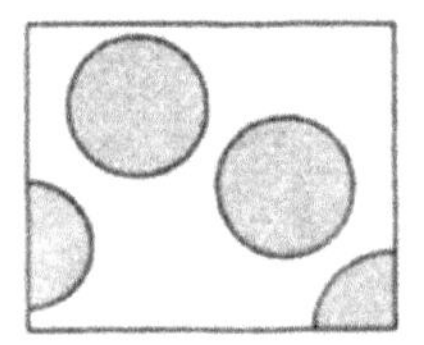

C = 0; isolated fibers, contiguous matrix

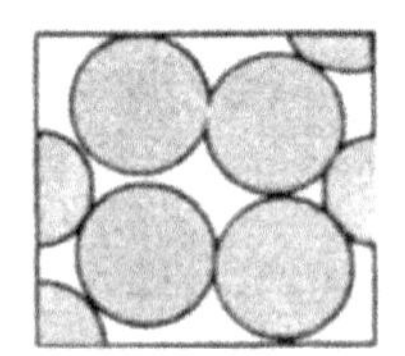

C = 1; isolated matrix, contiguous fibers

Figure 3.27 Models for extremes in contiguity factor C.

where $0.9 < k < 1.0$. The remaining elastic constants can similarly be defined. These expressions, which include the contiguity factor C, are

$$E_2 = A^*[(1-C)B^* + CC^*] \tag{3.44}$$

$$\nu_{12} = (1-C)D^* + CE^* \tag{3.45}$$

$$G_{12} = (1-C)F^* + CG^* \tag{3.46}$$

where the constants A^* through G^* are given as

$$A^* = 2[1 - \nu_f + v_m(\nu_f - \nu_m)]$$

$$B^* = \frac{K_f(2K_m + G_m) - G_m(K_f - K_m)v_m}{(2K_m + G_m) + 2(K_f - K_m)v_m}$$

$$C^* = \frac{K_f(2K_m + G_f) + G_f(K_m - K_f)v_m}{(2K_m + G_f) - 2(K_m - K_f)v_m}$$

$$D^* = \frac{K_f\nu_f(2K_m + G_m) + K_m\nu_m(2K_f + G_m)v_m}{K_f(2K_m + G_m) - G_m(K_f - K_m)v_m}$$

$$E^* = \frac{K_m\nu_m(2K_f + G_f)v_m + K_f\nu_f(2K_m + G_f)v_f}{K_f(2K_m + G_m) + G_f(K_m - K_f)v_m}$$

$$F^* = \frac{G_m[2G_f - (G_f - G_m)\nu_m]}{2G_m + (G_f - G_m)v_m}$$

$$G^* = \frac{G_f[(G_f + G_m) - (G_f - G_m)\nu_m]}{(G_f + G_m) + (G_f - G_m)v_m}$$

These approximations do not generally yield better results than previous cases considered and typically represent bounds on the true modulus.

3.4.1.5 Halpin–Tsai equations

The Halpin–Tsai equations [23] are an interpolative procedure for approximating elastic moduli and Poisson's ratio. They are considered to be accurate for many cases and are

$$E_1 = E_f v_f + E_m v_m \tag{3.47}$$

$$\nu_{12} = \nu_f v_f + \nu_m v_m \tag{3.48}$$

$$\frac{M}{M_{\mathrm{m}}}=\frac{1+\xi\eta v_{\mathrm{f}}}{1-\eta v_{\mathrm{f}}} \tag{3.49}$$

where

$$\eta=\frac{(M_{\mathrm{f}}/M_{\mathrm{m}})-1}{(M_{\mathrm{f}}/M_{\mathrm{m}})+\xi} \tag{3.50}$$

In these equations, M is the composite modulus (E_2, G_{12} or ν_{12}), M_{f} the fiber modulus ($E_{\mathrm{f}}, G_{\mathrm{f}}, \nu_{\mathrm{f}}$), and M_{m} the matrix modulus ($E_{\mathrm{m}}, G_{\mathrm{m}}, \nu_{\mathrm{m}}$). The parameter ξ is a measure of fiber reinforcement in the composite and depends on various conditions such as loading, fiber and packing geometries. The value of ξ is obtained by comparing Equations (3.49) and (3.50) with exact elasticity solutions and is not constant for a given material. It may change values depending upon the modulus being evaluated. The upper and lower limits imposed on ξ are $0 \le \xi \le \infty$. If $\xi = 0$, the lower bound solution for modulus is obtained and

$$\frac{1}{M}=\frac{v_{\mathrm{f}}}{M_{\mathrm{f}}}+\frac{v_{\mathrm{m}}}{M_{\mathrm{m}}}$$

For $\xi = \infty$ the upper bound solution for modulus is obtained and $M = M_{\mathrm{f}} v_{\mathrm{f}} + M_{\mathrm{m}} v_{\mathrm{m}}$. The limiting values of η are

$$\eta=\begin{cases} 1 & \text{Rigid inclusion} \\ 0 & \text{Homogenous material} \\ -1/\xi & \text{Voids} \end{cases}$$

Results from the Halpin–Tsai equations show acceptable correlation to actual data [23] for certain values of ξ. As in the case of approximations to E_2 from (3.41), the accuracy of correlation depends on the composite parameter being evaluated. The value of ξ giving the best correlation for E_2 may not be the same for G_{12}. Determination of an appropriate value for ξ requires fitting experimental data to the theoretical predictions.

3.4.1.6 Additional techniques

The techniques already presented for predicting mechanical properties of lamina do not constitute the entire range of possible models. Predictions of E_1 and ν_{12} are generally less model dependent than E_2 and G_{12} predictions. Classical approaches to micromechanics modeling consider constituent properties ($E_{\mathrm{f}}, E_{\mathrm{m}}, \nu_{\mathrm{f}}$, etc.) and volume fractions (v_{f} and v_{m}). With the exception of contiguity considerations, these models do not account for the fiber packing geometry, which can influence the predicted properties. Chamis [24,25] has developed a set of relationships which incorporate fiber spacing as part of the model and predicts mechanical, thermal, and hygral properties. The model is based on an assumed square array of fiber packing, in which interfiber spacing (δ), fiber diameter (d_{f}), ply thickness (t_{p}), and RVE cell size (s) are incorporated into the model, as schematically shown in Figure 3.28. In this model the number

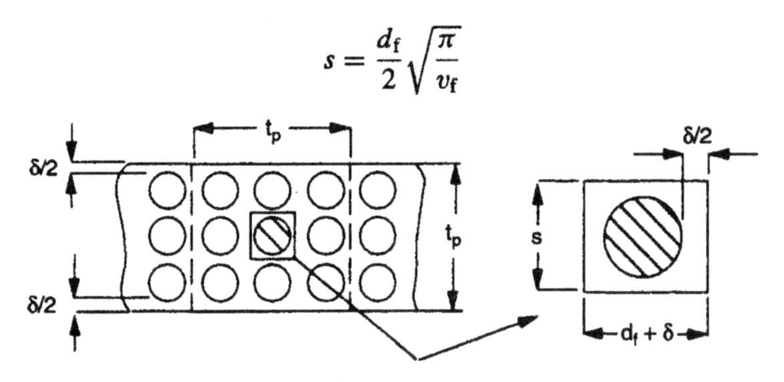

Figure 3.28 Geometric relationships for unified model.
After Chamis [24].

of fibers through the ply thickness can be used to estimate fiber volume fraction (v_f). For a square array of fibers, it has been established that the cell size is related to v_f and d_f by $s=(d_f/2)\sqrt{\pi/v_f}$. Using this relationship, the number of fibers (N_f) through the lamina can be easily estimated from $N_f = t_p/s$.

Relationships between volume fractions, constituent densities (ρ_f, ρ_m), and weight ratios (λ_f, λ_m) are useful in estimating intermediate relationships. Constituent densities are usually available from material suppliers and volume fractions can be experimentally determined. The relations cited above are expressed as

$$\begin{aligned} &v_f + v_m + v_v = 1 \quad \lambda_f + \lambda_m = 1 \\ &v_m = \frac{1 - v_v}{1 + \frac{\rho_m}{\rho_f}\left(\frac{1}{\lambda_m - 1}\right)} \quad v_f = \frac{1 - v_v}{1 + \frac{\rho_f}{\rho_m}\left(\frac{1}{\lambda_f - 1}\right)} \end{aligned} \tag{3.51}$$

The relationships developed by Chamis [25] treat the matrix as isotropic and homogenous, where E_m, G_m, and ν_m are related by $G_m = E_m/2(1+\nu_m)$. The fibers are assumed to have direction dependency. The longitudinal and transverse elastic moduli for the fibers are defined as E_{f11} and E_{f22}. Similarly, longitudinal and transverse shear moduli, and Poisson's ratio are expressed as G_{f11}, G_{f22}, ν_{f11}, and ν_{f22}. The micromechanics relations between constituent properties, volume fractions, and composite properties in Ref. [24] are

$$\begin{aligned} &E_1 = E_{f11}v_f + E_m v_m, \quad \nu_{12} = \nu_{f12}v_f + \nu_m v_m \\ &E_2 = E_3 = \frac{E_m}{1-\sqrt{v_f}(1-E_m/E_{f22})}, \quad G_{12} = G_{13} = \frac{G_m}{1-\sqrt{v_f}(1-G_m/G_{f12})} \\ &G_{23} = \frac{G_m}{1-\sqrt{v_f}(1-G_m/G_{f23})}, \quad \nu_{23} = \frac{E_2}{2G_{23}} - 1 \end{aligned} \tag{3.52}$$

Aside from the $\sqrt{v_f}$ in some of these expressions, the major notable difference between them and previously developed relations is that the fiber is treated as an orthotropic material with direction dependent material properties. Material properties for various fibers and matrices are cited in Ref. [24].

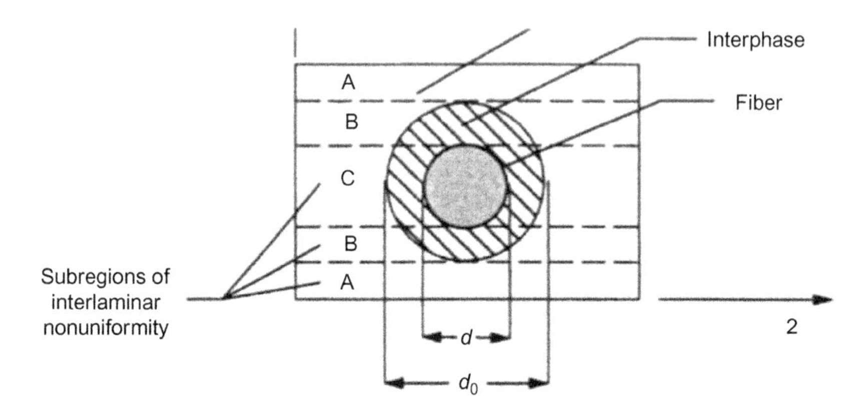

Figure 3.29 RVE of fiber subregions.
After Hopkins and Chamis [26].

Hopkins and Chamis [26] developed a procedure similar to that defined in Ref. [24] for high temperature metal matrix composites. The equations are derived from a mechanics of materials formulation, where a single fiber (in a square array) is assumed to be surrounded by an interphase region (to account for chemical reactions which commonly occur between fiber and matrix), and matrix, as shown in Figure 3.29. The fiber, interphase, and matrix regions are assumed to be transversely isotropic (isotropic behavior in the 2–3 plane is assumed). Each modulus is defined from a model of either constant strain or stress. The size of fiber, interphase, and matrix regions (s_f, s_d, and s_m, respectively) is such that we can define $s_f = d\sqrt{\pi/4}$, $s_d = (d_o - d)\sqrt{\pi/4}$, and $s_m = s - d_o\sqrt{\pi/4}$, where $s = d\sqrt{\pi/4v_f}$.

For a uniaxial load in the transverse direction (2-direction in Figure 3.29), and neglecting Poisson's ratio, the displacement compatibility of subregion C is $s\varepsilon_2 = s_f\varepsilon_f + s_d\varepsilon_d + s_m\varepsilon_m$. A force balance for equilibrium requires $\sigma_2 = \sigma_f = \sigma_d = \sigma_m$ and leads to a definition of E_2 for region C (E_2^C). Introducing the definitions of s_f, s_d, and s_m into the expression for E_2^C produces an equation in terms of diameters d_o and d as well as elastic moduli. By setting $d/d_o = 0$, the equivalent modulus for subregion B (E_2^B) is obtained. Since region A is the matrix, $E_2^A = E_m$. Combining the moduli from each subregion produces an estimate for by solving the equation $E_2 s = E_2^C s_f + E_2^B s_d + E_2^A s_m$. The result is

$$E_2 = E_m\left\{(1 - v_f). + \frac{\sqrt{v_f}[1 - (d/d_o)]}{1 - \sqrt{v_f}[1 - (E_m/E_d)]}\right.$$
$$\left. + \frac{\sqrt{v_f}(d/d_o)}{1 - \sqrt{v_f}[1 - \{1 - (d/d_o)\}(E_m/E_d) - (d/d_o)(E_m/E_d)]}\right\}$$

Following similar procedures, Hopkins and Chamis [26] developed a complete set of high temperature metal matrix composite micromechanics relations for predicting elastic moduli. These are

$$
\begin{aligned}
E_1 &= v_m E_{m11} + v_f\left\{\left[1-(d/d_o)^2\right]E_{d11} + (d/d_o)^2 E_{f11}\right\}\\
E_2 = E_3 &= E_{m22}\left\{(1-v_f) + \frac{\sqrt{v_f}[1-(d/d_o)]}{1-\sqrt{v_f}[1-(E_{m22}/E_{d22})]}\right.\\
&\quad \left. + \frac{\sqrt{v_f}(d/d_o)}{1-\sqrt{v_f}[1-\{1-(d/d_o)\}(E_{m22}/E_{d22})-(d/d_o)(E_{m22}/E_{f22})]}\right\}\\
G_{12} = G_{13} &= G_{m12}\left\{(1-v_f) + \frac{\sqrt{v_f}[1-(d/d_o)]}{1-\sqrt{v_f}[1-(G_{m12}/G_{d12})]}\right.\\
&\quad \left. + \frac{\sqrt{v_f}(d/d_o)}{1-\sqrt{v_f}[1-\{1-(d/d_o)\}(G_{m12}/G_{d12})-(d/d_o)(G_{m12}/G_{f12})]}\right\}\\
G_{23} &= G_{m23}\left\{(1-v_f) + \frac{\sqrt{v_f}[1-(d/d_o)]}{1-\sqrt{v_f}[1-(G_{m23}/G_{d23})]}\right.\\
&\quad \left. + \frac{\sqrt{v_f}(d/d_o)}{1-\sqrt{v_f}[1-\{1-(d/d_o)\}(G_{m23}/G_{d23})-(d/d_o)(G_{m23}/G_{f23})]}\right\}\\
\nu_{12} = \nu_{13} &= v_m \nu_{m12} + v_f\left\{\left[1-(d/d_o)^2\right]\nu_{d12} + (d/d_o)^2 \nu_{f12}\right\}\\
\nu_{23} &= \left(\frac{E_{22}}{2G_{23}}\right) - 1
\end{aligned}
\tag{3.53}
$$

Although one may attempt to use these in correlating moduli predictions from (3.53) to those of the other procedures, some difficulties arise. For example, defining the actual interphase zone size (d_o) or the elastic moduli associated with it is not a well established procedure. Therefore, it is generally more convenient to express (3.53) in a form more compatible with other procedures. In doing this, it is first assumed that $d/d_o = 1.0$, and the matrix is isotropic with $E_{m11} = E_{m22} = E_m$, $G_{m12} = G_{m23} = G_m$, and $\nu_{m12} = \nu_m$. Similarly, the interphase properties become the fiber properties, so that $E_{d11} = E_{f11}$, etc. Using these assumptions leads to

$$
\begin{aligned}
E_1 &= v_m E_m + v_f E_{f11}\\
E_2 = E_3 &= E_m\left\{(1-v_f) + \frac{\sqrt{v_f}}{1-\sqrt{v_f}(1-E_m/E_{f22})}\right\}\\
G_{12} = G_{13} &= G_m\left\{(1-v_f) + \frac{\sqrt{v_f}}{1-\sqrt{v_f}(1-G_m/G_{f12})}\right\}\\
G_{23} &= G_m\left\{(1-v_f) + \frac{\sqrt{v_f}}{1-\sqrt{v_f}(1-G_m/G_{f23})}\right\}\\
\nu_{12} = \nu_{13} &= v_m\nu_m + v_f\nu_f \quad \nu_{12} = \nu_{13} = v_m\nu_m + v_f\nu_f\\
\nu_{23} &= \left(\frac{E_{22}}{2G_{23}}\right) - 1
\end{aligned}
\tag{3.54}
$$

The expressions in (3.54) are similar to those defined in (3.52) and yield similar results.

An additional micromechanical model, which takes the form of the Halpin–Tsai relationships, has been developed by Spencer [27] for estimating E_2 and G_{12}. The model is developed based upon a square array of fibers and includes the effects of strain concentrations at points of minimum clearance between fibers in the RVE. Spencer assumed that only the matrix is isotropic and homogenous. The fiber separation, expressed as $\gamma = s/d$ (not to be confused with shear strain), can be established for three packing arrangements (triangular, square, and hexagonal). Each packing arrangement results in a different numerical relationship between γ and the fiber volume fraction v_f, which is expressed as an index (I), given as $I = 1/(\gamma^2 v_f)$. These indexes are identical to those developed at the beginning of Section 3.4 and are 1.103, 1.272, and 1.654 for triangular, square, and hexagonal packing arrangements, respectively. Although each of these results in a different variation of γ with v_f, Spencer has approximated all three arrangements by defining a modified index as $I = 1.1v_f^2 - 2.1v_f + 2.2$. Using this index, γ becomes

$$\gamma = \frac{1}{\sqrt{\left(1.1v_f^2 - 2.1v_f + 2.2\right)v_f}} \tag{3.55}$$

The semiemperical relationship developed in Ref. [27] for predicting elastic modulus is

$$\frac{M_c}{M} = \frac{\gamma - 1}{\gamma} + \frac{1}{k}\left(-\frac{\pi}{2} + \frac{2\gamma}{\sqrt{\gamma^2 - k^2}}\tan^{-1}\sqrt{\frac{\gamma + k}{\gamma - k}}\right) \tag{3.56}$$

where M_c is the composite modulus (either E_2 or G_{12}), M is the appropriate matrix modulus (either E_m or G_m), and

$$k = \begin{cases} 1 - E_m/E_{f22} \\ 1 - G_m/G_{f12} \end{cases}$$

These expressions, although not correlated with experimental data in Ref. [27], are considered to provide accurate approximations to E_2 and G_{12}. Numerical problems arise for special cases in which $k = \gamma$, and the approximation is no longer valid. Expressions for E_1 and ν_{12} were not developed since it is commonly felt that existing expressions provide sufficient accuracy for most applications.

3.4.1.7 Predictive technique summary

From these discussions, it is obvious that no single model or procedure exists which can be classified as the best approximation. Simple rule of mixtures approximations for E_1 and ν_{12} based on strength of materials techniques are generally reliable for the range of fiber volume fractions typically encountered. The prediction of E_2 and G_{12} is not as reliable.

Each of the relationships presented in the previous section is based on a micromechanical model developed from the study of interacting periodic cells. Two-phase cells have been the primary focus, but a three-phase cell model for metal matrix composites has also been presented, which can be degraded to a two-phase model when required. The text by Aboudi [28] is dedicated to the study of a unified approach to micromechanics focusing on periodic cell models. Elastic as well as nonelastic constituents (e.g., viscoelastic, elastoplastic, and nonlinear elastic) are discussed. In addition, a comprehensive reference for micromechanical models of continuous, particulate, and discontinuous fiber composites is available [29]. Included in this work are models for viscoelastic response and transport properties.

Example 3.6 This example presents a comparison of predicted moduli from several procedures. Assume the material is *S*-glass fibers ($E_f = 12.4 \times 10^6$ psi, $G_f = 5.77 \times 10^6$ psi, $\nu_f = 0.20$) in a PMR matrix.

$E_m = 0.47 \times 10^6$ psi, $G_m = 0.173 \times 10^6$ psi, $\nu_m = 0.32$). The largest variation in predicted moduli between the procedures considered is in E_2 and G_{12}. For the purpose of discussion, only E_2 is considered. The five relationships between E_2, volume fractions, and constituent moduli given in Equations (3.37), (3.40), (3.52), (3.54), and (3.56) are considered. For comparative purposes, we will assume a representative volume fraction of fibers and matrix to be $v_f = 0.6$ and $v_m = 0.4$. The first four equations can be numerically summarized in terms of the constituent volume fractions and elastic moduli. The remaining equation, (3.56), requires an intermediate calculation using (3.55), which results in $\gamma = 1.117$. In addition, $k = 0.962$. Using the material properties cited above with $v_f = 0.6$ and $v_m = 0.4$, we can estimate E_2 for each predictive equation under consideration. For these estimates, we note that $E_f = E_{f22}$.

$$\text{Equation (3.37): } E_2 = \frac{E_f E_m}{E_m v_f + E_f v_m} \Rightarrow E_2 = 1.112 \times 10^6 \text{psi}$$

$$\text{Equation (3.40): } \frac{1}{E_2} = \frac{v_f}{E_f} + \frac{v_m}{E_m} - v_f v_m \frac{\nu_f^2 E_m/E_f + \nu_m^2 E_f/E_m - 2\nu_f \nu_m}{v_f E_f + v_m E_m} \Rightarrow E_2 = 1.298 \times 10^6 \text{psi}$$

$$\text{Equation (3.52): } E_2 = \frac{E_m}{1 - \sqrt{v_f}(1 - E_m/E_{f22})} \Rightarrow E_2 = 1.844 \times 10^6 \text{psi}$$

$$\text{Equation (3.54): } E_2 = E_m \left\{ (1 - v_f) + \frac{\sqrt{v_f}}{1 - \sqrt{v_f}(1 - E_m/E_{f22})} \right\} \Rightarrow E_2 = 1.535 \times 10^6 \text{psi}$$

$$\text{Equation (3.56): } E_2 = E_m \left\{ \frac{\gamma - 1}{\gamma} + \frac{1}{k}\left(-\frac{\pi}{2} + \frac{2\gamma}{\sqrt{\gamma^2 - k^2}} \tan^{-1}\sqrt{\frac{\gamma + k}{\gamma - k}} \right) \right\} \Rightarrow E_2 = 1.775 \times 10^6 \text{psi}$$

The estimate given by (3.37) is from the simple rule of mixtures approximation and is known to underestimate the true value of E_2. The remaining estimates are within approximately 30% of each other and should be viewed as nothing more than estimates.

3.4.2 Physical properties and strength estimates

Physical parameters defining thermal and hygral behavior of lamina can be estimated in a rule of mixtures manner. The models are similar to those used for the elastic moduli. As a result, the topic is not fully developed herein, and only the results from selected references are presented [24–26,30,31]. Micromechanics expressions for composites with isotropic constituents are given in Ref. [30], and the case of orthotropic constituents is discussed in Refs. [25,26,31]. For isotropic constituents, the thermal expansion coefficients are represented as

$$\begin{aligned}\alpha_1 &= \frac{\bar{E}\alpha}{E} \\ \alpha_2 &= \alpha_f v_f(1+\nu_f) + \alpha_m v_m(1+\nu_m) - \alpha_1(v_f\nu_f + v_m\nu_m)\end{aligned} \tag{3.57}$$

where $\overline{E\alpha} = E_f\alpha_f v_f + E_m\alpha_m v_m$ and $E = E_f v_f + E_m v_m$.

Chamis [24] assumed the fibers experience orthotropic thermal expansion with α_{f1} and α_{f2} representing their longitudinal and transverse coefficients of thermal expansion, respectively. The corresponding relationships for α_1 and α_2 are

$$\begin{aligned}\alpha_1 &= \frac{v_f\alpha_{f1}E_{f11} + v_m\alpha_m E_m}{E_1} \\ \alpha_2 &= \alpha_{f2}\sqrt{v_f} + \alpha_m\left(1-\sqrt{v_f}\right)\left(1+\frac{v_f\nu_m E_{f11}}{E_1}\right)\end{aligned} \tag{3.58}$$

where E_1 is given by (3.52). Although these expressions do not reflect fiber orthotropic behavior, the expressions for thermal conductivity in Ref. [24] do.

Expressions involving interphase properties are presented in Ref. [26]. Imposing the same assumptions and limitations on d/d_o, E_d, etc. as done in the corresponding elastic modulus estimates results in

$$\begin{aligned}\alpha_1 &= \frac{v_f\alpha_{f1}E_{f11} + v_m\alpha_m E_m}{E_1} \\ \alpha_2 &= \frac{E_m}{E_2}\left\{\alpha_m(1-\sqrt{v_f}) + \frac{\alpha_m\sqrt{v_f} - v_f(\alpha_m - \alpha_{f2})}{1-\sqrt{v_f}(1-E_m/E_{f22})}\right\}\end{aligned} \tag{3.59}$$

In (3.59) the expressions for E_1 and E_2 are given by (3.54). The expressions for α_1 in (3.58) and (3.59) are identical, while those for α_2 are different. All three expressions for α_1 and α_2 yield reasonable results.

The procedures used to develop coefficients of hygral expansion are analogous to those for thermal expansion coefficients. The primary difference is that $\beta_f = 0$ for both isotropic or orthotropic fibers since they are generally not sensitive to moisture absorption. The coefficients of moisture expansion in each direction are, from Ref. [30],

$$\begin{aligned}\beta_1 &= \frac{\beta_m E_m v_m}{E_f v_f + E_m v_m} \\ \beta_2 &= \beta_m v_m\left\{\frac{(1+v_m)E_f v_f + [1 - v_f(1+\nu_f)]E_m}{E_f v_f + E_m v_m}\right\}\end{aligned} \tag{3.60}$$

The analogous expressions defined in Ref. [25] are

$$\begin{aligned}\beta_1 &= \frac{\beta_m E_m v_m}{E_1} \\ \beta_2 &= \beta_m\left(1-\sqrt{v_f}\right)\left\{1+\frac{E_m\sqrt{v_f}\left(1-\sqrt{v_f}\right)}{E_2\sqrt{v_f}+E_m\left(1-\sqrt{v_f}\right)}\right\}\end{aligned} \tag{3.61}$$

The expressions for β_1 in each equation are identical. The two equations yield different results since the E_1 and E_2 terms in (3.61) are assumed to be defined by (3.52).

The relationships between constituent strengths, moduli, and volume fractions can be obtained from various sources, including Refs. [26,32,33]. A wide range of approaches are possible when attempting to develop a micromechanics model for strength predictions, several of which are presented in this section.

The expressions presented herein assume tensile and compressive failure strengths of fibers and matrix, represented as S_{fT}, S_{fC}, S_{mT}, and S_{mC}, respectively. The fiber is assumed to be insensitive to shear and has no denotable shear strength. The matrix can experience shear failure, which is denoted as S_{mS}. Experimentally determined failure strengths for continuous fiber lamina are typically established under conditions of longitudinal, transverse, and shear loadings in the 1–2 plane and are expressed as S_1, S_2, and S_{12}, respectively. Both S_1 and S_2 can have a tensile (T) and compressive (C) subscript, while S_{12} is invariant to positive or negative shear. One should not confuse the S_{12} defined in this section with the S_{12} compliance term previously defined.

The simplest expression available is based on a simple mechanics of materials model of Chamis [32] which does not define an S_{12}. These failure strengths are given to be

$$S_{1T,C} = S_{fT,C}\left[v_f + \frac{v_m E_m}{E_{f11}}\right], \quad S_{2T,C} = S_{mT,C}\left[v_m + \frac{v_f E_{f11}}{E_m}\right] \tag{3.62}$$

The first of these expressions is sometimes reduced by the assumption that since $E_m \ll E_{f1}$ the second term can be omitted without a significant loss in accuracy. In some cases this assumption may be valid, while in others it may not. The compressive form of the expression for S_1 does not account for possible fiber buckling and is therefore not as reliable as the tensile form of the expression.

Expressions for failure strengths defined by Chamis [33] consider shear failure as well as failures in the 1- and 2-directions and are

$$\begin{aligned}S_{1T,C} &\cong S_{fT,C} \\ S_{2T,C} &= S_{mT,C}\left[1-\left(\sqrt{v_f}-v_f\right)\left(1-\frac{E_m}{E_{f22}}\right)\right] \\ S_{12} &= S_{mS}\left[1-\left(\sqrt{v_f}-v_f\right)\left(1-\frac{G_m}{G_{f12}}\right)\right]\end{aligned} \tag{3.63}$$

The S_1 expression in (3.63) is identical to that in (3.62) under the assumption that the v_m term can be neglected, as previously discussed.

The relationships developed in Ref. [26] are more elaborate than either of the previous relationships. For a tensile stress in the 1-direction the strength is predicted by

$$S_{1T} = S_{fT}\left[v_f + \frac{v_m E_m}{E_{f11}}\right] \tag{3.64}$$

For a compressive stress there are three options. The appropriate one being that which produces the minimum value. Therefore, for compression

$$S_{fC} = \min\begin{cases} S_{fC}\left[v_f + \dfrac{v_m E_m}{E_{f11}}\right] \\ S_{mC}\left[v_m + \dfrac{v_f E_f}{E_m}\right] \\ G_{m12}\left[v_m + \dfrac{v_f G_{m12}}{G_{f12}}\right] \\ S_{mC} + S_{12} \end{cases} \tag{3.65}$$

The expressions for $S_{2T,C}$ and S_{12} are similar and are

$$\begin{aligned} S_{2T,C} &= \frac{S_{mT,C}}{\{1-\sqrt{v_f}[1-E_m/E_{f22}]\}\sqrt{1+\phi(\phi-1)+(\phi-1)^2/3}} \\ S_{12} &= \frac{S_{mS}}{\{1-\sqrt{v_f}[1-G_m/G_{f22}]\}\sqrt{1+\phi(\phi-1)+(\phi-1)^2/3}} \end{aligned} \tag{3.66}$$

where ϕ corresponding to $S_{2T,C}$ is

$$\phi = \left[\frac{1}{\sqrt{\dfrac{\pi}{4v_f}}-1}\right]\left\{\sqrt{\frac{\pi}{4v_f}} - \frac{E_m/E_{f2}}{1-\sqrt{v_f}(1-E_m/E_{f2})}\right\} \tag{3.67}$$

The expression for ϕ corresponding to S_{12} is

$$\phi = \left[\frac{1}{\sqrt{\dfrac{\pi}{4v_f}}-1}\right]\left\{\sqrt{\frac{\pi}{4v_f}} - \frac{G_m/G_{f2}}{1-\sqrt{v_f}(1-G_m/G_{f2})}\right\} \tag{3.68}$$

The strength predictions presented above generally contain E_{f11}, E_{f22}, and E_m. These moduli are assumed to be independent of load direction, and a tension test is assumed to produce the same modulus as a compression test. Although this may be true for some materials, it is not always true. In order to account for this possible bi-modular behavior one could approximate the compressive modulus by using constituent properties based on a compression test.

3.5 Problems

3.1 For the state of plane stress shown, determine the stress in the principal material directions in terms of σ_o.

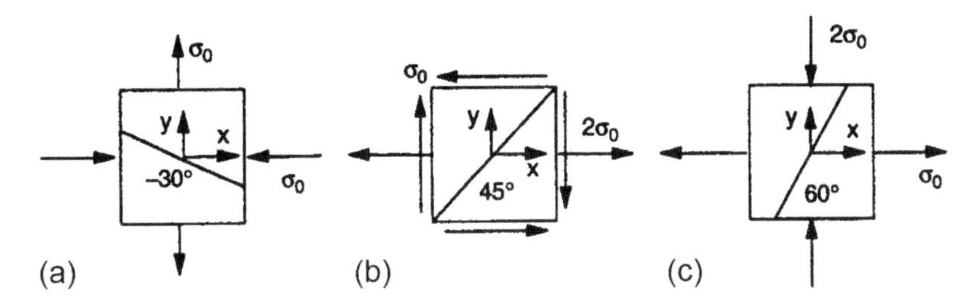

3.2 A unidirectional lamina with dimensions shown is stretched into the deformed shape indicated by the dashed lines. Determine the state of stress in the x–y plane required to produce this deformation. Knowing that $E_1 = 30.3 \times 10^6$ psi, $E_2 = 2.80 \times 10^6$ psi, $G_{12} = 0.93 \times 10^6$ psi, and $\nu_{12} = 0.21$.

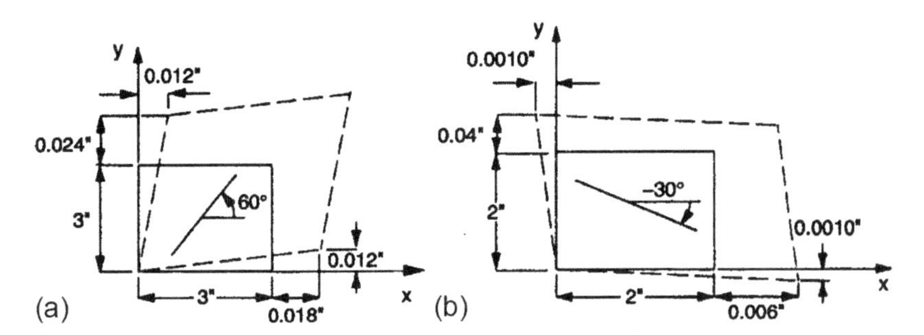

3.3 Work problem 3.2 with $E_1 = 7.0 \times 10^6$ psi, $E_2 = 2.10 \times 10^6$ psi, $G_{12} = 0.80 \times 10^6$ psi, and $\nu_{12} = 0.26$.

3.4 A unidirectional lamina with $E_1 = 30.3 \times 10^6$ psi, $E_2 = 2.80 \times 10^6$ psi, $G_{12} = 0.93 \times 10^6$ psi, and $\nu_{12} = 0.21$ is subjected to the stresses shown. Find the corresponding displacement field.

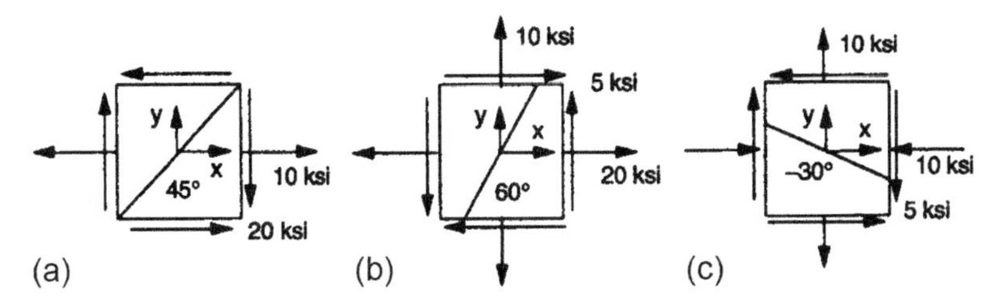

3.5 Work problem 3.4 assuming $E_1 = 20.02 \times 10^6$ psi, $E_2 = 1.30 \times 10^6$ psi, $G_{12} = 1.03 \times 10^6$ psi, and $\nu_{12} = 0.30$.

3.6 A unidirectional lamina with the material properties of problem 3.5 is subjected to the normal stress shown. Determine the apparent Poisson's ratio ν_{xy}.

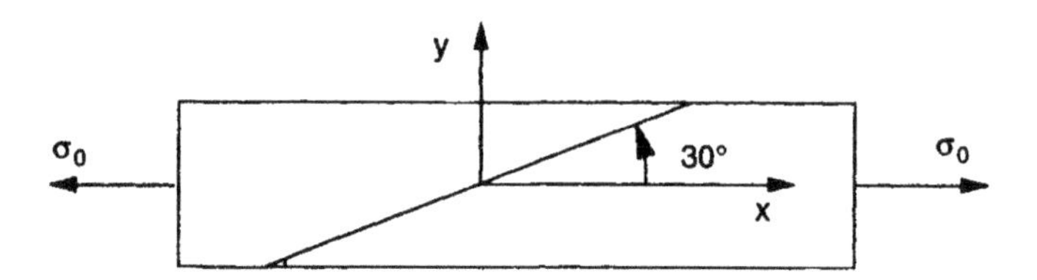

3.7 A unidirectional boron/epoxy lamina is cured at 370 °F and allowed to return to room temperature at 70 °F. The coefficients of thermal expansion are $\alpha_1 = 1.5 \times 10^{-6}$ in/in/°F, $\alpha_2 = 12.2 \times 10^{-6}$ in/in/°F. The lamina is loaded as shown. Find the strain in the principal material directions knowing $E_1 = 30.3 \times 10^6$ psi, $E_2 = 2.80 \times 10^6$ psi, $G_{12} = 0.93 \times 10^6$ psi, and $\nu_{12} = 0.21$.

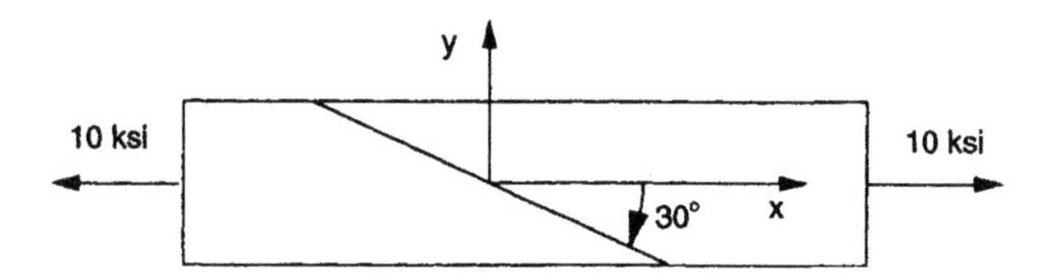

3.8 The lamina of problem 3.7 is subjected to an environment of 95 °F and a relative humidity of 95%. The coefficients of moisture absorption as well as the initial and equilibrium moisture contents and the appropriate equation for relating moisture content and time are $\beta_1 = 0.01$, $\beta_2 = 0.01$, $M_0 = 0.005$, $M_\infty = 0.0171$.

$$\frac{\bar{M} - M_0}{M_\infty - M_0} = 1.0 - 0.8105e^{-\left(8.27 \times 10^{-8} t\right)}$$

where t is the time measured in seconds. Determine the strain in the principal material direction after 6 h.

3.9 A lamina with mechanical properties given in problem 3.7 is placed between two rigid walls in the stress free state (370 °F). As the lamina cools to 70 °F its overall length (the distance between the walls) remains unchanged. Therefore a stress in the x-direction is present. Determine the resulting Cartesian and principal material direction stresses and strains.

(A) $\theta = 30°$
(B) $\theta = 45°$
(C) $\theta = -60°$

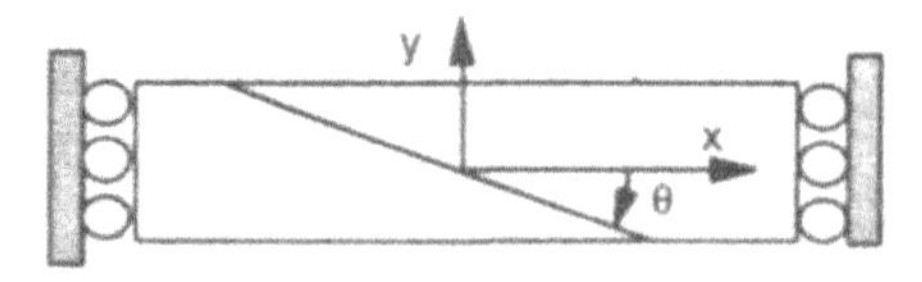

3.10 The lamina of problem 3.9 is also subjected to hygral effects for 6 h. The hygral properties in problem 3.8 are applicable. Determine the Cartesian and principal direction stresses and strains for this lamina for $\theta = 30°$.

3.11 Rigid beam *AB* is pinned at *C* and supported at *A* by a pinned composite column. The dimensions of the composite column and beam *AB* are shown. In order for the entire system to function as designed, member *AE* must be allowed to displace 0.025-in. when beam *AB* is subjected to the loading shown. Determine the required fiber orientation for this to happen knowing $E_1 = 20.0 \times 10^6$ psi, $E_2 = 1.30 \times 10^6$ psi, $G_{12} = 1.03 \times 10^6$ psi, and $\nu_{12} = 0.30$.

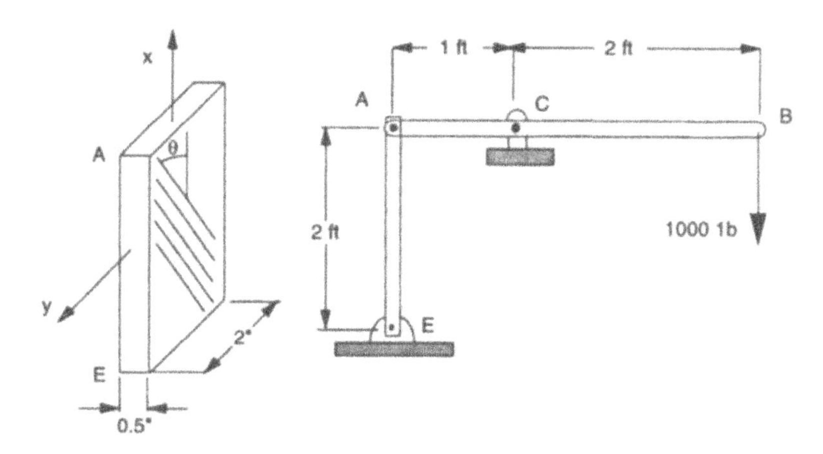

3.12 Assume triangular and regular hexagonal fiber packing arrangements as shown. The fibers have a diameter *d* and a separation distance *s*. Prove that

$$v_f = \frac{\pi d^2}{2\sqrt{3}s^2} \text{ (triangular)}$$

$$v_f = \frac{\pi d^2}{3\sqrt{3}s^2} \text{ (regular hexagonal)}$$

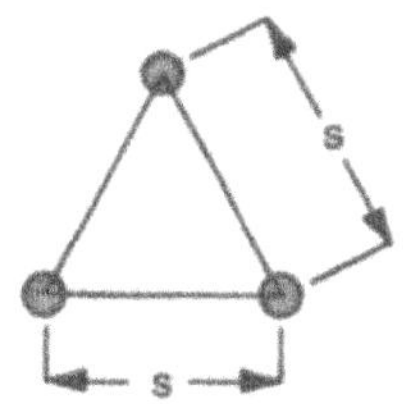

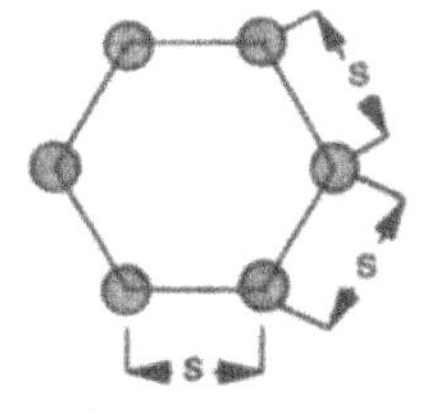

3.13 Assume a rectangular fiber of dimensions *a* and *b* is embedded in a matrix material. Determine v_f for each packing geometry.

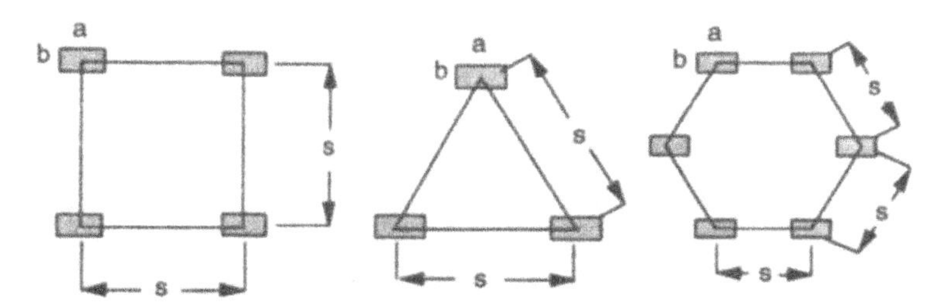

3.14 A continuous fiber "hybrid" composite is assumed to be modeled as shown. The fibers are made of two different materials. The fibers and matrix are isotropic and homogenous with elastic constants $(E_f, G_f, \nu_f)_1$, $(E_f, G_f, \nu_f)_2$, and (E_m, G_m, ν_m). The subscripts 1 and 2 refer to fibers 1 and 2. An RVE of the hybrid lamina is also shown. Use this RVE shown to:

(A) Derive expressions for E_1 and E_2 by appropriate use of the Voigt and Reuss models using the model shown. Be aware that fibers 1 and 2 are different and that $E_1^f \neq E_2^f$.

(B) Assume the total volume of the material modeled is $V = 1.0$, and the volume fraction of matrix in the lamina of 40%. Assume the elastic modulus of fiber 1 and fiber 2 can be related by $E_1^f = nE_2^f$, where $1 \leq n \leq 5$. If $E_1^f = 20E_m$, plot E_1/E_m and E_2/E_m vs. n.

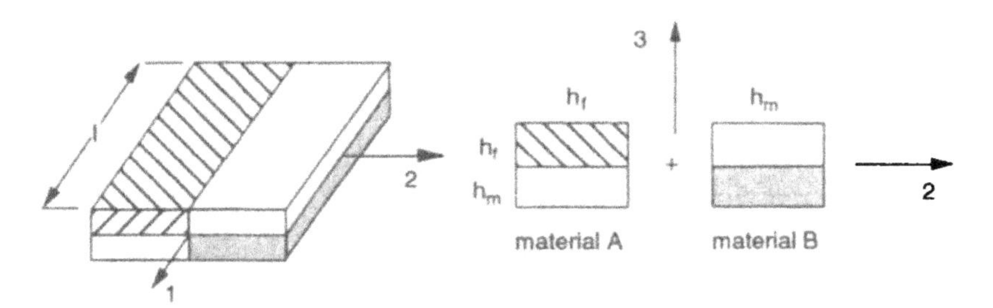

3.15 A layered medium consists of three alternating layers of dissimilar materials. All materials can be assumed to be elastic, isotropic, and homogenous, with properties $[E_A, G_A, \nu_A]$, $[E_B, G_B, \nu_B]$, and $[E_C, G_C, \nu_C]$. Clearly state assumptions regarding constant strain, etc., and use simple rule of mixture assumptions to

(A) Estimate the effective elastic moduli E_1, E_2, and E_3.

(B) Based upon the results of part (A), approximate G_{12} and G_{13}.

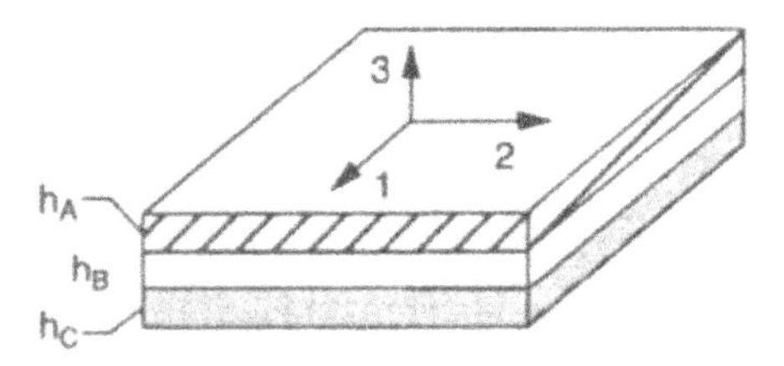

3.16 Assume a graphite/epoxy lamina can be modeled as shown.

(A) Derive expressions for E_1 and E_2 using Voigt and Reuss models.

(B) Assume a 70% fiber volume fraction and material properties of $E_f = 40 \times 10^6$ psi, $\nu_f = 0.25$, $E_m = 0.5 \times 10^6$ psi, and $\nu_m = 0.35$. Compute E_1 and E_2 using the expressions derived in part (A).

(C) Compute E_1 and E_2 using the Halpin–Tsai equations with $\xi = 1.0$.

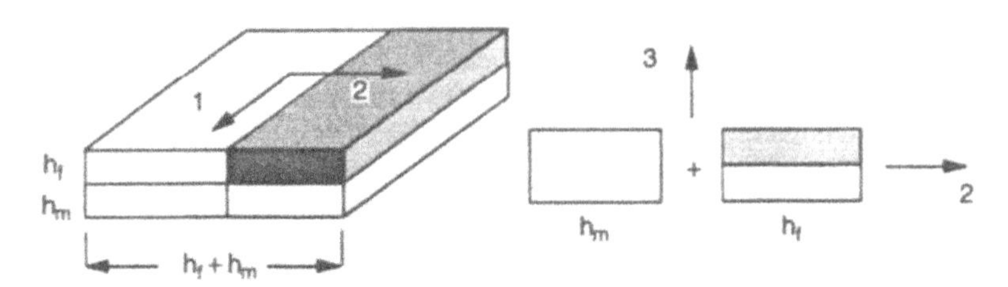

3.17 The elastic moduli for a material are defined by Equations (3.36)–(3.39). The material properties for each constituent are $E_f = 30 \times 10^6$ psi, $G_f = 12 \times 10^6$ psi, $\nu_f = 0.25$, $E_m = 1.0 \times 10^6$ psi, $G_m = 0.385 \times 10^6$ psi, and $\nu_m = 0.30$. Plot ε_1/σ_x and ε_2/σ_x vs. v_f for the state of stress shown. Allow the volume fraction of fibers to be in the range $0.40 \leq v_f \leq 0.70$. Assume a fiber orientation of

(A) $\theta = 30°$

(B) $\theta = 45°$

(C) $\theta = 60°$

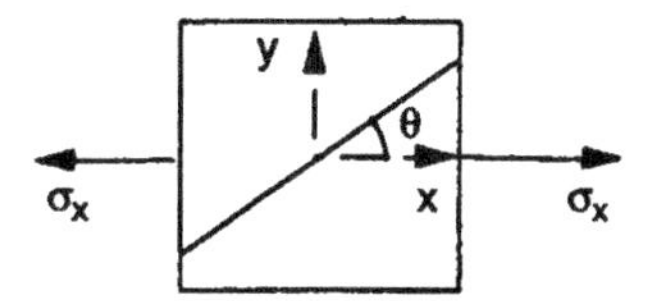

References

[1] Lempriere BM. Poisson's ratio in orthotropic materials. AIAA J 1968;6:2226–7.

[2] Jones RM. Mechanics of composite materials. New York: Hemisphere Publishing; 1975.

[3] Tsai SW, Pagano NJ. Invariant properties of composite materials. AFML-TR 67-349, March; 1968.

[4] Lekhniski SG. Theory of elasticity of an anisotropic elastic body. San Francisco: Holden-Day; 1963.

[5] Tsai SW, Hahn HT. Introduction to composite materials. Lancaster, PA: Technomic Publishing Co. Inc.; 1980.

[6] Tsai SW. Composite design. Dayton, OH: Think Composites; 1987.

[7] Beland S. High performance thermoplastic resins and their composites. Park Ridge, NJ: Noyes Data Corporation; 1990.

[8] Shen C-H, Springer GS. Moisture absorption and desorption of composite materials. J Compos Mater 1976;10:2–20.

[9] Springer GS, editor. Environmental effects on composite materials, vol. 1. Lancaster, PA: Technomic Publishing Co. Inc.; 1981.

[10] Springer GS, editor. Environmental effects on composite materials, vol. 2. Lancaster, PA: Technomic Publishing Co. Inc.; 1984.

[11] Hashin Z. The elastic moduli of heterogeneous materials. J Appl Mech 1962;29:143–50.

[12] Hashin Z. Analysis of composite materials—a survey. J Appl Mech 1983;50:481–505.

[13] Chamis CC, Sendeckyj GP. Critique on theories predicting thermoelastic properties of fibrous composites. J Compos Mater 1968;2:332–58.

[14] Kachanov LM. On the creep fracture time. Izv AN SSSR, Ofd TeckhnNovk 1958;8:26–31 [in Russian].

[15] Krajcinovic D. Continuum damage mechanics. Appl Mech Rev 1984;37:1–6.

[16] Haritos GK, et al. Mesomechanics: the microstructure—mechanics connection. Int J Solids Struct 1988;24:1084–96.

[17] Kachanov LM. Introduction to continuum damage mechanics. The Netherlands: Martinus Nijhoff; 1986.

[18] Gibson RF. Principles of composite material mechanics. Singapore: McGraw-Hill; 1994.

[19] Paul B. Predictions of elastic constants of multiphase materials. Trans Metall Soc AIME 1960;218:36–41.

[20] Hashin Z, Rosen BW. The elastic moduli of fiber-reinforced materials. J Appl Mech 1964;31:223–32, Errata, March 1965, p. 219.
[21] Tsai SW. Structural behavior of composite materials. NASA CR-71; 1964.
[22] Noyes JV, Jones BH. Analytical design procedures for the strength and elastic properties of multilayer fiber composites, In: Proceedings of the AIAA/ASME 9th structures, dynamics and materials conference, paper 68–336; 1968.
[23] Halpin JC, Tsai SW. Effects of environmental factors on composite materials. AFML-TR 67-423; 1969.
[24] Chamis CC. Simplified composite micromechanics equations for hygral, thermal and mechanical properties. SAMPE Q 1984;15:14–23.
[25] Chamis CC. Simplified composite micromechanics equations for mechanical, thermal and moisture-related properties. In: Weeton JW, Peters DM, Thomas KL, editors. Engineer's guide to composite materials. Materials Park, OH: American Society for Metals; 1987.
[26] Hopkins DA, Chamis CC. A unique set of micromechanical equations for high temperature metal matrix composites. In: DiGiovanni PR, Adsit NR, editors. Testing technology of metal matrix composites. ASTM STP, vol. 964. Philadelphia, PA: ASTM International; 1988.
[27] Spencer A. The transverse moduli of fibre composite material. Compos Sci Technol 1986;27:93–109.
[28] Aboudi J. Mechanics of composite materials, a unified micromechanical approach. The Netherlands: Elsevier; 1991.
[29] Whitnel JM. Deleware composites design encyclopedia. Micromechanical materials modeling, vol. 2. Lancaster, PA: Technomic Publishing; 1990.
[30] Schapery RA. Thermal expansion coefficients of composite materials based on energy principles. J Compos Mater 1968;2:380–404.
[31] Hashin Z. Analysis of properties of fiber composites with anisotropic constituents. J Appl Mech 1979;46:543–50.
[32] Chamis CC. Micromechanics strength theories. In: Broutman LS, editor. Composite materials, volume 5: fracture and fatigue. New York: Academic Press; 1974.
[33] Chamis CC. Simplified composite micromechanics equations for strength, fracture toughness and environmental effects. SAMPE Q 1984;15:41–55.

Mechanical test methods for lamina

4

4.1 Introduction

Birefringent coatings, holography, anisotropic photoelasticity, and Moire' have been successfully used in experimentally evaluating composite materials. Topics relating to experimental procedures and laminate test methods are available in texts [1–3], or from periodic publications. The discussions presented herein focus on methods used to establish mechanical and physical properties of orthotropic lamina.

Many test procedures and specimen geometries used with isotropic materials are not applicable to composites. For composites, one is generally concerned with defining load and displacement (or strain) histories throughout a specific test sequence using LVDT's, extensometers, or strain gages. An LVDT or extensometer (using optical or electrical resistance strain gages) measures the relative displacement between reference points on a specimen, and the sensing elements of either device are not directly applied to the specimen. An electrical-resistance strain gage can be applied directly to the specimen. Information from each of these devices is processed to define the parameter(s) of interest. Procedures for accomplishing this are discussed in texts such as Dally and Riley [4] and are not presented herein. Strain gages are perhaps the most commonly used strain measuring device and are briefly discussed.

4.2 Strain gages applied to composites

The concept behind electrical-resistance strain gages is simple and is based on the original findings of Lord Kelvin in 1856 [4], who found that the resistance of copper and iron wires increased as tensile loads were applied to each. Since the applied loads caused changes in the original length of each wire, which are expressible as strains, a direct correlation between strain and resistance change is obtainable. The evolution of strain gage technology from the first practical application in 1938 by the separate efforts of Ruge and Simmons has been substantial. There are many factors which can cause errors in correlating resistance change to strain, and they can be grouped into six categories [3]:

(a) The wire must be firmly bonded to the specimen so that its deformation accurately represents the deformation in the specimen.
(b) The wire must not locally reinforce the structure. If it does, the deformation of the wire does not accurately reflect specimen deformation.
(c) The wire must be electrically insulated from the structure.

Laminar Composites. http://dx.doi.org/10.1016/B978-0-12-802400-3.00004-0

(d) The change in wire resistance per unit microstrain is generally small but must be accurately measured.
(e) Deformation of the structure via mechanisms other than applied loads (such as temperature) must be accounted for.
(f) Aggressive environments may cause oxidation of the wire and lead to resistance changes of the wire which cause erroneous results.

The selection of an appropriate strain gage for a specific application is not a trivial matter, and issues such as temperature compensation, working environment, appropriate strain measuring circuits, etc. must all be considered for accurate collection and evaluation of data. These topics are beyond the scope of this text but are addressed in various references such as Refs. [3–9].

4.2.1 General interpretation of strain gage data

A single-element strain gage applied to a uniaxial tension specimen is represented in Figure 4.1. The longitudinal axis of the gage defines the direction in which strains are measured. Although a uniaxial state of stress exists, a state of biaxial strain results. Both axial and transverse specimen strains affect the strain measured by the gage. The relation between resistance change and a general state of strain is written as [4]:

$$\frac{\Delta R}{R} = S_a \varepsilon_a + S_t \varepsilon_t + S_s \gamma_{st} \tag{4.1}$$

where ΔR and R are the change in and original gage resistance, respectively.

S_a, S_t, and S_s are the axial, transverse, and shear sensitivities of the gage, respectively.

$\varepsilon_a, \varepsilon_t$, and γ_{st} are the axial, transverse, and shear strains of the gage, respectively.

In general, the sensitivity of a strain gage to shear strain is small, and therefore, S_s is neglected. A parameter called the transverse sensitivity factor is introduced into Equation (1) and is

$$K = S_t / S_a \tag{4.2}$$

The numerical values for K generally range from −0.05 to 0.05, and manufacturers report these numbers as percent. Using this definition of transverse sensitivity, and setting $S_s = 0$, Equation (4.1) becomes

$$\frac{\Delta R}{R} = S_a(\varepsilon_a + K\varepsilon_t) \tag{4.3}$$

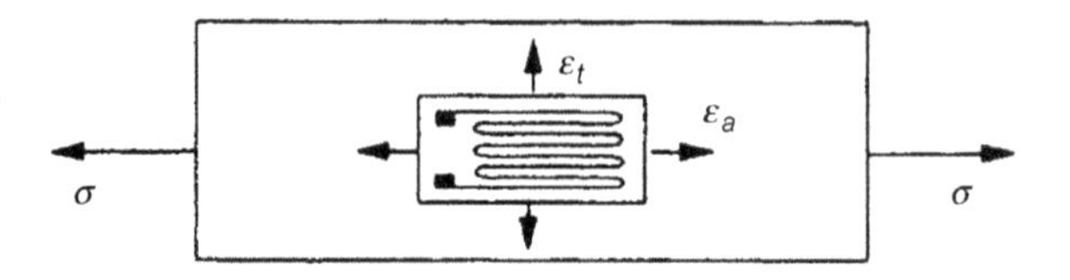

Figure 4.1 Single element strain gage on a uniaxial tension specimen.

A calibration constant known as the gage factor S_g (supplied with each strain gage) relates the resistance change to the axial strain by $\Delta R/R = S_g \varepsilon_a$. The calibration constant is determined from a test performed on each lot of gages being produced. This test is conducted by mounting a gage on a standardized beam so that its longitudinal axis coincides with the direction of maximum normal strain when the beam is deflected a specified amount. The state of stress at the gage location is uniaxial tension, but a state of biaxial strain exists in which $\varepsilon_t = -\nu_0 \varepsilon_a$. Where ν_0 is Poisson's ratio of the calibration beam (generally, $\nu_0 = 0.285$). Substituting this into Equation (4.3) results in

$$\frac{\Delta R}{R} = S_a \varepsilon_a (1 - \nu_0 K) \tag{4.4}$$

From this expression, the gage factor can be defined as

$$S_g = \frac{\Delta R/R}{\varepsilon_a} = S_a(1 - \nu_0 K) \tag{4.5}$$

The strain experienced by the gage (in its longitudinal direction) is related to the resistance change by

$$\varepsilon_a = \frac{\Delta R/R}{S_g} \tag{4.6}$$

Equation (4.6) is based on the following assumptions: (1) the gage is subjected to a uniaxial stress field; (2) the gage grid is parallel to the direction of the stress field; and (3) the gage is mounted on a material for which $\nu = \nu_0$ used in the calibration test. If all of these assumptions are satisfied, the measured strain (ε_m) is identical to ε_a. This ideal situation seldom exists, and in general, Equation (4.6) should not be used directly. Although erroneous results obtained by direct application of Equation (4.6) are not severe in many situations involving isotropic materials, they can be when considering composite materials. The percent error associated with using Equation (4.6) directly has been established in Ref. [4] to be

$$E = \frac{K(\varepsilon_t/\varepsilon_a + \nu_0)}{1 - \nu_0 K}(100) \tag{4.7}$$

The ratio of transverse to axial strain ($\varepsilon_t/\varepsilon_a$) is established from the loading conditions which produce the strain to be measured. The errors should be estimated from Equation (4.7) with an estimated ratio of $\varepsilon_t/\varepsilon_a$ based on the material being tested.

In order to compensate for the effects of transverse sensitivity, a minimum of two strain measurements is required. Assuming a biaxial strain gage rosette is used (two strain gages mounted on a specimen so that two orthogonal strains are measured), the true strain in the x and y directions can be established from the measured strains [4]. Denoting ε_x and ε_y as the true strains, and ε_{mx} and ε_{my} as the measured strains, respectively, the following relationships are obtained.

$$\varepsilon_x = \frac{(1-\nu_0 K)(\varepsilon_{mx} - K\varepsilon_{my})}{1-K^2} \quad \varepsilon_y = \frac{(1-\nu_0 K)(\varepsilon_{my} - K\varepsilon_{mx})}{1-K^2} \tag{4.8}$$

These expressions for the biaxial strain gage rosette are only applicable for that particular type of rosette. A variety of other types of strain gage rosettes are available. Two of these often used with composites are the rectangular and delta rosettes (each containing three strain gage elements). These are schematically shown in Figure 4.2. Relationships between measured and true strains for the rectangular rosette are given by Pendleton and Tuttle [3] as

$$\varepsilon_x = \frac{(1-\nu_0 K)(\varepsilon_{mx} - K\varepsilon_{my})}{1-K^2} \quad \varepsilon_y = \frac{(1-\nu_0 K)(\varepsilon_{my} - K\varepsilon_{mx})}{1-K^2} \tag{4.9}$$

$$\varepsilon_{45} = \frac{(1-\nu_0 K)}{1-K^2}\left\{\varepsilon_{m45} - K(\varepsilon_{mx} + \varepsilon_{my} - \varepsilon_{m45})\right\}$$

where ε_{m45} is the measured strain of the 45° strain gage.

For the delta rosette, the relationships between true and measured strains are

$$\varepsilon_x = \frac{(1-\nu_0 K)}{1-K^2}\left\{\left(1+\frac{K}{3}\right)\varepsilon_{mx} - 2K\frac{\varepsilon_{m60} + \varepsilon_{m120}}{3}\right\}$$

$$\varepsilon_{60} = \frac{(1-\nu_0 K)}{1-K^2}\left\{\left(1+\frac{K}{3}\right)\varepsilon_{m60} - 2K\frac{\varepsilon_{mx} + \varepsilon_{m120}}{3}\right\} \tag{4.10}$$

$$\varepsilon_{120} = \frac{(1-\nu_0 K)}{1-K^2}\left\{\left(1+\frac{K}{3}\right)\varepsilon_{m120} - 2K\frac{\varepsilon_{mx} + \varepsilon_{m60}}{3}\right\}$$

Equations (4.8)–(4.10) are only valid if K is identical for each gage. For situations in which K varies from gage to gage, the appropriate relations can be found in a technical note [10]. The Cartesian strain components associated with each of these true strains are obtained from the strain transformation equation in Chapter 2.

The typically low values of K which strain gage manufactures can obtain indicate that measured strain gage data generally yield accurate results. This is true in many cases involving isotropic materials, and it is not generally true for composite

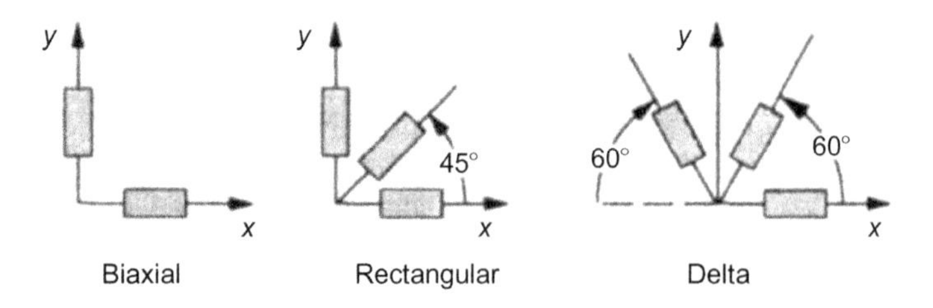

Figure 4.2 Biaxial, rectangular, and delta strain gage rosettes.

materials. Transverse sensitivity effects for composites are typically enhanced since the apparent Poisson's ratio for an orthotropic material is generally different from the ν_0 of the gage calibration material. The evaluation of transverse sensitivity effects for strain gages on composites presented by Tuttle [11] is summarized in Ref. [3]. Tests were conducted on unidirectional carbon/epoxy specimens subjected to uniaxial tension. Both axial and transverse strain gages with $K = 0.03$ were used, and fiber orientations varied. The axial gage had at most a 1% error between measured and corrected data for the worst case (fiber orientation greater than approximately 70°). The transverse gage, however, was shown to exhibit severe errors when K was not used to correct the measured strain gage data. Results of the analysis are described by Pendleton and Tuttle [3].

Example 4.1 Assume a delta rosette ($K = 0.05$ for each gage) is applied to a unidirectional composite. The strain gages shown in Figure E4.1 indicate strains of $\varepsilon_A = 23{,}015\,\mu\text{in./in.}$, $\varepsilon_B = 22{,}307\,\mu\text{in./in.}$, and $\varepsilon_C = -9936\,\mu\text{in./in.}$ Neglecting transverse sensitivity ($K = 0$), the strains for each gage are related through $\varepsilon_\theta = \varepsilon_x \cos^2\theta + \varepsilon_y \sin^2\theta + \gamma_{xy} \sin\theta\cos\theta$. This results in

$$\begin{Bmatrix} \varepsilon_A \\ \varepsilon_B \\ \varepsilon_C \end{Bmatrix} = \begin{Bmatrix} 23{,}015 \\ 22{,}307 \\ -9936 \end{Bmatrix} \times 10^{-6} = \begin{bmatrix} 1 & 0 & 0 \\ \cos^2 60^\circ & \sin^2 60^\circ & \sin 60^\circ \cos 60^\circ \\ \cos^2 120^\circ & \sin^2 120^\circ & \sin 120^\circ \cos 120^\circ \end{bmatrix} \begin{Bmatrix} \varepsilon_x \\ \varepsilon_y \\ \gamma_{xy} \end{Bmatrix}$$

Solving this expression yields

$$\begin{Bmatrix} \varepsilon_x \\ \varepsilon_y \\ \gamma_{xy} \end{Bmatrix} = \begin{Bmatrix} 23{,}015 \\ 576 \\ 37{,}232 \end{Bmatrix} \times 10^{-6}$$

Using the measured strain (ε_A, etc.) and Equation (4.10) with $K = 0.05$ and $\nu_0 = 0.285$, the true strain for each gage is

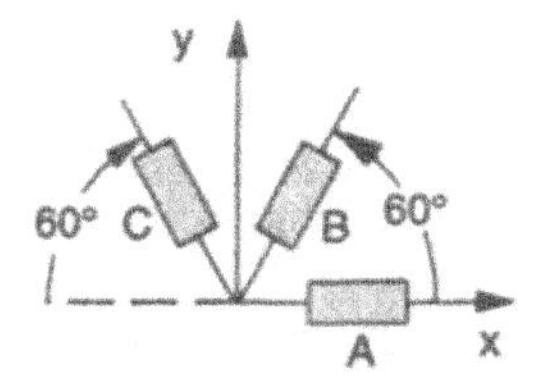

Figure E4.1 Delta strain gage rosette orientation.

$$\begin{Bmatrix} \varepsilon_x \\ \varepsilon_{60} \\ \varepsilon_{120} \end{Bmatrix} = \begin{bmatrix} 1.0040 & -0.0329 & -0.0329 \\ -0.0329 & 1.0040 & -0.0329 \\ -0.0329 & -0.0329 & 1.0040 \end{bmatrix} \begin{Bmatrix} 23{,}015 \\ 22{,}307 \\ -9936 \end{Bmatrix}$$
$$= \begin{Bmatrix} 22{,}700 \\ 21{,}966 \\ -11{,}467 \end{Bmatrix} \times 10^{-6}$$

These strains, when used to evaluate the Cartesian strain components, result in

$$\begin{Bmatrix} \varepsilon_x \\ \varepsilon_y \\ \gamma_{xy} \end{Bmatrix} = \begin{bmatrix} 1 & 0 & 0 \\ 0.25 & 0.75 & 0.433 \\ 0.25 & 0.75 & -0.433 \end{bmatrix} \begin{Bmatrix} 22{,}700 \\ 21{,}966 \\ -11{,}467 \end{Bmatrix} = \begin{Bmatrix} 22{,}700 \\ 17{,}184 \\ 27{,}115 \end{Bmatrix} \times 10^{-6}$$

A comparison between these results and those for $K = 0$ shows a substantial error exists for ε_y and γ_{xy}.

4.2.2 Strain gage misalignment

For an isotropic material, the effects of poor alignment are not as critical as for a composite. For example, using a steel specimen and allowing a set of biaxial gages to be misaligned by some angle β with respect to the loading axis results in errors for both the axial and transverse gages. In Ref. [3], results for a gage misalignment of $-4^\circ \leq \beta \leq -4^\circ$ indicate that the maximum errors in axial and transverse gages were -0.63% and -2.20%, respectively.

For a composite, both the misalignment (β) and fiber orientation angle (θ) contribute to the errors in strain gage measurement. In Tuttle and Brinson [12], tests on the effects of strain gage misalignment as a function of fiber orientation were investigated for the specimen shown schematically in Figure 4.3. The percentage error in axial and transverse gages which resulted was reported in Ref. [3] to be similar to those shown in Figures 4.4 and 4.5, respectively. The transverse gage experienced the most severe errors. Situations often arise in which the complete state of strain is required, a single-element strain gage, or a biaxial rosette are not adequate. A strain gage rosette in which three normal strains are determined is recommended for such cases. The transverse sensitivity corrections for rectangular and delta rosettes expressed by Equations (4.9) and (4.10), respectively, are applicable in this case. Proper alignment of strain gages on composite can be critical when defining a complete state of strain, and care should be taken to ensure proper alignment.

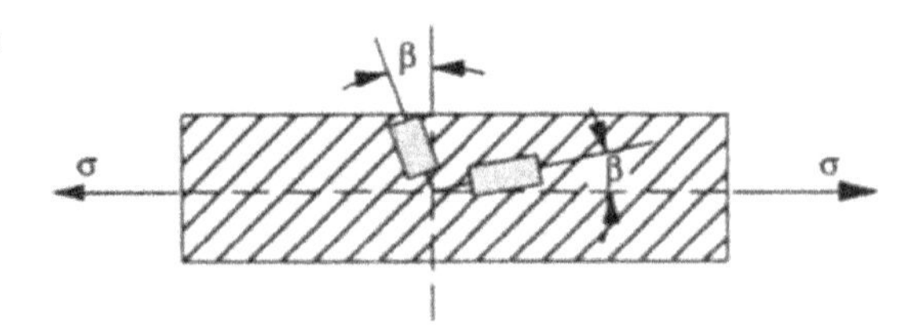

Figure 4.3 Gage misalignment on a composite specimen.

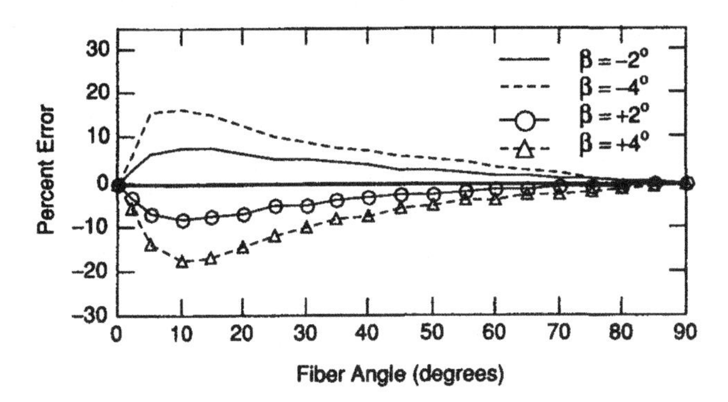

Figure 4.4 Percent error in axial gage (after [3]).

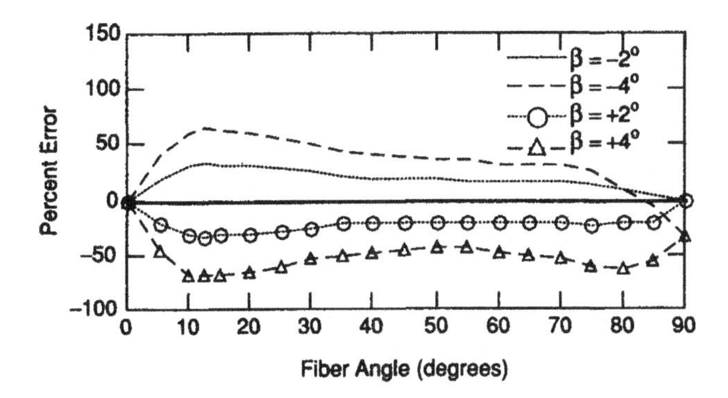

Figure 4.5 Percent error in transverse gage (after [3]).

4.2.3 Strain gage reinforcing effects

Strain gages applied directly to a specimen have been shown to produce reinforcement crrors in tcsts with low modulus materials such as plastics [13–16]. It is possible that similar reinforcement effects occur in composites. Strain gage reinforcement is most likely to occur in regions where the geometric cross-section of the specimen is thin and the elastic modulus in the gage direction is low. The normalized variation of E_x, as defined by Equation (3.19), with respect to the minimum elastic modulus of the material (E_2) decreases rapidly with increasing fiber orientation as shown in Figure 4.6 for Scotchply 1002 glass/epoxy. The magnitudes of E_x/E_2 differ for other material systems, but the trend is the same. For fiber orientations of $-30° \leq \theta \leq 30°$, $E_x > 2E_2$, while for all other fiber orientations, $E_x < 2E_2$. This can influence the degree to which a strain gage will reinforce a specimen and provides inaccurate measures of the actual strain.

In many practical situations, the strain gage is unlikely to significantly reinforce the composite. Given the wide range of possible material properties available with composites, there is a possibility that strain gage reinforcement can affect test results.

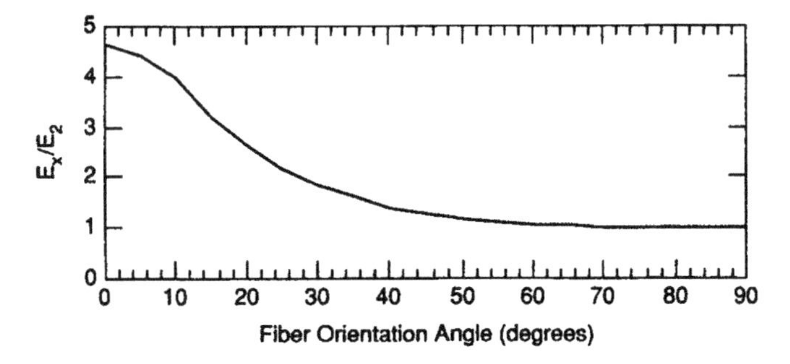

Figure 4.6 Variation of E_x/E_2 with fiber orientation.

Methods of accounting for the reinforcing effect of strain gages are available [3]. The procedure consists of establishing a set of correlation parameters from tests on a calibration specimen, and subsequently relating them to the measured strains on the actual specimen. A simple model for estimating the amount of fiber reinforcement can be defined following the procedures in Dally and Riley [4] for birefringent coatings.

The simplest case to evaluate is uniaxial tension. Assume a unidirectional lamina has a single-element strain gage applied to its surface in the direction of the applied state of stress (σ_x) as shown in Figure 4.7a. Assume the gage is perfectly bonded to the specimen, there are no stress concentrations at the specimen/gage interface, and the load sharing between the gage and specimen can be modeled as shown in Figure 4.7b.

In this figure, σ_{xu} is the axial stress in the ungaged region of the specimen. Stress σ_{xs} and σ_{xg} represent the axial stresses in the specimen and gage, respectively. The thickness of the specimen and gage are represented by h_s and h_g. Assuming the representative volume element associated with Figure 4.7b has a width dy, a force balance in the x direction results in $h_s\sigma_{xu}\mathrm{d}y = h_s\sigma_{xs}\mathrm{d}y + h_g\sigma_{xg}\mathrm{d}y$. Therefore,

$$\sigma_{xu} = \sigma_{xs} + \sigma_{xg}\left(\frac{h_g}{h_s}\right) \tag{4.11}$$

The relationship between axial strain and stress is established from Equation (3.15). Although reinforcement effects may occur transverse to the applied load, the major

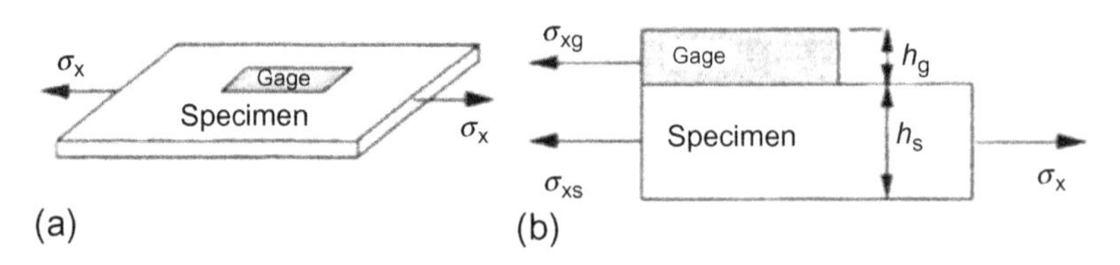

Figure 4.7 Model of load sharing between specimen and strain gage for uniaxial tension.

reinforcement will be in the direction of load application. The relationship between applied stress and strain in the ungaged region of the specimen is

$$\sigma_{xu} = \frac{\varepsilon_{xu}}{\bar{S}_{11}} = E_{xs}\varepsilon_{xs}$$

The specimen and gage stresses are defined by $\sigma_{xs} = E_{xs}\varepsilon_{xs}$ and $\sigma_{xg} = E_g\varepsilon_{xg}$, respectively. Using the assumption that the specimen and gage experience the same strain ($\varepsilon_{xg} = \varepsilon_{xs}$), Equation (4.11) becomes

$$\varepsilon_{xu} = \left[1 + \frac{h_g}{h_s}E_g\bar{S}_{11}\right]\varepsilon_{xg}$$

The ratio of specimen to gage thickness can be represented as $n = h_s/h_g$. Introducing this into the equation above, and rearranging it in order to relate the axial strains in the ungaged region of the specimen to those in the gaged region results in

$$\frac{\varepsilon_{xu}}{\varepsilon_{xg}} = \left(1 + \frac{E_g\bar{S}_{11}}{n}\right) \tag{4.12}$$

The greatest reinforcement effect results when $\bar{S}_{11}$ is a maximum, which generally occurs at fiber orientations of 90°. For this case, Equation (4.12) can be expressed as

$$\frac{\varepsilon_{xu}}{\varepsilon_{xg}} = \left(1 + \frac{E_g}{nE_2}\right) \tag{4.13}$$

The degree of reinforcement ($\varepsilon_{xu}/\varepsilon_{xg}$) depends on the composite material, strain gage material, and the ratio of specimen to gage thickness ($n = h_s/h_g$). A strain gage can be characterized as a plastic, so E_g is typically less than E_2. Evaluating Equation (4.13) for various E_g/E_2 ratios as a function of h_s/h_g results in the distribution shown in Figure 4.8, where it is obvious that the effect of strain gage reinforcement decreases

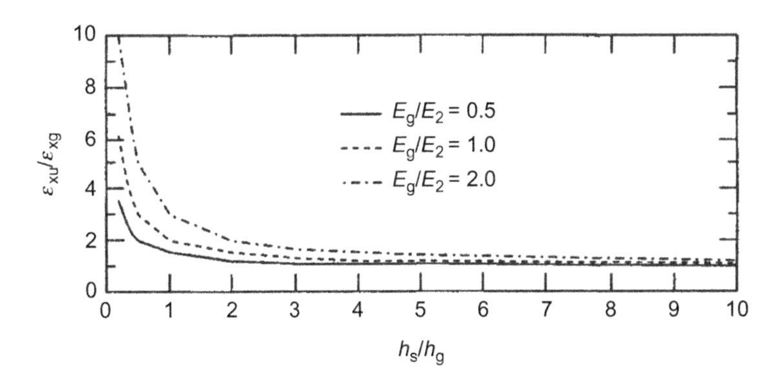

Figure 4.8 Strain gage reinforcement effect for uniaxial tension.

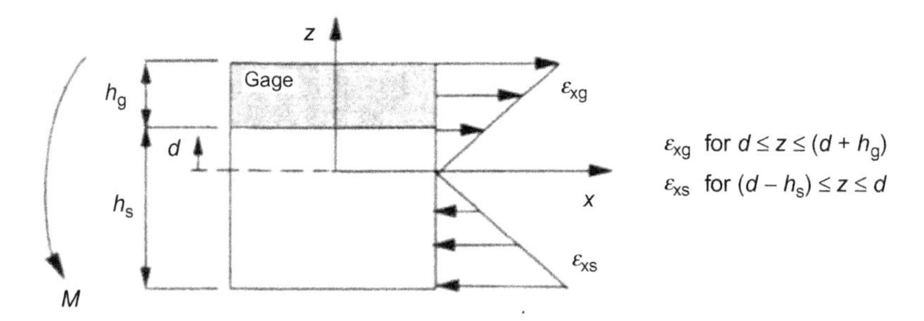

Figure 4.9 Model for strain gage reinforcement due to pure bending.

rapidly with increasing h_s/h_g. It appears as if substantial reinforcement is present for a large range of h_s/h_g, but strain gages are generally on the order of 0.0035 inch thick. Therefore, at $h_s/h_g = 10$, the specimen would be 0.035 in thick. One should evaluate $\varepsilon_{xu}/\varepsilon_{xg}$ from either Equation (4.12) or Equation (4.13) to determine if possible errors warrant compensation.

Strain gage reinforcement is more pronounced when flexure stresses exist. Assume a unidirectional composite lamina is subjected to pure bending. The strain variation through the specimen (ε_{xs}) and strain gage (ε_{xg}) in the region containing the gage are modeled in Figure 4.9.

The distance d is used to define the location of the neutral bending axis. The strain gage is assumed to be isotropic, and the stress in the gage and specimen are expressed as $\sigma_{xg} = E_g\varepsilon_{xg}$ and $\sigma_{xs} = E_{xs}\varepsilon_{xs} = \varepsilon_{xs}/\bar{S}_{11}$, respectively. Since the specimen is subjected to pure bending, the strains are related to the radius of curvature by

$$\varepsilon_{xs} = \frac{z}{\rho} \text{ for } d - h_s \le z \le d \text{ and } \varepsilon_{xg} = \frac{z}{\rho} \text{ for } d \le z \le d + h_g \tag{4.14}$$

Satisfying the condition of equilibrium of forces in the x direction requires

$$\int_{d-h_s}^{d} \sigma_{xs} \mathrm{d}z + \int_{d}^{d+h_g} \sigma_{xg} \mathrm{d}z = 0$$

Using Equation (4.14) in the expressions for σ_{xs} and σ_{xg} results in

$$E_{xs} \int_{d-h_s}^{d} (z/\rho)\mathrm{d}z + E_g \int_{d}^{d+h_g} (z/\rho)\mathrm{d}z = 0$$

Evaluation of these integrals yields an explicit definition of the neutral bending axis location d as a function of material properties and thickness of constituent materials.

$$d = \frac{E_{xs}h_s^2 - E_g h_g^2}{2\left[E_{xs}h_s + E_g h_g\right]} \tag{4.15}$$

The ratio of specimen to gage thickness ($n = h_s/h_g$) can be introduced into this expression as it was for the case of uniaxial tension, which allows (4.15) to be expressed as

$$d = C_1 h_g \tag{4.16}$$

where $C_1 = \dfrac{E_{xs}n^2 - E_g}{2[E_{xs}n + E_g]}$

The radius of curvature is determined by establishing the standard conditions of equilibrium for moments from $M = \int_{d-h_s}^{d} z\sigma_{xs}\mathrm{d}z + \int_{d}^{d+h_g} z\sigma_{xg}\mathrm{d}z = 0$. Using Equation (4.14) in the expressions for σ_{xs} and σ_{xg} results in $M = \frac{1}{\rho}\left[E_{xs}\int_{d-h_s}^{d} z^2\mathrm{d}z + E_g\int_{d}^{d+h_g} z^2\mathrm{d}z\right] = 0$. Upon evaluation of these integrals, it is convenient to define two additional terms

$$C_2 = 3\mathrm{d}^2 h_s - 3\mathrm{d}h_s^2 + h_s^3 \quad \text{and} \quad C_3 = 3\mathrm{d}^2 h_g + 3\mathrm{d}h_g^2 + h_g^3 \tag{4.17}$$

where d is defined by Equation (4.16). The curvature and bending moment in the strain-gaged area are related by $(1/\rho)_g = 3M/(E_{xs}C_2 + E_g C_3)$. For an unreinforced specimen, the relationship between curvature and moment is easily established since $h_g = 0$, $d = h_s/2$, and $\sigma_{xg} = 0$. The resulting relationship is $(1/\rho)_u = 12M/E_{xs}h_s^3$. Since $\varepsilon_{xg} = z/\rho_g$ and $\varepsilon_{xu} = z/\rho_u$, the ratio of strains in the ungaged region to those in the gaged region is $\varepsilon_{xu}/\varepsilon_{xg} = \rho_g/\rho_u$. Using the relationships for $(1/\rho)_g$ and $(1/\rho)_u$ results in

$$\frac{\varepsilon_{xu}}{\varepsilon_{xg}} = \frac{4\{E_{xs}C_2 + E_g C_3\}}{E_{xs}h_s^2} \tag{4.18}$$

where C_2 and C_3 are defined by Equation (4.17).

Equation (4.18) is more complex than either Equation (4.12) or Equation (4.13) due to the relationships between C_2, C_3, d, and the relative thickness of both specimen and gage. The maximum reinforcing ($\varepsilon_{xu}/\varepsilon_{xg}$) will generally occur for fiber orientations of 90° with respect to the x-axis (in the x–y plane). Expressing Equation (4.18) in terms of the elastic constants E_g and E_2 is not warranted for this case due to coupling of terms related through d. Evaluating Equation (4.18) for various ratios of h_s/h_g and E_g/E_2 produces the results shown in Figure 4.10. At $h_s/h_g = 10$, the best case shown ($E_g/E_2 = 0.5$) predicts

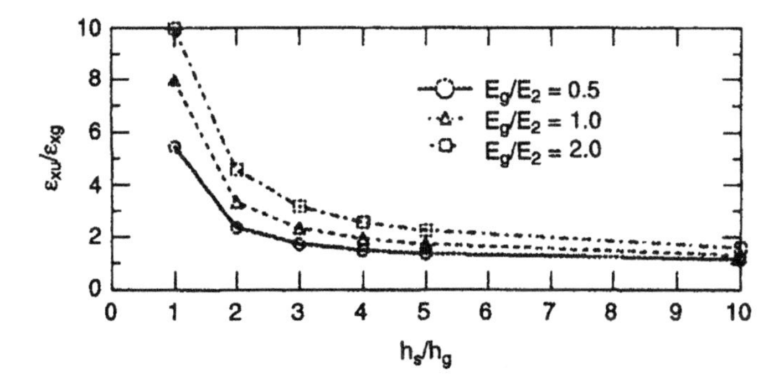

Figure 4.10 Strain gage reinforcing effects for flexure in the range $0 \le h_s/h_g \le 10$.

$\varepsilon_{xu}/\varepsilon_{xg} = 1.17$, which implies a 17% stiffening effect. As with uniaxial tension, the possibility of gage reinforcement from bending should be evaluated prior to strain gage application.

The amount of reinforcing associated with a specific set of gage and specimen properties can be estimated from Equation (4.12) or Equation (4.18). If reinforcement is considered a potential problem, the procedures in Ref. [3] can be used.

4.3 Experimental determination of mechanical properties

Nine independent elastic constants are required to define the mechanical response of an orthotropic lamina. In many cases of practical importance, a state of plane stress exists and the out of plane material properties are not required since they are often approximated by in-plane properties ($E_2 = E_3$, etc.). The mechanical properties generally considered to be of greatest interest are $E_1, E_2, G_{12}, \nu_{12}$, and ν_{21}. Lamina failure strengths can be established as part of the experimental procedures used for determining elastic properties. A unidirectional laminate is typically used since individual lamina is too thin and weak in the transverse direction to sustain sufficient load for determining elastic moduli and failure strength. The procedures discussed herein are those most commonly used in establishing the properties identified above and those which are typically easiest to implement.

4.3.1 Tensile testing

Each previously cited material property can be established from uniaxial tension tests of unidirectional laminates. Although G_{12} can be established from uniaxial tension tests, discussion of shear modulus determination is reserved for a later section. The recommended test procedures which should be followed in establishing these properties are described in ASTM D3039-76.

A dogbone specimen used in uniaxial tension tests of flat coupons for isotropic materials is not acceptable for laminates. Establishing E_1 and E_2 (along with ν_{12} and ν_{21}) requires test specimens with fiber orientations of 0° and 90°, respectively. A dogbone-shaped specimen with a 0° fiber orientation will result in the formation of matrix cracks parallel to the fibers, and an eventual failure in the region indicated in Figure 4.11. A stress–strain curve generated from such a specimen may contain a region of valid data (up to the point where the matrix cracks begin to develop) but will generally not yield an accurate modulus prediction or failure strength.

The 90° specimen will not fail in this manner. Damage induced by machining the specimen into a dogbone shape may weaken the matrix to the extent that invalid

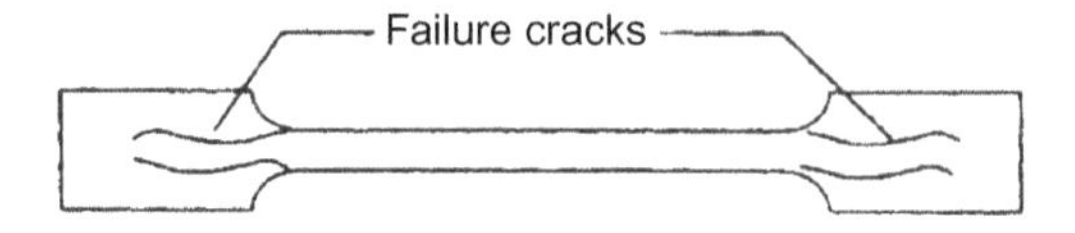

Figure 4.11 Failure mechanism for a composite dogbone specimen.

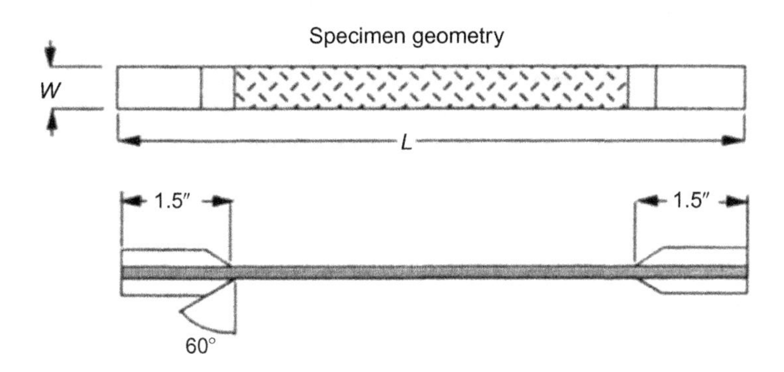

Figure 4.12 Geometry and dimensions of 0° and 90° tensile specimens.

predictions of E_2 result. A more appropriate specimen is a flat coupon with end tabs. The end tabs help reduce the probability of failure in the grip region where the applied loads are transferred from the testing machine to the specimen. Without end tabs, the normal force between the specimen and grips could crush the fibers in the grip region and produce premature failure.

The standard dimensions of a test specimen depend on fiber orientation. The geometry and dimensions for 0° and 90° tensile coupons are given in Figure 4.12. The total thickness of a specimen depends on the number of plies in the laminate. The end tabs are typically 1.50 in long and 0.125 in thick. The tabs are beveled to allow for a more uniform load transfer from the grips to the specimen.

Strain gages or extensometers are often used to determine the stress–strain history of a specimen. Using both longitudinal and transverse gages, a single test can produce either E_1 ν_{12} or E_2 ν_{21}, depending upon fiber orientation, as illustrated in Figure 4.13. Strain gages placed on the front and back of each specimen (in a full Wheatstone bridge) negate the effects of bending due to eccentricity of the load line [4]. A biaxial extensometer could also be used.

Specimen dimensions			
Fiber orientation	**Width (in.)**	**Number of plies**	**Length (in.)**
0°	0.50	6–8	9.00
90°	1.00	8–16	9.00

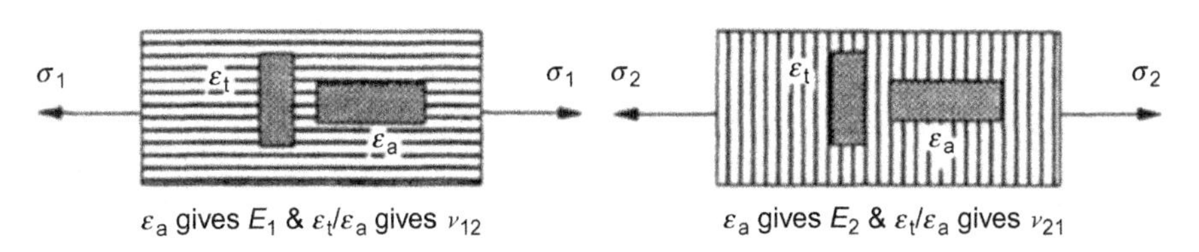

Figure 4.13 Schematic of fiber orientations and strain gage positioning for determining E_1, E_2, ν_{12} and ν_{21}.

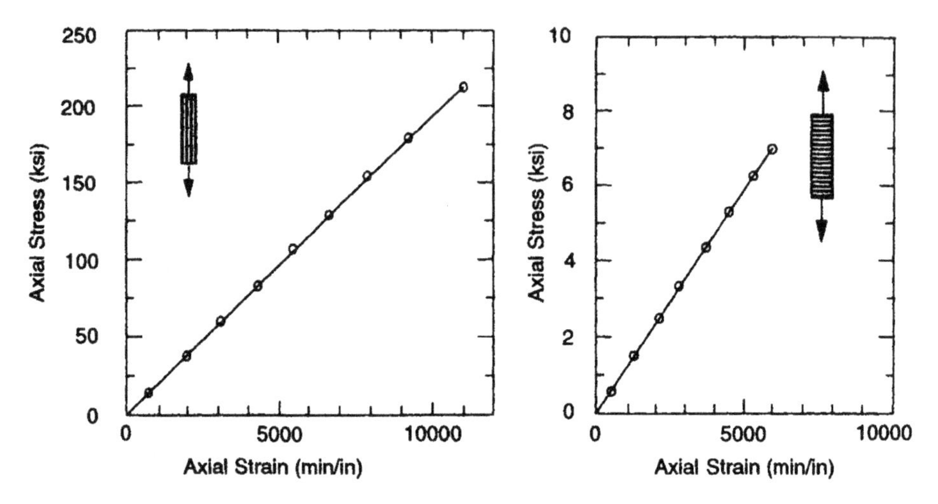

Figure 4.14 Stress-strain curves for AS/3502 graphite/epoxy.

A typical set of stress–strain curves for 0° and 90° AS/3502 graphite/epoxy tensile specimens are presented in Figures 4.14. Failure of the 90° specimen is matrix dominated. The fibers are capable of sustaining a far greater load than the matrix; therefore, the 0° specimen fails in a significantly different manner. Along the failure surface, there are jagged edges indicating that failure was not instantaneous. As the ultimate load is approached, individual fibers begin to fail with an audible "ping" sound. Individual fibers will fail at slightly different load levels.

4.3.2 Compression testing

In conducting compression tests, it has been noted that a composite material may exhibit different tensile and compressive moduli (E_1,E_2, etc.) and is termed bimodular. The influence of bimodularity on analysis techniques and failure analysis can be significant [17]. The failure strength is generally considered more significant than modulus when comparing tensile and compressive behavior. Some of the differences between tensile and compressive behavior can be attributed to the difficulty of compression testing. Slight geometric variations in the specimen may result in eccentric loads, which enhance the possibility of failure due to instability as opposed to stress. There are three accepted test methods which reduce this possibility as described by Whitney et al. [1]. Each is briefly outlined, and schematic diagrams of grip arrangements are presented. Strain gages are generally used for each of these test methods. Additional compression test methods are available, which are presented in a survey article pertaining to compression testing of composites [18].

TYPE I. This method is characterized by having a completely unsupported specimen with a relatively short test section length. Several types of fixtures exist for this method. The Celanese (ASTM D-3410-75) test fixture and associated specimen

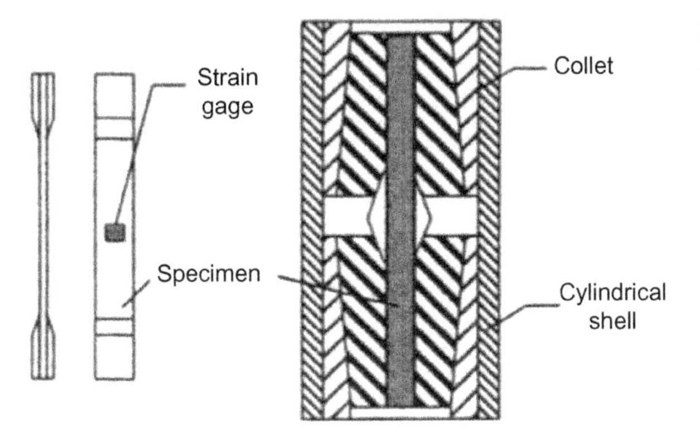

Figure 4.15 Celanese test fixture and specimen (ASTM D 3410-75).

geometry are shown in Figure 4.15. The Illinois Institute of Technology Research Institute (IITRI) [19] test fixture uses a test specimen identical to the Celanese fixture and is shown in Figure 4.16. Strain gages are mounted on the specimen, which is loaded through serrated wedges constrained by solid steel bases. The Northrop test fixture [20] is simpler than the Celanese or IITRI fixtures and is shown in Figure 4.17. The final example of Type I compression testing is the NBS (National Bureau of Standards) test fixture [21]. This fixture combines aspects of the Celanese and IITRI fixtures and adds features which allow for tensile tests. The NBS fixture is shown in Figure 4.18. All four of the Type I test methods yield acceptable results but are difficult to conduct due to load line eccentricity.

TYPE II. In this class of tests, the specimens are characterized as having a relatively long test section which is fully supported. The SWRI (Southwest Research Institute) [22] and the Lockheed type fixtures [23] are schematically shown in Figures 4.19 and 4.20, respectively. Results from experiments using these grips are comparable to

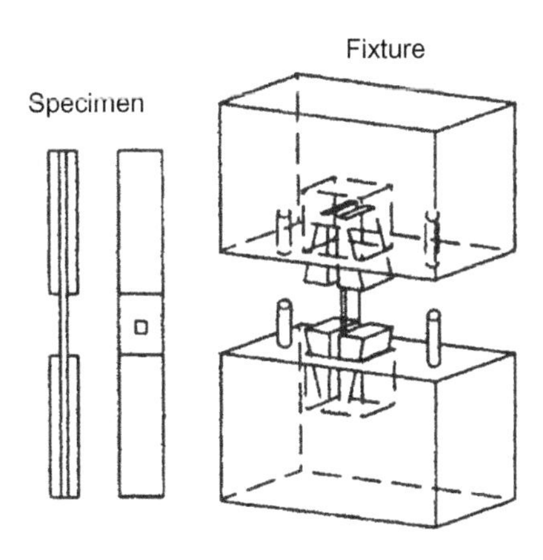

Figure 4.16 Modified grips for IITRI compression test (after [19]).

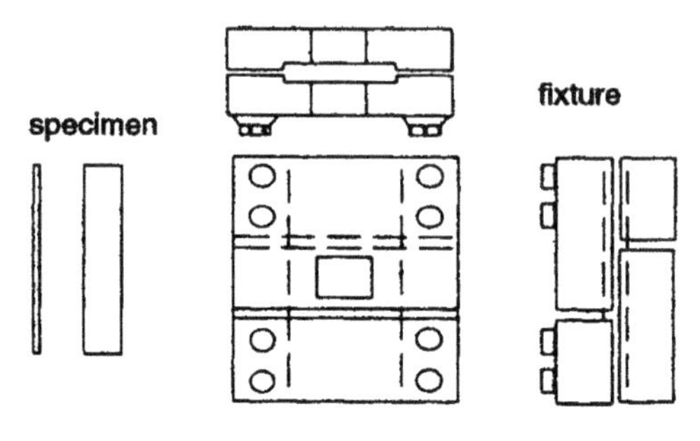

Figure 4.17 Northrop compression test specimen and fixture (after [20]).

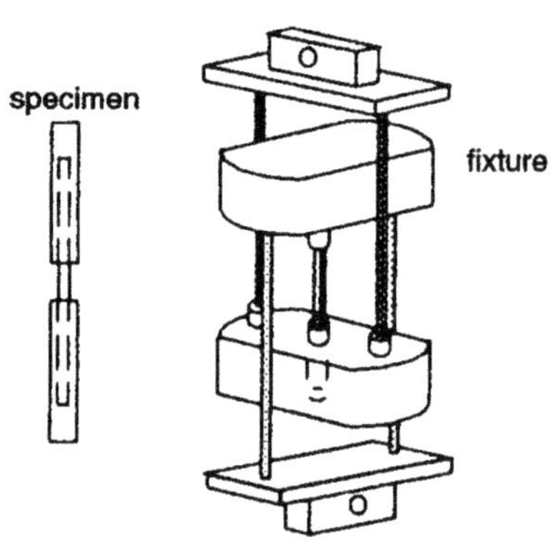

Figure 4.18 NBS compression test specimen and fixture (after [21]).

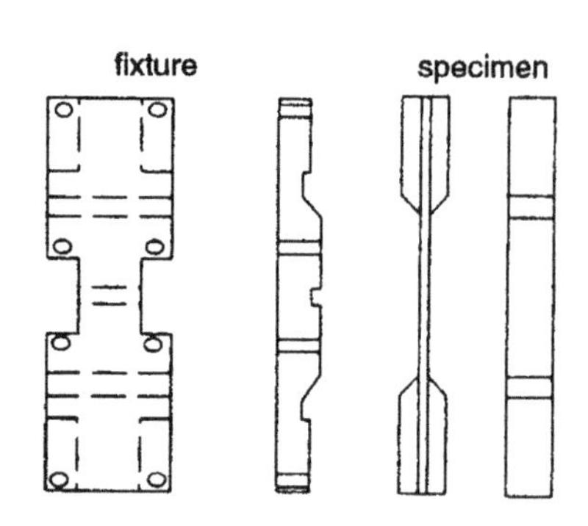

Figure 4.19 SWRI compression test fixture (after [22]).

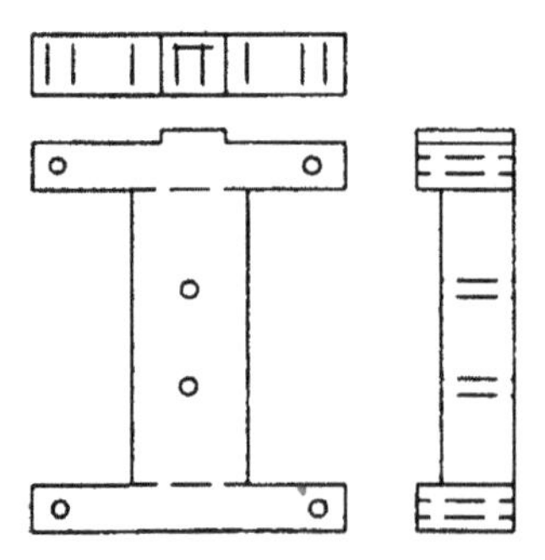

Figure 4.20 Lockheed compression test fixture (after [23]).

data from Type I tests. The SWRI grip has a cut in one support to allow for a transverse gage to measure Poisson's ratio in compression. Longitudinal strain is measured by an extensometer or strain gage placed on the edge of the specimen. The specimen is a modified tensile specimen in which the overall length is reduced while the end tab lengths are increased. The entire specimen length is supported by the fixture. The Lockheed fixture uses side supports only over the gage section of the specimen, which is the primary difference between it and the SWRI fixture.

TYPE III. The final class of compression test methods involves two sandwich beam specimen configurations. In each case, straight-sided coupons are bonded to a honeycomb core, which supplies lateral support. The elastic moduli and Poisson's ratio are determined from relationships between applied loads and strain gage readings taken from the specimen [1]. Results of failure strengths from this method are usually higher than from the other methods. The sandwich beam method can also be used to determine tensile properties [24]. The two specimen configurations are shown schematically in Figures 4.21 and 4.22. The specimen in Figure 4.21 is referred to as

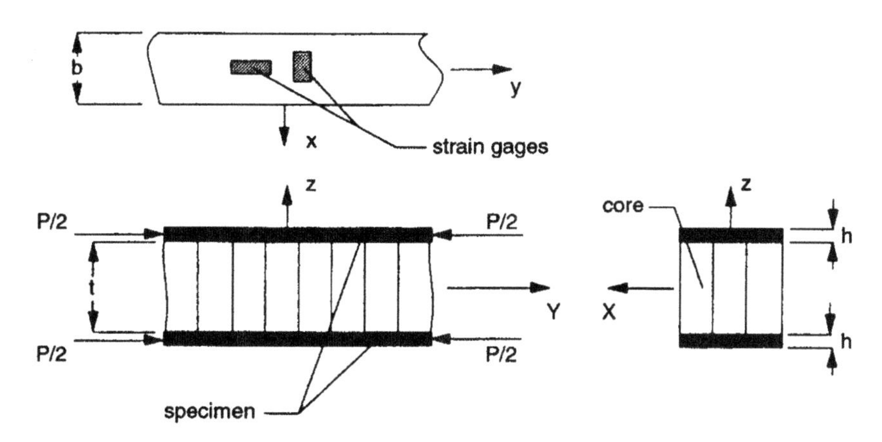

Figure 4.21 Sandwich beam edgewise compression test configuration (after [1]).

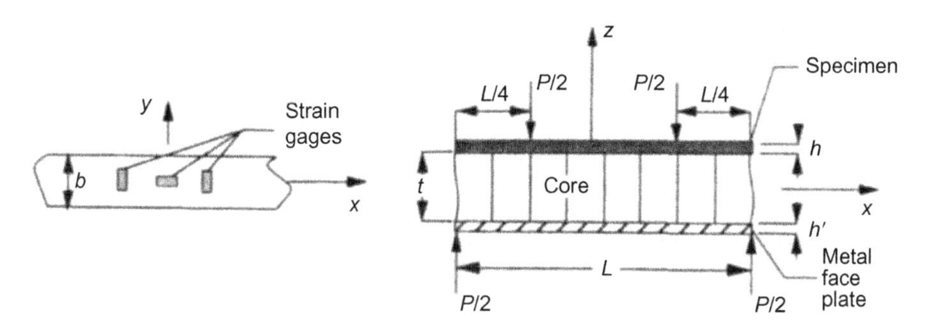

Figure 4.22 Four-point bend sandwich beam compression test (after [1]).

the *edgewise compression test specimen* and is used to determine E_y and ν_{yx} from the initial linear portion of the load–displacement curves generated during testing. The applied load is assumed to be distributed equally between the top and bottom specimens. The core is assumed to carry no in-plane load and is intended to supply lateral stability so that the potential for buckling is reduced. The elastic modulus and Poisson's ratio are determined from strain gage readings to be

$$\nu_{yx} = -\frac{\varepsilon_x}{\varepsilon_y} \quad E_y = \frac{\sigma_y}{\varepsilon_y} = \frac{P}{2A\varepsilon_y} = \frac{P}{2bh\varepsilon_y} \tag{4.19}$$

The specimen in Figure 4.22 is somewhat different since it is loaded in four-point bending. The specimen is the top sheet, which experiences compression. The bottom face sheet is in tension and is metal. Since the sandwich beam is subjected to flexure, various parameters (metal face sheet strength, core cell size, etc.) can be changed to achieve the desired compression failure of the specimen [24,25]. Poisson's ratio for this specimen is determined from direct strain gage readings to be $\nu_{xy} = -\varepsilon_y/\varepsilon_x$. The elastic modulus E_x is somewhat harder to establish since it requires an assumption of uniform deformation in each face sheet while bending stresses in the core are neglected. The approximation of E_x is

$$E_x = \frac{PL}{4bh\varepsilon_x(2t + h + h')} \tag{4.20}$$

4.3.3 Shear tests

The material properties in the plane of lamination (1-2) are commonly termed in-plane, while those in the 1-3 and 2-3 planes are known as interlaminar properties. As with extensional and compressive properties, the in-plane (1-2) properties are generally of more interest for classical laminate analysis than interlaminar properties. Five commonly accepted methods for in-plane shear testing are presented below. One of these procedures contains discussions applicable to interlaminar properties. The short-beam shear test, commonly used to define interlaminar shear strength, is discussed in Section 4.3.4. Discussions regarding the cross-beam sandwich, picture frame panel, and slotted tension test procedures for establishing shear properties are not considered herein but may be found in various articles, including Ref. [26]. In each of the test methods discussed, strain gages are typically used.

1. *Torsion*. Torsion of round specimens produces a state of pure shear, which is optimum for determining the in-plane shear modulus. Two types of round specimens, either a solid rod or a hollow tube, can be used.

   Solid Rod. This specimen consists of a unidirectional rod, generally machined from a square bar. The shear stress distribution in this type of specimen is known to vary linearly with distance from the center of the specimen according to $\tau = T\rho/J$. Where T, ρ, and J are the applied torque, distance from the rod center, and polar area moment of inertia, respectively. Knowing the applied torque allows for a simple prediction of stress on the outside surface of the rod. Using strain gages to determine the shear strain as a function of applied load gives a simple procedure for developing a $\tau-\gamma$ curve. The solid rod configuration is not used too often for two reasons: (1) a typical load–displacement ($T-\Phi$) diagram has a large region of nonlinear response; therefore, only a limited region of the curve provides useful data and (2) a solid specimen is expensive and difficult to produce.

Hollow (Thin-Walled) Tube. This specimen configuration is the most desirable from a mechanics viewpoint, since the shear stress is approximately constant over the wall thickness [27]. The actual variation of shear strain from the inner to outer surfaces of the specimen can be evaluated using strain gages applied to both surfaces. Proper application of a gage to the inside surface is difficult. A single-element gage oriented at 45° to the axis of the specimen (Figure 4.23) provides a simple analysis tool for defining shear strain since $\gamma_{xy}=2\varepsilon_{45}$. As an alternative to the single-element gage, a biaxial rosette can be used. The biaxial rosette should be applied so that each sensing element is at 45° to the axis of the specimen (Figure 4.23). The strains indicated by the $+45^{\circ}$ and -45° gages should be equal in magnitude and opposite in sign. The shear strain is the summation of the individual readings. A third possibility is a rectangular rosette applied so that the gages are oriented as indicated in Figure 4.23. The $\pm45^{\circ}$ gages provide the same information as the biaxial gage, while the gage aligned with the axis of the tube provides a measure of the extent to which pure torsion is achieved. This gage should indicate no strain, or a very small strain which remains constant with increasing torque. Although axial strains are not uncommon due to compressive end forces exerted on the specimen by the torsion machine, they should be small and relatively insignificant when compared to the shear strains.

The approximately constant through the wall shear stress in the hollow tube specimen produces a good $\tau-\gamma$ curve. Two problems associated with this test procedure are the expense of producing a hollow thin-walled tube, and a tube can be crushed by the end loads required to secure it to the torsion machine. One acceptable approach to gripping has been presented by Hahn and Erikson [28] and is schematically shown in Figure 4.24. This configuration provides a rigid base for clamping and adequate load transfer through pins and glue to the thin-walled torsion specimen. Variations to this adaptive end configuration are easily devised. The wall thickness to diameter ratio should generally be less than 0.030 to ensure a uniform stress distribution.

2. *Shear Rail Test*. The shear rail test is easier to prepare and conduct than the torsion tests. There are two acceptable configurations for the shear rail test: two-rail and three-rail. A schematic of the load fixture for each is shown in Figures 4.25 and 4.26, respectively. Both configurations are attributed to the ASTM D-30 committee. The specimen is simple to construct and machine. The suggested overall dimensions and hole sizes for both the two- and three-rail specimens are shown in Figure 4.27.

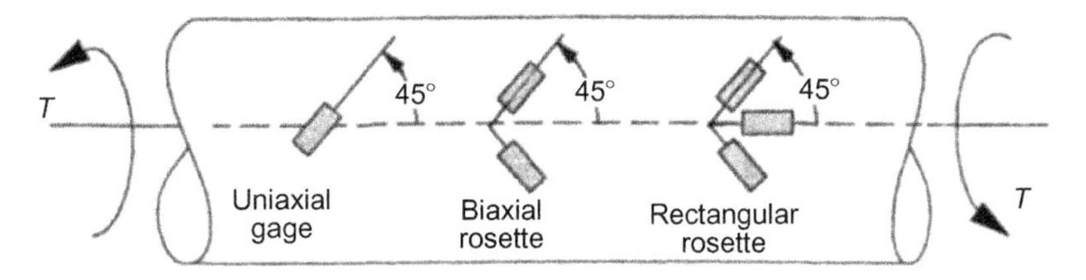

Figure 4.23 Strain gage orientation on torsion specimens.

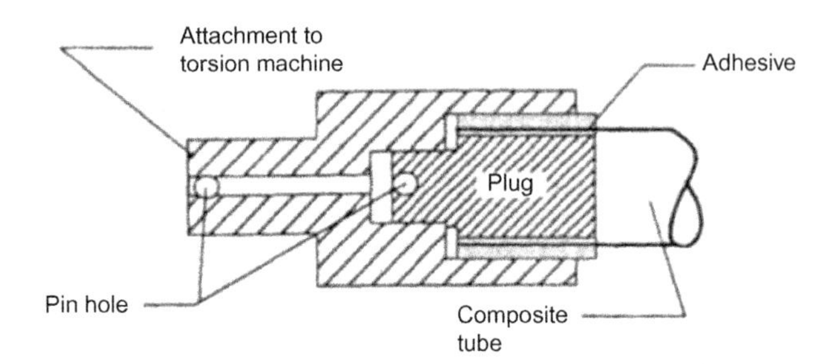

Figure 4.24 Grips or torsion testing composite tubes (after [28]).

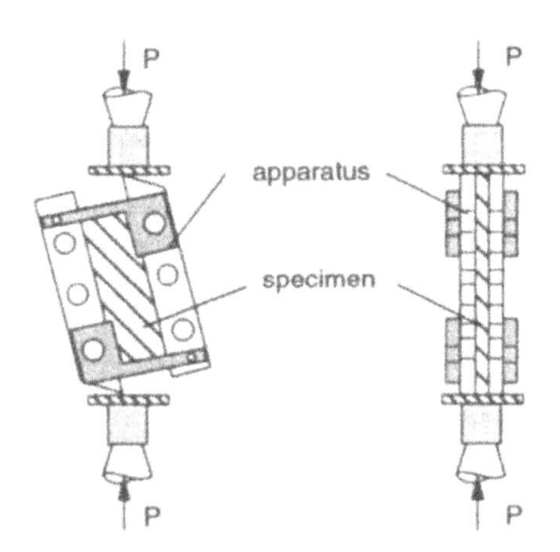

Figure 4.25 Two-rail shear test apparatus and specimen.

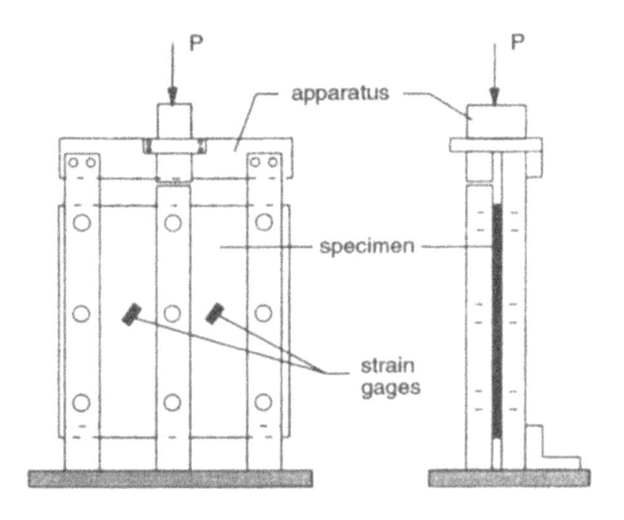

Figure 4.26 Three-rail shear test apparatus and specimen (after [1]).

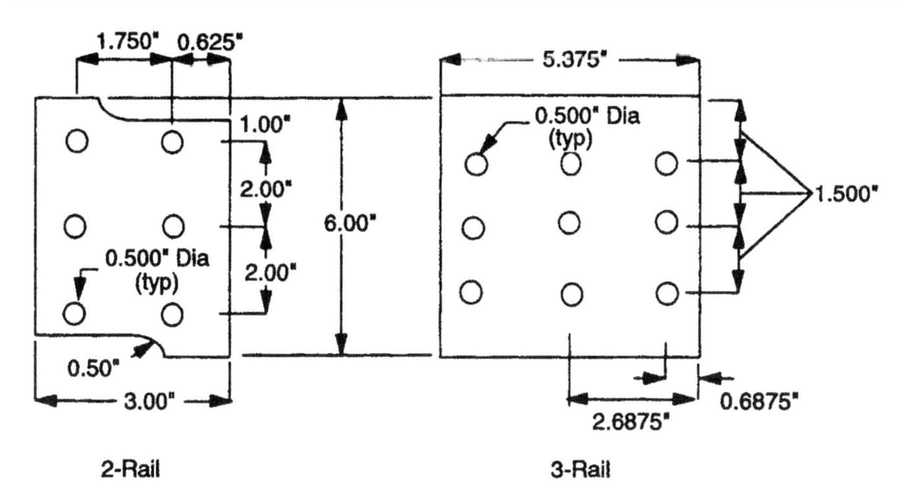

Figure 4.27 Specimen geometry for two- and three-rail shear tests (after [26]).

For both the two- and three-rail configurations, the shear stress in the strain-gaged region of each specimen is defined in terms of the applied load P and the specimen thickness (h), as well as the distance between each vertical rail (b). The shear stress for each configuration is approximated by

Two-rail: $\tau_{xy} = \frac{P}{bh}$ Three-rail: $\tau_{xy} = \frac{P}{2bh}$

Due to the method of load application, free surfaces at the top and bottom of each specimen experience large normal stresses concentrated at the corners [29]. A length-to-width ratio of 10:1 has been shown to approximate a state of pure shear stress provided the edges are perfectly clamped. The requirement of perfect clamping can be met if the bolts in the rails each apply the same clamping pressure to the edges. Since a state of pure shear is only approximated with the two- and three-rail configurations, a single-element strain gage oriented at 45° to the load axis may not adequately define the true state of strain.

3. *10° Off-Axis Test.* An off-axis test is generally performed in order to establish stress–strain responses in directions other than the principal material directions. The off-axis test is a tension test, and no special fixtures or specimen preparation is required. Consider the unidirectional test coupon loaded as shown in Figure 4.28. The rectangular rosette in this figure is not required for establishing G_{12}. Its presence is solely for the purpose of indicating that an off-axis test can be used for defining more than one parameter. The strains indicated by each

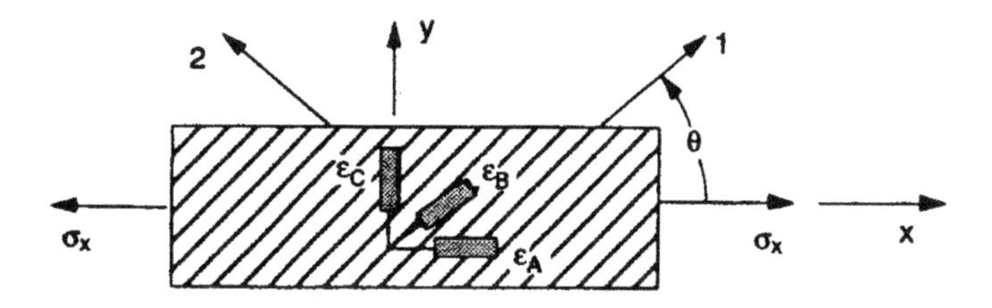

Figure 4.28 Off-axis test specimen.

gage in the rosette are related to Cartesian (x–y) strains by the strain transformation relations in Chapter 2. The relations between gage strain and the Cartesian strains are $\varepsilon_x = \varepsilon_A$, $\varepsilon_y = \varepsilon_C$, and $\gamma_{xy} = 2\varepsilon_B - \varepsilon_A - \varepsilon_C$, where ε_A, ε_B, and ε_C are the strains indicated by gages A, B, and C, respectively.

The normal stress σ_x and strain ε_x (from strain gage measurements) are related by $E_x = \sigma_x / \varepsilon_x$. E_x is a function of fiber orientation, shown in Equation (3.19) as

$$\frac{1}{E_x} = \frac{m^4}{E_1} + \left(\frac{1}{G_{12}} - \frac{2\nu_{12}}{E_1}\right) m^2 n^2 + \frac{n^4}{E_2}$$

Assuming E_1, E_2, ν_{12} are known, and E_x is defined from testing the specimen of Figure 4.28, the only remaining unknown is G_{12}, which can be determined from the equation above.

The uniaxial state of stress results in a biaxial state of strain in the specimen. Chamis and Sinclair [30] deduced from theoretical and experimental results that the best angle for establishing G_{12} is 10°. The 10° angle was chosen since it minimizes the effect of longitudinal and transverse tensile stress components σ_1 and σ_2 on the shear response. A comparison of the 10° off-axis procedure with other approaches has shown it to produce reasonable results for in-plane shear properties [31]. The simplicity of the 10° off-axis test for establishing G_{12} should not be taken for granted, since problems can result due to the specimen being orthotropic.

A uniaxial tensile stress in an orthotropic specimen can result in a shear-coupling deformation as shown in Figure 4.29a. Constraints imposed on the specimen by rigid clamping forces at the ends (Figure 4.29b) impose other testing difficulties [32]. Clamping at the end of the specimen prohibits localized rotation and produces a nonuniform strain field. A uniform strain field can be developed at the center of the specimen provided L/w is sufficiently large [32]. The specimen length is considered to be the region between end tabs. The effect of shear coupling can be defined by the shear coupling ratio, $\eta_{xy} = \gamma_{xy}/\varepsilon_x$. It can be shown that η_{xy} is a function of the stiffness matrix such that

$$\eta_{xy} = \frac{\bar{S}_{16}}{\bar{S}_{11}} \left[1 + \frac{3}{2}\left(\frac{w}{L}\right)^2 \left(\frac{\bar{S}_{66}}{\bar{S}_{11}} - \left(\frac{\bar{S}_{16}}{\bar{S}_{11}}\right)^2 \right) \right]^{-1}$$

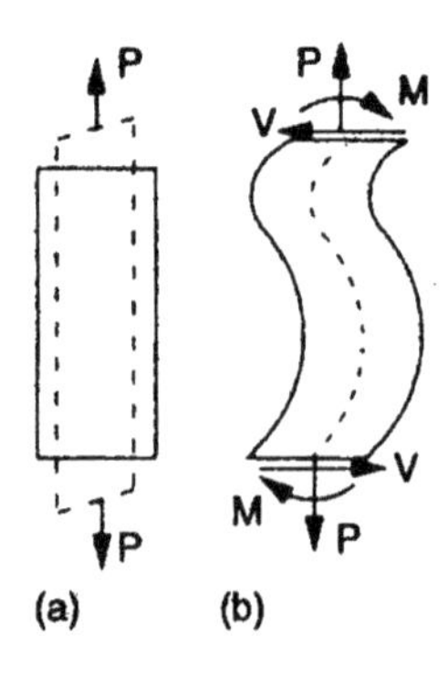

Figure 4.29 Effects of end constraints on off-axis tensile specimens.

The relationship between the apparent modulus E_x^* (established during testing) and the actual modulus E_x is $E_x = (1-\eta)E_x^*$, where

$$\eta = \frac{1}{\bar{S}_{11}}\left[\frac{3\bar{S}_{16}^2}{3\bar{S}_{66}+2\bar{S}_{11}(L/w)^2}\right]$$

As L/w increases, η decreases and E_x approaches E_x^*. The actual value of L/w at which one can assume the shear coupling to be negligible is dependent on the material system and fiber orientation being considered, as well as the tolerable error.

4. *Iosipescu Shear Test.* The Iosipescu shear test [33] is similar to an antisymmetric four-point bend (AFPB) test method for composites [34]. The major difference is that for the Iosipescu test, the shear force through the test section is equal to the applied load. The Iosipescu test fixture and specimen are shown in Figure 4.30. This test procedure can be applied to composites for determining material properties in the 1-2, 2-3, and 1-3 directions [35]. The appropriate fiber orientations for determining in-plane and interlaminar properties are shown in Figure 4.31. This test method is versatile and allows for determination of a wider variety of material properties than other procedures. Analysis of the procedure has lead to the evolution of several specimen and fixture geometries. The University of Wyoming Iosipescu test specimen and fixture [36] is commonly accepted as producing reliable results. Techniques for specimen preparation and modified testing procedures to eliminate variability of results have been introduced by Lee and Munro [37].
5. $[\pm 45]_{2s}$ *Coupon Test.* This procedure involves a uniaxial tension test of a $[\pm 45]_{2s}$ laminate, with strain gages. Although a biaxial rosette is sufficient, a three element rosette provides additional information which can be used as a verification of the state of stress in the specimen. Specimen preparation and testing are identical to a conventional tension test. A complete discussion of this procedure is not presented at this point since the specimen is a laminate. Further discussions of this procedure are deferred until laminate analysis procedures are established in Chapter 6. Results from the $[\pm 45]_{2s}$ test are in good agreement with those from other procedures, and it is considered to be a reliable test configuration.

4.3.3.1 Summary of shear test methods

A definitive conclusion as to which of the available procedures for establishing shear properties is "best" would be difficult to defend, since some procedures work better with one type of material than another. Evaluations of several procedures [38] indicate that more than one procedure can be categorized as appropriate for defining in-plane

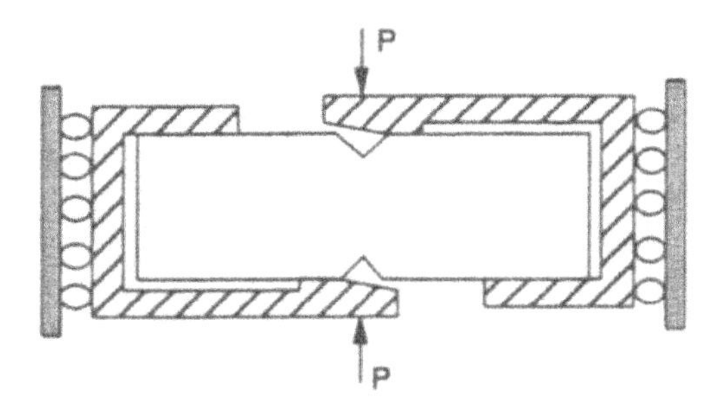

Figure 4.30 Schematic of Iosipescu test fixture and specimen (after [35]).

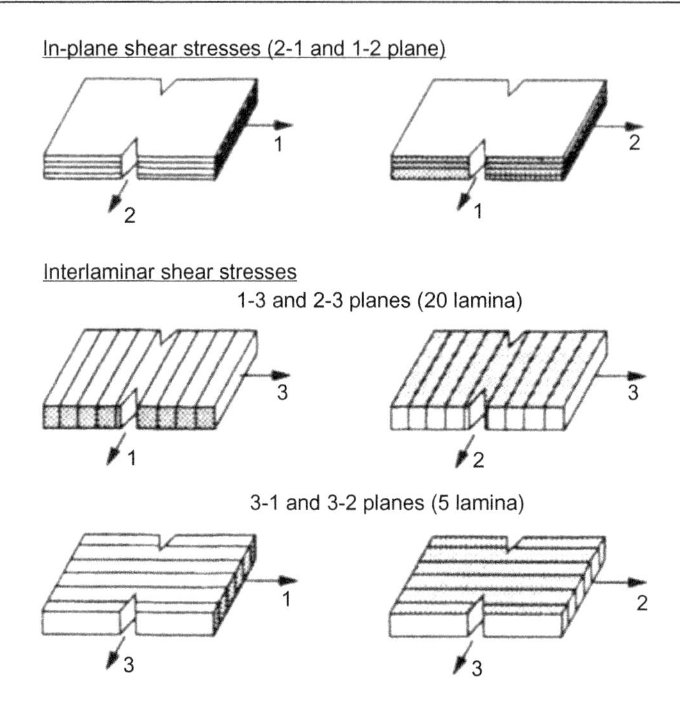

Figure 4.31 Specimen configurations for determination of shear properties from the Iosipescu test procedures (after [35]).

shear properties. Lee and Munro [37] attempted to evaluate nine experimental procedures for determining in-plane shear properties. They established 11 criteria within 4 broad categories relating to a successful experiment. These categories are fabrication cost, testing cost, data reproducibility, and accuracy. Each criterion was rated from 0 to 10, with 10 the highest, and a weighting function was applied to each. Both the rating and weighting functions of each criterion are subject to author preference. Table 4.1 presents the raw score and overall rating for each method discussed in

Table 4.1 Evaluation of in-plane shear test methods (after [37])

Test procedure	Raw score	Rating
Two-rail shear	6880	6
Three-rail shear	7200	4
$[\pm 45]_{2s}$	8030	1
10° off-axis	7860	3
Cross-beam	5890	9
Picture frame	6650	7
Thin-walled tube	6530	8
Slotted-tenisle	6990	5
Iosipescu	8030	1

Ref. [36]. Three of the procedures presented in Table 4.1 have not been discussed herein but can be found in the archival literature relating to composite materials testing.

4.3.4 Flexure tests

There are two commonly used loading conditions for flexure testing, three-point and four-point bending. Each is shown schematically in Figure 4.32 with the general specimen geometry. The *L*/4 load reaction position in the four-point bend configuration is sometimes replaced by an *L*/3 reaction position. The *L*/4 location is generally used with high modulus materials (graphite/epoxy, boron/epoxy, etc.). The objective of these tests is to determine flexure strength and material modulus in specific directions. These tests are not recommended for generating design data. The flexure test can be used to determine interlaminar shear properties. Requirements for specific types of flexure tests (specimen dimensions, loading rate, etc.) are given in ASTM D790-71. Unidirectional specimens with fibers oriented at either 0° or 90° to the beam axis can be used to determine the elastic modulus of a material. The modulus along the beam axis of the specimen is generally designated as E_x and depending upon fiber orientation correspond to E_1 or E_2. The appropriate fiber orientations corresponding to estimates of E_1 and E_2 for flexure specimens are shown in Figure 4.32. The recommended span to depth ratio, *L*/*h*, for a flexure specimen depends on the ratio of tensile strength parallel to the beam axis and interlaminar shear strength. For 0° specimens with a strength ratio less than 8:1, the recommended ratio is $L/h = 16$. For high modulus materials (such as graphite/epoxy or boron/epoxy), an $L/h = 32$ is suggested. The requirement on a 90° lamina is less severe, and $L/h = 16$ is generally acceptable for all materials.

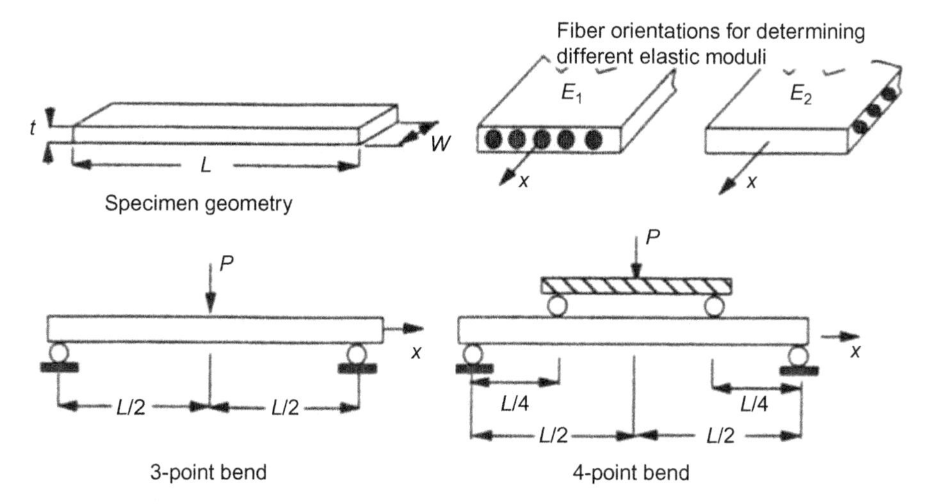

Figure 4.32 Schematic of three- and four-point bend tests.

Estimating the modulus for 0° or 90° specimens requires knowing the deflection (w) of the center of the beam. Given the beam width (d), center span deflection (w), applied load (P), and beam length (L), the elastic modulus along the beam axis (E_x) for each configuration can be determined from strength of materials techniques. For the four-point bend specimen, the deflection at $x = L/2$ is $E_x I w = 11PL^3/768$. For a beam with cross-sectional dimensions as shown, $I = bh^3/12$. This results in the elastic modulus for the beam being expressed as $E_x = 11PL^3/(64bh^3w)$. A similar approach is used for the three-point bend specimen. If shear deformation is considered, these equations contain an additional term. The predicted elastic modulus for each configuration (including shear deformation) is as defined in Ref. [1]

$$\text{3-Point bend}: E_x = \frac{PL^3}{4bh^3w}(1+S) \quad \text{4-Point bend}: E_x = \frac{PL^3}{64bh^3w}(11+8S) \tag{4.21}$$

where S is the shear correction factor which is a function of specimen geometry and deformation. General definitions of the shear correction factor are found in strength of many materials texts. For a rectangular cross section, the expression for S is $S = 3h^2E_x/(2L^2G_{xz})$. In this expression, G_{xz} is the shear modulus in the longitudinal plane through the thickness of the specimen. It is difficult to determine and, in many cases, is not adequately known. It can be neglected by allowing the shear correction factor (S) to be set equal to zero, thus reducing the above equations to a simpler form. With $S = 0$, each of the above expressions is a simple function of $(L/h)^3$. In order to estimate E_x, a series of tests with increasing L/h ratios of the specimen are conducted. The modulus E_x for each test is computed, and when a constant E_x is obtained between several test specimens, the modulus is considered to have been determined.

An additional type of flexure test is the short-beam shear test. Unlike the flexure tests, this test is designed to estimate interlaminar shear strength only. There are difficulties associated with it, and its overall value is questionable. The procedure and specimen dimensions for this test are discussed in ASTM D2344-76. The specimen should be designed so that shear deformation effects are as large as possible, and failure results from interlaminar shear stresses rather than normal stresses. The ratio L/h should be small ($L/h = 4$ is suggested for graphite/epoxy), and a three-point bend test is used. The specimen has a parabolic shear stress distribution through the thickness, and the maximum shear stress may not be at the mid-surface of the beam. For the short-beam shear test, the interlaminar shear strength is expressed as

$$\tau_m = \frac{3P}{4bh} \tag{4.22}$$

where τ_m is the maximum interlaminar shear stress, P is the applied load, and b and h are the beam width and thickness, respectively.

4.3.5 Failure strengths

The failure strengths of unidirectional composites (lamina) can be determined from the same tests used to estimate moduli. The failure strengths in the principal fiber directions require both tensile and compressive tests. Shear strength, on the other hand, is independent of load direction.

4.4 Physical properties

The physical properties of a composite material system can be as important as mechanical properties in assessing suitability for a particular application. The properties of most practical interest from a stress analysis point of view are density, fiber volume fraction, and coefficients of thermal and hygral expansion. The general procedures used to estimate these properties are highlighted below.

4.4.1 Density

The density can be determined by first preparing a specimen with a volume on the order of $V \geq 1\,\text{cm}^3$ $(0.061\,\text{in.}^3)$. The procedure for estimating density is:

1. Accurately determine the specimen dimensions.
2. Weigh the specimen in air. The weight of the specimen in air is designated as "*a*."
3. Weigh the specimen in water while suspending it by a wire. As part of this step, the weight of the wire (used to suspend the specimen) and the sinker (used to assure the specimen is submerged) must be taken into account. Therefore, two additional terms must be accounted for:
   w—weight of sinker and immersed wire
   b—specimen weight + *w*
4. Calculate the lamina density in mg/m^3 from

$$\rho = \frac{0.9975a}{a + w - b}$$

where 0.9975 is a conversion factor from specific gravity to density. A complete description of this procedure can be found in ASTM D792-66.

4.4.2 Fiber volume fraction

In order to estimate the volume fraction of fibers in a given specimen, a series of tests is required. Results from some of these tests are sensitive to measurements taken during the experiment. The procedures to be followed are based on the assumption that a small specimen has been prepared from a larger sample of the lamina under investigation. The general procedures are:

1. Determine the weight of the composite sample (W_c) and its density (ρ_c) as described in the previous section.

2. Allow the matrix to be digested by either an acid bath or burning it away. The appropriate procedure is dictated by the fiber material. Graphite fibers, for example, require acid digestion, while glass fibers can be burned.
3. After the matrix is removed, only the fiber remains. The fiber weight (W_f) and density (ρ_f) must then be determined from procedures established in Section 4.4.1. The volume fraction of fiber is then estimated from:

$$v_f = \frac{W_f/\rho_f}{W_c/\rho_c}$$

Similar procedures can be used to determine the volume fraction of voids. The mass fractions of fiber and matrix and their respective densities are required to compute v_v. Slight errors in measurements can lead to significant errors in estimates of volume fractions. In estimating v_v, the relationship developed in Section 3.4 is useful.

$$v_v = 1 - \rho\left[\frac{m_f}{\rho_f} + \frac{m_m}{\rho_m}\right]$$

Example 4.2 A 1-g glass/epoxy composite specimen is being tested. The matrix is burned away, and the remaining fibers are found to weigh 0.5 g. The density of the glass fiber and the epoxy matrix are known to be $\rho_{\text{fiber}} = 2.60\ \text{kg/m}^3$ and $\rho_{\text{matrix}} = 1.25\ \text{kg/m}^3$. Assuming the volume fraction of voids is zero ($v_v = 0$), determine the density of the composite. Then, assume the actual density of the composite is $\rho_{\text{composite}} = 1.55\ \text{kg/m}^3$ and determine the volume fraction of voids.

From the date given, we determine $M = 1$, $M_f = 0.5$, $m_f = 0.5$, $m_f + m_m = 1 \Rightarrow m_m = 0.5$

Then, applying the equation above, we have

$$\begin{aligned} v_v &= 1 - \rho\left[\frac{m_f}{\rho_f} + \frac{m_m}{\rho_m}\right] = 1 - 0.5\rho\left[\frac{1}{\rho_f} + \frac{1}{\rho_m}\right] = 1 - 0.5\rho\left[\frac{1}{2.6} + \frac{1}{1.25}\right] \\ &= 1 - 0.5\rho[1.1846] \end{aligned}$$

Next, we set $v_v = 0$, resulting in $\rho = \rho_{\text{composite}} = 1.6883\ \text{kg/m}^3$.

Given $\rho_{\text{composite}} = 1.55\ \text{kg/m}^3$, we determine (from the same equation)

$$v_v = 1 - 0.5(1.55)[1.1846] = 0.081935$$

4.4.3 *Thermal expansion and moisture swelling coefficients*

Coefficients of thermal expansion can be determined by using a dilatometer, or strain gages. A dilatometer measures the elongation of a specimen subjected to either high or low temperatures. The strains in the x and y directions (which could represent the 1 and

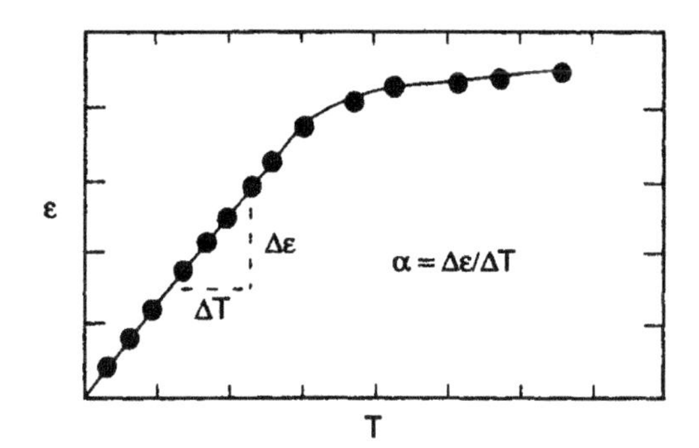

Figure 4.33 Schematic for determination of α.

2 directions of a lamina) can be determined by $\varepsilon_x = \Delta x / b$ and $\varepsilon_x = \Delta y / a$, where Δx and Δy are specimen deformations recorded by deflection gages, and a and b are the lengths of each side. The strains could alternately be measured from strain gages. Each strain is plotted as a function of temperature. The slope of the resulting curve in the linear portion of the graph is taken to represent the coefficient of thermal expansion α as illustrated in Figure 4.33. Different expansion coefficients can be observed during heating and cooling cycles of a test sequence.

The moisture swelling coefficients are determined in a manner analogous to that for estimating thermal expansion coefficients. By measuring the volume change in a specimen as a function of moisture weight gain, a plot similar to Figure 4.33 can be established for moisture. The slope of the resulting curve represents the coefficient of expansion. Compilation of sufficient data to establish swelling coefficients can be a long process, since temperature and humidity are coupled.

4.5 Material properties of selected composites

Tables 4.2 and 4.3 present material properties for selected composite materials in both English and SI units, respectively. These tables are by no means comprehensive. Many possible materials, such as thermoplastic, metal, and ceramic matrix systems have been omitted. Material properties for these types of composites can be found in numerous references, including Refs. [41–43]. Some of the more common epoxy resin material systems are presented. The primary references from which most data are taken are cited. Entries marked by an "*" are obtained from either Refs. [39,40,44] or vendor supplied data sheets. The notations used for strength properties are:

- X, X' tensile and compressive strength in 1-direction, respectively;
- Y, Y' tensile and compressive strength in 2-direction, respectively; and
- S shear strength.

In Table 4.2, the notation Msi replaces the more familiar 10^6 psi notation commonly associated with elastic modulus.

Table 4.2 **Material properties for selected composite material systems (English units)**

Material [reference]		v_f	E_1 (Msi)	E_2 (Msi)	G_{12} (Msi)	ν_{12}	X (ksi)	X' (ksi)	Y (ksi)	Y' (ksi)	S (ksi)	α_1	α_2	β_1	β_2
Graphite/Epoxy															
T300/5208	[39]	0.70	26.27	1.49	1.04	0.28	217.7	217.7	5.80	35.7	9.86	0.011	12.47	0.00	0.44
	[40]	0.70	22.20	1.58	1.03	0.30	100.0	110.0	4.00	13.9	9.00	—	—	—	—
T300/934	[40]	0.60	23.70	1.70	0.93	0.30	107.0	105.0	—	—	14.8	—	—	—	—
T300/SP-286	[40]	0.60	21.90	1.53	0.96	0.31	185.6	362.6	8.80	44.7	15.2	—	—	—	—
AS/3501	[39]	0.66	19.90	1.29	1.03	0.30	210.0	210.0	7.50	29.9	13.5	–0.17	15.57	—	—
	[40]	0.67	20.02	1.30	1.03	0.30	209.9	209.9	7.50	29.9	13.5	—	—	—	—
Glass/Epoxy															
Scotchply: type 1002	[39]	0.54	5.60	1.20	0.60	0.26	154.1	88.5	4.50	17.1	10.5	4.77	12.24	—	—
type SP-250-S29*		0.54	7.00	2.10	0.80	0.26	260.0	145.0	6.20	29.0	14.0	—	—	—	—
E-Glass/Epoxy	[40]	0.72	8.80	3.60	1.74	0.23	187.0	119.0	6.70	25.3	6.50	3.50*	11.4*	—	—
S-Glass/XP-251		0.67	8.29	2.92	0.86	0.262	289.0	170.0	11.00	29.0	9.00	3.60*	11.1*	—	—
Boron/Epoxy															
B(4)/5505	[39]	0.50	29.60	2.69	0.811	0.23	183.0	363.0	8.90	29.3	9.70	3.38	16.79	—	—
	[40]	0.67	30.30	2.80	0.930	0.21	185.6	362.6	8.80	44.7	15.2	3.40*	16.9*	—	—
Aramid/Epoxy															
Kevlar 49/Epoxy	[39]	0.60	11.03	0.798	0.333	0.34	203.2	34.1	1.74	7.69	4.93	–2.22	43.77	—	—
	[40]	0.60	11.02	0.798	0.334	0.34	203.1	34.1	1.74	7.69	4.93	–2.2*	39.0*	—	—

Note: The units on the coefficients of thermal expansion are μin./in./°F.

Table 4.3 Material properties for selected composite material systems (SI units)

Material [reference]		υ_f	E_1 (GPa)	E_2 (GPa)	G_{12} (GPa)	ν_{12}	X (MPa)	X' (MPa)	Y (MPa)	Y' (MPa)	S (MPa)	α_1	α_2	β_1	β_2
Graphite/Epoxy															
T300/5208	[39]	0.70	181.0	10.30	7.17	0.28	1500.0	1500.0	40.0	246.0	68.0	0.020	22.5	0.00	0.44
	[40]	0.70	153.0	10.90	5.60	0.30	689.5	758.5	27.6	96.5	62.1	—	—	—	—
T300/934	[40]	0.60	163.0	11.90	6.50	0.30	738.0	724.0	—	—	102.0	—	—	—	—
T300/SP-286	[40]	0.60	151.0	10.60	6.60	0.31	1401.0	1132.0	54.0	211.0	72.0	—	—	—	—
AS/3501	[39]	0.66	138.0	8.96	7.10	0.30	1447.0	1447.0	51.7	206.0	93.0	–0.31	28.1	—	—
	[40]	0.67	138.0	8.96	7.10	0.30	1447.0	1447.0	51.7	206.0	93.0	—	—	—	—
Glass/Epoxy															
Scotchply: type 1002	[39]	0.54	38.6	8.27	4.14	0.26	1062.0	610.0	31.0	118.0	72.0	8.6	22.1	—	—
type SP-250-S29*		0.54	48.3	14.50	5.50	0.26	1790.0	1000.0	43.0	200.0	97.0	—	—	—	—
E-Glass/Epoxy	[40]	0.72	60.7	24.80	11.99	0.23	1288.0	820.5	45.9	174.4	44.8	6.5*	20.6*	—	—
S-Glass/XP-251		0.67	57.2	20.10	5.90	0.262	1993.0	1172.0	76.0	200.0	62.0	6.6*	19.7*	—	—
Boron/Epoxy															
B(4)/5505	[39]	0.50	204.0	18.50	5.59	0.23	1260.0	2500.0	61.0	202.0	67.0	6.1*	30.3*	—	—
	[40]	0.67	209.0	19.00	6.40	0.21	1280.0	2500.0	61.0	308.0	105.0	6.1*	30.3*	—	—
Aramid/Epoxy															
Kevlar 49/Epoxy	[39]	0.60	76.0	5.50	2.30	0.34	1400.0	235.0	12.0	53.0	34.0	–4.0	79.0	—	—
	[40]	0.60	76.0	5.50	2.30	0.34	1400.0	235.0	12.0	53.0	34.0	–4.0*	70.0*	—	—

Note: The units on the coefficients of thermal expansion are μm/m/°C.

The entries in each table illustrate the variations which exist in reported properties for materials. This can, to some extent, be explained as a result of specimens having different fiber volume fractions. Similarly, specimen preparation and testing procedures can affect results. Material moduli, failure strength, and coefficients of thermal and hygral expansion are dependent upon testing environments. Many material manufacturers supply mechanical data based on different test temperatures and expansion coefficients for a range of temperatures. The data presented in the tables herein are for room temperature test conditions. The thermal expansion coefficients are based on a temperature range from room temperature to an appropriate elevated temperature (typically around $200\,^{\circ}F$). The lack of data for thermal and hygral coefficients is apparent but does not imply that this information is unimportant.

4.6 Testing lamina constituents

Complete characterization of a composite material includes fiber and matrix properties. Both physical and mechanical properties are determinable in many cases. Since the constituent materials are typically isotropic, some of the complexities of testing orthotropic lamina are reduced, while some tests applicable to lamina are not appropriate for the constituents. For example, compression testing of single fibers cannot be accomplished. Tensile testing of fibers, however, is a well-established procedure as discussed in detail in ASTM D3379-75. Special test fixtures, such as the one schematically shown in Figure 4.34, as well as data reduction procedures are required for testing fibers.

Since fibers can only be tested in tension, the basic properties generally established for resin systems are tensile. The procedures for testing different polymeric resin systems depend on the availability of the material as either a thick sheet or a thin film. The test procedures and specimen configurations for thick sheet forms of material are detailed in ASTM D638-72. Similar information corresponding to a thin film material is available in ASTM D882-73. An appropriate procedure for characterizing them depends on their general classification as elastic, plastic, viscoelastic, etc. Tables 4.4 and 4.5 present a range of values for some of the physical and mechanical properties of selected fibers and resins systems in English and SI units, respectively.

Figure 4.34 Schematic of test fixture for fiber testing.

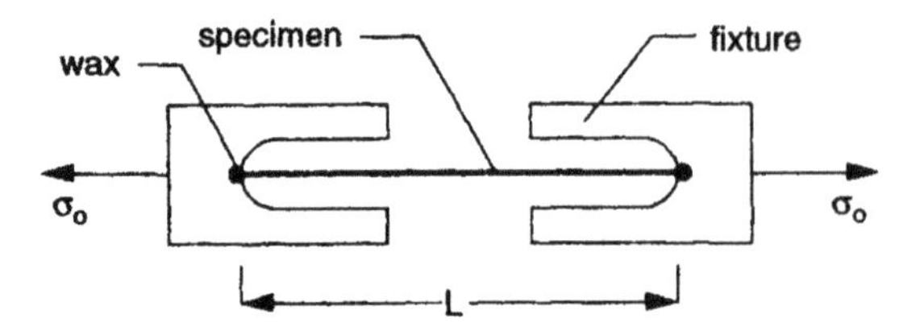

Table 4.4 **Material properties for selected fibers and resins (English units)**

Material [reference]		Diameter (in. ×10^{-3})	Density (1b/in.^{3})	Elastic Modulus (Msi)	ν	Tensile Strength (ksi)	α(μin./in./°F)
Fibers							
E-Glass	[45]	0.40	0.092	10.5	—	450.0	2.8
	[46]	0.30–0.55	0.093	11.0	—	203.0–263.0	2.7
S-Glass	[45]	0.40	0.090	12.3	—	650.0	—
Graphite	[45]	0.20–0.40	0.053–0.065	35.0–100.0	—	250.0–500.0	1.5
Carbon (Pan-based)	[46]	0.28–0.38	0.063–0.070	36.0–57.0	—	319.0–392.0	–0.05–0.67
Kevlar 49	[46]	0.468	0.052	18.0	—	406.0–522.0	–1.1
	[47]	0.470	0.052	18.0	—	406.0	—
Kevlar 29	[47]	0.470	0.052	9.5	—	406.0	—
Boron	[39]	4.00	0.094	59.5	—	500.0	—
	[45]	4.00	0.095	60.0	—	400.0	2.8
Resins							
Epoxy	[45]	—	—	0.50	0.35	—	32.0
	[46]	—	0.039–0.051	0.44–0.88	0.38–0.40	5.1–14.5	33.0
	[47]	—	0.059	—	—	5.1–12.3	44.0–55.0
Polyimid	[45]	—	—	0.40	0.33	—	28.0–35.0
	[47]	—	0.053	—	—	17.4	50.0
Polyester	[46]	—	0.430–0.540	0.35–0.65	0.37–0.39	5.8–13.0	55.0–110.0
Phenolic	[47]	—	0.047	—	—	7.2–8.0	2.5–6.1

Table 4.5 Material properties for selected fibers and resins (SI units)

Material [reference]		Diameter (μm)	Density (kg/m^3)	Elastic Modulus (Gpa)	ν	Tensile Strength (Mpa)	α (μm/m/°C)
Fibers							
E-Glass	[45]	10.0	2547	72.3	—	3102	5.1
	[46]	8.0–14.0	2560	76.0	—	1400–2500	4.9
S-Glass	[45]	10.0	2491	84.8	—	4481	—
Graphite	[45]	5.0–10.0	1467–1799	241.0–690.0	—	1720–3345	2.7
Carbon (PAN-based)	[46]	7.0–9.7	1750–1950	250.0–390.0	—	2200–2700	– 0.10– –1.20
Kevlar 49	[46]	11.9	1450	125.0	—	2800–3600	–2.0
	[47]	12.0	1440	125.0	—	2800	—
Kevlar 29	[47]	12.0	1440	65.0	—	2800	—
Boron	[39]	100.0	2600	410.0	—	3450	—
	[45]	100.0	2630	414.0	—	2758	5.1
Resins							
Epoxy	[45]	—	—	3.5	0.35	—	58.0
	[46]	—	1100–1400	3.0–6.0	0.38–0.40	35–100	60.0
	[47]	—	1380	—	—	35–85	80.0–110.0
Polyimid	[45]	—	—	2.75	0.33	—	51.0–63.0
	[47]	—	1460	—	—	120	90.0
Polyester	[46]	—	1200–1500	2.0–4.5	0.37–0.39	40–90	100.0–200.0
Phenolic	[47]	—	1300	—	—	50–55	4.5–11.0

4.7 Problems

4.1 Assume a uniaxial strain gage is mounted on a unidirectional tensile specimen made of T300/5208 graphite/epoxy. The gage is oriented as shown, and the specimen is subjected to an applied load σ_0 as indicated. Determine the percent error resulting from neglecting transverse sensitivity for

$(A)\ K = 0.40\%,\ \ (B)\ K = -1.5\%$

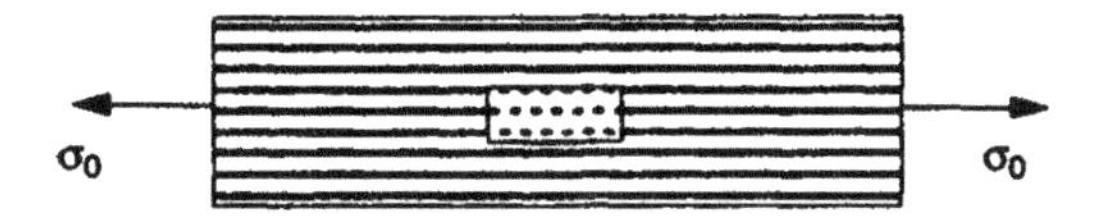

4.2 The strain gage of problem 4.1 is applied to a T300/5208 graphite/epoxy specimen as indicated. Determine the percent error which results from neglecting the transverse sensitivity for

$(A)\ K = 0.40\%,\ \ (B)\ K = -1.5\%$

4.3 A uniaxial strain gage is used to measure the maximum tensile strain on the outside surface of a closed end pressure vessel subjected to an internal pressure P. The vessel is made from a unidirectional graphite/epoxy (AS/3501) with its fibers oriented along the longitudinal axis of the vessel. Determine the percent error resulting if transverse sensitivity is neglected. [Hint: determine the state of stress in the vessel using classical thin-walled pressure vessel theory.]

$(A)\ K = 0.40\%,\ \ (B)\ K = -1.5\%$

4.4 The results from a uniaxial tension test on AS/3501 graphite/epoxy are presented below. Find $E_1, E_2, \nu_{12}, \nu_{21}, X$, and Y.

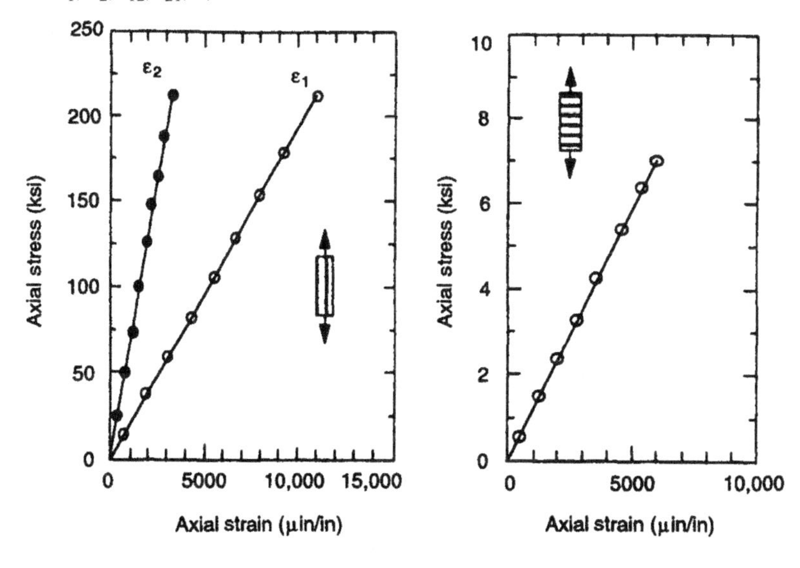

4.5 The four-point bend sandwich beam compression test is schematically shown in Figure 4.22. In the region of the beam where the bending moment is constant, the forces in both top and bottom layers can be modeled as shown. The bending moment at section A-A in this section can be expressed in terms of the applied load P and the beam length L.

(A) Determine the relationship between F_1, F_2, t, h, h' and the bending moment at section A-A expressed in terms of P and L.

(B) Assuming that $F_1 = F_2$, determine the expression for E_x which is known to be

$$E_x = \frac{PL}{4bh(2t + h + h')\varepsilon_x}$$

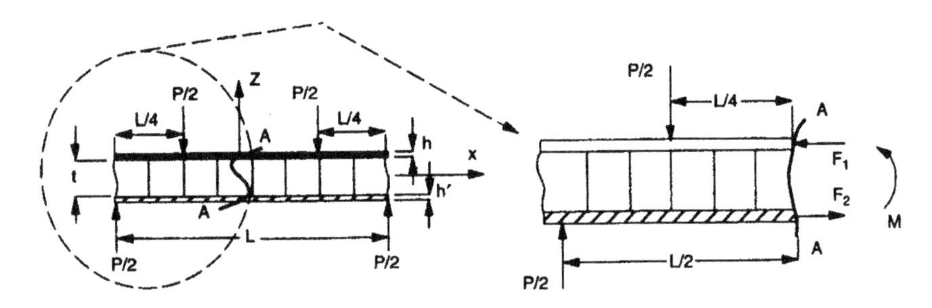

4.6 The Iosipescu shear test is schematically shown in Figure 4.30. A schematic of the loads which the fixture transmits to the specimen is shown in Fig. A, while a schematic of the loads experienced by the specimen are shown in Fig. B

(A) Use fig. A to verify that $P_1 = Pa/(a - b)$ and $P_2 = Pb/(a - b)$.

(B) Use fig. B to show that the specimen test section (the vertical plane through the notch) experiences a shear force equal to the applied load P and no bending.

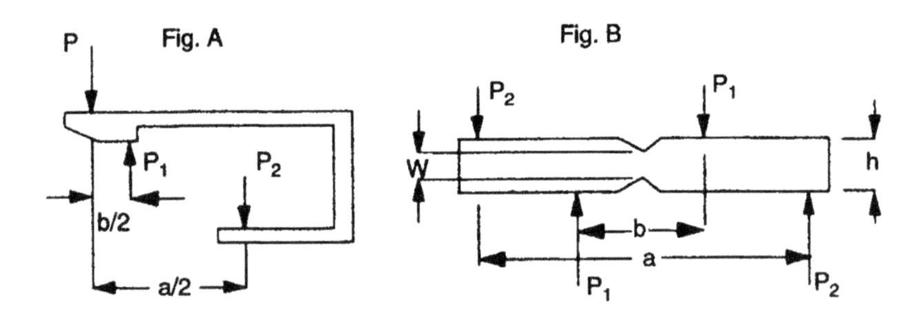

4.7 A schematic diagram for the asymmetrical four-point bend (AFPB) shear loading fixture [34] is shown below. For this test configuration, verify that the shear force in the specimen test section (the vertical plane through the notch) is given as

$$V = P\frac{a - b}{a + b}$$

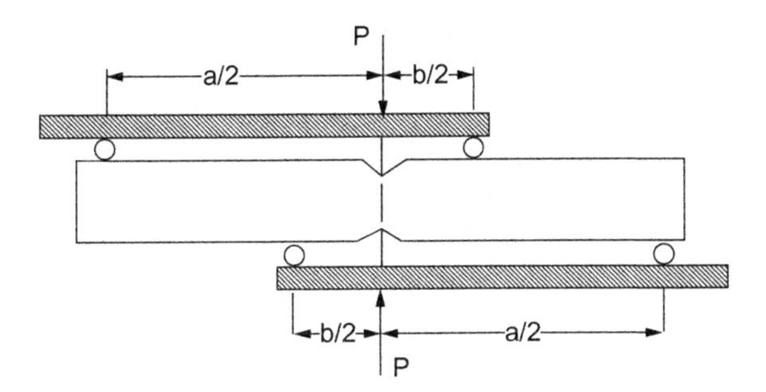

4.8 A unidirectional lamina is being tested for a specific material property using what is called a cross-beam test procedure. Two composite specimens with fibers oriented at an angle θ are made into the shape of a cross. The two specimens are separated by a core. The core has an elastic modulus approximately 2 orders of magnitude less than the specimens. The "sandwich" beam (the core with specimens bonded to the top and bottom) has a strain gage applied to the top surface in the region indicated. The arms of the cross are then loaded with concentrated forces of magnitude P directed as shown.

(A) Draw an element showing the state of stress in the area containing the strain gage. Be as specific as possible in identifying the components of stress (for example, $\sigma_x = 2\sigma_y$, or whatever the relationship may be).

(B) Use the element defined in (A) to identify each of principal direction stresses in terms of m and n and the stresses defined in (A) and define the single material property (E_1 or E_2 or G_{12}, etc.) that can be determined form this test and the required angle θ.

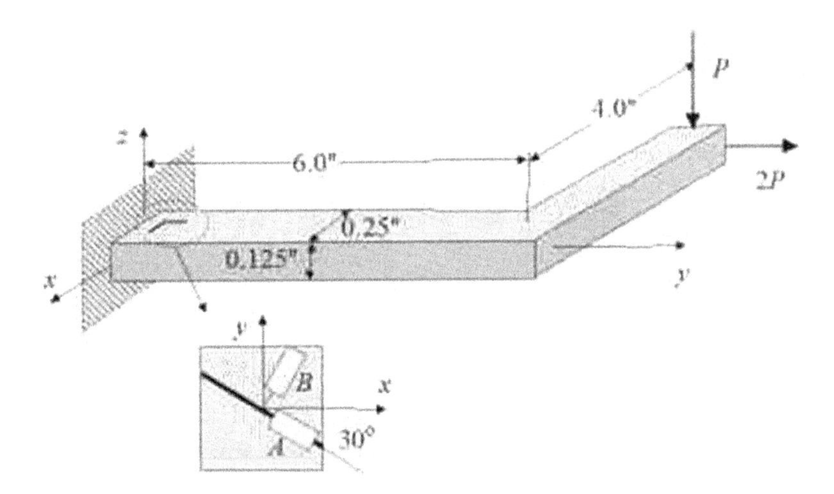

4.9 Experimental data from thermal expansion tests are presented below for three types of materials. From these data, determine the coefficients of thermal expansion for each material.

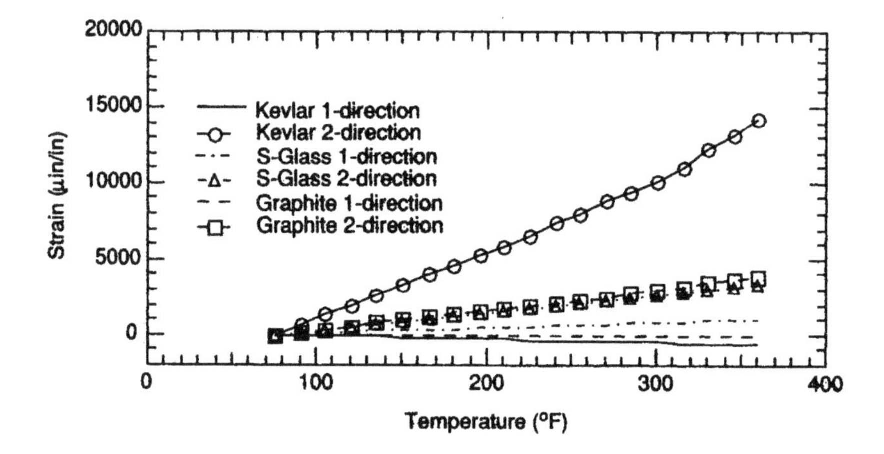

References

[1] Whitney JM, Daniel IM, Pipes RB. Experimental mechanics of fiber reinforced composite materials, SESA Monograph No. 4; 1982.

[2] Carlson LA, Pipes RB. Experimental characterization of advanced composite materials. NJ: Prentice-Hall; 1987.

[3] Pendleton RL, Tuttle ME. Manual on experimental methods of mechanical testing of composites. Bethel, CT: SEM; 1990.

[4] Dally JW, Riley WF. Experimental stress analysis. NY: McGraw-Hill; 1991.

[5] Student manual for strain gage technology, Bulletin 309, Measurements Group Inc., Raleigh, NC; 1992.

[6] Surface preparation for strain gage bonding, Bulletin B-129-5, Measurements Group Inc., Raleigh, NC; 2014.

[7] Perry CC, Lissner HR. The strain gage primer. 2nd ed. NY: McGraw-Hill; 1962.

[8] Optimizing strain gage excitation levels, M-M Tech Note TN-502, Measurements Group Inc., Raleigh, NC; 2010.

[9] Standard test methods for performance characteristics of bonded resistance strain gages, ASTM Standard E251-67.

[10] Errors due to transverse sensitivity in strain gages, M-M Tech Note TN-509, Measurements Group Inc., Raleigh, NC.

[11] Tuttle ME. Error in strain measurement obtained using strain gages on composites. In: Proc. 1985 SEM Fall Mtg., Nov. 17–20; 1985. p. 170–9.

[12] Tuttle ME, Brinson HF. Resistance-foil strain gage technology as applied to composite materials. Exp Mech 1984;24:54–65.

[13] McCalvey LF. Strain measurements on low modulus materials. In: Presented at BSSM conf., Univ. of Surrey, U.K., Sept; 1982.

[14] White RN. Model Study of The Failure of a Steel Structures. In: Presented at ASCE/SESA Exchange Session on Physical Modeling of Shell and Space Structures, ASCE Annual Convention, New Orleans, LA; 1982.

[15] Steklin P. Strain distribution in and around strain gages. J Strain Anal 1972;7:228–35.

[16] Beatty MF, Chewning SW. Nuerical analysis of the reinforcement effect of a strain gage applied to a soft material. Int J Eng Sci 1979;17:907–15.

[17] El-Tahan WW, Staab GH, Advani SH, Lee JK. Structural analysis of bimodular materials. ASCE J Eng Mech 1989;963–81.
[18] Schoeppner GA, Sierakowski RL. A review of compression test methods for organic matrix composites. J Compos Technol Res 1990;V12:3–12.
[19] Hofer Jr KE, Rao N Larsen D, Development of engineering data on mechanical properties of advanced composite materials, AFML-TR-72-205, Part I, Sept. 1972.
[20] Verette RM, Labor JD. Structural criteria for advanced composites, AFFDL-TR-76-142, Vol. 1, Summary, March 1977.
[21] Kasen MB, Schramm RE, Read DT. Fatigue of composites at cryogenic temperatures. In: Reifsnider KL, Lauraitis KN, editors. Fatigue of Filamentary Composites, ASTM STP 636; 1977. p. 141–51.
[22] Grimes GC, Frances PH, Commerford GE, Wolfe GK. An Experimental Investigation of the Stress Levels at Which Significant Damage Occurs in Graphite Fiber Plastic Composites, AFML-TR-72-40; 1972.
[23] Ryder JT, Black ED. Compression testing of large gage length composite coupons. In: Composite materials: testing and design (Fourth Conference), ASTM STP 617; 1977. p. 170–89.
[24] Lantz RB, Baldridge KG. Angle-plied boron/epoxy test method—a comparison of beam-tension and axial tension coupon testing. In: Composite materials: testing and design; 1969, ASTM STP 460, p. 94–107.
[25] Uniaxial compression. section 4.2.4, Advanced composites design guide. 3rd ed. Materials, vol. 4. Air Force Flight Dynamics Laboratoy; 1975.
[26] Lee S, Munro M. Evaluation of in-plane shear test method for advanced composite materials by the decision analysis technique. Composites 1986;17:13–22.
[27] Whitney JM, Halpin JC. Analysis of laminated anisotropic tubes under combined loading. J Comp Mater 1968;2:360–7.
[28] Hahn HT, Erikson J. Characterization of composite laminates using tubular specimens, AFML-TR-77-144, Aug. 1977.
[29] Whitney JM, Stansbarger DL, Howell HB. Analysis of the rail shear test—applications and limitations. J Comp Mater 1971;5:24–35.
[30] Chamis CC, Sinclair TH. Ten-degree off-axis test for shear properties in fiber composites. Exp Mech 1977;17(9):339–46.
[31] Yeow YT, Brinson HF. A comparison of simple shear characterization methods for composite laminates. Composites 1978;9:49–55.
[32] Halpin JC, Pagano NJ. Influence of end constraint in the testing of anisotropic bodies. J Comp Mater 1968;2:18–31.
[33] Iosipescu N. New accurate procedure for single shear testing of metals. J Mater 1967;2(3):537–66.
[34] Slepetz JM, Zagaeski TF, Novello RF. In-plane shear test for composite materials. Watertown, MA: Army Materials and Mechanics Research Center; 1978, AMMRC TR 78-30.
[35] Walrath DE, Adams DF. The Iosipescu shear test as applied to composite materials. Exp Mech 1983;23(1):105–10.
[36] Adams DF, Walrath DE. Further development of the Iosipescu shear test method. Exp Mech 1987;27:113–9.
[37] Lee S, Munro M. Evaluation of testing techniques for the Iosipescu shear test for advanced composite materials. J Comp Mater 1990;24:419–40.
[38] Lee S, Munro M, Scott RF. Evaluation of three in-plane shear test methods for advanced composite materials. Composites 1990;21:495–502.

[39] Tsai SW, Hahn HT. Introduction to composite materials. Montreal: Technomec Publishing Inc.; 1980.
[40] Vinson JR, Sierakowski RL. The behavior of structures composed of composite materials. Netherlands: Martinus Nijhoff; 1987.
[41] Reinhart TJ, et al, editors. Engineered materials handbook. In: Composites, vol. 1. Pennsylvania: ASTM International; 1989.
[42] Schwartz MM. Composite materials handbook. NY: McGraw-Hill; 1984.
[43] Beland S. High performance thermoplastic resins and their composites. NJ: Noyes Data Corporation; 1990.
[44] Jones RM. Mechanics of composite materials. New York: Hemisphere Publishing; 1975.
[45] Ashton JE, Halpin JC, Petit PH. Primer on composite materials: analysis. Pennsylvania: Technomis; 1969.
[46] Hull D. An introduction to composite materials. Cambridge: Cambridge University Press; 1981.
[47] Chawal KK. Composite materials science and engineering. New York: Springer-Verlag; 1987.

Lamina failure theories

5

5.1 Introduction

The bond between adjacent lamina in a laminated composite is assumed to be perfect and is not considered when discussing in-plane failure theories. Special consideration is given to failures involving interlaminar stress. Models and analysis techniques used to address interlaminal failure are summarized by Pagano [1].

The mechanisms for complete laminate failure are best understood by first considering lamina failure. Fiber orientations of adjacent lamina in a laminate may be different; thus, the apparent stiffness in specific directions may vary through the laminate. The state of stress experienced by individual lamina can be correlated to the effective stiffness of the lamina. Early efforts by the paper products industry to predict failures in orthotropic materials led to current failure theories for composite materials. Detailed reviews of many failure theories have been presented [2–8]. None of the orthotropic failure theories currently available are considered accurate enough to be used as a sole performance predictor in design. They all tend to be phenomenological and empirical in nature rather than mechanistic.

Failure of a unidirectional laminate begins on the microscopic level. Initial microscopic failures can be represented by local failure modes, such as:

- Fiber failure—breakage, microbuckling, dewetting
- Bulk matrix failure—voids, crazing
- Interface/flaw dominated failures—crack propagation and edge delamination

Microscopic failures can become macroscopic and result in catastrophic failure. The general nature of failure for orthotropic materials is more complicated than for an isotropic material.

Analysis techniques valid for isotropic materials are not adequate for composites.

A lamina is stronger in the fiber direction than in the transverse direction. The largest stress on the lamina may not be the one that causes failure. Assume the failure stresses are as follows:

X = Maximum failure strength in 1-direction = 50 ksi
Y = Maximum failure strength in 2-direction = 1 ksi
S = Maximum shear failure strength = 2 ksi

The state of stress is assumed to be

$$\begin{Bmatrix} \sigma_1 \\ \sigma_2 \\ \tau_{12} \end{Bmatrix} = \begin{Bmatrix} 45 \\ 2 \\ 1 \end{Bmatrix} \text{ksi}$$

Laminar Composites. http://dx.doi.org/10.1016/B978-0-12-802400-3.00005-2

It is obvious that failure in the 1-direction (maximum stress direction) will not occur, but in the 2-direction it will.

There are numerous theories for predicting lamina failure, a summary of which is given in Rowlands [2]. In this text, two failure theories are considered in detail. In the following discussions, the notation is used for identifying failure strength in various directions:

X,X': Maximum tensile and compressive failure strengths in the longitudinal (fiber) direction
Y,Y': Maximum tensile and compressive failure strengths in the transverse (X_2) direction
S: Maximum shear failure strength

The prime ($'$) denotes compression. Tensile and compressive failure strengths of continuous fiber laminates are generally different. The sign of an applied shear stress does not influence the failure strength in shear but can affect predicted failure loads.

5.2 Maximum stress theory

This theory is commonly attributed to Jenkins [9] and is an extension of the maximum normal stress theory for the failure of orthotropic materials (such as wood). Consider a lamina subjected to uniaxial tension as shown in Figure 5.1. For a failure to occur according to the maximum stress theory, one of three possible conditions must be met:

$$\sigma_1 \geq X, \quad \sigma_2 \geq Y, \quad \tau_{12} \geq S \tag{5.1}$$

In this case, the stresses in the principal material directions are

$$\begin{Bmatrix} \sigma_1 \\ \sigma_2 \\ \tau_{12} \end{Bmatrix} = [T_\sigma] \begin{Bmatrix} \sigma_x \\ 0 \\ 0 \end{Bmatrix} = \begin{bmatrix} m^2 & n^2 & 2mn \\ n^2 & m^2 & -2mn \\ -mn & mn & m^2-n^2 \end{bmatrix} \begin{Bmatrix} \sigma_x \\ 0 \\ 0 \end{Bmatrix} = \begin{Bmatrix} m^2 \\ n^2 \\ -mn \end{Bmatrix} \sigma_x$$

where $m = \cos\theta$ and $n = \sin\theta$. To ensure that failure does not occur under the conditions represented in Equation (5.1), the stresses in the principal material directions must be less than the respective strengths in those directions such that

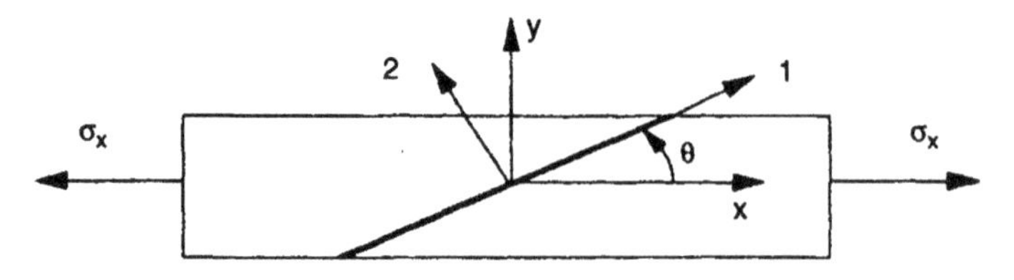

Figure 5.1 Uniaxial tension for a unidirectional lamina.

$$\sigma_x \le \frac{X}{\cos^2\theta} \quad \sigma_x \le \frac{Y}{\sin^2\theta} \quad \sigma_x \le \frac{S}{\sin\theta\cos\theta}$$

If the applied stresses were compressive, the X and Y would be replaced by X' and Y' so that the failure conditions become

$$\sigma_1 > X', \quad \sigma_2 > Y' \tag{5.2}$$

The failure criterion for shear remains unchanged, since S is independent of the sign of the applied shear stress.

If any one of the inequalities is not satisfied, it is assumed that failure occurs. In cases of multiaxial stress, the simple relationships just given are no longer valid. The relationship between applied stress components and principal material direction stresses must be determined through stress transformation, while the inequalities in Equations (5.1) and (5.2) remain valid. For example, assume a general state of stress in which σ_x, σ_y, and τ_{xy} exist. The principal direction stresses and condition for failure to occur for this case are

$$\sigma_1 = m^2\sigma_x + n^2\sigma_y + 2mn\tau_{xy} \ge X$$
$$\sigma_2 = n^2\sigma_x + m^2\sigma_y - 2mn\tau_{xy} \ge Y$$
$$\tau_{12} = (\sigma_y - \sigma_x)mn + (m^2 - n^2)\tau_{xy} \ge S$$

5.3 Maximum strain theory

The maximum strain failure criteria is an extension of St. Venant's maximum strain theory to accommodate orthotropic material behavior. The maximum strain failure theory is expressed as

$$\varepsilon_1 \le X_\varepsilon, \varepsilon_1 > X'_\varepsilon \quad \varepsilon_2 \le Y_\varepsilon, \varepsilon_2 > Y'_\varepsilon, \quad \gamma_{12} \le S_\varepsilon, \tag{5.3}$$

where

$$X_\varepsilon = \frac{X}{E_1} \quad Y_\varepsilon = \frac{Y}{E_2} \quad S_\varepsilon = \frac{S}{G_{12}} \quad X'_\varepsilon = \frac{X'}{E_1} \quad Y'_\varepsilon = \frac{Y'}{E_2} \tag{5.4}$$

For a case of uniaxial tension as in Figure 5.1,

$$\varepsilon_1 = \frac{1}{E_1}(\sigma_1 - \nu_{12}\sigma_2) \quad \varepsilon_2 = \frac{1}{E_2}(\sigma_2 - \nu_{21}\sigma_1) \quad \gamma_{12} = \frac{\tau_{12}}{G_{12}}$$

The principal material direction stresses are $\sigma_1 = m^2\sigma_x$, $\sigma_2 = n^2\sigma_x$, and $\tau_{12} = -mn\sigma_x$. Therefore, it is a simple matter to show that

$$\varepsilon_1 = \frac{m^2 - \nu_{12}n^2}{E_1}\sigma_x \quad \varepsilon_2 = \frac{n^2 - \nu_{21}m^2}{E_2}\sigma_x \quad \gamma_{12} = \frac{-mn}{G_{12}}\sigma_x$$

From these relationships, it is easy to show that in order to avoid failure for a condition of uniaxial tension the following conditions must be checked:

$$\sigma_x \leq \frac{X}{m^2 - \nu_{12}n^2} \quad \sigma_x \leq \frac{Y}{n^2 - \nu_{21}m^2} \quad \sigma_x \leq \frac{S}{mn}$$

The maximum stress and strain failure theories generally yield different results and are not extremely accurate. They are often used because of their simplicity. As with the maximum stress theory, a more complex state of stress results in different expressions. As discussed in the previous section, a more general state of stress results in a more complex representation of ε_1, ε_2, and γ_{12}.

Example 5.1 The composite reinforced pressure vessel in Example 3.3 is considered again. The analysis presented here incorporates actual material properties as defined in Table 4.2. The analysis procedure in Example 3.3 relating normal stress (σ_x) to reinforcement spacing (s) is used in this example with $\sigma_x = 7200s$. It is assumed that the vessel has failed, and the reinforcements sustain all circumferential loads originally carried by the vessel as illustrated in Figure E5.1-1. Both the maximum stress and strain failure theories are investigated. Assume the reinforcements are E-glass/epoxy. Because of the state of stress, only tensile failure strengths are considered, with $X = 187$, $Y = 6.7$, and $S = 6.5$ ksi. Corresponding to these are the failure strengths for maximum strain failure established from Equation (5.4). For some fiber orientations, the normal strain may be compressive, so both tensile and compressive properties are required.

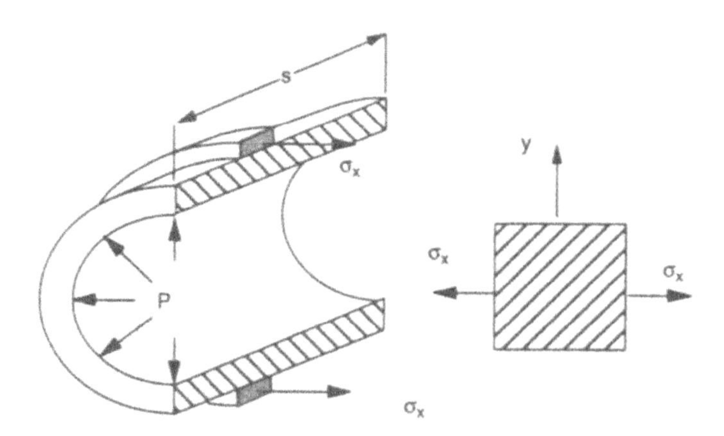

Figure E5.1-1 Assumed stress in pressure vessel reinforcements.

$X_\varepsilon = 21{,}250\ \mu\text{in./in.}\quad Y_\varepsilon = 1861\ \mu\text{in./in.}\quad S_\varepsilon = 3767\ \mu\text{in./in.}$
$X'_\varepsilon = 13{,}523\ \mu\text{in./in.}\quad Y'_\varepsilon = 7027\ \mu\text{in./in.}$

The analysis for failure due to the maximum stress theory is simple since only the principal direction stresses are required. Using the stress transformations from Chapter 2, the failure criteria for the maximum stress theory are

$$\begin{Bmatrix}\sigma_1\\ \sigma_2\\ \tau_{12}\end{Bmatrix} = \begin{Bmatrix}187{,}000\\ 6700\\ 6500\end{Bmatrix} = [T_\sigma]\begin{Bmatrix}7200s\\ 0\\ 0\end{Bmatrix} = \begin{Bmatrix}m^2\\ n^2\\ -mn\end{Bmatrix}(7200s)$$
$$\Rightarrow \begin{Bmatrix}25{,}970\\ 0.931\\ 0.903\end{Bmatrix} = \begin{Bmatrix}m^2\\ n^2\\ -mn\end{Bmatrix}s$$

The solution of each possible failure depends on fiber orientation, so no unique solution exists. The spacing depends on which failure criterion is met first at a particular fiber orientation. Recalling that $m = \cos\theta$ and $n = \sin\theta$, we denote the fiber spacing according to which stress component satisfies the failure condition by

σ_1 controlled: $s = 25{,}970/\cos^2\theta$
σ_2 controlled: $s = 0.931/\sin^2\theta$
τ_{12} controlled: $s = 0.903/\sin\theta\cos\theta$

Figure E5.1-2 shows the fiber spacing as a function of fiber orientation for each failure mode. The spacing which could be used is obtained by taking the minimum spacing for each angle considered, as shown in Figure E5.1-3. At 0°, the spacing associated with σ_1 is required, since all other solutions predict that $s = \infty$. Up to 45°, the τ_{12}-controlled failure predicts the appropriate spacing, and from 45° to 90°, the σ_2-controlled spacing is appropriate.

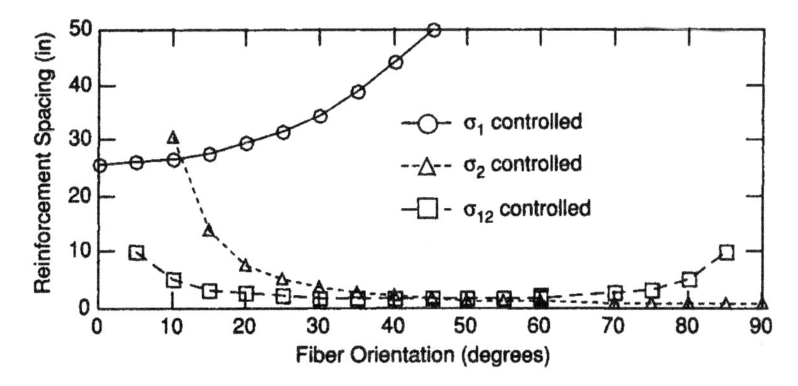

Figure E5.1-2 Reinforcement spacing for each component of the maximum stress failure theory.

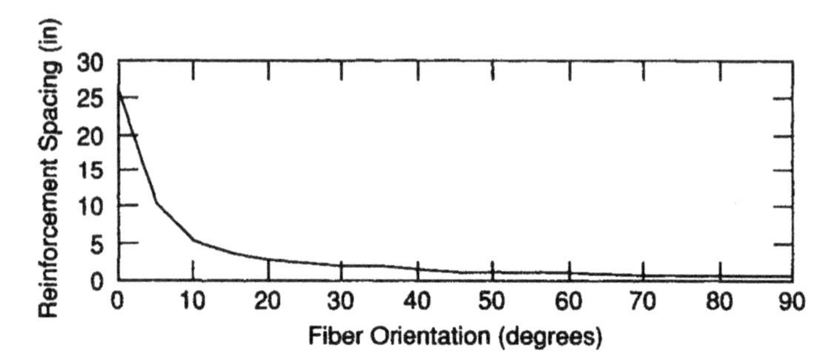

Figure E5.1-3 Reinforcement spacing from the maximum stress failure theory.

The strains in the principal material directions are related to the stresses through the relationships in Chapter 2. Recall that

$$\begin{Bmatrix} \varepsilon_1 \\ \varepsilon_2 \\ \gamma_{12} \end{Bmatrix} = [S] \begin{Bmatrix} \sigma_1 \\ \sigma_2 \\ \tau_{12} \end{Bmatrix}$$

where $[S]$ is a function of fiber orientation. The principal material direction stresses are related to the applied stresses through the stress transformation relations. Coupling this with the preceding relationship, the maximum strain failure theory is written as

$$\begin{Bmatrix} X_\varepsilon \\ Y_\varepsilon \\ S_\varepsilon \end{Bmatrix} \leq \begin{Bmatrix} \varepsilon_1 \\ \varepsilon_2 \\ \gamma_{12} \end{Bmatrix} = [S] \begin{Bmatrix} m^2 \\ n^2 \\ -mn \end{Bmatrix} \sigma_x = [S] \begin{Bmatrix} m^2 \\ n^2 \\ -mn \end{Bmatrix} (7200s)$$

For some fiber orientations, Y_ε must be replaced with Y_ε' since the strain in the 2-direction is compressive. Reinforcement spacing as a function of fiber orientation is shown in Figure E5.1-4. This curve is similar to that presented for the maximum

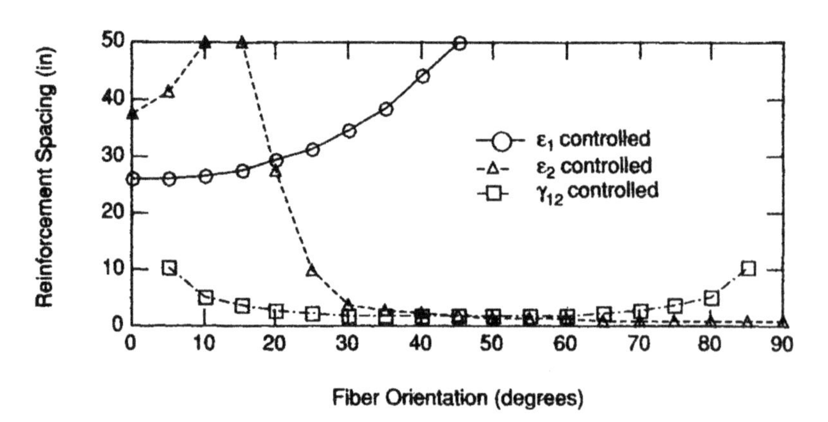

Figure E5.1-4 Reinforcement spacing for each component of the maximum strain failure theory.

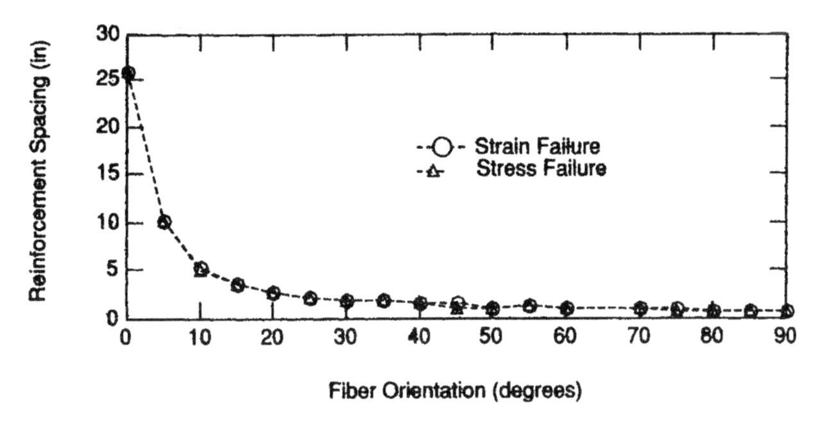

Figure E5.1-5 Reinforcement spacing from the maximum strain failure theory.

stress theory. A direct correlation of the maximum stress and strain theories indicates that they predict virtually identical results, as shown in Figure E5.1-5.

This close correlation is not always observed when comparing the two theories. The state of stress in the component plays a more significant role in defining the state of strain that may be evident from this example.

Example 5.2 Assume the lamina shown in Figure E5.2 is subjected to the multiaxial state of stress indicated.

The stresses in the principal material directions are

$$\begin{Bmatrix} \sigma_1 \\ \sigma_2 \\ \tau_{12} \end{Bmatrix} = \begin{bmatrix} m^2 & n^2 & 2mn \\ n^2 & m^2 & -2mn \\ -mn & mn & m^2-n^2 \end{bmatrix} \begin{Bmatrix} -10 \\ 15 \\ 10 \end{Bmatrix}$$

$$= \begin{bmatrix} 0.25 & 0.75 & 0.866 \\ 0.75 & 0.25 & -0.866 \\ -0.433 & 0.433 & -0.5 \end{bmatrix} \begin{Bmatrix} -10 \\ 15 \\ 10 \end{Bmatrix} = \begin{Bmatrix} 17.41 \\ -12.41 \\ 5.825 \end{Bmatrix} \text{ksi}$$

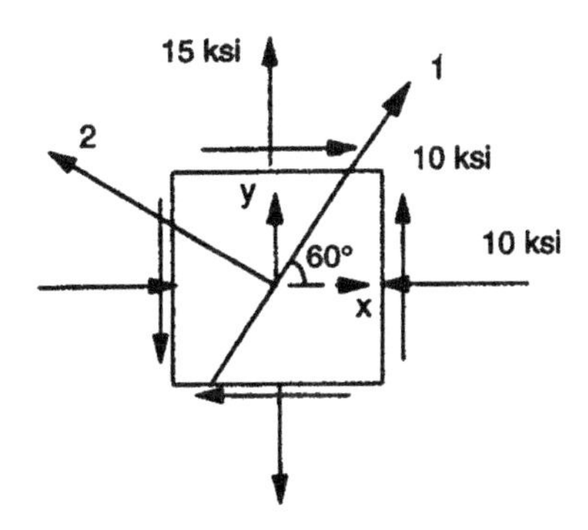

Figure E5.2 Multiaxial state of stress.

The material properties and failure strengths are assumed to be

$$E_1 = 8.42 \times 10^6\,\text{psi} \quad E_2 = 2.00 \times 10^6\text{psi} \quad G_{12} = 0.77 \times 10^6\,\text{psi} \quad \nu_{12} = 0.293$$
$$X = 136\,\text{ksi} \quad X' = 280\,\text{ksi} \quad Y = 4\,\text{ksi} \quad Y' = 20\,\text{ksi}$$
$$S = 6\,\text{ksi}$$

When the preceding stresses are compared to these failure strengths, it is evident that no failure has occurred according to the maximum stress theory. For completeness, however, the failure strains should also be checked. The stresses just given can be used to determine the principal direction strains by using the compliance matrix so that

$$\begin{Bmatrix} \varepsilon_1 \\ \varepsilon_2 \\ \gamma_{12} \end{Bmatrix} = \begin{bmatrix} S_{11} & S_{12} & 0 \\ S_{12} & S_{22} & 0 \\ 0 & 0 & S_{66} \end{bmatrix} \begin{Bmatrix} 17.41 \\ -12.41 \\ 5.825 \end{Bmatrix}$$

$$= \begin{bmatrix} 1.2 & -0.35 & 0 \\ -0.35 & 5 & 0 \\ 0 & 0 & 12.9 \end{bmatrix} \times 10^{-7} \begin{Bmatrix} 17.41 \\ -12.41 \\ 5.825 \end{Bmatrix} \times 10^3 = \begin{Bmatrix} 2536 \\ -6814 \\ 7518 \end{Bmatrix} \mu\text{in./in.}$$

The failure strains associated with this material are

$$X_\varepsilon = \frac{X}{E_1} = 16{,}152\,\mu\text{in./in.} \quad Y'_\varepsilon = \frac{Y}{E_2} = 10{,}000\,\mu\text{in./in.}$$

$$S_\varepsilon = \frac{S}{G_{12}} = 7792\,\mu\text{in./in.}$$

A comparison of the failure strains with those resulting from the applied state of stress shows that failure does not occur. In each case, the shear term (either strain or stress) was the closest to failure. This result brings to light some interesting aspects of shear failures and the general importance of shear stresses.

5.4 The significance of shear stress

Unlike tension or compression, shear failures are not distinguishable as being either tensile or compressive since the shear failure strength is independent of the sign of τ. Consider the lamina shown in Figure 5.2a, with fibers oriented at $\theta = -45°$, and a positive shear stress τ applied. The stresses in the 1–2 material plane are as shown in Figure 5.2b and are

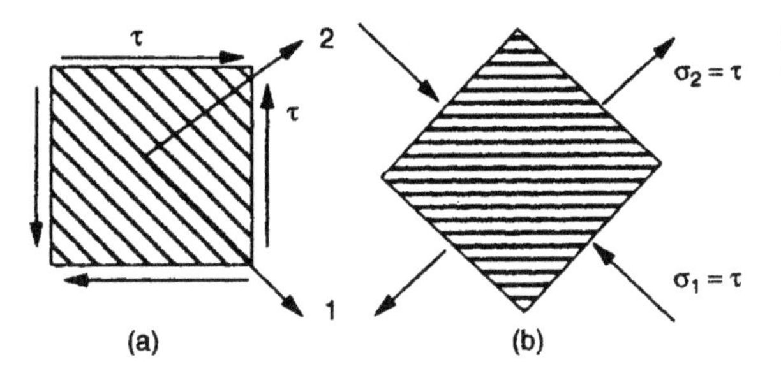

Figure 5.2 Pure shear with $+\tau$ and $\theta = -45°$.

$$\begin{Bmatrix} \sigma_1 \\ \sigma_2 \\ \tau_{12} \end{Bmatrix} = [T_\sigma] \begin{Bmatrix} \sigma_x \\ \sigma_y \\ \tau_{xy} \end{Bmatrix} = \begin{bmatrix} 0.5 & 0.5 & -1 \\ 0.5 & 0.5 & 1 \\ 0.5 & -0.5 & 0 \end{bmatrix} \begin{Bmatrix} 0 \\ 0 \\ \tau \end{Bmatrix} = \begin{Bmatrix} -\tau \\ \tau \\ 0 \end{Bmatrix}$$

The signs of these stresses indicate that failure is likely to be matrix dominated and occur in the 2-direction since its failure strength is much lower than that in the 1-direction. If the material is the same as that used in Example 5.2, the shear stress that would cause failure is $\tau > 4$ ksi.

If the direction of the applied shear stress is reversed so that $\tau = -\tau$ (as shown in Figure 5.3a), and the angle $\theta = -45°$ is maintained, the principal material direction stresses illustrated in Figure 5.3b and are

$$\begin{Bmatrix} \sigma_1 \\ \sigma_2 \\ \tau_{12} \end{Bmatrix} = [T_\sigma] \begin{Bmatrix} \sigma_x \\ \sigma_y \\ \tau_{xy} \end{Bmatrix} = \begin{bmatrix} 0.5 & 0.5 & -1 \\ 0.5 & 0.5 & 1 \\ 0.5 & -0.5 & 0 \end{bmatrix} \begin{Bmatrix} 0 \\ 0 \\ -\tau \end{Bmatrix} = \begin{Bmatrix} \tau \\ -\tau \\ 0 \end{Bmatrix}$$

For this case, there is a tensile stress in the 1-direction and a compressive stress in the 2-direction. Referring to the failure stresses of Example 5.2, this indicates that failure occurs for $\tau = 20$ ksi.

Shear stress can play a significant role in lamina failure, since it contributes to both the magnitude and sign of stresses in various directions. Although the sign of the applied shear stress does not affect the shear failure strength (S), there is interaction between stress components and the associated failure of a composite lamina.

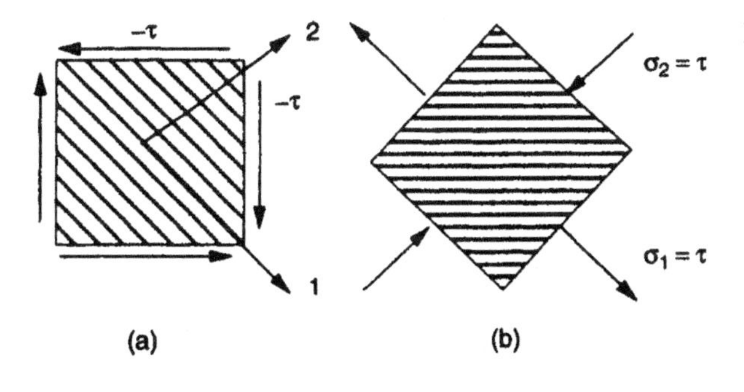

Figure 5.3 Pure shear with $-\tau$ and $\theta = -45°$.

Example 5.3 Assume a state of pure shear stress on an element of unidirectional lamina for which the fiber orientation is arbitrary (either $+\theta$ or $-\theta$) as illustrated in Figure E5.3-1. The applied stress is assumed to be either positive as shown or negative. We wish to determine the stress required to cause failure as a function of θ.

Assume this lamina is made of the material in Example 5.2. The stresses in the principal material directions are established from Equation (2.3) to be

$$\begin{Bmatrix} \sigma_1 \\ \sigma_2 \\ \tau_{12} \end{Bmatrix} = [T_\sigma] \begin{Bmatrix} 0 \\ 0 \\ \tau_{xy} \end{Bmatrix} \begin{Bmatrix} 2mn \\ -2mn \\ m^2 - n^2 \end{Bmatrix} \tau_{xy}$$

Analysis of $+\theta$ fiber orientations: A $+\theta$ fiber orientation results in a tensile σ_1 and compressive σ_2. Using the maximum stress theory with the failure strengths of the previous example, three possible failure conditions result:

$$\tau_{xy} \le \begin{cases} \dfrac{X}{2mn} = \dfrac{68}{mn} \\ \dfrac{Y}{2mn} = \dfrac{10}{mn} \\ \dfrac{S}{m^2 - n^2} = \dfrac{6}{m^2 - n^2} \end{cases}$$

Analysis of $-\theta$ fiber orientations: A $-\theta$ fiber orientation results in a compressive σ_1 and tensile σ_2. The three resulting failure conditions are

$$\tau_{xy} \le \begin{cases} \dfrac{140}{mn} \\ \dfrac{2}{mn} \\ \dfrac{6}{m^2 - n^2} \end{cases}$$

Substituting various values of θ into these equations results in failures that are predominantly controlled by the shear stress failure condition $S/(m^2 - n^2)$. For some fiber

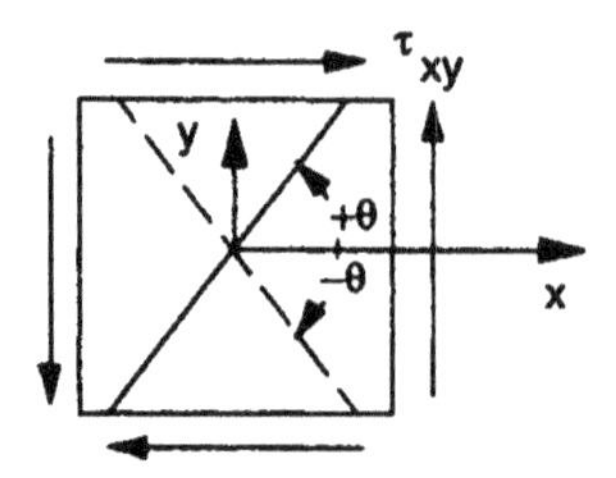

Figure E5.3-1 Pure shear on an element with arbitrary fiber orientation.

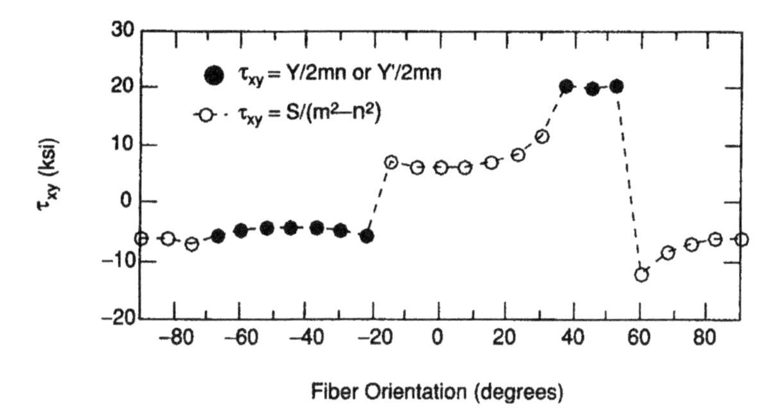

Figure E5.3-2 Applied shear stress vs. θ required to produce failure of a lamina with arbitrary fiber orientations.

orientations, however, failure is controlled by the appropriate σ_2 failure condition of either $Y/2\,mn$ or $Y'/2\,mn$. This is illustrated in Figure E5.3-2 by a plot of the shear stress required to produce failure as a function of θ. The applied shear stress required to produce failure is dependent on both the direction of the applied shear and fiber orientation θ. A similar set of results can be obtained using the maximum strain failure theory.

5.5 Interactive failure theories

The interaction of stress components was partially illustrated in Example 5.3. Initial efforts to formulate an interactive failure criterion are credited to Hill [10] in 1950. Since then, others have proposed modifications to the initial theory. Some interactive failure theories are complicated, and in certain cases, the amount of work needed to define critical parameters for a specific theory exceeds the benefit of using it. The general form of a majority of these theories can be put into one of two categories, each of which has a different form. These criteria and their associated failure theories are expressed in tensor notation below.

Criterion	Theory
$F_{ij}\sigma_i\sigma_j=1$	Ashkenazi [11], Chamis [12], Fischer [13], Tsai–Hill [14], Norris [15]
$F_{ij}\sigma_i\sigma_j+F_i\sigma_i=1$	Cowin [16], Hoffman [17], Malmeister [18], Marin [19], Tsai–Wu [20], Gol'denblat–Kopnov [21]

Each theory is summarized in Tables 5.1 and 5.2. The fundamental difference in each is the extent to which the terms F_i and F_{ij} are defined. The interactive term F_{12} can be influential in predicting failure and is often difficult to experimentally define.

Table 5.1 Summary of interactive failure theories governed by $F_{ij}\sigma_i\sigma_j =$ (after [2])

Theory	F_{11}	F_{22}	F_{12}	F_{66}
Ashkenazi [11]	$\frac{1}{X^2}$	$\frac{1}{Y^2}$	$\frac{1}{2}\left[\frac{4}{U^2}-\frac{1}{X^2}-\frac{1}{Y^2}-\frac{1}{S^2}\right]$	$\frac{1}{S^2}$
Chamis [12]	$\frac{1}{X^2}$	$\frac{1}{Y^2}$	$-\frac{K_1K_1'}{2XY}$	$\frac{1}{S^2}$
Fisher [13]	$\frac{1}{X^2}$	$\frac{1}{Y^2}$	$-\frac{K}{2XY}$	$\frac{1}{S^2}$
Tsai–Hill [14]	$\frac{1}{X^2}$	$\frac{1}{Y^2}$	$-\frac{1}{2X^2}$	$\frac{1}{S^2}$
Norris[1] [15]	$\frac{1}{X^2}$	$\frac{1}{Y^2}$	$-\frac{K}{2XY}$	$\frac{1}{S^2}$

$K_1 = \frac{(1+4\nu_{12}-\nu_{13})E_{22}+(1-\nu_{23})E_{11}}{\sqrt{E_{11}E_{22}(2+\nu_{12}+\nu_{13})(2+\nu_{21}+\nu_{23})}}$ $K = \frac{(1+\nu_{12})E_{11}+(1-\nu_{12})E_{22}}{2\sqrt{E_{11}E_{22}(1+\nu_{12})(1+\nu_{21})}}$

K_1' = experimentally determined correlation coefficient.
U = longitudinal strength of a 45° off-axis coupon.
[1]An additional condition in the Norris theory is that $\sigma_1^2 = X^2$ and $\sigma_2^2 = Y^2$.

The interactive theories presented here are not the only ones available for composites. Energy formulations as well as complete laminate theories (as opposed to isolating individual lamina) have been proposed. For example, Petit and Waddoups [22] extended the conventional maximum strain theory to include nonlinear terms. Sandhu [23] formulated a parallel failure theory based on energy methods. Puppo and Evensen [24] postulated a failure theory directly applicable to the entire laminate. Similarly, Wu and Scheriblein [25] formulated a tensor polynomial for direct laminate failure evaluation. A more detailed discussion of many of these theories is presented in Rowlands [2]. The Tsai–Hill and Tsai–Wu interactive failure theories considered to be representative of those from each category and are discussed herein.

5.5.1 Tsai–Hill (maximum work) theory

The Tsai–Hill theory [14] is considered an extension of the Von Mises failure criterion. The failure strengths in the principal material directions are assumed to be known. The tensor form of this criterion is $F_{ij}\sigma_i\sigma_j = 1$. If this expression is expanded and the F_{ij} terms replaced by letters, the failure criterion is

$$F(\sigma_2-\sigma_3)^2+G(\sigma_3-\sigma_1)^2+H(\sigma_1-\sigma_2)^2+2\left(L\tau_{23}^2+M\tau_{13}^2+N\tau_{12}^2\right)=1$$

Expanding and collecting terms,

$$(G+H)\sigma_1^2+(F+H)\sigma_2^2+(F+G)\sigma_3^2-2[H\sigma_1\sigma_2+F\sigma_2\sigma_3+G\sigma_1\sigma_3] \\ +2\left[L\tau_{23}^2+M\tau_{13}^2+N\tau_{12}^2\right]=1$$

where F, G, H, L, M, and N are anisotropic material strength parameters.

The failure strength in the principal material directions are represented by X, Y, and Z. Application of a uniaxial tensile stress in each of the three principal material directions while keeping all other stresses zero (i.e., $\sigma_1 \neq 0$, $\sigma_2 = \sigma_3 = \tau_{12} = \tau_{13} = \tau_{23} = 0$) yields

Table 5.2 Summary of interactive failure thoeories governed by $F_i\sigma_i + F_{ij}\sigma_i\sigma_j = 1$ (after [2])

Theory	F_1	F_2	F_{11}	F_{22}	F_{12}	F_{66}
Cowin [16]	$\frac{1}{X}-\frac{1}{X'}$	$\frac{1}{Y}-\frac{1}{Y'}$	$\frac{1}{XX'}$	$\frac{1}{YY'}$	$\frac{2S^2\sqrt{F_{11}F_{22}}-1}{2S^2}$	$\frac{1}{S^2}$
Hoffman [17]	$\frac{1}{X}-\frac{1}{X'}$	$\frac{1}{Y}-\frac{1}{Y'}$	$\frac{1}{XX'}$	$\frac{1}{YY'}$	$-\frac{1}{2XX'}$	$\frac{1}{S^2}$
Malmeister [18]	$\frac{1}{X}-\frac{1}{X'}$	$\frac{1}{Y}-\frac{1}{Y'}$	$\frac{1}{XX'}$	$\frac{1}{YY'}$	$\frac{S'_{45}\left[(F_1-F_2)+(S'_{45})(F_{11}+F_{22})\right]-1}{2(S'_{25})^2}$	$\frac{1}{S^2}$
Marin [19]	$\frac{1}{X}-\frac{1}{X'}$	$\frac{1}{Y}-\frac{1}{XX'}$	$\frac{1}{XX'}$	$\frac{1}{YY'}$	$\frac{1}{XX'}-\frac{XX'-S[X'-X-X'(X/Y)+Y]}{2S^2XX'}$	—
Tsai–Wu [20]	$\frac{1}{X}-\frac{1}{X'}$	$\frac{1}{Y}-\frac{1}{Y'}$	$\frac{1}{XX'}$	$\frac{1}{YY'}$	$\leq \pm\sqrt{F_{11}F_{22}}$ and is determined under biaxial stress	$\frac{1}{S^2}$
Gol'denblat–Kopnov [21][1]	$\frac{1}{2}\left(\frac{1}{X}-\frac{1}{X'}\right)$	$\frac{1}{2}\left(\frac{1}{Y}-\frac{1}{Y'}\right)$	$\frac{1}{4}\left(\frac{1}{X}-\frac{1}{X'}\right)^2$	$\frac{1}{4}\left(\frac{1}{Y}-\frac{1}{Y'}\right)^2$	$\frac{1}{8}\left[\left(\frac{1}{X}+\frac{1}{X'}\right)^2+\left(\frac{1}{Y}+\frac{1}{Y'}\right)^2-\left(\frac{1}{S_{45}}+\frac{1}{S'_{45}}\right)^2\right]$	$\frac{1}{S^2}$

[1] S_{45} and S'_{45} are the shear strength of the 45° coupon subjected to positive and negative shear.

$$G+H=\frac{1}{X^2} \quad F+H=\frac{1}{Y^2} \quad G+F=\frac{1}{Z^2}$$

These expressions can be solved for the unknowns G, H, and F:

$$2H=\frac{1}{X^2}+\frac{1}{Y^2}-\frac{1}{Z^2} \quad 2G=\frac{1}{X^2}-\frac{1}{Y^2}+\frac{1}{Z^2} \quad 2F=-\frac{1}{X^2}+\frac{1}{Y^2}+\frac{1}{Z^2}$$

Assuming a state of plane stress ($\sigma_3=\tau_{13}=\tau_{23}=0$), the failure theory is written as

$$\begin{aligned}
&(G+H)\sigma_1^2+(F+H)\sigma_2^2-2H\sigma_1\sigma_2+2N\tau_{12}^2=1\\
&G\sigma_1^2+H\sigma_1^2+F\sigma_2^2+H\sigma_2^2-2H\sigma_1\sigma_2+2N\tau_{12}^2=1\\
&G\sigma_1^2+F\sigma_2^2+H(\sigma_1-\sigma_2)^2+2N\tau_{12}^2=1
\end{aligned}$$

Application of a pure shear stress τ_{12}, with $\sigma_1=\sigma_2=0$, results in an expression for the only remaining parameter, N:

$$2N=\frac{1}{S^2}$$

Substitution of the failure parameters F, G, H, and N into the plane stress failure criterion yields

$$\left(\frac{\sigma_1}{X}\right)^2-\left(\frac{1}{X^2}+\frac{1}{Y^2}-\frac{1}{Z^2}\right)\sigma_1\sigma_2+\left(\frac{\sigma_2}{Y}\right)^2+\left(\frac{\tau_{12}}{S}\right)^2=1$$

The primary load-resisting constituent in the 2 and 3 directions is the matrix, as illustrated in Figure 5.4.

Therefore, $Y=Z$, and the above expression simplifies to the plane stress form of the Tsai–Hill failure theory

$$\frac{\sigma_1^2}{X^2}-\frac{\sigma_1\sigma_2}{X^2}+\frac{\sigma_2^2}{Y^2}+\frac{\tau_{12}^2}{S^2}=1 \tag{5.5}$$

Sometimes it is convenient to express the stress and strength in terms of stress and strength ratios. The stress and strength ratios for plane stress can be written as

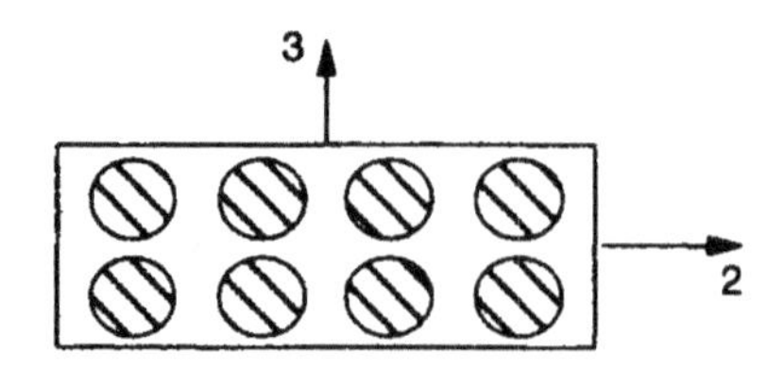

Figure 5.4 Schematic of 2- and 3-directions in a lamina.

Stress ratios: $p=\sigma_y/\sigma_x \quad q=\tau_{xy}/\sigma_x$
Strength ratios: $r=X/Y \quad S=X/S$

Using stress ratios, the Tsai–Hill failure theory is written as

$$\sigma_1^2-\sigma_1\sigma_2+r^2\sigma_2^2+s^2\tau_{12}^2=X^2 \tag{5.6}$$

It is generally assumed that the strength and stress ratio remains constant as the lamina loads increase. Consider the simple case of an off-axis lamina subjected to an axial stress σ_x (refer to Figure 5.1). The parameters p and q are both zero, since only an axial stress σ_x is applied. Therefore, with $p=q=0$, the stresses in the principal material directions are

$$\begin{Bmatrix}\sigma_1\\ \sigma_2\\ \tau_{12}\end{Bmatrix}=\begin{Bmatrix}m^2\\ n^2\\ -mn\end{Bmatrix}\sigma_x$$

Substituting these into Equation (5.5) and using the strength ratios results in

$$m^4+(s^2-1)m^2n^2+r^2n^4=\left(\frac{X}{\sigma_x}\right)^2$$

Solving for the applied stress, σ_x yields

$$\sigma_x=\frac{X}{\sqrt{m^4+(s^2-1)m^2n^2+r^2n^4}}$$

For the special case of $\theta=0°$, this reduces to $\sigma_x=X$. In a similar manner, if $\theta=90°$, this expression becomes $\sigma_x=Y$.

Example 5.4 The maximum stress and Tsai–Hill theories are investigated for pure shear. The lamina under consideration is assumed to have an arbitrary fiber orientation of either $-\theta$ or $+\theta$, as shown in Figure E5.4-1. The material is glass/epoxy with $E_1=7.8\times10^6$ psi, $E_2=2.6\times10^6$ psi, $G_{12}=1.25\times10^6$ psi, $\nu_{12}=0.25$, and failure strengths $X=X'=150$ ksi, $Y=4$ ksi, $y'=20$ ksi, and $S=8$ ksi. The stresses in the 1–2 plane based on an applied shear stress of $-\tau$ are

$$\begin{Bmatrix}\sigma_1\\ \sigma_2\\ \tau_{12}\end{Bmatrix}=[T_\sigma]\begin{Bmatrix}0\\ 0\\ -\tau\end{Bmatrix}=\begin{Bmatrix}-2mn\\ 2mn\\ n^2-m^2\end{Bmatrix}\tau$$

The tensile and compressive components of stress change with 0. For a positive angle, σ_1 is compressive and σ_2 is tensile. For a negative angle, σ_1 is tensile and σ_2 is compressive. The shear stress τ_{12} will not change sign as θ changes from positive to negative, but it will change sign based on the angle itself. Since $X=X'$, the sign of σ_1 is not

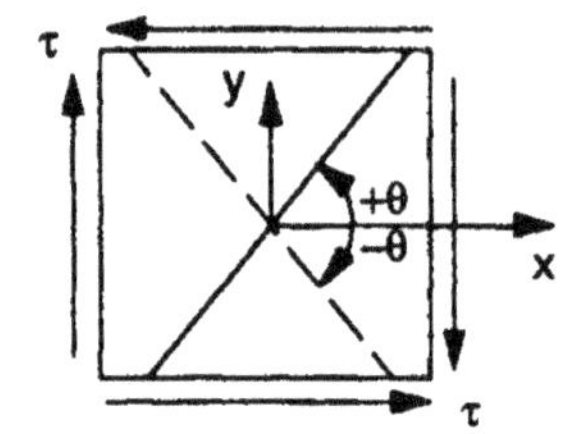

Figure E5.4-1 Pure shear with an arbitrary fiber orientation.

significant, but the sign of σ_2 will dictate which failure strength (Y or Y') is used for the 2-direction.

Maximum Stress Criterion. In the maximum stress criterion, four failure conditions must be checked. For materials in which $X \neq X'$, a fifth condition is required. Each condition is a function of θ:

$$\begin{array}{lll} \sigma_1 = 150{,}000 = -2mn\tau & \tau = 75{,}000/mn & (\text{for all } \theta) \\ \sigma_2 = 4000 = 2mn\tau & \tau = 2000/mn & (\text{for } +\theta) \\ \sigma_2 = 20{,}000 = 2mn\tau & \tau = 10{,}000/mn & (\text{for } -\theta) \\ \tau_{12} = 8000 = (n^2 - m^2) & \tau = 8000/(n^2 - m^2) & (\text{for all } \theta) \end{array}$$

Tsai–Hill Criterion. The Tsai–Hill criterion requires only one equation to establish failure. Substituting the stresses and failure strengths for this case into Equation (5.5) yields the failure criteria. The governing equation depends on the sign of σ_2, since it is the only stress component having two failure strengths (for this material).

$+\theta$: σ_1 is compressive and σ_2 is tensile, and the failure equation is

$$\left(\frac{(-2mn\tau)}{150}\right)^2 - \frac{(2mn\tau)(-2mn\tau)}{150^2} + \left(\frac{2mn\tau}{4}\right)^2 + \left(\frac{(n^2 - m^2)^2}{8}\right)^2 \tau^2 = 1$$

$$(-2mn\tau)^2 - (2mn\tau)(-2mn\tau) + \left(\frac{150}{4}\right)^2 (2mn\tau)^2 + \left(\frac{150}{8}\right)^2 (n^2 - m^2)^2 \tau^2 = (150 \times 10^3)^2$$

$$\tau^2 \left[5633m^2n^2 + 351.6(n^2 - m^2)^2\right] = 2.25 \times 10^{10}$$

$-\theta$: σ_1 is tensile and σ_2 is compressive, and the failure equation is

$$\left(\frac{(2mn\tau)}{150}\right)^2 - \frac{(2mn\tau)(-2mn\tau)}{150^2} + \left(\frac{-2mn\tau}{20}\right)^2 + \left(\frac{(n^2 - m^2)^2}{8}\right)^2 \tau^2 = 1$$

$$(2mn\tau)^2 - (2mn\tau)(-2mn\tau) + \left(\frac{150}{20}\right)^2 (-2mn\tau)^2 + \left(\frac{150}{8}\right)^2 (n^2 - m^2)^2 \tau^2 = (150 \times 10^3)^2$$

$$\tau^2 \left[233m^2n^2 + 351.6(n^2 - m^2)^2\right] = 2.25 \times 10^{10}$$

Solutions for the maximum stress criteria result in sign changes for τ similar to those in Example 5.2. Solutions for the Tsai–Hill criteria yield two roots for τ for each angle. To compare these theories, absolute value $|\tau|$ vs. θ is plotted in Figure E5.4-2. The Tsai–Hill theory produces a more uniform curve of $|\tau|$ vs. θ than the maximum stress theory. For negative fiber angles, the stress required to produce failure is greater than for positive angles. Depending on the fiber orientation angle, the maximum stress criterion will be controlled by either σ_2 or τ. The regions in which either shear or normal stress control failure for the maximum stress criterion are established by examination of the failure criteria at each angle. For example, $\theta = 10°$, for which cos $10° = 0.9848$ and sin $10° = 0.1736$. Comparing the σ_2 and the shear, it is easy to see that

$$\sigma_2 = \frac{2000}{(0.9848)(0.1736)} = 11,100\text{psi} \quad \tau = \frac{8000}{(0.030 - 0.9698)} = 8500\text{psi}$$

Therefore, the failure is shear controlled at this angle.

From Table 5.1, the primary difference between failure theories is the form of the interactive term, F_{12}. The Ashkenazi [11] and Chamis [12] theories require experimentally determined parameters not generally defined when X, X', Y, Y', and S are established. In the case of uniaxial tension applied to a lamina with a fiber orientation of 45°, it is easily shown that very little difference in predicted failure load exists between the theories. This is due primarily to the magnitude of F_{12} as compared to the other terms. For fiber orientations other than 45°, the same conclusion may not be valid. The influence of an interactive term on lamina failure is better observed in the Tsai–Wu theory developed in the next section.

The Tsai–Hill theory can also be formulated based on strains, by incorporating the appropriate relationships between principal direction strains and stresses and the failure strains ($X_\varepsilon = X/E_1$, etc.), and substituting into Equation (5.5).

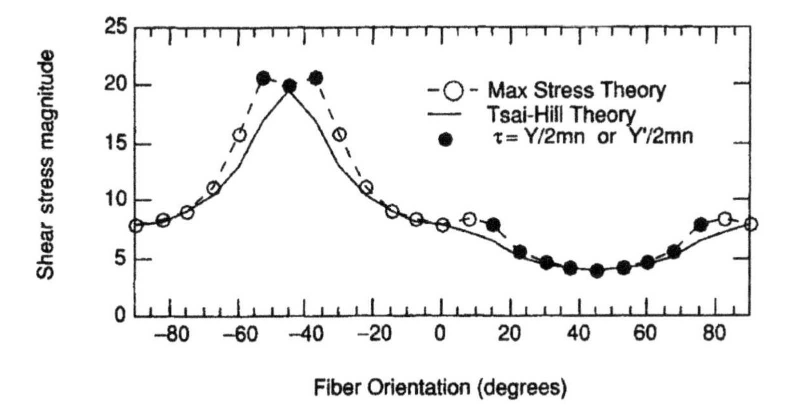

Figure E5.4-2 Comparison of maximum stress and Tsai–Hill failure criteria for pure shear.

5.5.2 Tsai–Wu tensor theory

The Tsai–Wu theory [20] has a form similar to that of several other interactive theories presented in Table 5.2. The most compact form for expressing this theory is through tensor notation:

$$F_i\sigma_i + F_{ij}\sigma_i\sigma_j = 1 \quad i,j = 1,2,\ldots,6,$$

where F_i and F_{ij} are strength tensors established through experimental procedures and are related to failure strengths in principal lamina directions. For an orthotropic lamina subjected to plane stress ($\sigma_3 = \tau_{13} = \tau_{23} = 0$), this reduces to

$$F_{11}\sigma_1^2 + 2F_{12}\sigma_1\sigma_2 + F_{22}\sigma_2^2 + F_{66}\sigma_6^2 + 2F_{16}\sigma_1\sigma_6 + 2F_{26}\sigma_2\sigma_6$$
$$+ F_1\sigma_1 + F_2\sigma_2 + F_6\sigma_6 = 1$$

The σ_6 term is the shear stress τ_{12}, as shown in Figure 5.5.

It is possible to define five of the six parameters from simple test procedures. For example, consider a load in the σ_1–ε_1 stress–strain plane as illustrated in Figure 5.6

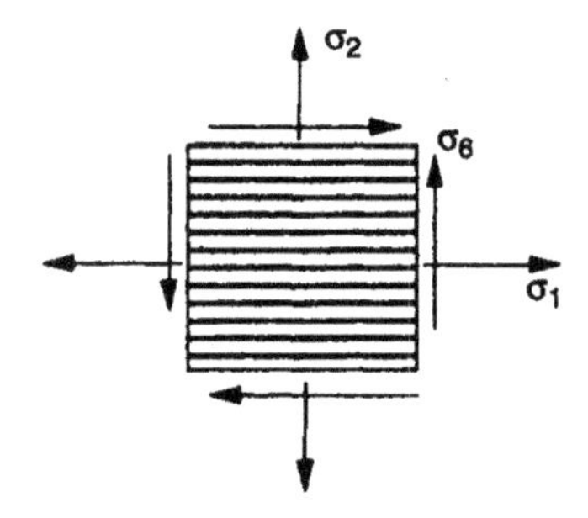

Figure 5.5 Plane stress components for failure analysis using Tsai–Wu tensor theory.

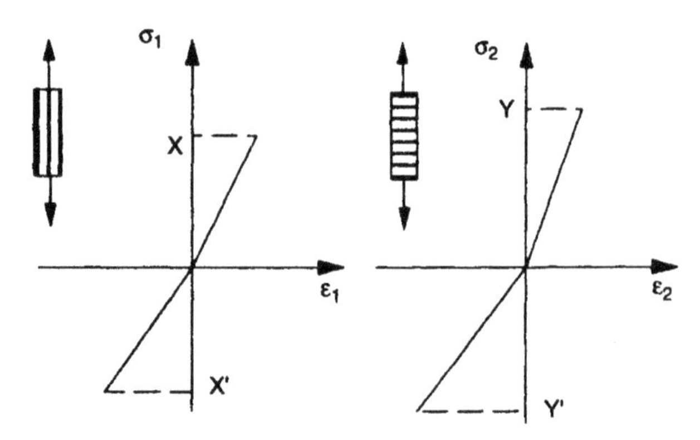

Figure 5.6 Uniaxial tension and compression tests to determine F_1, F_{11}, F_2, and F_{22}.

Assuming all other stress components are zero, the failure criterion reduces to $F_{11}X^2+F_1X=1$ for tension and $F_{11}X'^2-F_1X'=1$ for compression. Solving these equations simultaneously results in

$$F_{11}=\frac{1}{XX'}\quad F_1=\frac{1}{X}-\frac{1}{X'}$$

Following a similar procedure for the σ_2–ε_2 stress–strain space as shown in Figure 5.6 yields similar results with $F_{22}Y^2+F_2Y=1$ for tension and $F_{22}Y'^2-F_2Y'=1$ for compression. Solving these two equations yields

$$F_{22}=\frac{1}{YY'}\quad F_2=\frac{1}{Y}-\frac{1}{Y'}$$

Similarly, application of a pure shear stress results in $F_{66}S^2+F_6S=1$ for $+\tau$ and $F_{66}S'^2-F_6S'=1$ for $-\tau$. Solving these equations yields expressions identical in form to those for F_{11}, F_{22}, F_1, and F_2. Since $S=S'$, these terms are

$$F_{66}=\frac{1}{S^2}\quad F_6=0$$

In a similar manner, evaluation of the remaining first-order terms ($F_{16}\sigma_1\sigma_6$ and $F_{26}\sigma_2\sigma_6$) yields $F_{16}=F_{26}=0$. Because of this, the Tsai–Wu failure reduces to

$$F_{11}\sigma_1^2+2F_{12}\sigma_1\sigma_2+F_{22}\sigma_2^2+F_{66}\sigma_6^2+F_1\sigma_1+F_2\sigma_2=1 \tag{5.7}$$

The only remaining term to be determined is F_{12}. Wu [26] argued that this can be accomplished by applying a biaxial state of stress so that $\sigma_1=\sigma_2=\sigma$. In this case, the failure criterion becomes $F_{11}\sigma^2+2F_{12}\sigma^2+F_{22}\sigma^2+F_1\sigma+F_2\sigma=1$. Collecting terms and rearranging, $(F_{11}+2F_{12}+F_{22})\sigma^2+(F_1+F_2)\sigma=1$. This expression can now be solved for F_{12}, with

$$F_{12}=\frac{1-(F_1+F_2)\sigma-(F_{11}+F_{22})\sigma^2}{2\sigma^2}$$

Thus, F_{12} depends upon the various engineering strengths and the biaxial tensile failure stress σ. A biaxial tension test can be difficult to perform and cannot generally be used to define F_{12}. Originally, Tsai and Wu [20] suggested that a 45° off-axis tension or shear test would be good for determining F_{12}. It was later reported [27] that slight variations in the fiber orientation would completely obscure the estimates of F_{12}. The off-axis test has been shown to produce poor results for predicting F_{12} [28,29]. A theoretical test procedure by Evans and Zhang [30] has been proposed for determining F_{12}. They suggest a series of tests in which deformations in the directions transverse to the applied load are zero, as shown schematically in Figure 5.7.

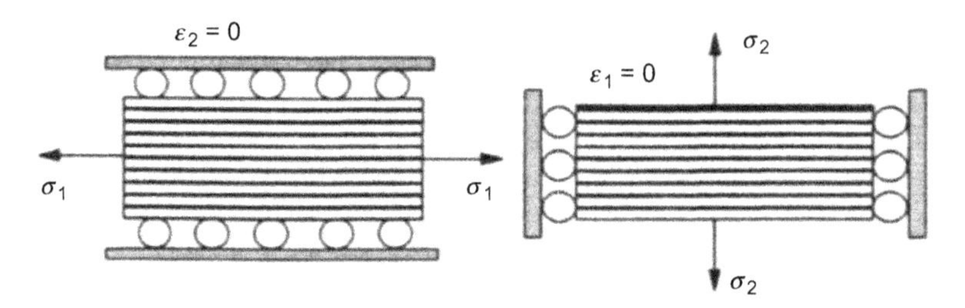

Figure 5.7 Schematic of test procedure proposed by Evans and Zhang [30] for determination of F_{12}.

The exact value of F_{12} cannot be explicitly determined, but there are bounds imposed on it by geometric considerations [31]. F_{12} can force the general form of the failure criterion to change from an ellipse to a parallel set of lines to a hyperbola, depending on its value. The interactive term is generally established from experiments in which $\tau = 0$. With $\tau = 0$, the failure criterion can be put into the form $(F_{11}\sigma_1 + F_{12}\sigma_2 + F_1)\sigma_1 + (F_{22}\sigma_2 + F_{12}\sigma_1 + F_2)\sigma_2 - 1 = 0$. A general second-degree expression of this type can define a quadratic surface. The type of surface defined depends on the sign of the determinant formed by the stress coefficients, called the discriminant [32]. The types of surface formed, as well as the discriminants, are

$$\text{Discriminant} = F_{11}F_{22} - F_{12}^2 \begin{cases} > 0 \text{ ellipse} \\ = 0 \text{ parallel lines} \\ < 0 \text{ hyperbola} \end{cases}$$

The appropriate form of a solution for defining a failure envelope is a closed surface, such as an ellipse. To determine F_{12}, a normalized interactive term is introduced:

$$F_{12}^* = \frac{F_{12}}{\sqrt{F_{11}F_{22}}}$$

The admissible range of values for F_{12}^* and its associated curve are

$-1 < F_{12}^* < 1$	(ellipse)
$F_{12}^* = 1$	(parallel lines)
$F_{12}^* < -1, \ 1 < F_{12}^*$	(hyperbola)

Using $F_{12} = F_{12}^*\sqrt{F_{11}F_{22}}$, the failure criterion can be written as

$$F_{11}\sigma_1^2 + 2F_{12}^*\sqrt{F_{11}F_{22}}\sigma_1\sigma_2 + F_{22}\sigma_2^2 + F_{66}\sigma_6^2 + F_1\sigma_1 + F_2\sigma_2 = 1$$

This equation describes a family of ellipses. F_{12}^* governs the slenderness ratio and the inclination of the major axis, which is +45° for $-F_{12}^*$ and −45° for $+F_{12}^*$. Assuming the orthotropic failure criterion described earlier is a generalization of the Von Mises failure criterion, the interactive term is best defined as $F_{12}^* = -1/2$.

The parameters selected to define F_{12} do not have to be those defined by Tsai and Hahn [31]. Wu and Stachorski [33] found that for materials such as thermoplastics and paper (which are less anisotropic than graphite/epoxy), good agreement between theory and experiment is achieved for an interactive term expressed as

$$F_{12} = -\frac{F_{11}F_{22}}{F_{11}+F_{22}}$$

This term is not the only one applicable to slightly anisotropic materials such as paper. A correlation of various strength theories with experimental results for paperboard is presented in Schuling et al. [34]. Some theories showed good correlations and some did not. In general, the form presented by Tsai and Hahn [31] appears to be better suited to more highly anisotropic materials such as graphite/epoxy than other representations.

Example 5.5 Consider a unidirectional lamina with failure strengths $X = X' = 217.7$ ksi, $Y = 5.8$ ksi, $Y' = 35.7$ ksi, and $S = 9.86$ ksi.

$$\begin{aligned}
F_{11} &= 1/XX' = 2.104 \times 10^{-11}(\text{psi})^{-2} \\
F_{22} &= 1/YY' = 4.833 \times 10^{-9}(\text{psi})^{-2} \\
F_{66} &= 1/S^2 = 1.029 \times 10^{-8}(\text{psi})^{-2} \\
F_1 &= 1/X - 1/X' = 0 \\
F_2 &= 1/Y - 1/Y' = 1.44 \times 10^{-4}(\text{psi})^{-1}
\end{aligned}$$

Assuming $F_{12}^* = -1/2, F_{12} = F_{12}^*\sqrt{F_{11}F_{22}} = -1.594 \times 10^{-10}$, the failure criterion becomes

$$\sigma_1^2 - 15.156\sigma_1\sigma_2 + 229.7\sigma_2^2 + 488.8\sigma_6^2 + 6,844,000\sigma_2 = 4.752 \times 10^{10}$$

The solution to this equation depends on the state of stress in the 1–2 plane. Figure E5.5-1 shows the resulting failure surface for cases in which $\tau = 0$ and $\tau = 1000$ psi. By varying one component of normal stress and solving the resulting equation for the other, the complete curve is generated. For the case in which $\tau = 0$, the ellipse crosses the axes at the four intercept points corresponding to X, X', Y, and Y'. This is not the case when $\tau \neq 0$. As the shear stress increases, the failure ellipse shrinks. The effect of F_{12} on predicting failure can be seen by altering F_{12}^*, as shown in Figure E5.5-2. The variety of possible failure surfaces shown here illustrates the importance of correct selection of F_{12}^* for a particular material.

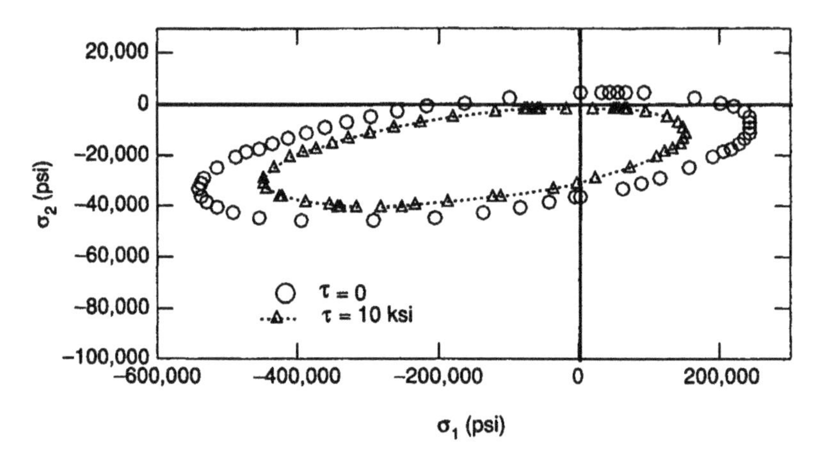

Figure E5.5-1 Failure ellipse for biaxial stress with $\tau = 0$ and $\tau \neq 0$.

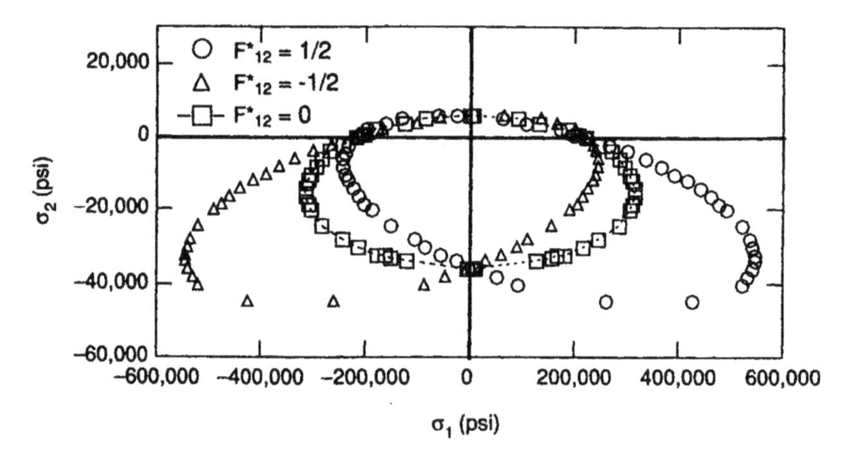

Figure E5.5-2 Effect of F^*_{12} on failure ellipses.

Example 5.6 As a variation to establishing failure ellipses, consider the unidirectional lamina in Figure E5.6. The failure strengths are the same as those previously given, and the stresses are

Figure E5.6 Unidirectional lamina subjected to σ_x.

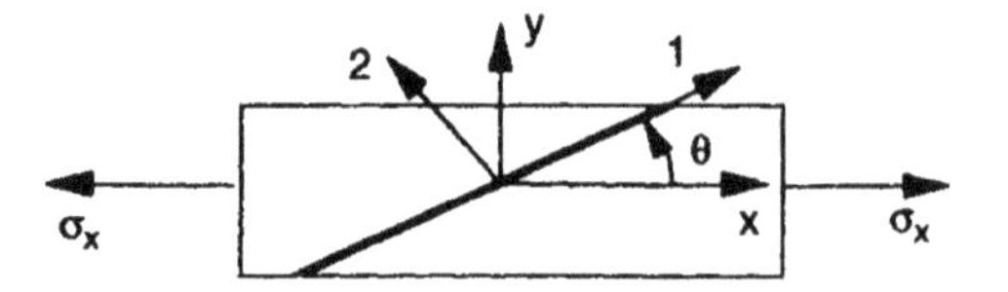

$$\begin{Bmatrix} \sigma_1 \\ \sigma_2 \\ \tau_{12} \end{Bmatrix} = \begin{bmatrix} m^2 & n^2 & 2mn \\ n^2 & m^2 & -2mn \\ -mn & mn & m^2-n^2 \end{bmatrix} \begin{Bmatrix} \sigma_x \\ 0 \\ 0 \end{Bmatrix} = \begin{Bmatrix} m^2 \\ n^2 \\ -mn \end{Bmatrix} \sigma_x$$

Substituting these stresses into the Tsai–Wu failure criterion gives $m^4F_{11}\sigma_x^2+2m^2n^2F_{12}\sigma_x^2+n^4F_{22}\sigma_x^2+m^2n^2F_{66}\sigma_x^2+n^2F_2\sigma_x=1$. Collecting terms:

$$\left(m^4F_{11}+2m^2n^2F_{12}+n^4F_{22}+m^2n^2F_{66}\right)\sigma_x^2+n^2F_2\sigma_x=1$$

For $\theta=45°$, σ_x is found by solving the preceding equation:

$$\begin{aligned} &0.25(F_{11}+2F_{12}+F_{22}+F_{66})\sigma_x^2+0.5F_2\sigma_x=1 \\ &3.705\times10^{-9}\sigma_x^2+7.2\times10^{-5}\sigma_x=1 \\ &\sigma_x^2+19.43\times10^3\sigma_x-2.699\times10^8=0 \end{aligned}$$

Solving this quadratic results in

$$\sigma_x=9.73\text{ksi},\quad -28.8\text{ksi}$$

It is instructive to compare these results with the failure stress predicted from the Tsai–Hill theory. For $\theta=45°$, Equation (5.5) becomes

$$0.25\left[\frac{1}{X^2}-\frac{1}{X^2}+\frac{1}{Y^2}+\frac{1}{S^2}\right]\sigma_x^2=1$$

Substituting the appropriate failure strengths and solving results in a predicted failure stress of $\sigma_x=9.99$ ksi. This is within 7% of the σ_x predicated from the Tsai–Wu theory.

The Tsai–Wu theory predicts two roots, one for a tensile stress and one for a compressive stress. For compression, the predicted failure stress (using X' and Y' instead of X and Y) is $\sigma_x=-19$ ksi.

The Tsai–Wu failure theory can also be expressed in terms of strain. The procedure for transformation into the strain space is identical to that described for the Tsai–Hill theory in the previous section.

5.5.2.1 *Strength ratios*

In the design and analysis of components using composite materials, the failure theory only answers the question of failure for a given state of stress. From a design viewpoint, it is equally important to identify the additional stress to which a component may be subjected prior to failure. The general form of the Tsai–Wu failure theory is

$$F_{11}\sigma_1^2+2F_{12}\sigma_1\sigma_2+F_{22}\sigma_2^2+F_{66}\sigma_6^2+F_1\sigma_1+F_2\sigma_2=1$$

It defines a go-no-go condition on failure for a specific state of stress. The left-hand side can be either

$$\text{Left-hand side}\begin{cases} <1 \text{ (no failure)} \\ =1 \text{ (criterion is met)} \\ >1 \text{ (not physically possible)} \end{cases}$$

Tsai and Hahn [31] extended the use of a failure theory by defining an additional variable, the strength ratio R, such that $\sigma_{i\text{a}} = R\sigma_i$. The subscript a means allowable or ultimate stress. The strength ratio R has features that make it a convenient parameter to incorporate into a failure theory: (1) if the applied stress or strain is zero, $R = \infty$; (2) the stress/strain level is safe if $R > 1$; (3) the stress/strain level is unsafe if $R = 1$; and (4) there is no physical significance if $R < 1$. An analogous development can be made for strain [31].

The strength ratio can be used to define the allowable stress or strain ($R = 1$), and a factor of safety. If, for example, $R = 2$, the applied stress may be double before failure.

For a specimen in uniaxial tension, the generalized criterion for failure involving strength ratios can be expressed as

$$\left[F_{11}\sigma_1^2 + 2F_{12}\sigma_1\sigma_2 + F_{22}\sigma_2^2 + F_{66}\sigma_6^2\right]R^2 + \left[F_1\sigma_1 + F_2\sigma_2\right]R = 1$$

The solution of this equation involves two roots, R and R', applicable to either the tensile (R) *or* the compressive (R') failure strength.

Example 5.7 Consider the specimen in Example 5.6. The only nonzero applied stress is assumed to be $\sigma_x = 5$ ksi. Failure strengths and strength coefficients (F_{11}, etc.) with $F_{12}^* = -1/2$ are

$$\begin{array}{llll} X = X' = 218\,\text{ksi} & Y = 5.8\,\text{ksi} & Y' = 35.7\,\text{ksi} & S = 9.86\,\text{ksi} \\ F_{11} = 2.104 \times 10^{-11} & F_{22} = 4.833 \times 10^{-9} & F_{12} = -1.598 \times 10^{-10} & \\ F_{66} = 1.0285 \times 10^{-8} & F_1 = 0 & F_2 = 1.440 \times 10^{-4} & \end{array}$$

The stresses in the principal material directions are defined by stress transformation as

$$\begin{Bmatrix} \sigma_1 \\ \sigma_2 \\ \sigma_6 \end{Bmatrix} = \begin{Bmatrix} m^2 \\ n^2 \\ -mn \end{Bmatrix} \sigma_x$$

Incorporating strength ratios into the conventional form of the Tsai–Wu failure criterion results in

$$\left[F_{11}m^4 + 2F_{12}m^2n^2 + F_{22}n^4 + F_{66}m^2n^2\right]\sigma_x^2R^2 + n^2F_2\sigma_xR = 1$$

The solution of this equation depends upon the angle θ. For example, consider the two possible angles of $\theta = 0°$ and $90°$:

$$\begin{aligned}\theta=0^\circ: &\quad R=43.6,\ \text{and}\ \sigma_a(\theta=0)=5R=5(43.6)=218\,\text{ksi}\\ \theta=90^\circ: &\quad R=1.16,\ -7.121,\ \text{which implies}:\\ \text{For tension}: &\quad \sigma_a=5(1.16)=5.8\,\text{ksi}\\ \text{For compression}: &\quad \sigma_a=5(-7.12)=-35.7\,\text{ksi}\end{aligned}$$

These results are comparable to those described in Example 5.5. They indicate the points on the 1–2 stress plane where the failure ellipse crosses an axis, which are the failure strengths for this material in the 1 or 2 directions. Assume the fibers are oriented at $\theta=45°$ (as in Example 5.6). For this case, $R=1.874$, -5.76, and the allowables are

For tension: $\sigma_x=5(1.874)=9.37$ ksi
For compression: $\sigma_x=5(-5.76)=-28.8$ ksi

These results are the same as those given in Example 5.6, and they illustrate the use of strength ratios.

5.6 Buckling

Most of the efforts associated with buckling of composites have centered around plates and shells. A review of the buckling of laminated composite plates and shells is given by Leissa [35]. Buckling failures associated with lamina have not been investigated to the same extent as those associated with laminates. A survey of fiber microbuckling is presented by Shuart [36]. The problem encountered in buckling is that it generally results from a geometric instability rather than a material failure due to overstressing. A failure theory based on stress (or strain) is not applicable for buckling analysis. Initial investigations of the fiber microbuckling problem were formulated by Rosen [37] and are based on the procedures established by Timoshenko [38] for columns on an elastic foundation. The procedures described in Rosen [37] are also presented in Jones [39] and form the basis for one of the discussions presented herein. The phenomenon of fiber buckling can be defined as *fiber instability followed by a decreased capability of the fibers to carry load, with the final result being matrix failure by overstressing*. The model from which an analysis procedure can be developed is shown in Figure 5.8. It is assumed that the fibers are equally spaced, and each is subjected to the same compressive load P. It is further assumed that the fibers can be modeled as plates of thickness h and an out-of-plane width of unity. The fibers are separated by matrix plates of thickness $2c$. There are two distinct modes of buckling in lamina analysis, and the same model is used for each. The out-of-plane dimensions are disregarded, making the model two-dimensional. In each mode, failure results from instability, which in turn causes the matrix to fail. The manner in which the matrix deforms motivates the failure modes termed *extension* and *shear* modes.

Extension mode. In the extension mode, all fibers are assumed to buckle with the same wavelength, but adjacent fibers are out of phase. The most prominent deformation is extension of the matrix seen in Figure 5.9. An energy approach is used to

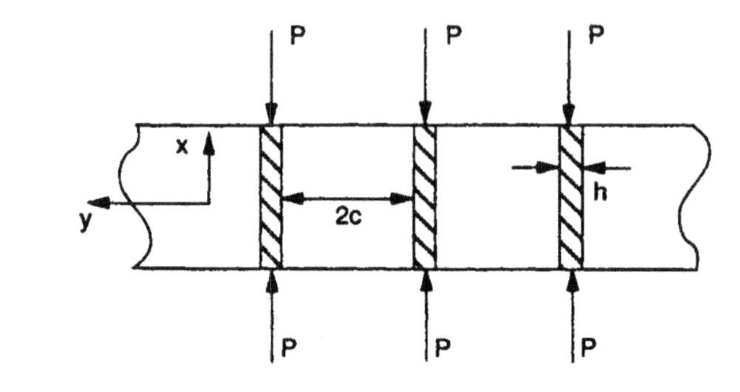

Figure 5.8 Fiber-buckling model.

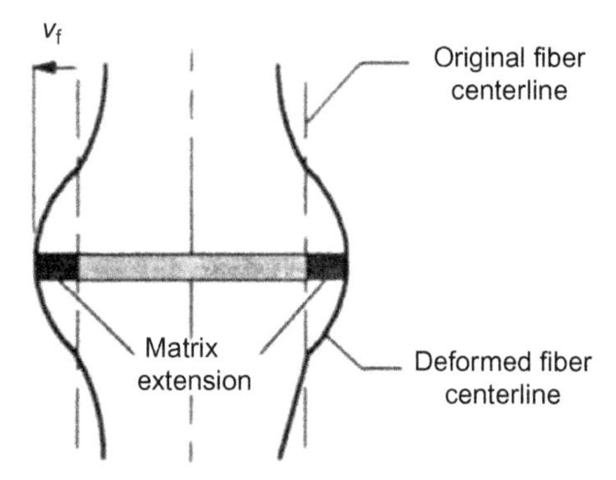

Figure 5.9 Model used to define the extension mode of fiber buckling.

develop a solution for this mode of failure. The work and energy terms required are ΔU_f = change in strain energy of the fiber, ΔU_m = change in strain energy of the matrix, and ΔW_e = change in work due to external loads.

These energies are related by $\Delta W_e = \Delta U_f + \Delta U_m$.

It is assumed that in its buckled state, the displacement of the fiber from its original position is expressed as

$$V_f = \sum_{n=1}^{\infty} a_n \sin\frac{n\pi x}{L}$$

The change in strain energy of the fiber is obtained by energy methods. Since the fiber is assumed to be a flat plate, the strain energy is written in terms of an elastic modulus and an area moment of inertia as

$$\Delta U_f = \frac{1}{2E_f I_f}\int_0^L (M)^2 \mathrm{d}x,$$

where $M = E_f I_f(\mathrm{d}^2V_f/\mathrm{d}x^2)$. Substituting this expression for M and noting

$$\frac{d^2V_f}{dx^2} = \sum -a_n \frac{n^2\pi^2}{L^2} \sin \frac{n\pi x}{L}$$

results in

$$\Delta U_f = \frac{E_f I_f}{2} \int_0^L \left(\frac{d^2V_f}{dx^2}\right)^2 dx = \frac{E_f I_f \pi^4}{4L^3} \sum n^4 a_n^2$$

The extensional strain and stress in the matrix (in the y-direction) are $\varepsilon_y^m = V_f/c$ and $\varepsilon_y^m = E_m V_f/c$. In the x-direction, the changes in strain energy are assumed negligible for the matrix. The total change in strain energy for the matrix is

$$\Delta U_m = \frac{1}{2}\int_A \varepsilon_y^m \sigma_y^m dA = \frac{1}{2}\int_A \left(\frac{E_m}{c^2}\right)[V_f]^2 dA = \left(\frac{E_m}{c}\right)\int_0^L (V_f)^2 dx = \frac{E_m L}{2c}\sum a_n^2$$

The external work is found by considering the total compressed state of the fiber. In the compressed state, it is assumed that the actual fiber length does not change. The end of the fiber travels a distance δ as shown in Figure 5.10. The work is $W_e = P\delta$.

The displacement δ is found by considering the length of the fiber, established from

$$L = \int ds = \int \sqrt{dx^2 + dy^2} = \int \sqrt{1 + (dy/dx)^2}dx$$

Expanding the radical in a binomial series results in

$$L \approx \int_0^{L-\delta} \left[1 + \frac{1}{2}\left(\frac{dy}{dx}\right)^2\right] dx \approx (L - \delta) + \frac{1}{2}\int_0^{L-\delta} \left(\frac{dy}{dx}\right)^2 dx$$

Solving for δ yields

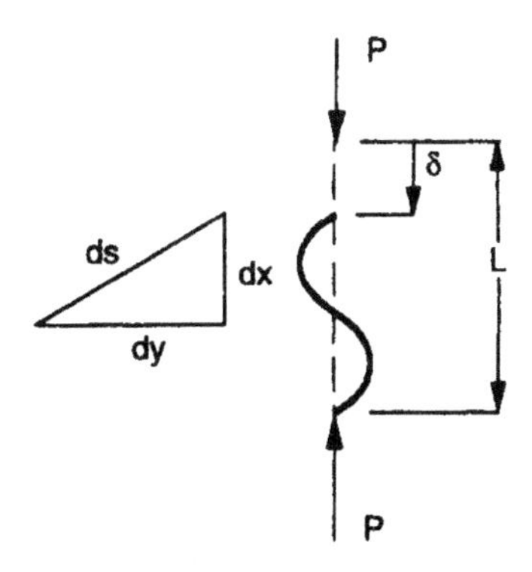

Figure 5.10 Displacement of fiber centerline in applied load direction.

$$\delta = \frac{1}{2}\int_0^{L-\delta} (y')^2 \mathrm{d}x$$

Since $L \gg \delta$, $L - \delta \rightarrow L$, and $\delta \approx 1/2\int_0^L (y')^2 \mathrm{d}x$, where $y' = \mathrm{d}V_\mathrm{f}/\mathrm{d}x$. The work due to external forces is

$$\Delta W_\mathrm{e} = \frac{P}{2}\int_0^L \left(\frac{\mathrm{d}V_\mathrm{f}}{\mathrm{d}x}\right)^2 \mathrm{d}x = \frac{P\pi^2}{4L}\sum n^2 a_n^2$$

Combining these energy expressions produces the general form of the buckling equation. The value of P can be established from that expression as

$$P = \frac{\frac{E_\mathrm{f} I_\mathrm{f} \pi^4}{L^2}\sum\left(n^2 a_n^2\right) + 2\frac{E_\mathrm{m} L^2}{c}\sum\left(a_n^2\right)}{\pi^2 \sum\left(n^2 a_n^2\right)}$$

The fibers are modeled as flat plates of unit width. The inertia term I_f can be replaced by $I_\mathrm{f} = h^3/12$. If it is now assumed that P reaches the minimum critical value required for a particular since wave, the mth wave, the preceding equation can be put into a different form. The expression for the critical load can be expressed as

$$P_\mathrm{CR} = \frac{\pi^2 E_\mathrm{f} h^3}{12L^2}\left\{m^2 + \left(\frac{24L^4 E_\mathrm{m}}{\pi^4 c h^3 E_\mathrm{f}}\right)\frac{1}{m^2}\right\} \tag{5.8}$$

The critical buckling load is a function of material properties, length, and m. The minimum wave number for buckling is determined from $\partial P_\mathrm{CR}/\partial m = 0$, subject to the condition $\partial^2 P_\mathrm{CR}/\partial m^2 > 0$. These operations result in $m^2 = \sqrt{24L^4 E_\mathrm{m}/\pi^4 c h^3 E_\mathrm{f}}$. For certain material combinations and geometries, this expression can yield unrealistically large values of m for the extension failure mode. Using the preceding expression for m^2, the critical buckling load is

$$P_\mathrm{CR} = \sqrt{\frac{2E_\mathrm{f} E_\mathrm{m} h^3}{3c}} \tag{5.9}$$

Equation (5.9) can be modified to reflect changes in volume fractions. The volume fraction of fibers can be modeled as $\upsilon_\mathrm{f} = h/(h + 2c)$. This can be rearranged so that $c = h(1 - \upsilon_\mathrm{f})/2\upsilon_\mathrm{f}$. Using this expression for C, Equation (5.9) becomes

$$P_\mathrm{CR} = \sqrt{\frac{\upsilon_\mathrm{f} E_\mathrm{f} E_\mathrm{m}}{3(1 - \upsilon_\mathrm{f})}} \tag{5.10}$$

Associated with the critical buckling load is a critical stress for the lamina, which is presented after the second mode of buckling is considered.

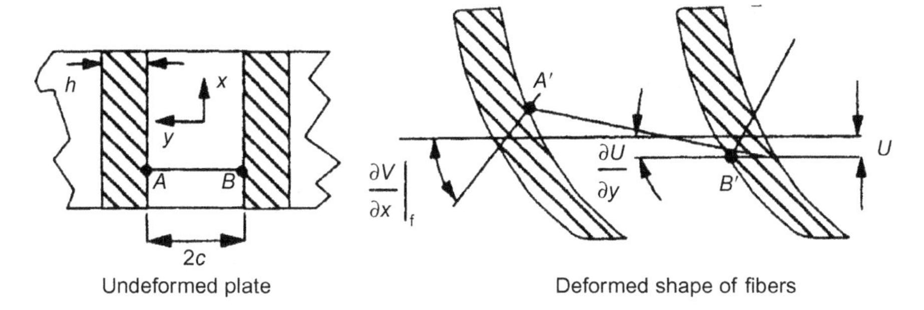

Figure 5.11 Model used to define the shear mode of fiber buckling.

Shear mode. In this mode, the fibers are assumed to buckle with the same wavelength and are considered to be in phase with each other. The matrix deformation is predominantly shear as illustrated in Figure 5.11. The buckling load is determined from energy methods. It is assumed that the matrix displacement in the y-direction is independent of y. The shear strain in the matrix is expressed as

$$\gamma_{xy}^{\mathrm{m}} = \partial V_{\mathrm{m}}/\partial x + \partial U_{\mathrm{m}}/\partial y.$$

Since the transverse displacement is independent of y, $\partial V_{\mathrm{m}}/\partial x = \partial V_{\mathrm{f}}/\partial x$. Therefore, $\partial U_{\mathrm{m}}/\partial y = U_{\mathrm{m}}(c) - U_{\mathrm{m}}(-c)/2c$, where $U_{\mathrm{m}}(c)$ and $U_{\mathrm{m}}(-c)$ are defined in terms of $\mathrm{d}V_{\mathrm{f}}/\mathrm{d}x$ by $U_{\mathrm{m}}(c) = (h/2)(\mathrm{d}V_{\mathrm{f}}/\mathrm{d}x)$ and $U_{\mathrm{m}}(-c) = -(h/2)(\mathrm{d}V_{\mathrm{f}}/\mathrm{d}x)$. This results in $\partial U_{\mathrm{m}}/\partial x = (h/2c)(\mathrm{d}V_{\mathrm{f}}/\mathrm{d}x)$. Using these expressions, the shear strain and stress in the matrix are $\gamma_{xy}^{\mathrm{m}} = (1 + h/2c)(\mathrm{d}V_{\mathrm{f}}/\mathrm{d}x)$ and $\tau_{xy}^{\mathrm{m}} = G_{\mathrm{m}}\gamma_{xy}^{\mathrm{m}}$. The change in strain energy in the matrix is due to matrix shear and is

$$\Delta U_{\mathrm{m}} = \frac{1}{2}\int_A \tau_{xy}^{\mathrm{m}}\gamma_{xy}^{\mathrm{m}}\,\mathrm{d}A = \frac{1}{2}\int_A G_{\mathrm{m}}\left(\gamma_{xy}^{\mathrm{m}}\right)^2\mathrm{d}A$$

Using the expression for the transverse displacement of the fiber from the extensional case, the shear strain is

$$\gamma_{xy}^{\mathrm{m}}\left(1 + \frac{h}{2c}\right)\sum a_n\left(\frac{n\pi}{L}\right)\cos\left(\frac{n\pi x}{L}\right)$$

The subsequent change in strain energy of the matrix is

$$\Delta U_{\mathrm{m}} = G_{\mathrm{m}}c\left(1 + \frac{h}{2c}\right)^2\left(\frac{\pi^2}{2L}\right)\sum a_n^2 n^2$$

The changes in strain energy for the fiber (ΔU_{f}) and the work due to external loads (ΔW_{e}) remain unchanged from the previous case. Following the same procedures as before, the critical buckling load is

$$P_{CR} = \frac{hG_m}{\upsilon_f \upsilon_m} + \frac{h\pi^2 E_f}{12}\left(\frac{mh}{L}\right)^2 \tag{5.11}$$

The term L/m is called the *buckle wavelength.* The second term in this expression is small when the buckle wavelength is large compared to the fiber width h. For this reason, the second term is generally neglected and the critical buckling load reduces to

$$P_{CR} = \frac{hG_m}{\upsilon_f \upsilon_m} \tag{5.12}$$

The critical stresses for the extensional and shear modes of buckling can be approximated from the critical buckling loads. Defining the stress as load divided by area, and recognizing that the area in question is that of the fiber, which is approximated as h, the critical stress is $\sigma_{CR} = P_{CR}/h$. For the extension and shear modes, the critical buckling stress is

$$\text{Extension}: \sigma_{CR} = 2\sqrt{\frac{\upsilon_f E_f E_m}{3(1-\upsilon_f)}}$$
$$\text{Shear}: \sigma_{CR} = \frac{G_m}{\upsilon_f \upsilon_m}$$

In order to assess which of these two modes is actually the most critical, both cases are examined for typical values of elastic moduli. The fiber modulus is generally greater than that of the matrix. Therefore, it is assumed that $E_f = 20E_m$. Since $v_m = 1 - \upsilon_f$, the critical stress for extension is

$$\sigma_{CR} = 2\sqrt{\frac{20\upsilon_f E_m^2}{3\upsilon_m}} = 5.16E_m\sqrt{\frac{\upsilon_f}{\upsilon_m}}$$

For the shear mode, the shear modulus of the matrix (assuming isotropic constituent material behavior) can be expressed as $G_m = E_m/2(1+\nu_m)$. Assuming $\nu_m = 0.35$, the critical buckling stress for shear becomes

$$\sigma_{CR} = \frac{0.37E_m}{\upsilon_f \upsilon_m}$$

From these expressions, it appears as if the shear mode is the most probable mode of buckling. For a glass/epoxy lamina with $E_f = 10.6 \times 10^6$ psi, $E_m = 0.5 \times 10^6$ psi, and $G_m = 0.185 \times 10^6$ psi, the critical buckling stresses for the extension and shear modes are

$$\text{Extension}: \sigma_{CR} = 2.645 \times 10^6\sqrt{\frac{\upsilon_f}{\upsilon_m}}$$
$$\text{Shear}: \sigma_{CR} = \frac{0.185 \times 10^6}{\upsilon_f \upsilon_m}$$

Plotting these two stresses as a function of υ_f yields the results shown in Figure 5.12.

The shear mode is seen to be the most critical for a large range of fiber volume fractions. At very low (and generally unrealistic) fiber volume fractions, the extension mode dominates. Although this illustration does not include experimental results, the dominance of the shear mode as a failure mechanism has been experimentally investigated and verified by Greszczuk [40]. From experimental and theoretical studies of graphite fibers in various matrices, he formulated the following conclusions: (1) microbuckling in the shear mode dominates for low-modulus resin systems, (2) transverse tension failures (including fiber "splitting") result with intermediate-modulus resins, and (3) the reinforcements fail in compression with high-modulus resins.

An expanded form of the critical stress for the shear failure mode is obtained by dividing the critical load in Equation (5.11) by the fiber area h. Comparing this form of σ_{CR} to that predicted by the Euler buckling formula [40] results in the following relationship between σ_{Euler} and $\sigma_{microbuckling}$:

$$\sigma_{Euler} = \frac{w^2}{\Lambda L^2 + h^2}\sigma_{microbuckling}, \tag{5.13}$$

where w, L, and h are the specimen width, length, and fiber diameter, respectively, and

$$\Lambda = \frac{12G_m}{\upsilon_f \upsilon_m \pi^2 E_f}$$

A more comprehensive discussion of the relationship between Euler buckling and microbuckling is presented by Greszczuk [41]. In many cases, $\sigma_{Euler} > \sigma_{microbuckling}$.

The model presented herein assumes the fibers to be initially straight. Predictions resulting from this model tend to be larger than experimental measurements, which prompted Davis [42] to develop a model incorporating fiber geometry allowing for initial fiber curvature. He investigated a boron/epoxy composite and evaluated both

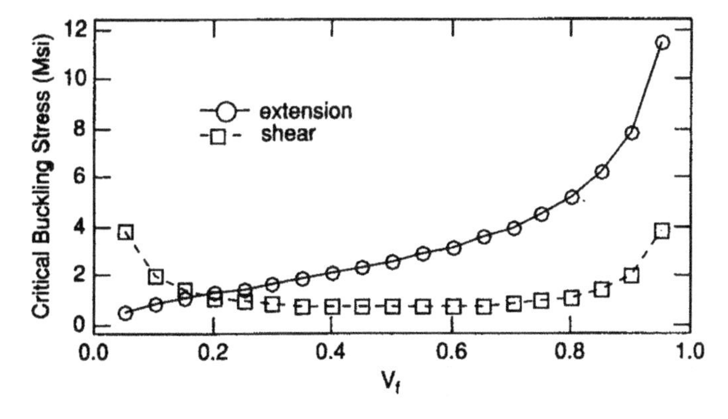

Figure 5.12 Critical buckling stress for extension and shear modes.

delamination and shear instability. The conclusion drawn from his study was that shear instability is critical.

An alternative viewpoint [43–48] is that compression failures may be the result of fiber kinking, which can be linked to kink bands formed along existing glide planes in the load direction. A mode of failure associated with fiber buckling and kinking is *shear crippling*. Shear crippling appears as a shear failure on a plane at some angle to the loading direction if viewed macroscopically. A microscopic investigation reveals that shear crippling is frequently the result of kink-band formation as schematically illustrated in Figure 5.13. The analysis of kink-related failures generally requires a nonlinear material model and analysis techniques beyond the scope of this text.

Experimental findings [49,50] strongly suggest the origin of failures for uniaxial specimens loaded in the fiber direction by compression is the free edge of the lamina. Based on these observations, Wass et al. [51] developed a model for incorporating the free edge as the origin of a buckling failure. The model used is shown in Figure 5.14. Analysis of this model consists of three parts, which require elasticity formulations beyond the scope of this text. The logical progression from a single-fiber composite to an isolated single-buckled fiber and its relationship to the matrix material during buckling are easily understood. The combination of several similar models (one of which is the free surface model in Figure 5.14) results in a complete solution. Results from Wass et al. [51] are lower than those presented by Rosen [37], and at high fiber volume fractions, the results violate some of the general beam theory assumptions used to establish the model.

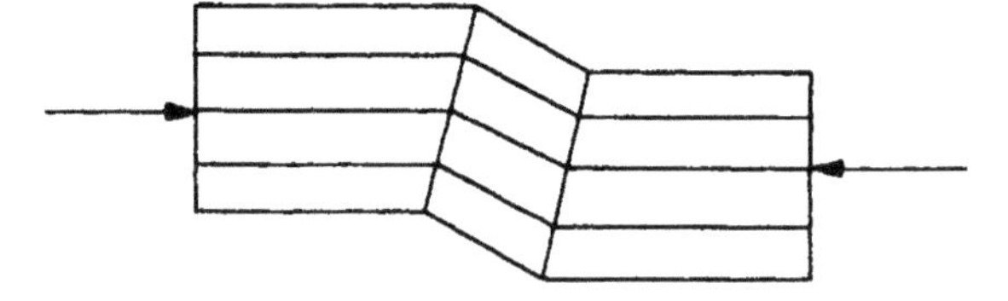

Figure 5.13 Schematic of kink-band geometry (after Hahn and Williams [49]).

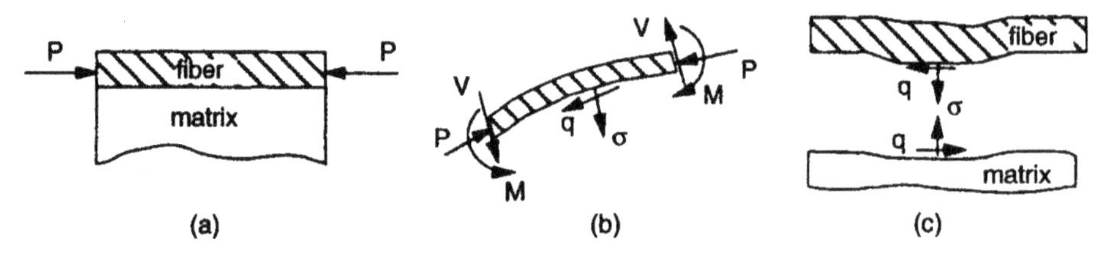

Figure 5.14 Model of (a) single fiber composite, (b) isolated buckled fiber, and (c) matrix configuration at buckling (after Wass et al. [51]).

5.7 Design examples incorporating failure analysis

As an example of design and failure analysis, consider the filament-wound, closed-end pressure vessel with an outside diameter d and an internal pressure P. The winding angle θ is assumed to be positive and is allowed to vary in the range $0° < \theta < 90°$. The required lamina thickness as a function of winding angle, $t = f(\theta)$, is to be determined so that failure does not result. Figure 5.15 illustrates the pressure vessel and the associated state of stress in the wall.

From classical thin-walled pressure vessel theory, the longitudinal and circumferential stresses are the only ones present in the x–y plane, and the state of stress for the vessel is shown in Figure 5.15, where $\sigma_x = Pd/4t$, $\sigma_y = Pd/2t$, and $\tau_{xy} = 0$.

Assume the material may be either boron/epoxy or glass/epoxy. Both the Tsai–Hill and Tsai–Wu failure theories are used for analysis. The internal operating pressure is $P = 100$ psi. Introducing a constant $a = Pd/2 = 1500$ psi-in., the stresses in the 1–2 plane are

$$\begin{Bmatrix} \sigma_1 \\ \sigma_2 \\ \tau_{12} \end{Bmatrix} = [T_\sigma] \begin{Bmatrix} a/2t \\ a/t \\ 0 \end{Bmatrix} = \begin{Bmatrix} (m^2/2) + n^2 \\ (n^2/2) + m^2 \\ mn/2 \end{Bmatrix} \left(\frac{a}{t}\right)$$

The failure strengths for each material are:

Material	X (ksi)	X′ (ksi)	Y (ksi)	Y′ (ksi)	S (ksi)
Boron/epoxy	185	363	8.8	44.7	15.2
Glass/epoxy	187	119	6.67	25.3	6.5

5.7.1 Tsai–Hill criterion

$$\sigma_1^2 - \sigma_1\sigma_2 + \sigma_2^2 r^2 + \tau_{12}^2 s^2 = X^2$$

Substituting the stress just given into this expression yields

$$\left[\left(\frac{m^2}{2} + n^2\right)^2 - \left(\frac{m^2}{2} + n^2\right)\left(\frac{n^2}{2} + m^2\right) + r^2\left(\frac{n^2}{2} + m^2\right)^2 + s^2\left(\frac{mn}{2}\right)^2\right]\left(\frac{a}{t}\right)^2 = X^2$$

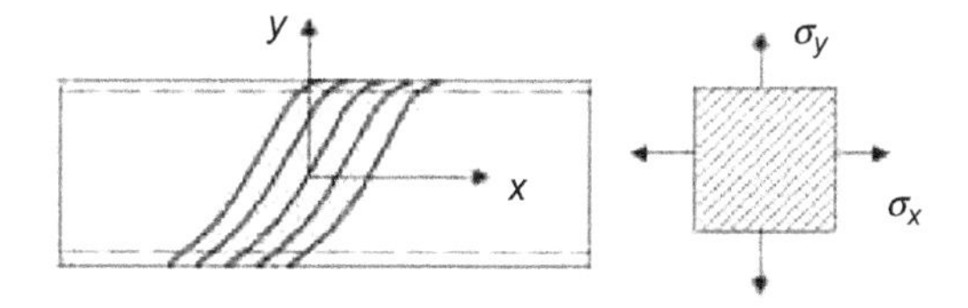

Figure 5.15 Filament wound pressure vessel and associated state of stress.

This reduces to

$$\left[\frac{m^4(4r^2-1)}{4}+\frac{n^4(r^2+2)}{4}+\frac{m^2n^2(4r^2+s^2-1)}{4}\right]\left(\frac{a}{t}\right)^2=X^2$$

Assuming only tensile stresses, and using boron/epoxy, the strength ratios are $r=X/Y=21.02$ and $s=X/S=12.2$. Therefore, the failure criterion is

$$\left[442m^2+111n^2+479m^2n^2\right]\left(\frac{a}{t}\right)^2=\left(185\times10^3\right)^2$$

Substituting $a=Pd/2$ and solving for t yields

$$t=8.1\times10^{-3}\sqrt{442m^4+111n^4+479m^2n^2}$$

For glass/epoxy, assuming only tension, with strength ratios of $r=28.03$ and $s=28.76$, the failure theory becomes

$$\left[7852m^2+197n^2+991m^2n^2\right]\left(\frac{a}{t}\right)^2=\left(187\times10^3\right)^2$$

Substituting $a=Pd/2$ and solving for t results in

$$t=8.02\times10^{-3}\sqrt{785m^4+197n^4+991m^2n^2}$$

5.7.2 Tsai–Wu criterion

$$F_{11}\sigma_1^2+2F_{12}\sigma_1\sigma_2+F_{22}\sigma_2^2+F_{66}\sigma_6^2+F_1\sigma_1+F_2\sigma_2=1$$

This criterion requires more coefficients. Substituting the preceding stresses into this equation and collecting terms yields, after some manipulation

$$\left\{m^4\left(\frac{F_{11}+4F_{12}+4F_{22}}{4}\right)+n^4\left(\frac{4F_{11}+4F_{12}+F_{22}}{4}\right)+m^2n^2\left(\frac{4F_{11}+10F_{12}+4F_{22}+F_{66}}{4}\right)\right\}a^2 + \left\{m^2\left(\frac{F_1+2F_2}{2}\right)+n^2\left(\frac{2F_1+F_2}{2}\right)\right\}at=t^2$$

For boron/epoxy, the strength coefficients with $F_{12}^*=-1/2$ are $F_{11}=1.489\times10^{-11}$, $F_{22}=2.54\times10^{-9}$, $F_{66}=4.328\times10^{-9}$, $F_{12}=-9.72\times10^{-11}$, $F_1=2.65\times10^{-6}$, and $F_2=9.13\times10^{-5}$. Using these, the failure equation becomes

$$\left\{m^4\left(2.44\times10^{-9}\right)+n^4\left(5.526\times10^{-10}\right)+m^2n^2\left(3.39\times10^{-9}\right)\right\}a^2 + \left\{m^2\left(9.26\times10^{-5}\right)+n^2\left(4.83\times10^{-5}\right)\right\}at=t^2$$

For glass/epoxy, the coefficients are $F_{11}=4.494\times10^{-11}$, $F_{22}=5.989\times10^{-9}$, $F_{66}=2.367\times10^{-8}$, $F_{12}=-2.594\times10^{-10}$, $F_1=-3.06\times10^{-6}$, and $F_2=1.104\times10^{-4}$, and the failure equation becomes

$$\left\{m^4\left(5.74\times10^{-9}\right)+n^4\left(1.275\times10^{-9}\right)+m^2n^2\left(1.13\times10^{-8}\right)\right\}a^2$$
$$+\left\{m^2\left(1.088\times10^{-4}\right)+n^2\left(5.214\times10^{-5}\right)\right\}at=t^2$$

The variation of wall thickness with θ is shown in Figure 5.16. As seen, there is little difference between the predicted wall thickness for either theory. The main difference is in the material selection.

As a second example, consider the support bracket shown in Figure 5.17. It is designed to safely sustain an applied load of 500 lb, directed as shown. The material is carbon/epoxy (T300/5208) with properties defined in Table 4.2 (reference [42] in Chapter 4). The bracket is assumed to have an arbitrary fiber orientation. The purpose of this analysis is to establish a design envelope for fiber orientations as a function of applied load. Although the design load is 500 lb, it is initially assumed that the applied force has an unknown magnitude F. The design envelope is to be established using both the Tsai–Wu and maximum strain failure theories. Assume that from previously

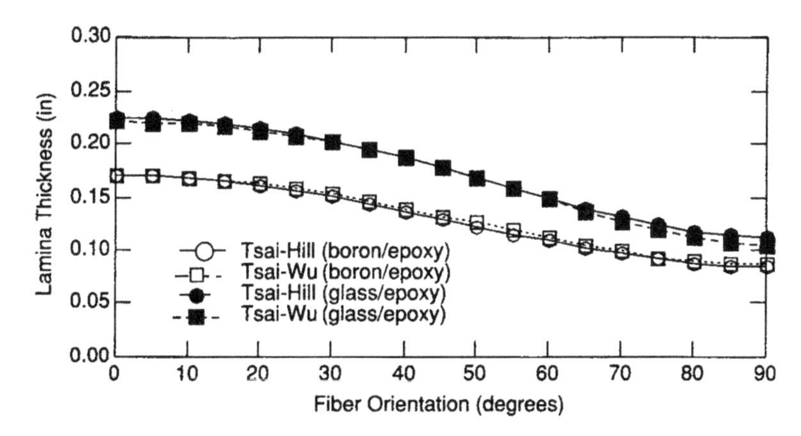

Figure 5.16 Variation of wall thickness as predicted by Tsai–Wu and Tsai–Hill failure theories.

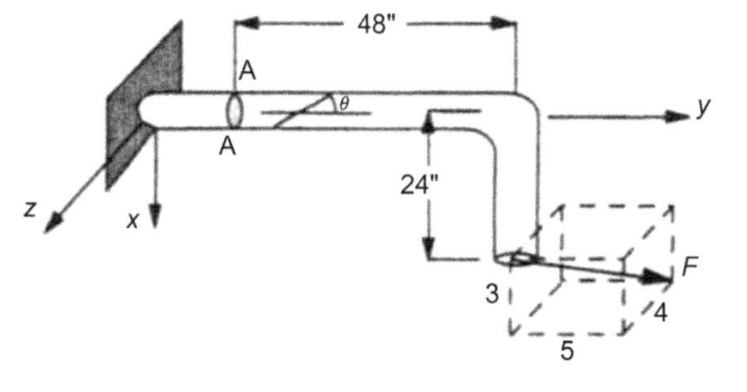

Figure 5.17 Support bracket with arbitrary fiber orientation.

designed components of this type, it is known that section AA in Figure 5.17 is the critical section.

The initial step in the analysis is to define the state of stress at critical points in cross-section A–A, and subsequently, the principal direction stresses and strains for failure assessment. In order to establish the state of stress, we first define the loads and moments acting on section A–A. Considering the free body diagram in Figure 5.18, we see that the applied force vector at the free end of the bracket will result in a force vector which must satisfy the condition of static equilibrium of forces,

$$\sum F = F_R + \frac{F}{\sqrt{50}}(3i + 5j - 4k) = 0$$

In a similar manner, the moment at section A–A must satisfy the condition of equilibrium of moments, given by

$$\sum M_A = M_R + (24i + 48j)x\frac{F}{\sqrt{50}}(3i + 5j - 4k) = 0$$

Solving these expressions for the vector components of F_R and M_R results in the cross-section loads and moments expressed in terms of the applied force F as shown in Figure 5.18.

The transverse shear forces (in the x- and z-directions) are assumed to have negligible effects upon the state of stress in section A–A. For convenience, the two bending moments, being vector components, are combined to yield a single moment oriented with respect to the x-axis. This resultant moment will produce tensile and compressive stresses at two points (B and C) as shown in Figure 5.19. From geometry, the cross-sectional area, second area moment of inertia, and polar moment of inertia are $A = 1.374$ in.^{2}, $I = 0.537$ in.^{4}, and $J = 1.074$ in.^{4}.

The state of stress at points B and C is determined by combining the stress components resulting from the individual loads and moments, as established from elementary strength of materials considerations.

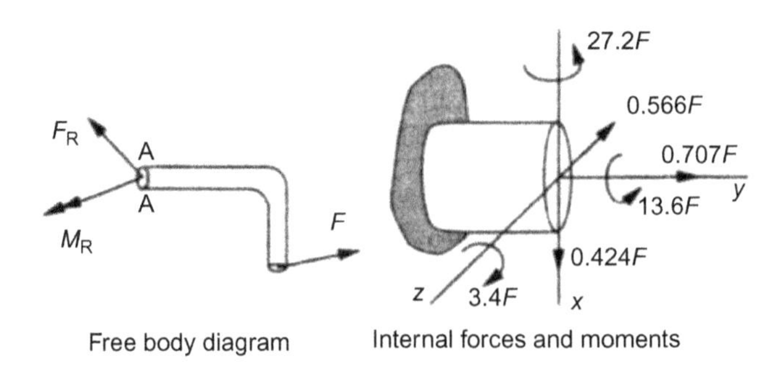

Figure 5.18 FBD and internal loads and moments at section A–A.

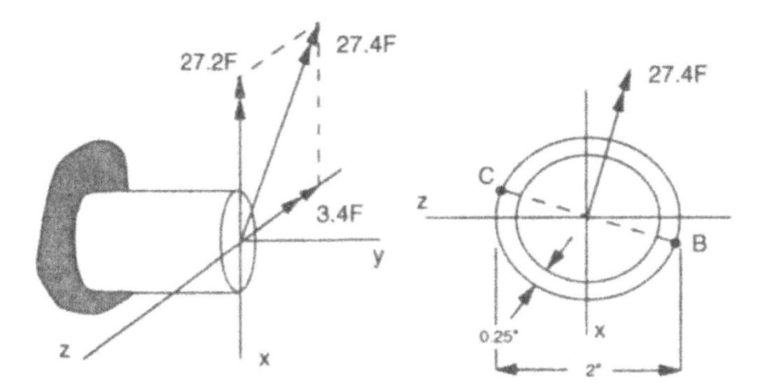

Figure 5.19 Resultant bending moment at section A–A.

$$\sigma_F = \frac{F_y}{A} = \frac{0.707}{1.374} = 0.515F \quad \sigma_M = \frac{Mr}{I} = \frac{27.4F(1.0)}{0.537} = 51.0F$$

$$\tau = \frac{Tr}{J} = \frac{13.6F(1.0)}{1.074} = 12.7F$$

Combining the normal stress components at points B and C and including the shear stress result in the state of stress at both points, as shown in Figure 5.20. The normal stress exists only in the y-direction. It is assumed that the x-axis has been redefined to coincide with the line connecting points B and C and is indicated as x'.

The state of stress in the principal material directions at either point is established by stress transformation:

$$\begin{Bmatrix} \sigma_1 \\ \sigma_2 \\ \tau_{12} \end{Bmatrix} = \begin{bmatrix} m^2 & n^2 & 2mn \\ n^2 & m^2 & -2mn \\ -mn & mn & m^2-n^2 \end{bmatrix} \begin{Bmatrix} 0 \\ \sigma_y \\ \tau_{xy} \end{Bmatrix} = \begin{Bmatrix} m^2\sigma_y + 2mn\tau_{xy} \\ n^2\sigma_y - 2mn\tau_{xy} \\ mn\sigma_y + (m^2-n^2)\tau_{xy} \end{Bmatrix}$$

Incorporating these stresses into the Tsai–Wu theory, assuming $F^*_{12} = -1/2$, and using the failure strengths in Table 4.2 result in

$$2.11\times10^{-11}(\sigma_1)^2 - 1.595\times10^{-10}(\sigma_1)(\sigma_2) + 4.83\times10^{-9}(\sigma_2)^2 + 1.03 \times 10^{-8}(\tau_{12})^2 + 1.44\times10^{-4}(\sigma_2) = 1$$

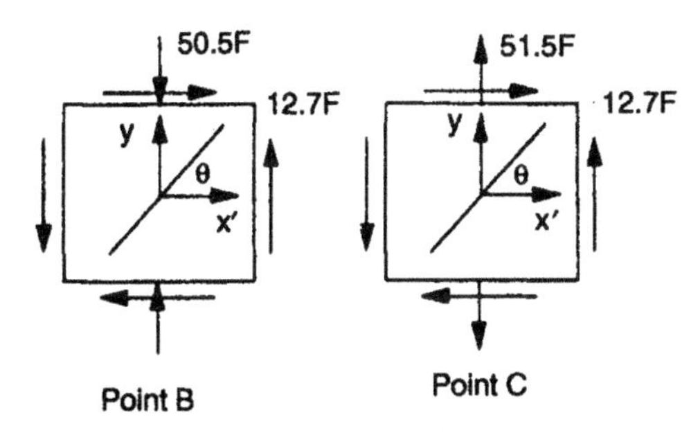

Figure 5.20 State of stress at points B and C.

An alternative form of this expression can be established by using explicit σ_1, σ_2, and τ_{12} terms given earlier. The failure theory becomes

$$2.11\times10^{-11}\left(m^2\sigma_y+2mn\tau_{xy}\right)^2-1.595\times10^{-10}\left(m^2\sigma_y+2mn\tau_{xy}\right)\left(n^2\sigma_y-2mn\tau_{xy}\right)$$
$$+4.83\times10^{-9}\left(n^2\sigma_y-2mn\tau_{xy}\right)^2+1.03\times10^{-8}\left(mn\sigma_y+(m^2-n^2)\tau_{xy}\right)^2$$
$$+1.44\times10^{-4}\left(n^2\sigma_y-2mn\tau_{xy}\right)=1$$

Either form of the failure theory can be expressed in terms of the applied load F, since σ_y and τ_{xy} are functions of F. An explicit solution for F is established by solving one of the foregoing relationships for a specific fiber orientation. The results differ from point B to C, since one experiences a normal tensile stress and the other a normal compressive stress. The principal direction stresses as a function of F and the actual failure load satisfying the preceding criterion for points B and C are summarized in Table 5.3 for selected fiber orientations.

The principal direction strains and stresses are related by

$$\begin{Bmatrix}\varepsilon_1\\ \varepsilon_2\\ \gamma_{12}\end{Bmatrix}=[S]\begin{Bmatrix}\sigma_1\\ \sigma_2\\ \tau_{12}\end{Bmatrix}$$

The stresses established from strength of materials approximations are related to loads and geometry, while strains are associated with deformations. Incorporating the material properties into the general expression just given results in the principal direction strains:

$$\begin{Bmatrix}\varepsilon_1\\ \varepsilon_2\\ \gamma_{12}\end{Bmatrix}=\begin{bmatrix}0.038 & -0.011 & 0\\ -0.011 & 0.689 & 0\\ 0 & 0 & 0.962\end{bmatrix}\begin{Bmatrix}\sigma_1\\ \sigma_2\\ \tau_{12}\end{Bmatrix}$$

when ε_1 is replaced with X_ε, etc., this expression becomes

$$\begin{Bmatrix}X_\varepsilon\\ Y_\varepsilon\\ S_\varepsilon\end{Bmatrix}=\begin{bmatrix}0.038 & -0.011 & 0\\ -0.011 & 0.689 & 0\\ 0 & 0 & 0.962\end{bmatrix}\begin{Bmatrix}\sigma_1\\ \sigma_2\\ \tau_{12}\end{Bmatrix}$$

Although X_ε and X'_ε are identical, Y_ε and Y'_ε are not. Therefore, the preceding expression must be examined according to the sign of ε_2. The sign of each strain component

Table 5.3 Summary of failure loads at points B and C

	Point B				Point C			
θ	σ_1	σ_2	τ_{12}	F (lb)	σ_1	σ_2	τ_{12}	F (1b)
0	0	$-50.5F$	$12.7F$	633	0	$51.5F$	$12.7F$	111
15	$2.9F$	$-53.5F$	$-1.6F$	664	$9.8F$	$41.7F$	$23.9F$	128
30	$-1.6F$	$-48.9F$	$-15.5F$	618	$23.9F$	$27.6F$	$30.3F$	164
45	$-12.6F$	$-37.9F$	$-25.3F$	542	$38.5F$	$13.1F$	$25.8F$	259
60	$-26.9F$	$-23.6F$	$-28.2F$	500	$49.6F$	$1.9F$	$15.9F$	564
75	$-40.7F$	$-9.7F$	$-23.6F$	532	$54.4F$	$-2.9F$	$1.9F$	4026
90	$-50.5F$	0	$-12.7F$	764	$51.5F$	0	$-12.7F$	763

is established from $\{\varepsilon\}=[S]\{\sigma\}$. Each component of strain, expressed in terms of the applied load F, is presented in Table 5.4 for various fiber orientation angles. The units of each strain component are μin./in.

The sign of ε_2 varies with fiber orientation. Taking this into account, the maximum strain failure criterion for each sign of ε_2 is

$$\begin{array}{cc} \varepsilon_2 > 0 & \varepsilon_2 < 0 \\ \begin{Bmatrix} \varepsilon_1 \\ \varepsilon_2 \\ \gamma 12 \end{Bmatrix} = \begin{Bmatrix} 8287 \\ 3893 \\ 9481 \end{Bmatrix} & \begin{Bmatrix} \varepsilon_1 \\ \varepsilon_2 \\ \gamma 12 \end{Bmatrix} = \begin{Bmatrix} 8287 \\ 23{,}960 \\ 9481 \end{Bmatrix} \end{array}$$

For $\varepsilon_2 < 0$, the negative sign is omitted, with the understanding that these components of strain are compressive. The load required to cause failure is established by correlating the information for individual strain components with the failure strain for that case. Take, for example, point B with a fiber orientation of 30°. The relationship between the normal strain (noting that $\varepsilon_2 < 0$) and failure strain for each component results in failure loads determined from

$$\begin{Bmatrix} 0.456F \\ 33.7F \\ 14.9F \end{Bmatrix} = \begin{Bmatrix} 8268 \\ 23{,}960 \\ 9481 \end{Bmatrix}$$

Solving each of these results in three possible failure loads. The lowest one is the obvious choice for this example. A summary of the failure load for each point, based on each condition, is given in Table 5.5. In this table, the load at which failure initially results is underlined.

From this, it is obvious that failure generally results from either a normal strain in the 2-direction or the shear strain. A direct comparison of results predicted by the maximum strain theory and the Tsai–Wu theory is presented in Figure 5.21. It is apparent that point C is the critical point within the cross section. For fiber orientations, less than approximately 60°, both theories predict failure loads below the design load of 500 lb. The fiber orientations corresponding to a safe operating condition are those for which the curves lie above the 500-lb threshold. Since the maximum strain theory

Table 5.4 **Summary of strains at points B and C**

	Point B			Point C		
θ	ε_1	ε_2	γ_{12}	ε_1	ε_2	γ_{12}
0	0.535F	−34.5F	12.2F	−0.55F	35.5F	12.2F
15	0.679F	−36.9F	−1.5F	−0.07F	28.6F	23.0F
30	0.456F	−33.7F	−14.9F	0.62F	18.8F	29.1F
45	−0.077F	−25.9F	−24.3F	1.32F	8.6F	24.8F
60	−0.772F	−15.9F	−27.1F	1.86F	0.78F	15.3F
75	−1.440F	−6.3F	−22.7F	2.09F	−2.6F	1.8F
90	−1.920F	0.53F	12.2F	1.96F	−0.55F	−12.2F

Table 5.5 Summary loads required to produce failure according to the maximum strain theory failure

	ε_1		ε_2		γ_{12}	
θ	**Point B**	**Point C**	**Point B**	**Point C**	**Point B**	**Point C**
0	15,500	15,070	694	110	777	777
15	12,200	118,400	649	136	6160	412
30	18,200	13,400	711	207	640	326
45	107,700	6300	925	453	390	382
60	10,740	4500	1507	4991	350	620
75	5760	3970	3803	9215	420	5180
90	4300	4200	7345	43,563	780	780

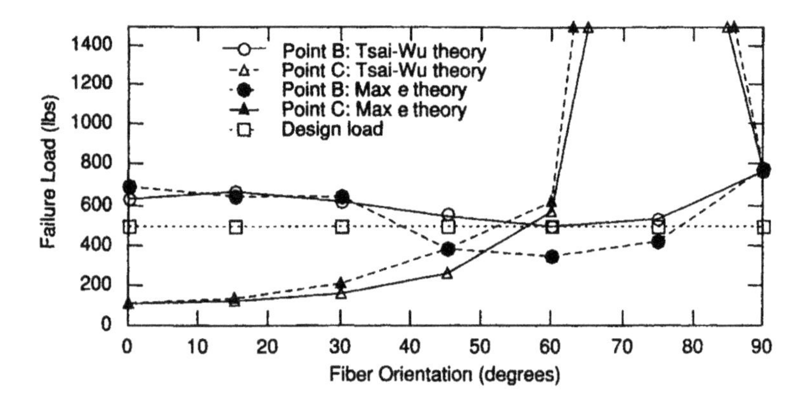

Figure 5.21 Failure envelope for support bracket established by Tsai–Wu and maximum strain theories.

predicts failure loads less than 500 lb at point B for $40° < \theta < 80°$, the safest fiber orientation would be $\theta > 80°$. The assessment of which failure theory is most acceptable will influence the final design choice.

In general, structural members made from unidirectional composite materials are initially analyzed in much the same manner as isotropic materials. The first step requires an explicit identification of the forces and moments which exist on a plane containing the points being assessed. Subsequently, the state of stress at each point must be identified. Failure analysis can then be conducted using any (or all) of the methods presented herein, or other suitable procedures.

In performing an analysis involving interactive failure theories, the strength parameters are essential. For convenience, the strength parameters for the Tsai–Hill and Tsai–Wu failure theories are presented in Tables 5.6 and 5.7 for each of the materials in Table 4.2 and/or Table 4.3. In these tables, the appropriate references *from* Chapter 4 are cited. The English units are presented along with the SI units (which appear in parentheses).

Material [Chapter 4 reference]	$1/X^2$	$1/(X')^2$	$1/Y^2$	$1/(Y')^2$	$1/S^2$
Carbon/Epoxy					
T300/5208 [42]	2.11×10^{-11}	2.11×10^{-11}	2.97×10^{-8}	7.85×10^{-10}	1.02×10^{-8}
	(4.44×10^{-19})	(4.44×10^{-19})	(6.25×10^{-16})	(1.65×10^{-17})	(2.16×10^{-16})
[43]	1.0×10^{-10}	8.2610^{-11}	6.25×10^{-8}	5.18×10^{-9}	1.24×10^{-8}
	(2.10×10^{-18})	(1.74×10^{-18})	(1.31×10^{-15})	(1.07×10^{-16})	(2.59×10^{-16})
T300/934 [43]	8.73×10^{-11}	9.07×10^{-11}	—	—	4.57×10^{-9}
	(1.84×10^{-18})	(1.91×10^{-18})			(9.61×10^{-17})
T300/SP-286	2.90×10^{-11}	7.61×10^{-12}	1.29×10^{-8}	5.00×10^{-10}	4.33×10^{-9}
[43]	(4.49×10^{-19})	(7.80×10^{-19})	(3.43×10^{-16})	(2.25×10^{-17})	(1.93×10^{-16})
AS/3501	2.27×10^{-11}	2.27×10^{-11}	1.78×10^{-8}	1.12×10^{-9}	5.49×10^{-9}
[42, 43]	(4.78×10^{-19})	(4.78×10^{-19})	(3.74×10^{-16})	(2.36×10^{-17})	(1.16×10^{-16})
Glass/Epoxy					
Scotchply: Type 1002 [42]	2.42×10^{-11}	1.28×10^{-10}	4.94×10^{-8}	3.42×10^{-9}	9.07×10^{-9}
	(8.87×10^{-19})	(2.69×10^{-18})	(1.04×10^{-15})	(7.18×10^{-17})	(1.93×10^{-16})
Type	1.48×10^{-11}	4.76×10^{-11}	2.60×10^{-8}	1.19×10^{-9}	5.10×10^{-9}
SP-250-S29	(3.12×10^{-19})	(1.00×10^{-18})	(5.41×10^{-16})	(2.5×10^{-17})	(1.06×10^{-16})
E-glass/epoxy	2.86×10^{-11}	7.06×10^{-11}	2.23×10^{-8}	1.56×10^{-9}	2.37×10^{-8}
[43]	(6.02×10^{-19})	(1.49×10^{-18})	(4.75×10^{-16})	(3.29×10^{-17})	(4.98×10^{-16})
S-glass/XP-251	1.20×10^{-11}	3.46×10^{-11}	8.26×10^{-9}	1.19×10^{-9}	1.24×10^{-8}
[43]	(2.52×10^{-19})	(7.28×10^{-19})	(1.73×10^{-16})	(2.50×10^{-17})	(2.60×10^{-16})
Boron/Epoxy					
B(4)/5505	2.99×10^{-11}	7.59×10^{-12}	1.26×10^{-8}	1.17×10^{-9}	1.06×10^{-8}
[42]	(6.30×10^{-19})	(1.60×10^{-19})	(2.69×10^{-16})	(2.450×10^{-17})	(2.23×10^{-16})
[43]	2.90×10^{-11}	7.61×10^{-12}	1.29×10^{-8}	5.00×10^{-10}	4.33×10^{-9}
	(6.10×10^{-19})	(1.60×10^{-19})	(2.69×10^{-16})	(1.05×10^{-17})	(9.07×10^{-17})
Aramid/Epoxy					
Kevlar 49/EP	2.42×10^{-11}	8.60×10^{-10}	3.30×10^{-7}	1.69×10^{-8}	4.11×10^{-8}
[42, 43]	(5.10×10^{-19})	(1.81×10^{-17})	(6.94×10^{-15})	(3.56×10^{-16})	(8.65×10^{-16})

Units are $(psi)^{-2}$ for English and $(Pa)^{-2}$ for SI (coefficients in parentheses).

Table 5.7 **Strength coefficients for Tsai–Wu theory**

Material [Chapter 4 reference]	F_1	F_2	F_{11}	F_{22}	$\sqrt{F_{11}F_{22}}$	F_{66}
Carbon/Epoxy						
T300/5208 [42]	0	1.44×10^{-4}	2.11×10^{-11}	4.83×10^{-9}	3.19×10^{-10}	1.02×10^{-8}
	(0)	(2.09×10^{-8})	(4.44×10^{-19})	(1.02×10^{-16})	(6.72×10^{-18})	(2.16×10^{-16})
[43]	9.09×10^{-7}	1.78×10^{-4}	9.09×10^{-11}	1.80×10^{-8}	1.28×10^{-9}	1.24×10^{-8}
	(1.32×10^{-10})	(2.59×10^{-8})	(1.91×10^{-18})	(3.76×10^{-16})	(2.68×10^{-17})	(2.59×10^{-16})
T300/934 [43]	-1.78×10^{-7}	—	8.90×10^{-10}	—	—	4.57×10^{-9}
	(-2.62×10^{-11})		(1.87×10^{-18})			(9.61×10^{-17})
T300/SP-286	2.63×10^{-6}	9.13×10^{-5}	1.49×10^{-11}	2.54×10^{-9}	1.94×10^{-10}	4.33×10^{-9}
[43]	(-1.70×10^{-10})	(1.38×10^{-8})	(6.31×10^{-19})	(8.78×10^{-17})	(7.44×10^{-18})	(1.93×10^{-16})
AS/3501	0	9.99×10^{-5}	2.27×10^{-11}	4.46×10^{-9}	3.18×10^{-10}	5.49×10^{-9}
[42, 43]	(0)	(1.45×10^{-8})	(4.78×10^{-19})	(9.39×10^{-17})	(6.70×10^{-18})	(1.16×10^{-16})
Glass/Epoxy						
Scotchply: Type	-4.81×10^{-6}	1.64×10^{-4}	7.33×10^{-11}	1.30×10^{-8}	9.76×10^{-10}	9.07×10^{-9}
1002 [42]	(-6.98×10^{-10})	(2.38×10^{-8})	(1.54×10^{-18})	(2.73×10^{-16})	(2.05×10^{-17})	(1.93×10^{-16})
Type SP-250-S29	-3.05×10^{-6}	1.27×10^{-4}	2.65×10^{-11}	5.56×10^{-9}	3.84×10^{-10}	5.10×10^{-9}
	(-4.41×10^{-10})	(1.83×10^{-8})	(5.59×10^{-19})	(1.16×10^{-16})	(8.06×10^{-18})	(1.06×10^{-16})
E-glass/epoxy	-3.06×10^{-6}	1.10×10^{-4}	4.49×10^{-11}	5.90×10^{-9}	5.15×10^{-10}	2.37×10^{-8}
[43]	(-4.42×10^{-10})	(1.61×10^{-8})	(9.45×10^{-19})	(1.25×10^{-16})	(1.09×10^{-17})	(4.98×10^{-16})
S-glass/XP-251	-2.42×10^{-6}	5.64×10^{-5}	2.04×10^{-11}	3.14×10^{-9}	2.53×10^{-10}	1.24×10^{-8}
[43]	(-3.52×10^{-10})	(8.16×10^{-9})	(4.28×10^{-18})	(6.58×10^{-17})	(5.31×10^{-18})	(2.60×10^{-16})
Boron/Epoxy						
B(4)/5505	2.71×10^{-6}	7.82×10^{-5}	1.51×10^{-11}	3.84×10^{-9}	2.40×10^{-10}	1.06×10^{-8}
[42]	(3.94×10^{-10})	(1.14×10^{-8})	(3.18×10^{-19})	(8.12×10^{-17})	(5.08×10^{-18})	(2.23×10^{-16})
[43]	2.63×10^{-6}	9.13×10^{-5}	1.49×10^{-11}	2.54×10^{-9}	1.94×10^{-10}	4.33×10^{-9}
	(3.81×10^{-10})	(1.32×10^{-8})	(3.13×10^{-19})	(5.32×10^{-17})	(4.08×10^{-18})	(9.07×10^{-17})
Aramid/Epoxy						
Kevlar 49/EP	-2.44×10^{-6}	4.45×10^{-4}	1.44×10^{-10}	7.47×10^{-8}	3.28×10^{-9}	4.11×10^{-8}
[42, 43]	(-3.54×10^{-9})	(6.45×10^{-8})	(3.04×10^{-18})	(1.57×10^{-15})	(6.91×10^{-17})	(8.65×10^{-16})

5.8 Problems

The material properties and failure strengths for all composite material systems in the following problems are found in Tables 4.2 and 4.3.

5.1 An E-glass/epoxy laminate is subjected to the state of stress shown. Determine if failure will occur according to the maximum stress, maximum strain, Tsai–Hill, and Tsai–Wu (with $F_{12}^{*}=-1/2$) theories, assuming a fiber orientation of
(A) 30
(B) −30
(C) 60°

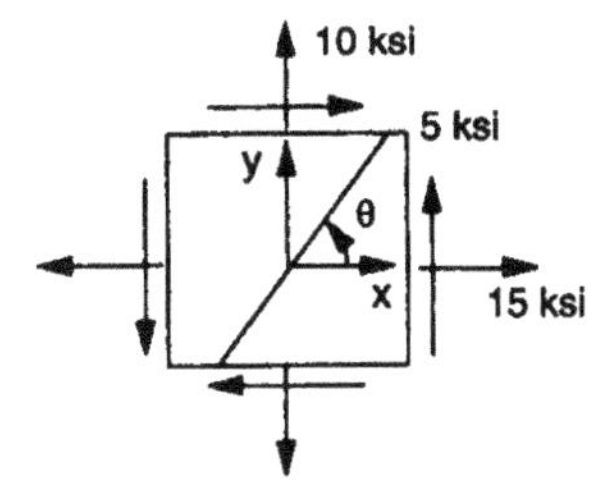

5.2 Assume that stress ratios p and q are defined as $p=\sigma_y/\sigma_x$ and $q=\tau_{xy}/\sigma_x$. A unidirectional laminate of arbitrary positive fiber orientation θ is subjected to the state of stress shown. Use the maximum stress criteria to determine the angle θ at which failure would occur for each possible failure condition (X,Y,S).

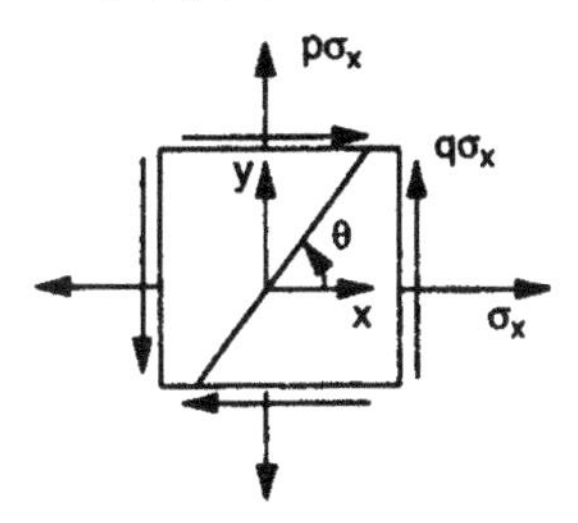

5.3 For each of the following states of stress, determine if failure will occur using the Tsai–Hill criteria for an S-glass/XP-251 lamina.

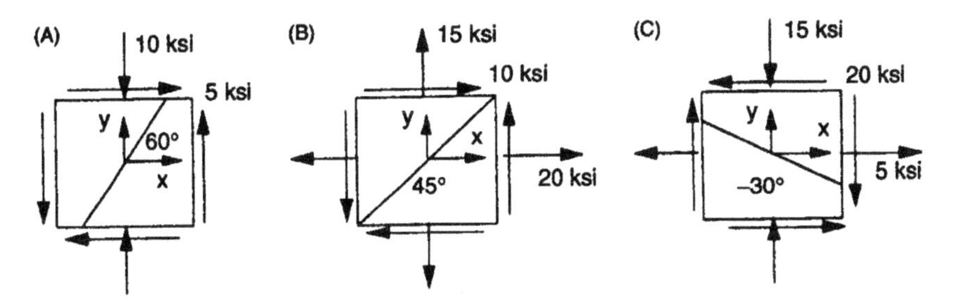

5.4 Work Problem 5.3 using $F_{12}^{*}=-1/2$ in the Tsai–Wu failure criteria.

5.5 For each of the following, the fiber orientation θ is arbitrary ($-90°\leq\theta\leq 90°$). Plot the failure stress σ_x as a function of θ using the Tsai–Wu failure theory with $F_{12}^{*}=-1$ and

$F^*_{12} = +1$. What conclusions regarding the interactive term can be drawn from this plot? Assume that the lamina is made from Scotchply-1002.

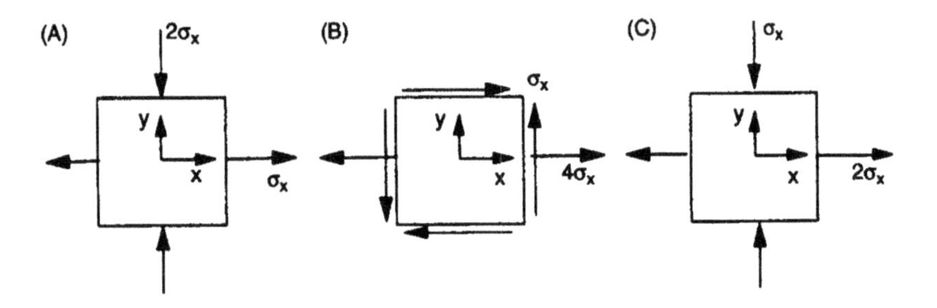

5.6 Use the Tsai–Wu theory to plot the failure stress σ as a function of the arbitrary fiber angle θ ($-90° \leq \theta \leq 90°$). Assume the material is Kevlar 49/epoxy, and $F^*_{12} = -1/2$.

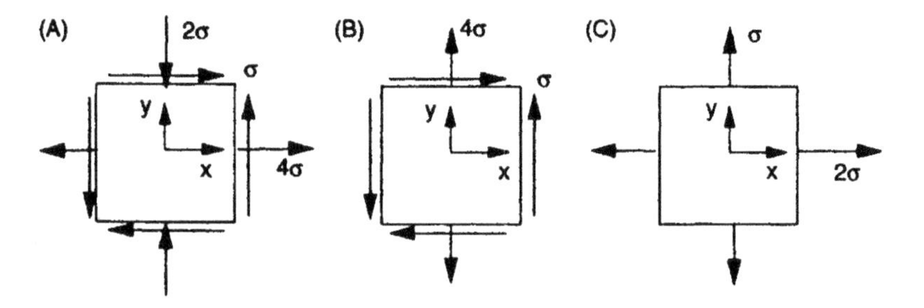

5.7 Determine the stress required to produce failure using the Tsai–Wu (with $F^*_{12} = -1/2$), Cowin, and Hoffman failure theories (see Table 5.2). Assume the material has properties from reference [42] in Chapter 4.

(A) T300/5208, $\theta = 30°$
(B) Boron/epoxy, $\theta = 60°$

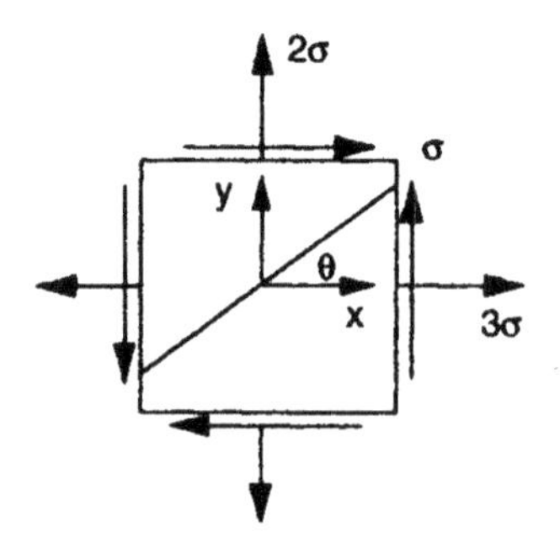

5.8 A filament-wound composite pressure vessel is to be made from unidirectional AS/3501 graphite/epoxy. The winding angle is $0° \leq \theta \leq 90°$. The pressure vessel has closed ends and an internal pressure P. In addition, a torque T is applied as indicated. The torque will result in a shear stress expressed as $\tau = 2a/t$, where $a = pd/2 = 1500$ psi-in. The diameter $d = 30$ in. Plot the required lamina thickness as a function of angle θ to ensure a safe design using

(A) Maximum stress theory
(B) Tsai–Hill theory

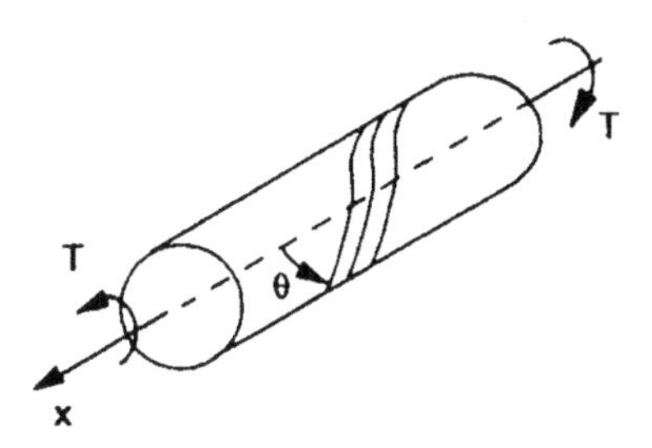

5.9 Assume the closed-end pressure vessel of Problem 5.8 is also subjected to an applied bending moment as shown. The bending stress is known to be $\sigma_B = 5a/6t$, where $a = 1500$ psi-in. Determine the variation of thickness with fiber orientation θ. Do not forget to take into account that the bending stress is either tensile or compressive, depending upon which circumferential position is being evaluated. Assess the failure using
(A) Tsai–Hill theory
(B) Tsai–Wu theory ($F_{12}^{*} = -1/2$)

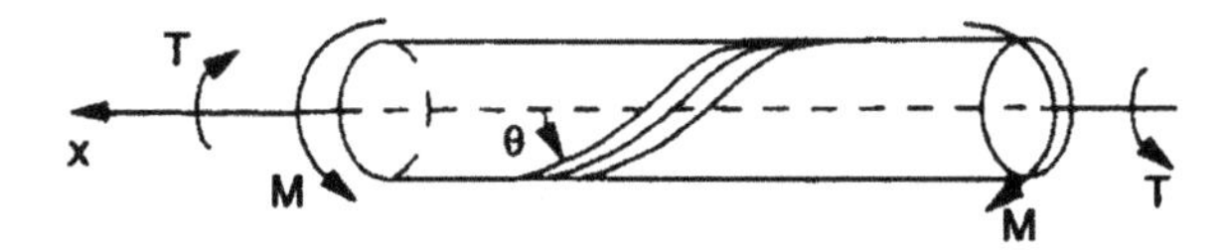

5.10 A unidirectional lamina with an arbitrary positive fiber orientation θ is placed between two rigid walls in its stress-free state. As the temperature is changed, the walls remain the same distance apart. The rollers will never lose contact with the walls. Determine and plot the temperature change ΔT required to produce failure as a function of the fiber angle θ using the material and failure theory given.
(A) T300/5208 (Chapter 4, reference [21]), Tsai–Hill
(B) E-glass/epoxy (Chapter 4, reference [43]), Tsai–Wu with $F_{12}^{*} = -1/2$

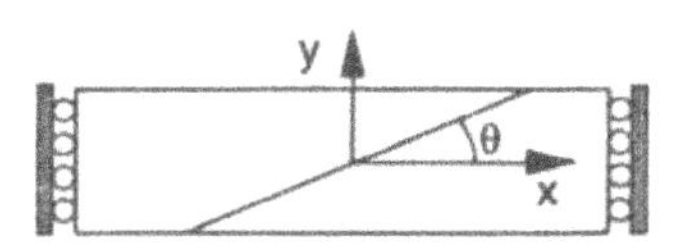

5.11 A support structure made of a unidirectional composite is subjected to the load shown. The fiber orientation is $\theta = 60°$ from the vertical y-axis. Use Tsai–Wu failure theory with $F_{12}^{*} = -1/2$ to determine the maximum applied load which the structure can support based on an analysis of a plane through section A–A. Use elementary strength of materials approximations to define the states on plane A–A prior to performing the failure analysis. Neglect transverse shear and assume the material is
(A) T300/5208 (Chapter 4, reference [42])
(B) E-glass/epoxy (Chapter 4, reference [43])

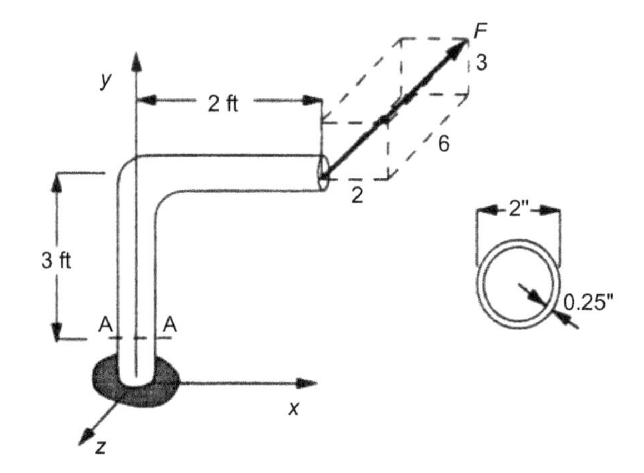

5.12 The support structure of Problem 5.11 is subjected to the applied load shown. The fiber orientation angle θ is allowed to be arbitrary but limited to positive angles. Determine the appropriate angle θ to assure that failure will not occur by plotting θ vs. the failure load F using the Tsai–Hill failure theory. On this plot, indicate the safe and unsafe regions assuming a load of $F = 500$ lb. The material is the same as in Problem 5.11. Neglect transverse shear effects.

5.13 A rectangular bracket is made from a boron/epoxy composite material with mechanical properties as defined by reference [43] of Chapter 4. The fibers are oriented at 30° to the x-axis, as shown. The bracket is cured at 370 °F, and the operating temperature is 70 °F. The bracket is cantilevered from a wall and loaded by forces of P and $2P$ as shown. At (very near) the wall, a biaxial strain gage is applied so that each gage is oriented in a principal material direction as indicated. For a rectangular section with the dimensions shown, the shear stress due to torsion (T) may be approximated by $\tau_T \approx 1000T$ and the shear stress due to transverse shear (V) may be approximated by $\tau_V \approx 48V$.

(A) Limiting considerations to the location at which the strain gage is applied, determine force P that will cause failure according to the maximum normal stress failure theory.

(B) Determine the strain indicated by each gage at the failure load determined in part A.

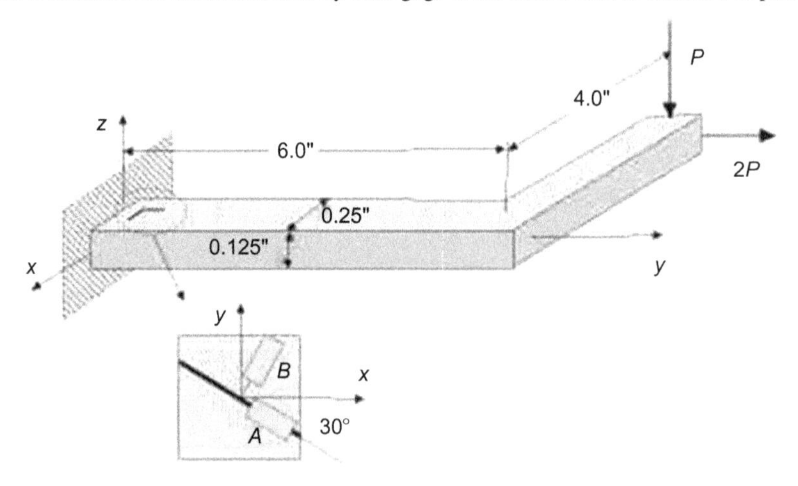

References

[1] Pagano NJ, editor. Interlaminar response of composite materials. Composite materials series, vol. 5. New York: Elsevier; 1989.
[2] Rowlands RE. Strength (failure) theories and their experimental correlation. In: Sih GC, Skudra AM, editors. Handbook of composites. Failure mechanics of composites, vol. 3. North-Holland: Amsterdam; 1984. p. 71–125.
[3] Wu EM. Phenomenological anisotropic failure criteria. In: Broutman LJ, Kroch RH, Sendeckyj GP, editors. Treatise on composite materials. New York: Academic Press; 1973. p. 353–432.
[4] Wu EM. Strength and fracture of composites. In: Broutman LJ, editor. Composite materials. Fatigue and fracture, vol. 5. New York: Academic Press; 1974. p. 191–248.
[5] Franklin HG. Classical theories of failure of anisotropic materials. Fiber Sci Technol 1968;1(2):137–50.
[6] Sandhu RS. A survey of failure theories of isotropic and anisotropic materials. In: AFFDL-TR-72-71. Dayton, OH: WPAFB; 1972.
[7] Sendeckyj GP. A brief survey of empirical multiaxial strength criteria for composites. In: ASTM STP 497. Philadelphia: ASTM; 1972, p. 41–51.
[8] Vicario AA, Toland RH. Failure criterial and failure analysis of composite structural components. In: Chamis CC, editor. Composite materials. Structural design and analysis, Vol. 7. New York: Academic Press; 1975. p. 52–97.
[9] Jenkins CF. Materials of constructions used in aircraft and aircraft engines, Report to Great Britain Aeronautical Research Committee; 1920.
[10] Hill R. The mathematical theory of plasticity. New York: Oxford University Press; 1950.
[11] Ashkenazi EK, Pekker FP. Experimental testing of the applicability of a fourth degree polynomial describing surfaces of critical planar stress distributions in glass-reinforced plastics. Meckhanika Polimerov 1970;6(2):284–94, Polym Mech, 6(2) (1970): 251–78.
[12] Chamis CC. Failure criteria for filamentary composites. Composite materials: testing and design, ASTM STP 460. Philadelphia: ASTM; 1969, p. 336–51.
[13] Fischer L. Optimization of orthotropic laminates. Eng Ind Ser B 1967;89(3):399–402.
[14] Azzi VD, Tsai SW. Anisotropic strength of composites. Exp Mech 1965;5(9):283–8.
[15] Norris CD, Strength of Orthotropic Materials Subjected to Combined Stress, U.S. Forest Products Laboratory Report #1816, 1950.
[16] Cowin SC. On the Strength Anisotropy of Bone and Wood. Trans ASME J Appl Mech 1979;46(4):832–8.
[17] Hoffman O. The brittle strength of orthotropic materials. J Compos Mater 1967;1:200–6.
[18] Malmeister AK. Geometry of theories of strength. Mekhanica Polimerov 1966;6 (2):519–34, Polym Mech, 2(4) (1966): 324–31.
[19] Marin J. Theories of strength for combined stress on non-isotropic materials, J Aeronautical Sci. 24(4) (1957) 265-8.
[20] Tsai SW, Wu EM. A general theory of strength for anisotropic materials. J Compos Mater 1971;24:58–80.
[21] Gol'denblat I, Kopnov VA. Strength of glass-reinforced plastic in the complex stress state. Mekhanika Polimerov 1965;1(2):70–8, Polym Mech, 1(2) (1965): 54–9.
[22] Petit PH, Waddoups ME. A method of predicting the nonlinear behavior of laminated composites. J Compos Mater 1969;3:2–19.
[23] Sandhu RS. Ultimate strength analysis of symmetric laminates, AFFDL-TR-73–137; 1974.

[24] Puppo AH, Evensen HA. Strength of anisotropic materials under combined stresses. AIAA J 1972;10:468–74.
[25] Wu EM, Scheriblein JK. Laminated strength—a direct characterization procedure. In: Composite materials testing and design (3rd Conf.), ASTM STP 546. Philadelphia: ASTM; 1974. p. 188–206.
[26] Wu EM. Optimal experimental measurements of anisotropic failure tensors. J Compos Mater 1972;6:472–89.
[27] Tsai SW, Wu EM. A general theory of strength for anisotropic materials, AFML-TR-71-12, 1971.
[28] Collins BR, Crane RL. A graphical representation of the failure surface of a composite. J Compos Mater 1971;5:408–13.
[29] Pipes RB, Cole BW. On the off-axis strength test for anisotropic materials. J Compos Mater 1973;7:246–56.
[30] Evans KE, Zhang WC. The determination of the normal interactive term in the Tsai–Wu polynomial strength criterion. Compos Sci Technol 1987;30:251–62.
[31] Tsai SW, Hahn TH. Introduction to composite materials. Westport, CT: Technomic; 1980.
[32] Korn GA, Korn TH. Mathematical handbook for scientists and engineers. 2nd ed. New York: McGraw-Hill; 1968.
[33] Wu RU, Stachorski Z. Evaluation of the normal stress interaction parameter in the tensor polynomial strength theory for anisotropic materials. J Compos Mater 1984;18:456–63.
[34] Schuling JC, Rowlands RE, Johnson HW, Gunderson DE. Tensorial strength analysis of paperboard. Exp Mech 1985;25:75–84.
[35] Leissa AW. Buckling of laminated composite plates and shell panels, AFWAL-TR-85-3069; 1985.
[36] Shuart MJ. Short wave-length buckling and shear failures for compression loaded composite laminates, NASA-TM87640; 1985.
[37] Rosen BW. Mechanics of composite strengthening. In: Fiber composite materials. Metals Park, OH: American Society for Metals; 1965. p. 37–75.
[38] Timoshenko SP, Gere JM. Theory of elastic stability. New York: McGraw-Hill; 1961.
[39] Jones RM. Mechanics of composite materials. New York: Hemisphere Publishing; 1975.
[40] Greszczuk LB. Compressive strength and failure modes of unidirectional composites. Analysis of the test methods for high modulus fibers and composites, ASTM-STP-521. Philadelphia: ASTM; 1973, p. 192–217.
[41] Greszczuk LB. Failure mechanics of composites subjected to compressive loading, AFML-TR-72-107; 1972.
[42] Davis JG. Compressive strength of fiber-reinforced composite materials. Composite reliability, ASTM-STP-580. Philadelphia: ASTM; 1975, p. 364–77.
[43] Dale WC, Bare E. Fiber buckling in composite systems: a model for the ultra-structure of uncalcified collagen tissue. J Mater Sci 1974;9:369–82.
[44] Weaver CW, Williams J. Deformation of a carbon-epoxy composite under hydrostatic pressure. J Mater Sci 1975;10:1323–33.
[45] Chaplin CR. Compressive fracture in unidirectional glass reinforced plastics. J Mater Sci 1977;12:347–52.
[46] Evans AG, Adler WF. Kinking as a mode of structural degradation in carbon fiber composites. Acta Metall 1983;26:725–38.
[47] Budiansky B. Micromechanics. Comput Struct 1983;16:3–12.
[48] Robinson IM, Yeung PHJ, Galiotis C, Young RJ, Batchelder DU. Stress induced twinning of polydiacetylene single crystal fibers in composites. J Mater Sci 1986;21:3440–4.

[49] Hahn HT, Williams JG. Compression failure mechanisms in unidirectional composites. Composite materials testing and design: ASTM STP-893. Philadelphia: ASTM; 1986, p. 115–39.
[50] Sohi MS, Hahn HT, Williams JG. The effect of resin toughness and modulus on compression failure modes of quasi-isotropic gr/epoxy laminates, NASA TM-87604; 1984.
[51] Wass AH, Babcock CD, Knauss WG. A mechanical model for elastic fiber microbuckling. In: Hui D, Kozik T, editors. Composite materials technology. New York: ASME; 1989. p. 203–15.

Laminate analysis

6.1 Introduction

A *laminate* is a collection of lamina arranged in a specified manner. Adjacent lamina may be of the same or different materials and their fiber orientations with respect to a reference axis (by convention, the x-axis) may be arbitrary. The principles developed in previous chapters are used to establish load–strain and stress–strain relations for continuous fiber laminated composite plates. Discussions are restricted to laminate stress analysis. The procedures established in this chapter for thin plates can be extended to other structural elements such as beams, columns, and shells, which are beyond the scope of this text.

6.2 Classical lamination theory

Classical lamination theory (CLT) as presented herein is applicable to orthotropic continuous fiber laminated composites only. Derivations in this section follow the classical procedures cited in earlier publications [1–5]. The approach used in formulating CLT is similar to that used in developing load–stress relationships in elementary strength of materials courses. An initial displacement field consistent with applied loads is assumed. Through the strain–displacement fields and an appropriate constitutive relationship, a state of stress is defined. By satisfying the conditions of static equilibrium, a load–strain relation is defined, and subsequently a state of stress is defined for each lamina.

6.2.1 Strain–displacement relations

Consider the plate shown in Figure 6.1a, in which the xy-plane coincides with the mid-plane of the plate. With application of a lateral load, reference point A located at a position defined by the coordinates (x_A, y_A) is displaced. The displacement W of this point, as well as an assumed deformed shape of the plate in the x–z plane, is shown in Figure 6.1b. The displacements of any point within the plate in the x, y, and z directions are denoted by U, V, and W, respectively. The manner in which these displacements are modeled dictates the complexity of the strain–displacement and eventually the load–strain relation. The displacements are initially expressed by a power series in z, which takes the form

$$U=\sum_{i=0}^{\infty} z^i \Phi(x,y) \quad V=\sum_{i=0}^{\infty} z^i \Psi(x,y) \quad W=\sum_{i=0}^{\infty} z^i \Theta(x,y) \tag{6.1}$$

Laminar Composites. http://dx.doi.org/10.1016/B978-0-12-802400-3.00006-4

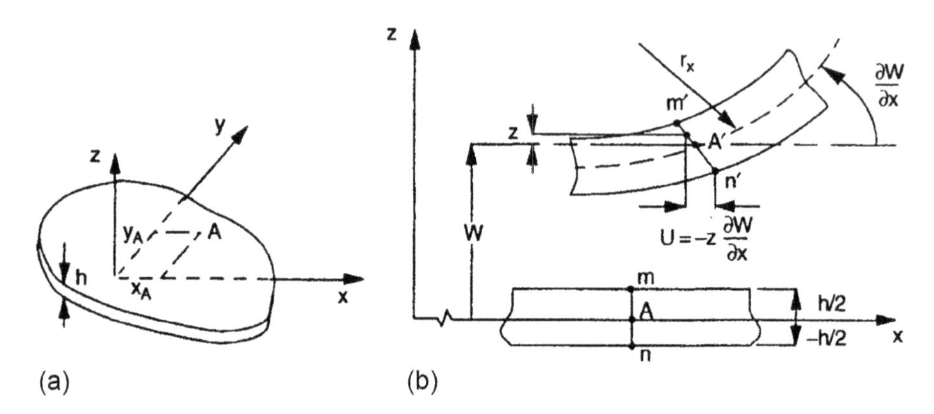

Figure 6.1 Plate geometry for classical lamination theory.

The number of terms retained, as well as assumptions made regarding permissible strain fields defines the form of Φ, Ψ, and Θ. The U, V, and W expressions in Equation (6.1) represent displacements resulting from both forces (normal and shear) and moments (bending and torsional). For thin plates subjected to small deformations, the fundamental assumptions are as follows:

1. Deflections of the mid-surface (geometric center of the plate) are small compared to the thickness of the plate, and the slope of the deflected plate is small.
2. The mid-plane is unstrained when the plate is subjected to pure bending.
3. Plane sections initially normal to the mid-plane remain normal to the mid-plane after bending. Shear strains γ_{yz} and γ_{xz} are assumed to be negligible ($\gamma_{yz} = \gamma_{xz} = 0$). Similarly, normal out-of-plane strains are assumed to be zero when plate deflections are due to bending.
4. The condition of $\sigma_z = 0$ is assumed to be valid, except in localized areas where high concentrations of transverse load are applied.

These assumptions are known as the Kirchhoff hypothesis for plates and the Kirchhoff–Love hypothesis for thin plates and shells. Different structural members such as beams, bars, and rods require alternate assumptions. For the case of thick plates (or short, deep beams), shear stresses are important and assumptions 3 and 4 are no longer valid, requiring a more general theory.

For thin laminated plates, the total laminate thickness h is usually small compared to other plate dimensions. A good approximation is achieved by retaining only the first few terms of U and V from Equation (6.1). The W displacement field is assumed to be constant, resulting in

$$U = U_0(x, y) + z\Phi(x, y) \quad V = V_0(x, y) + z\Psi(x, y) \quad W = W_0(x, y)$$

The terms U_0, V_0, and W_0 are the mid-surface displacements. They are not the same as the neutral bending axis displacements presented in strength of materials discussions for beams made of isotropic materials. The displacements of the plate with respect to the mid-surface are illustrated in Figure 6.1b for the x-direction. Similar relations can be established for the y-direction. Using the definitions of strain from Chapter 2, and the assumptions just given, the strains are

$$\varepsilon_x = \frac{\partial U}{\partial x} = \frac{\partial U_0}{\partial x} + z\frac{\partial \Phi}{\partial x}$$
$$\varepsilon_y = \frac{\partial V}{\partial y} = \frac{\partial V_0}{\partial x} + z\frac{\partial \Psi}{\partial y}$$
$$\varepsilon_z = \frac{\partial W}{\partial z} = 0$$
$$\gamma_{xz} = \frac{\partial U}{\partial z} + \frac{\partial W}{\partial x} = \left(\frac{\partial W_0}{\partial x} + \Phi\right) = 0$$
$$\gamma_{yz} = \frac{\partial V}{\partial z} + \frac{\partial W}{\partial y} = \left(\frac{\partial W_0}{\partial y} + \Psi\right) = 0$$
$$\gamma_{xy} = \frac{\partial U}{\partial y} + \frac{\partial V}{\partial x} = \left(\frac{\partial U_0}{\partial y} + \frac{\partial V_0}{\partial x}\right) + z\left(\frac{\partial \Phi}{\partial y} + \frac{\partial \Psi}{\partial x}\right)$$

The nonzero mid-surface strains are defined as

$$\{\varepsilon^0\} = \begin{Bmatrix} \varepsilon_x^0 \\ \varepsilon_y^0 \\ \gamma_{xy}^0 \end{Bmatrix} = \begin{Bmatrix} \partial U_0/\partial x \\ \partial V_0/\partial y \\ \partial U_0/\partial y + \partial U_0/\partial x \end{Bmatrix} \tag{6.2}$$

The mid-surface may experience curvatures related to the radius of curvature of the mid-surface. The curvatures are related to the displacement functions Ψ and Φ by

$$\begin{Bmatrix} \kappa_x \\ \kappa_y \\ \kappa_{xy} \end{Bmatrix} = \begin{Bmatrix} \partial \Phi/\partial x \\ \partial \Psi/\partial y \\ \partial \Phi/\partial y + \partial \Psi/\partial x \end{Bmatrix} \tag{6.3}$$

Each term in Equation (6.3) can be related to a radius of curvature of the plate. Each curvature and its associated relationship to Ψ and Φ is illustrated in Figure 6.2.

The strain variation through a laminate is expressed by a combination of Equations (6.2) and (6.3) as

$$\begin{Bmatrix} \varepsilon_x \\ \varepsilon_y \\ \gamma_{xy} \end{Bmatrix} = \begin{Bmatrix} \varepsilon_x^0 \\ \varepsilon_y^0 \\ \varepsilon_{xy}^0 \end{Bmatrix} + z\begin{Bmatrix} \kappa_x \\ \kappa_y \\ \kappa_{xy} \end{Bmatrix} \tag{6.4}$$

The strains in Equation (6.4) are valid for conditions of plane stress ($\gamma_{yz} = \gamma_{xz} = \sigma_z = 0$). In cases where the condition on the two shear strains being zero is relaxed, the strain relationship given by Equation (6.4) contains additional terms (γ_{xz} and γ_{yz}), which are not functions of the curvatures κ_x, κ_y, and κ_{xy}.

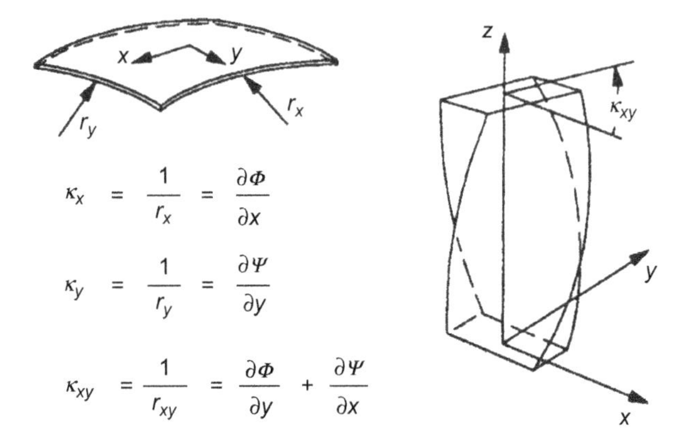

Figure 6.2 Plate curvatures for classical lamination theory.

With γ_{xz} and γ_{yz} zero, Ψ and Φ can be explicitly defined in terms of W_0 from the strain–displacement relations as

$$\Phi = -\frac{\partial W_0}{\partial_x} \quad \Psi = -\frac{\partial W_0}{\partial_y} \tag{6.5}$$

It follows directly from Equation (6.3) that the curvatures are

$$\begin{Bmatrix} \kappa_x \\ \kappa_y \\ \kappa_{xy} \end{Bmatrix} = \begin{Bmatrix} -\partial^2 W_0/\partial x^2 \\ -\partial^2 W_0/\partial y^2 \\ -2\partial^2 W_0/\partial x \partial y \end{Bmatrix} \tag{6.6}$$

Using these definitions of curvature and Equation (6.2), the strain variation through the laminate as represented by Equation (6.4) can be expressed in terms of displacements. This form of the strain variation is convenient for problems in which deflections are required. Examples of such problems are generally found with beams, plate and shell vibrations, etc., where Φ and Ψ are obtained from boundary and initial conditions. They are not considered herein.

6.2.2 Stress–strain relationships

The strain variation through a laminate is a function of both mid-surface strain and curvature and is continuous through the plate thickness. The stress need not be continuous through the plate. Consider the plane stress relationship between Cartesian stresses and strains,

$$\begin{Bmatrix} \sigma_x \\ \sigma_y \\ \tau_{xy} \end{Bmatrix} = [\bar{Q}] \begin{Bmatrix} \varepsilon_x \\ \varepsilon_y \\ \gamma_{xy} \end{Bmatrix}$$

Each lamina through the thickness may have a different fiber orientation and consequently a different $[\bar{Q}]$. The stress variation through the laminate thickness is

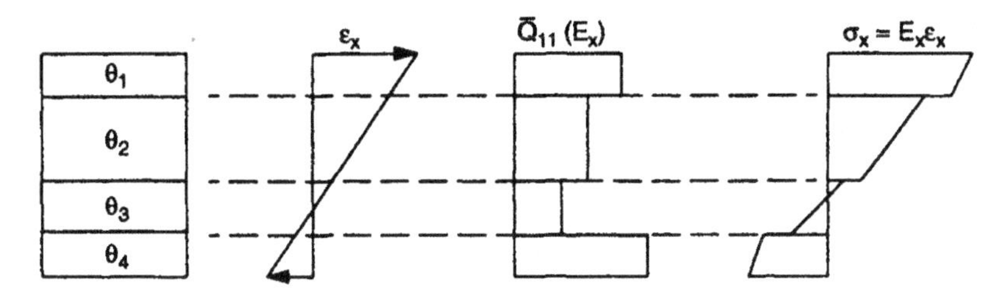

Figure 6.3 Stress variation in a variable-modulus material.

therefore discontinuous. This is illustrated by considering a simple one-dimensional model. Assume a laminate is subjected to a uniform strain ε_x (with all other strains assumed to be zero). The stress in the x-direction is related to the strain by $\bar{Q}_{11}$. This component of $[\bar{Q}]$ is not constant through the laminate thickness. The magnitude of $\bar{Q}_{11}$ is related to the fiber orientation of each lamina in the laminate. As illustrated in Figure 6.3, the linear variation of strain combined with variations of $\bar{Q}_{11}$ (which can be treated as a directional modulus designated as E_x) gives rise to a discontinuous variation of stress, described by $\sigma_x = E_x\varepsilon_x$.

In order to establish the state of stress at a point in a laminate, the state of strain at the point must first be defined. Combining this state of strain with an appropriate constitutive relation yields the stress. For general loading conditions, it is convenient to work with Cartesian components of stress and strain. For a specific lamina (termed the "kth" lamina), the appropriate constitutive relationship is $[\bar{Q}]_k$. Under conditions of plane stress, the Cartesian components of stress in the kth layer are

$$\begin{Bmatrix} \sigma_x \\ \sigma_y \\ \tau_{xy} \end{Bmatrix}_k = [\bar{Q}]_k \left(\begin{Bmatrix} \varepsilon_x^0 \\ \varepsilon_y^0 \\ \gamma_{xy}^0 \end{Bmatrix} + z \begin{Bmatrix} \kappa_x \\ \kappa_y \\ \kappa_{xy} \end{Bmatrix} \right) \tag{6.7}$$

This relationship is assumed valid for any layer of the laminate.

6.2.3 Laminate load–strain and moment–curvature relations

Formulating a simple working relationship between load, strain, and stress requires appropriate load–displacement relationships for the entire laminate. The admissible loads are assumed to be a set of *resultant* forces and moments, defined for a representative section of the laminate. The resultant forces have units of force per unit length of laminate (N/m or lb/in.) and are shown in Figure 6.4. Thin-plate theory omits the effect of shear strains γ_{xz} and γ_{yz}, but shear forces Q_x and Q_y are considered. They have the same order of magnitude as surface loads and moments, and are used in developing the equations of equilibrium [6]. The dimensions for these terms are force per unit length of laminate.

To satisfy conditions of equilibrium, resultant laminate forces must be balanced by the integral of stresses over the laminate thickness. The balance of forces (assuming $\sigma_z = 0$) is

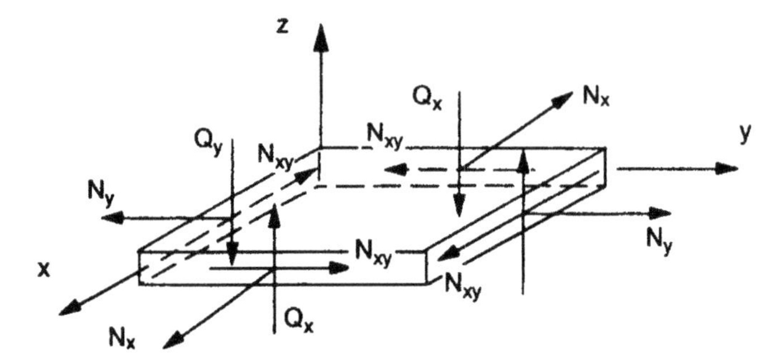

Figure 6.4 Positive sign convention for laminate loads.

$$\begin{Bmatrix} N_x \\ N_y \\ N_{xy} \\ Q_x \\ Q_y \end{Bmatrix} = \int_{-h/2}^{h/2} \begin{Bmatrix} \sigma_x \\ \sigma_y \\ \tau_{xy} \\ \tau_{xz} \\ \tau_{yz} \end{Bmatrix} \mathrm{d}z \tag{6.8}$$

The resultant moments assumed to act on the laminate have units of length times force per unit length of laminate (N-m/m or in.-lb/in.) and are shown in Figure 6.5. In a manner similar to that for resultant forces, the resultant moments acting on the laminate must satisfy the conditions of equilibrium. The out-of-plane shear stresses τ_{xz} and τ_{yz} do not contribute to these moments. Moments are related to forces through a simple relationship of force times distance. The balance of the resultant moments yields

$$\begin{Bmatrix} M_x \\ M_y \\ M_{xy} \end{Bmatrix} = \int_{-h/2}^{h/2} z \begin{Bmatrix} \sigma_x \\ \sigma_y \\ \tau_{xy} \end{Bmatrix} \mathrm{d}z \tag{6.9}$$

A general laminate consists of an arbitrary number of layers (N). The Cartesian stress components within any one of these layers, say the kth layer, are defined by Equation (6.7). The fiber orientation of each lamina is arbitrary, so the variation of stress through the

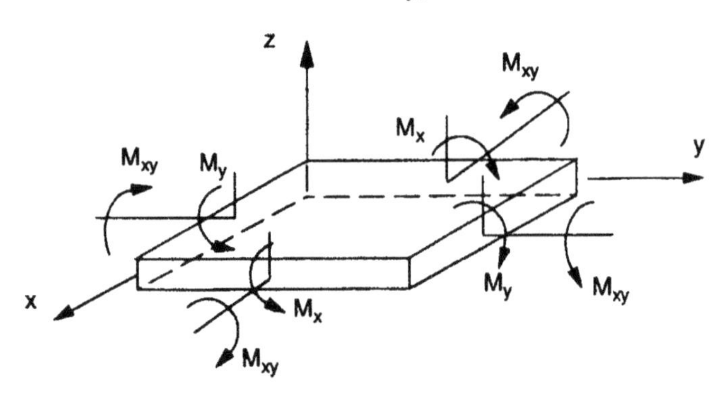

Figure 6.5 Positive sign convention for laminate moments.

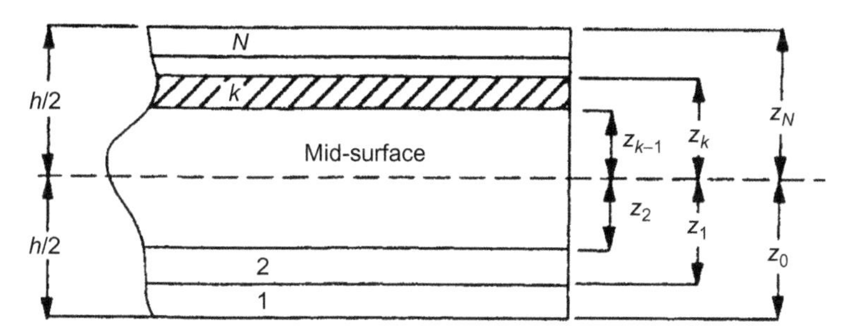

Figure 6.6 Laminate stacking sequence nomenclature.

laminate thickness cannot be expressed by a simple function. Each of the N layers is assigned a reference number and a set of z-coordinates to identify it, as illustrated in Figure 6.6. The location of each layer is important in defining the governing relations for laminate response. Using the topmost lamina (labeled N in Figure 6.6) as the first lamina is contrary to CLT procedures and affects numerical results.

Equations (6.8) and (6.9) can be expressed in terms of the stresses in each layer. Since the kth layer is assumed to occupy the region between z_k and z_{k-1}, it follows directly that Equations (6.8) and (6.9) can be expressed as

$$\begin{Bmatrix} N_x \\ N_y \\ N_{xy} \end{Bmatrix} = \sum_{k=1}^{N} \int_{z_{k-1}}^{z_k} \begin{Bmatrix} \sigma_x \\ \sigma_y \\ \tau_{xy} \end{Bmatrix}_k \mathrm{d}z \tag{6.10}$$

$$\begin{Bmatrix} Q_x \\ Q_y \end{Bmatrix} = \sum_{k=1}^{N} \int_{z_{k-1}}^{z_k} \begin{Bmatrix} \tau_{xz} \\ \tau_{yz} \end{Bmatrix}_k \mathrm{d}z \tag{6.11}$$

$$\begin{Bmatrix} M_x \\ M_y \\ M_{xy} \end{Bmatrix} = \sum_{k=1}^{N} \int_{z_{k-1}}^{z_k} z \begin{Bmatrix} \sigma_x \\ \sigma_y \\ \tau_{xy} \end{Bmatrix}_k \mathrm{d}z \tag{6.12}$$

where N is the total number of lamina in the laminate. The Q_x and Q_y have been segregated from the other load terms, for reasons to be subsequently discussed.

Substituting Equation (6.7) into the preceding equations, the laminate loads and moments are expressed in terms of the mid-surface strains and curvatures as

$$\begin{Bmatrix} N_x \\ N_y \\ N_{xy} \end{Bmatrix} = \sum_{k=1}^{N} [\bar{Q}]_k \left(\int_{z_{k-1}}^{z_k} \begin{Bmatrix} \varepsilon_x^0 \\ \varepsilon_y^0 \\ \varepsilon_{xy}^0 \end{Bmatrix} \mathrm{d}z + \int_{z_{k-1}}^{z_k} z \begin{Bmatrix} \sigma_x \\ \sigma_y \\ \tau_{xy} \end{Bmatrix}_k \mathrm{d}z \right) \tag{6.13}$$

$$\begin{Bmatrix} M_x \\ M_y \\ M_{xy} \end{Bmatrix} = \sum_{k=1}^{N} [\bar{Q}]_k \left(\int_{z_{k-1}}^{z_k} z \begin{Bmatrix} \varepsilon_x^0 \\ \varepsilon_y^0 \\ \varepsilon_{xy}^0 \end{Bmatrix} \mathrm{d}z + \int_{z_{k-1}}^{z_k} z^2 \begin{Bmatrix} \kappa_x \\ \kappa_y \\ \kappa_{xy} \end{Bmatrix} \mathrm{d}z \right) \tag{6.14}$$

Since the mid-surface strains $\{\varepsilon^0\}$ and curvatures $\{k\}$ are independent of the z-coordinate, the integration is simplified. The integrals in Equations (6.13) and (6.14) become simple integrals of $(1, z, z^2)$. The loads and moments can be expressed in matrix form, after integration, as

$$\begin{Bmatrix} N_x \\ N_y \\ N_{xy} \\ --- \\ M_x \\ M_y \\ M_{xy} \end{Bmatrix} = \left[\begin{array}{ccc|ccc} A_{11} & A_{12} & A_{16} & B_{11} & B_{12} & B_{16} \\ A_{12} & A_{22} & A_{26} & B_{12} & B_{22} & B_{26} \\ A_{16} & A_{26} & A_{66} & B_{16} & B_{26} & B_{66} \\ \hline B_{11} & B_{12} & B_{16} & D_{11} & D_{12} & D_{16} \\ B_{12} & B_{22} & B_{26} & D_{12} & D_{22} & D_{26} \\ B_{16} & B_{26} & B_{66} & D_{16} & D_{26} & D_{66} \end{array}\right] \begin{Bmatrix} \varepsilon_x^0 \\ \varepsilon_y^0 \\ \gamma_{xy}^0 \\ --- \\ \kappa_x \\ \kappa_y \\ \kappa_{xy} \end{Bmatrix} \tag{6.15}$$

Each component of the $[A]$, $[B]$, and $[D]$ matrices is defined by

$$\begin{aligned} [A_{ij}] &= \sum_{k=1}^{N} [\bar{Q}_{ij}]_k (z_k - z_{k-1}) \\ [B_{ij}] &= \frac{1}{2}\sum_{k=1}^{N} [\bar{Q}_{ij}]_k (z_k^2 - z_{k-1}^2) \\ [D_{ij}] &= \frac{1}{3}\sum_{k=1}^{N} [\bar{Q}_{ij}]_k (z_k^3 - z_{k-1}^3) \end{aligned} \tag{6.16}$$

The subscripts i and j are matrix notation, not tensor notation. The form of Equation (6.15) is often simplified to

$$\begin{Bmatrix} N \\ -- \\ M \end{Bmatrix} = \left[\begin{array}{c|c} A & B \\ \hline D & D \end{array}\right] \begin{Bmatrix} \varepsilon^0 \\ -- \\ \kappa \end{Bmatrix} \tag{6.17}$$

When using this abbreviated form of the laminate load–strain relationship, one must be aware that $\{N\}$, $\{M\}$, $\{\varepsilon^0\}$, and $\{\kappa\}$ are off-axis quantities. They define laminate behavior with respect to the Cartesian (x–y) coordinate system. Each of the 3×3 matrices in either Equation (6.15) or (6.17) has a distinct function identified by examination of Equation (6.17). These matrices are termed:

$[A_{ij}]$ = extensional stiffness matrix
$[B_{ij}]$ = extension-bending coupling matrix
$[D_{ij}]$ = bending stiffness matrix

The resultant shear forces Q_x and Q_y are treated differently since they are not expressed in terms of mid-surface strains and curvatures. From Equation (3.7), we see that the stress–strain relationship for shear in the kth layer is expressed as

$$\begin{Bmatrix} \tau_{xz} \\ \tau_{yz} \end{Bmatrix}_k = \begin{bmatrix} \bar{Q}_{55} & \bar{Q}_{45} \\ \bar{Q}_{45} & \bar{Q}_{44} \end{bmatrix}_k \begin{Bmatrix} \gamma_{xz} \\ \gamma_{yz} \end{Bmatrix}_k \tag{6.18}$$

From Equation (6.8), the expressions for Q_x and Q_y can be written as

$$\begin{Bmatrix} Q_x \\ Q_y \end{Bmatrix}_k = \sum_{k=1}^{N} \int_{z_{k-1}}^{z_k} \begin{bmatrix} \bar{Q}_{55} & \bar{Q}_{45} \\ \bar{Q}_{45} & \bar{Q}_{44} \end{bmatrix}_k \begin{Bmatrix} \gamma_{xz} \\ \gamma_{yz} \end{Bmatrix}_k \mathrm{d}z \tag{6.19}$$

It is generally assumed that the transverse shear stresses are parabolically distributed over the laminate thickness. This distribution is consistent with Reisner [7] and can be represented by a weighting function $f(z)=c\ [1-(2z/h)^2]$. The coefficient c is commonly termed the shear correction factor. The numerical value of c depends upon the cross-sectional shape of the laminate. For a rectangular section, generally of interest in laminate analysis, $c=6/5$ (1.20). The derivation of this can be found in many strength of materials texts. The expression for Q_x and Q_y can be written in a manner analogous to Equation (6.15) or (6.17):

$$\begin{Bmatrix} Q_x \\ Q_y \end{Bmatrix} = \begin{bmatrix} A_{55} & A_{45} \\ A_{45} & A_{44} \end{bmatrix}_k \begin{Bmatrix} \gamma_{xz} \\ \gamma_{yz} \end{Bmatrix}_k \tag{6.20}$$

Following the same procedures as before, it can be shown that

$$A_{ij} = c \sum_{k=1}^{N} \left[\bar{Q}_{ij}\right]_k \left\{ (z_k - z_{k-1}) - \frac{4}{3h^2}\left(z_k^3 - z_{k-1}^3\right) \right\} \tag{6.21}$$

where $i, j=4, 5$ and h is the total laminate thickness.

In general, the shear terms are seldom used in beginning laminate analysis. They are, however, useful in the formulation of plate analysis, as well as beam deflection problems. The stiffness terms ($\bar{Q}_{44}$, $\bar{Q}_{55}$, etc.) associated with Q_x and Q_y can be difficult to experimentally determine; therefore, they are often approximated.

6.2.3.1 Alternate formulation of A, B, D

A convenient form for the [A], [B], and [D] matrices can be established by examining the position of the kth lamina in Figure 6.7. Recall that each ply of the laminate is confined to the limits z_k and z_{k-1}. Using these bounds, the following definitions result:

$t_k = z_k - z_{k-1}$ = thickness of the kth lamina

$\bar{z}_k = \dfrac{z_k + z_{k-1}}{2}$ = location of the centroid of the kth lamina from the mid-plane of the laminate.

Note that Z_k can be either positive or negative.

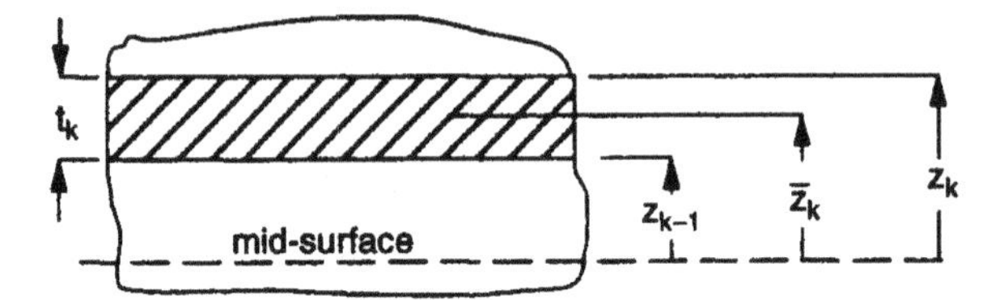

Figure 6.7 Relationship of $\bar{z}_k$ and t_k to z_k and z_{k-1}.

From Equation (6.16), the terms containing z_k and z_{k-1} can be written as

$$\frac{z_k^2 - z_{k-1}^2}{2} = \frac{(z_k - z_{k-1})(z_k + z_{k-1})}{2} = t_k \bar{z}_k$$
$$\frac{z_k^3 - z_{k-1}^3}{3} = t_k \bar{z}_k^2 + \frac{z_k^3}{12}$$

Substitution of these into Equation (6.16) results in

$$\begin{aligned}
[A_{ij}] &= \sum_{k=1}^{N} [\bar{Q}_{ij}]_k t_k \\
[B_{ij}] &= \sum_{k=1}^{N} [\bar{Q}_{ij}]_k t_k \bar{z}_k \\
[D_{ij}] &= \sum_{k=1}^{N} [\bar{Q}_{ij}]_k \left(t_k \bar{z}_k^2 + \frac{z_k^3}{12} \right)
\end{aligned} \tag{6.22}$$

For transverse shear, the analogous expression is

$$[A_{ij}] = c \sum_{k=1}^{N} [\bar{Q}_{ij}]_k t_k \left\{ 1 + \frac{4}{h^2} \left(\bar{z}_k^2 + \frac{t_k^3}{12} \right) \right\} \tag{6.23}$$

where c is the shear correction factor previously defined.

The $[A]$, $[B]$, and $[D]$ matrices defined by Equation (6.16) or (6.22) are the primary relations between load and strain for laminate analysis. These can be manipulated to define mid-surface strains and curvatures as a function of applied loads. It is obvious from examination of either equation used to define each component of $[A]$, $[B]$, and $[D]$ that they are functions of both lamina material and the location of each lamina with respect to the mid-surface. Consider, for example, a four-layer laminate consisting of only 0° and 90° plies, which can be arranged in one of two manners, as shown in Figure 6.8. All plies are assumed to have the same thickness (t_k), and the entire laminate has a thickness of $4t_k$. The $[A]$ matrix is a function of $[\bar{Q}]$, and ply thickness for each lamina is not influenced by position since t_k is a constant.

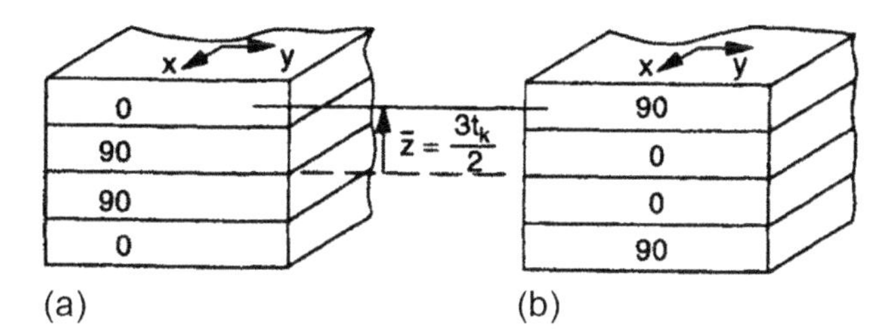

Figure 6.8 Two possible laminate stacking arrangements resulting in identical $[A]$ matrices.

The $[D]$ matrix is influenced by t_k, $\bar{z}_k$, and $[\bar{Q}]$ for each lamina. Considering only D_{11}, Equation (6.22) for this term is

$$[D_{11}] = \sum_{k=1}^{4} [\bar{Q}_{11}]_k \left(t_k \bar{z}_k^2 + \frac{t_k^3}{12} \right)$$

D_{11} for the laminate of Figure 6.8a and b is

$$\underset{[0/90]_s}{[D_{11}] = \sum \frac{t_k^3}{6} \left\{ 28 [\bar{Q}_{11}]_0 + [\bar{Q}_{11}]_{90} \right\}} \quad \underset{[90/0]_s}{[D_{11}] = \sum \frac{t_k^3}{6} \left\{ 28 [\bar{Q}_{11}]_{90} + [\bar{Q}_{11}]_0 \right\}}$$

Since $[\bar{Q}_{11}]_0 > [\bar{Q}_{11}]_{90}$, the laminate in Figure 6.8a has a larger D_{11} than that of Figure 6.8b. The $[B]$ matrix for both laminates in Figure 6.8 is zero.

6.3 Thermal and hygral effects

Thermal and hygral effects are dilatational (associated with dimension changes only) and influence only the strains. In this section, thermal effects are discussed and equations relating resultant thermal loads and moments to the load–strain relationship of Equation (6.17) are derived. Hygral effects are expressed in a similar manner. The combined effects of temperature and moisture are incorporated into a general governing equation for plane stress analysis.

6.3.1 Thermal effects

For some laminate stacking arrangements, thermal effects can result in large residual strains and curvatures. In order to illustrate residual (sometimes called curing) strains due to thermal effects, a cross-ply laminate (only 0° and 90° lamina) is considered. A one-dimensional model is used, and only the strains in the x-direction are considered. The coefficients of thermal expansion in the x-direction are different in each lamina. Figure 6.9 illustrates the deformations and residual strains developed in one direction for a cross-ply laminate.

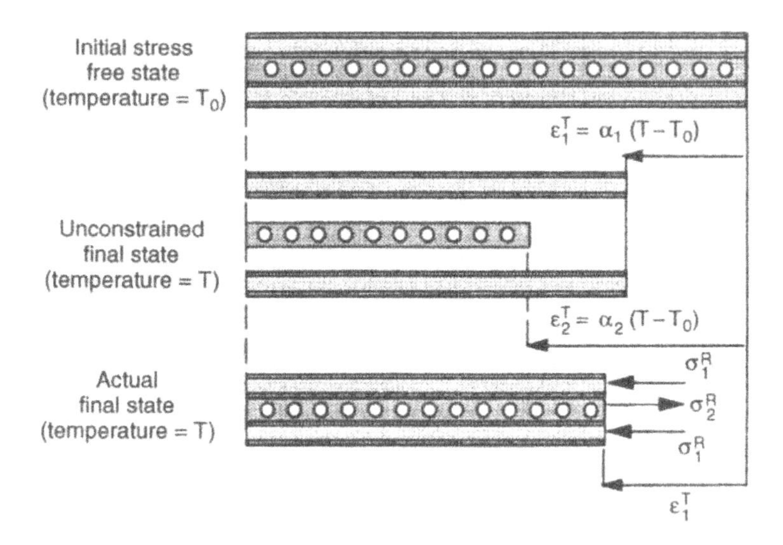

Figure 6.9 Residual stresses in one direction of a symmetric laminate.

Laminates fabricated by curing a stack of lamina at elevated temperatures are considered to be in a stress-free state during curing. In the initial stress-free state (at temperature T_0), all laminas are of the same length. After the cure cycle is complete, the laminate temperature returns to room temperature. If the laminas are unconstrained, they can assume the relative dimensions indicated in the middle sketch of Figure 6.9, resulting in the strains shown. Each ply is constrained to deform with adjacent plies resulting in a uniform residual strain, but different residual stresses in each lamina, as indicated in Figure 6.9. These stresses can be thought of as the stresses required to either pull or push each lamina into a position consistent with a continuous deformation (strain). The residual stresses (and strains) can be either tensile or compressive and depend on temperature difference, material, stacking sequence, etc. If the symmetric cross-ply laminate of Figure 6.9 were antisymmetric (consisting of two lamina, oriented at 0° and 90°), the residual strains would be accompanied by residual curvatures.

The general form of the stress in the kth layer of the laminate given by Equation (6.7) must be appended to account for temperature affects. Following the discussions of Section 3.3.1, the stress components in the kth lamina are

$$\begin{Bmatrix} \sigma_x \\ \sigma_y \\ \tau_{xy} \end{Bmatrix}_k = [\bar{Q}]_k \left(\begin{Bmatrix} \varepsilon_x^0 \\ \varepsilon_y^0 \\ \varepsilon_{xy}^0 \end{Bmatrix} + z \begin{Bmatrix} \kappa_x \\ \kappa_y \\ \kappa_{xy} \end{Bmatrix} - \begin{Bmatrix} \alpha_x \\ \alpha_y \\ \alpha_{xy} \end{Bmatrix}_k \Delta T \right) \tag{6.24}$$

The $\alpha \Delta T$ terms have the effect of creating thermal loads and moments, which must satisfy the conditions of equilibrium. The thermal stresses in the kth lamina are

$$\begin{Bmatrix} \sigma_x \\ \sigma_y \\ \tau_{xy} \end{Bmatrix}_k^T = [\bar{Q}]_k \begin{Bmatrix} \alpha_x \\ \alpha_y \\ \alpha_{xy} \end{Bmatrix} \Delta T$$

In a manner identical to that of Section 6.2.3, the thermal loads and moments are

$$\begin{aligned} \{N^{\mathrm{T}}\} = \begin{Bmatrix} N_x^{\mathrm{T}} \\ N_y^{\mathrm{T}} \\ N_{xy}^{\mathrm{T}} \end{Bmatrix} &= \int_{-h/2}^{h/2} \begin{Bmatrix} \sigma_x \\ \sigma_y \\ \tau_{xy} \end{Bmatrix}_k^{\mathrm{T}} \mathrm{d}z = \int_{-h/2}^{h/2} [\bar{Q}]_k \begin{Bmatrix} \alpha_x \\ \alpha_y \\ \alpha_{xy} \end{Bmatrix}_k \Delta T \, \mathrm{d}z \\ &= \sum_{k=1}^{N} [\bar{Q}]_k \begin{Bmatrix} \alpha_x \\ \alpha_y \\ \alpha_{xy} \end{Bmatrix}_k \Delta T \, t_k \end{aligned} \tag{6.25}$$

$$\begin{aligned} \{M^{\mathrm{T}}\} = \begin{Bmatrix} M_x^{\mathrm{T}} \\ M_y^{\mathrm{T}} \\ M_{xy}^{\mathrm{T}} \end{Bmatrix} &= \int_{-h/2}^{h/2} z \begin{Bmatrix} \sigma_x \\ \sigma_y \\ \tau_{xy} \end{Bmatrix}_k^{\mathrm{T}} \mathrm{d}z = \int_{-h/2}^{h/2} [\bar{Q}]_k z \begin{Bmatrix} \alpha_x \\ \alpha_y \\ \alpha_{xy} \end{Bmatrix}_k \Delta T \, \mathrm{d}z \\ &= \sum_{k=1}^{N} [\bar{Q}]_k \begin{Bmatrix} \alpha_x \\ \alpha_y \\ \alpha_{xy} \end{Bmatrix}_k \Delta T (t_k \bar{z}_k) \end{aligned} \tag{6.26}$$

Formulation of $\{N^{\mathrm{T}}\}$ and $\{M^{\mathrm{T}}\}$ is analogous to formulating the $[A]$ and $[B]$ matrices, provided ΔT is independent of z, $[N^{\mathrm{T}}]$ and $[M^{\mathrm{T}}]$ are related to integrals of $(1, z)\mathrm{d}z$, respectively. Thermal loads and moments are formed in a manner analogous to the formulation of $[A]$ and $[B]$. As a result, a symmetric laminate with a uniform temperature results in $\{M^{\mathrm{T}}\} = 0$. Incorporating thermal loads and moments into Equation (6.17) results in

$$\begin{Bmatrix} N \\ -- \\ M \end{Bmatrix} = \begin{bmatrix} A & | & B \\ -- & -- & -- \\ B & | & D \end{bmatrix} \begin{Bmatrix} \varepsilon^0 \\ -- \\ \kappa \end{Bmatrix} - \begin{Bmatrix} N^{\mathrm{T}} \\ -- \\ M^{\mathrm{T}} \end{Bmatrix} \tag{6.27}$$

6.3.2 Hygral effects

Hygral effects are similar to thermal effects in that moisture absorption introduces dilational strains into the analysis. Discussions of hygral effects parallel those of thermal effects. Hygral strains produce swelling, and the stresses due to these strains, when coupled with mid-surface strains and curvatures, are expressed as

$$\begin{Bmatrix} \sigma_x \\ \sigma_y \\ \tau_{xy} \end{Bmatrix}_k = [\bar{Q}]_k \left(\begin{Bmatrix} \varepsilon_x^0 \\ \varepsilon_y^0 \\ \gamma_{xy}^0 \end{Bmatrix} + z \begin{Bmatrix} \kappa_x \\ \kappa_y \\ \kappa_{xy} \end{Bmatrix} - \begin{Bmatrix} \beta_x \\ \beta_y \\ \beta_{xy} \end{Bmatrix}_k \bar{M} \right) \tag{6.28}$$

In Equation (6.28), $\bar{M}$ is the average moisture content, as discussed in Section 3.3.2.1. These equations are more complex if moisture gradients are considered, since the variation of moisture concentration through the laminate is a function of z. Moisture can produce both hygral loads and moments. These are found by considering equilibrium conditions. The hygral stresses in the kth lamina are

$$\begin{Bmatrix} \sigma_x \\ \sigma_y \\ \tau_{xy} \end{Bmatrix}_k^{\mathrm{H}} = [\bar{Q}]_k \begin{Bmatrix} \beta_x \\ \beta_y \\ \beta_{xy} \end{Bmatrix}_k \bar{M}$$

The hygral loads and moments are

$$\{N^{\mathrm{H}}\} = \begin{Bmatrix} N_x^{\mathrm{H}} \\ N_y^{\mathrm{H}} \\ N_{xy}^{\mathrm{H}} \end{Bmatrix} = \int_{-h/2}^{h/2} \begin{Bmatrix} \sigma_x \\ \sigma_y \\ \tau_{xy} \end{Bmatrix}_k^{\mathrm{H}} \mathrm{d}z = \int_{-h/2}^{h/2} [\bar{Q}]_k \begin{Bmatrix} \beta_x \\ \beta_y \\ \beta_{xy} \end{Bmatrix}_k \bar{M} \mathrm{d}z = \sum_{k=1}^{N} [\bar{Q}]_k \begin{Bmatrix} \beta_x \\ \beta_y \\ \beta_{xy} \end{Bmatrix}_k \bar{M} t_k \tag{6.29}$$

$$\{M^{\mathrm{H}}\} = \begin{Bmatrix} M_x^{\mathrm{H}} \\ M_y^{\mathrm{H}} \\ M_{xy}^{\mathrm{H}} \end{Bmatrix} = \int_{-h/2}^{h/2} \begin{Bmatrix} \sigma_x \\ \sigma_y \\ \tau_{xy} \end{Bmatrix}_k^{\mathrm{H}} \mathrm{d}z = \int_{-h/2}^{h/2} [\bar{Q}]_k z \begin{Bmatrix} \beta_x \\ \beta_y \\ \beta_{xy} \end{Bmatrix}_k \bar{M} \mathrm{d}z = \sum_{k=1}^{N} [\bar{Q}]_k \begin{Bmatrix} \beta_x \\ \beta_y \\ \beta_{xy} \end{Bmatrix}_k \bar{M}(t_k \bar{z}_k) \tag{6.30}$$

6.3.3 Combined effects

The combined effects of thermal and hygral considerations on a laminate are expressible in compacted form as

$$\begin{Bmatrix} N \\ -- \\ M \end{Bmatrix} = \begin{bmatrix} A & | & B \\ -- & -- & -- \\ B & | & D \end{bmatrix} \begin{Bmatrix} \varepsilon^0 \\ -- \\ k \end{Bmatrix} - \begin{Bmatrix} N^{\mathrm{T}} \\ -- \\ M^{\mathrm{T}} \end{Bmatrix} - \begin{Bmatrix} N^{\mathrm{H}} \\ -- \\ M^{\mathrm{H}} \end{Bmatrix} \tag{6.31}$$

where $\{N^{\mathrm{T}}\}$, $\{M^{\mathrm{T}}\}$, $\{M^{\mathrm{H}}\}$, and $\{M^{\mathrm{H}}\}$ are defined by Equations (6.25), (6.26), (6.29) and (6.30), respectively, and $[A]$, $[B]$, and $[D]$ are given by Equation (6.16) or (6.22).

Mid-surface strains and curvatures are defined by Equations (6.2) and (6.3). Equation (6.31) can be cast into a more compact form by defining

$$\{\hat{N}\} = \{N\} + \{N^{\mathrm{T}}\} + \{N^{\mathrm{H}}\} \quad \{\hat{M}\} = \{M\} + \{M^{\mathrm{T}}\} + \{M^{\mathrm{H}}\} \tag{6.32}$$

Using (6.32), Equation (6.31) is expressed as

$$\begin{Bmatrix} \hat{N} \\ -- \\ \hat{M} \end{Bmatrix} = \begin{bmatrix} A & | & B \\ -- & -- & -- \\ B & | & D \end{bmatrix} \begin{Bmatrix} \varepsilon^0 \\ -- \\ k \end{Bmatrix} \tag{6.33}$$

Two distinct cases are associated with Equation (6.33):

Case 1: $[B]=0$ is the simplest case since normal strains and curvatures are uncoupled. Two equations, $\{\hat{N}\}=[A]\{\varepsilon^0\}$ and $\{\hat{M}\}=[D]\{k\}$, must be solved.

Case 2: $[B]\neq 0$ is more complicated since the strains and curvatures are coupled, and Equation (6.33) must be solved.

6.4 Laminate codes

A code is generally used to identify the lamina stacking sequence in a laminate. Knowing the code can aid in identifying the form of $[A]$, $[B]$, and $[D]$ prior to analysis, which can simplify the procedure.

6.4.1 Single-layered laminates

A single-layered laminate is a unidirectional lamina with multiple layers. If the total thickness of the laminate is t, the $[A]$, $[B]$, and $[D]$ matrices are

$$[A_{ij}] = \bar{Q}_{ij}t \quad [B_{ij}] = 0 \quad [D_{ij}] - \bar{Q}_{ij}t^3/12$$

These forms of each matrix can be verified by examining Equation (6.16) or (6.22). Since $[B]=0$, bending and extension are uncoupled, allowing simplified solution procedures. The mid-surface strains and curvatures are obtained by solving

$$\begin{Bmatrix} \hat{N} \\ -- \\ \hat{M} \end{Bmatrix} = \begin{bmatrix} A & | & 0 \\ -- & -- & -- \\ 0 & | & D \end{bmatrix} \begin{Bmatrix} \varepsilon^0 \\ -- \\ k \end{Bmatrix}$$

For fiber orientations of 0° and 90°, $\bar{Q}_{16}=\bar{Q}_{26}=0$. This results in $A_{16}=A_{26}=D_{16}=D_{26}=0$. If the fiber orientation is anything other than 0° or 90°, both $[A]$ and $[D]$ are fully populated.

6.4.2 Symmetric laminates

A symmetric laminate has both geometric and material symmetries with respect to the mid-surface. Geometric symmetry results from having identical lamina orientations above and below the mid-surface. Material symmetry can result from either having all laminas of the same material or requiring different laminas to be symmetrically disposed about the mid-surface. A result of symmetry is $[B]=0$. In order to have a symmetric laminate, there may be either an *even* or an *odd* number of layers. Examples of symmetric laminate stacking sequences and notation are shown in Figure 6.10. Similarly, their notations can be used for describing laminates composed of different materials and/or ply orientations.

6.4.3 Antisymmetric laminates

This laminate is characterized by having its layers arranged in an antisymmetric fashion with respect to the mid-surface. There must be an *even* number of plies for a laminate to be antisymmetric. The $[B]$ matrix is not zero. An example of an antisymmetric laminate (fiber orientation, ply number, and laminate code) is given in Figure 6.11. This laminate

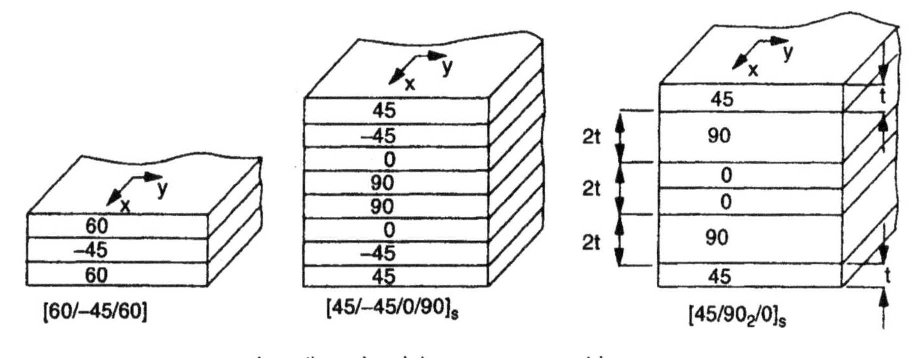

Figure 6.10 Examples of symmetric laminate ply orientations and codes.

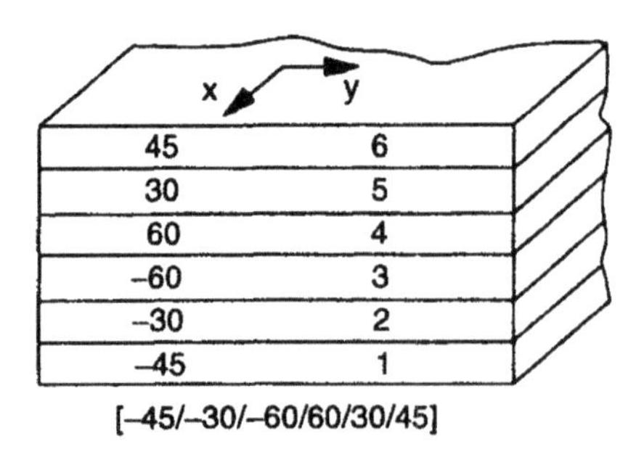

Figure 6.11 Antisymmetric laminate and code.

has several notable features. First, each $+\theta$ is accompanied by a $-\theta$. There is a distinct relationship between components of $[\bar{Q}]$ for $+\theta$ and $-\theta$ angles: $(\bar{Q}_{16})_{+\theta} = -(\bar{Q}_{16})_{-\theta}$ and $(\bar{Q}_{26})_{+\theta} = -(\bar{Q}_{26})_{-\theta}$, resulting in $A_{16} = A_{26} = D_{16} = D_{26} = 0$.

6.4.4 Cross-ply laminates

A cross-ply laminate contains an arbitrary number of plies, each with a fiber orientation of either 0° or 90°, and it can be either symmetric or antisymmetric. For the symmetric cross-ply, $[B] = 0$, but for an antisymmetric cross-ply laminate, $[B]$ can be shown to be

$$[B] = \begin{bmatrix} B_{11} & 0 & 0 \\ 0 & -B_{11} & 0 \\ 0 & 0 & 0 \end{bmatrix}$$

Since fiber orientations are either 0° or 90°, $Q_{16} = Q_{26} = 0$ for both plies. Therefore, $A_{16} = A_{26} = B_{16} = B_{26} = D_{16} = D_{26} = 0$, which is true for all cross-ply laminates.

6.4.5 Angle-ply laminates

Angle-ply laminates have an arbitrary number of layers (n). Each ply has the same thickness and is the same material. The plies have alternating fiber orientations of $+\theta$ and $-\theta$. An angle-ply laminate can be either symmetric or antisymmetric, and [0] is fully populated. Depending on whether the laminate is symmetric or antisymmetric, certain simplifications can be made in identifying components of $[A]$, $[B]$, and $[D]$. Examples of both symmetric and antisymmetric angle-ply laminates are shown in Figure 6.12.

Symmetric. The symmetric angle-ply laminate has an odd number of layers and $[B] = 0$. Since the plies alternate between $+\theta$ and $-\theta$, the shear terms $\bar{Q}_{16}$ and $\bar{Q}_{26}$ change sign between layers, while all other components of $[\bar{Q}]$ remain unchanged. For an n-ply laminate of total thickness h, the $[A]$ and $[D]$ terms that are easily related to $[\bar{Q}]$ are

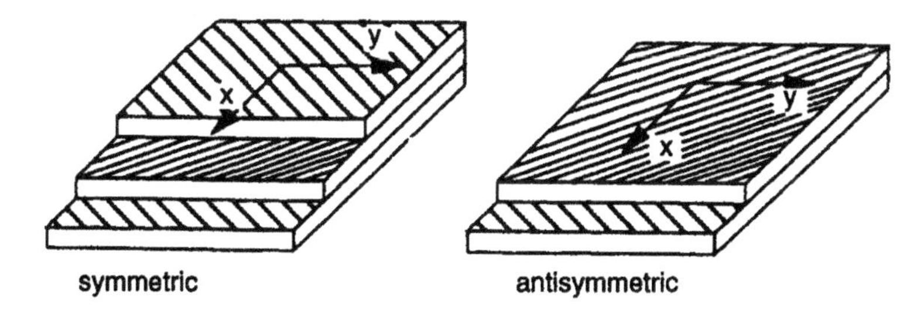

Figure 6.12 Symmetric and antisymmetric angle-ply laminates.

$$(A_{11}, A_{22}, A_{12}, A_{66}) = h(\bar{Q}_{11}, \bar{Q}_{22}, \bar{Q}_{12}, \bar{Q}_{66})$$

$$(A_{16}, A_{26}) = \left(\frac{h}{n}\right)(\bar{Q}_{16}, \bar{Q}_{26})$$

$$(D_{11}, D_{22}, D_{12}, D_{66}) = \left(\frac{h^3}{12}\right)(\bar{Q}_{11}, \bar{Q}_{22}, \bar{Q}_{12}, \bar{Q}_{66})$$

$$(D_{16}, D_{26}) = \left(\frac{h^3(3n^2 - 2)}{12n^3}\right)(\bar{Q}_{16}, \bar{Q}_{26})$$

Antisymmetric. The antisymmetric angle-ply laminate has an even number of layers (n). It is unique in that only the shear terms of the [B] matrix are present (B_{16} and B_{26}). The shear terms of [A] and [D] change from the symmetric representations just shown to $A_{16} = A_{26} = D_{16} = D_{26} = 0$. The nonzero shear terms of [B] are

$$(B_{16}, B_{26}) = -\left(\frac{h^2}{2n}\right)(\bar{Q}_{16}, \bar{Q}_{26}).$$

As the number of layers becomes larger, B_{16} and B_{26} approach zero.

6.4.6 Quasi-isotropic laminates

A quasi-isotropic laminate results when individual lamina are oriented in such a manner as to produce an isotropic [A] matrix. This means that extension and shear are uncoupled ($A_{16} = A_{26} = 0$), the components of [A] are independent of laminate orientation, and for the quasi-isotropic laminate

$$[A] = \begin{bmatrix} A_{11} & A_{12} & 0 \\ A_{12} & A_{11} & 0 \\ 0 & 0 & A_{66} \end{bmatrix}$$

The conditions of isotropic response only apply to the [A] matrix. The [B] and [D] matrices may or may not be fully populated, and extension–shear coupling is possible. Several rules apply for a quasi-isotropic laminate:

1. The total number of layers must be $n \geq 3$.
2. All layers must have identical orthotropic elastic constants (they must be the same material) and identical thickness.
3. The orientation of the kth layer of an n-layer laminate is

$$\theta_k = \frac{\pi(k-1)}{n}$$

Examples of lamina ply orientations which produce an isotropic [A] matrix are shown in Figure 6.13. These two examples can be altered by making each laminate symmetric. For example, instead of [60/0/−60], the laminate could be [60/0/−60]$_s$. This

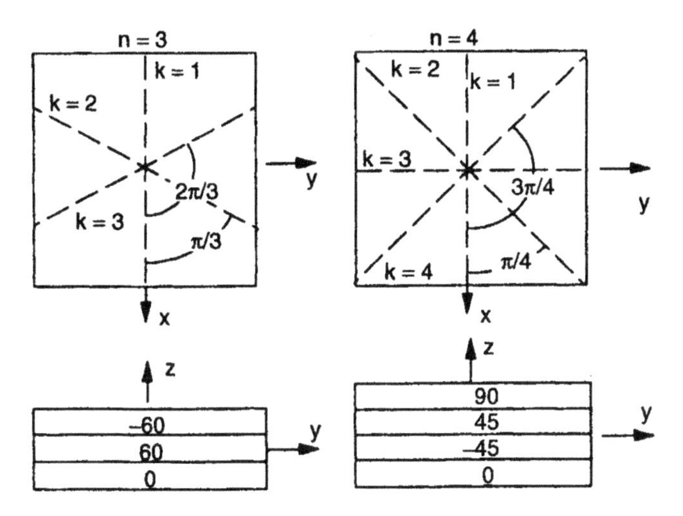

Figure 6.13 Examples of ply orientations for quasi-isotropic laminates.

condition would not alter the fact that $[A]$ is isotropic. The components of $[A]$ are different in each case, and $[B]$ and $[D]$ change.

A laminate may be quasi-isotropic and not appear to follow the rules cited. For example, a [0/−45/45/90] laminate is quasi-isotropic, and if each layer were oriented at some angle (assume +60°) from its original orientation, $[A]$ remains isotropic. The reorientation of each lamina (or the laminate as a whole) by +60° results in a [60/15/−75/−30] laminate.

6.4.7 General laminates

A general laminate does not conveniently fit into any of the specifically designated categories previously discussed. They consist of an arbitrary number of layers (either an even or odd number) oriented at selected angles with respect to the x-axis. Simplified rules for estimating components of $[A]$, $[B]$, and $[D]$ do not exist as they do for other types of laminates. Each of these matrices is typically fully populated, and residual curvatures are generally present. A [0/∓45/90] appears to be a general laminate, but is actually quasi-isotropic. Examples of several general laminate stacking arrangements are shown in Figure 6.14.

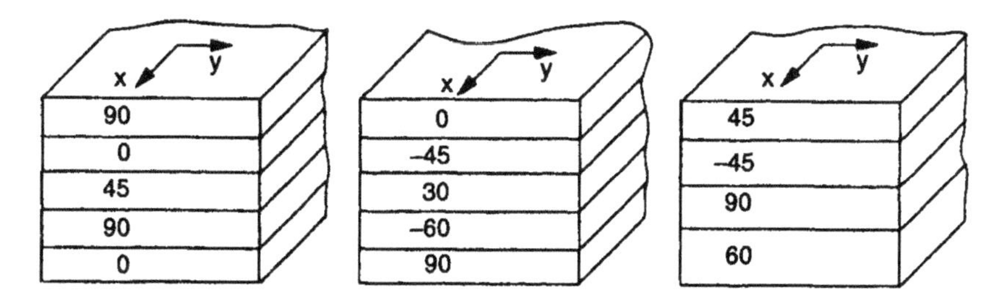

Figure 6.14 Examples of general laminate stacking sequences.

6.5 Laminate analysis

Laminate analysis requires the solution of Equation (6.33) for the mid-surface strains and curvatures. Numerical difficulties may arise when symmetric laminates ($[B]=0$) or other laminates with a large number of zeros are encountered. Rather than inverting the entire 6×6 matrix relating strain and curvature to loads, an alternative procedure can be used. Equation (6.33) can be written as two equations, one for loads and the other for moments. Through matrix manipulation, an intermediate set of matrix relationships can be defined as

$$\begin{array}{ll} [A^*]=[A]^{-1} & [B^*]=[A]^{-1}[B] \\ [C^*]=[B][A]^{-1} & [D^*]=[D]-[B][A]^{-1}[B] \\ [A']=[A^*]-[B^*][D^*]^{-1}[C^*] & [B']=[B^*][D^*]^{-1} \\ [C']=-[D^*]^{-1}[C^*] & [D']=[D^*]^{-1} \end{array} \tag{6.34}$$

The resulting solutions for $\{\varepsilon^0\}$ and $\{k\}$ as a function of $\{\hat{N}\}$ and $\{\hat{M}\}$ can be written as

$$\begin{Bmatrix} \varepsilon^0 \\ -- \\ k \end{Bmatrix} = \begin{bmatrix} A' & | & B' \\ -- & -- & -- \\ C' & | & D' \end{bmatrix} \begin{Bmatrix} \hat{N} \\ -- \\ \hat{M} \end{Bmatrix} \tag{6.35}$$

where $[C']=[B']^{\mathrm{T}}$. This relationship between $[C']$ and $[B']$ is verified by standard matrix techniques applied to the equations of (6.34).

After the mid-surface strains and curvatures are determined, the laminate can be analyzed for stresses, and subsequent failure. The failure analysis can be directed toward either a single lamina or the entire laminate. Failure depends on the state of stress in each lamina, which depends on the $[\bar{Q}]$ for each lamina. If thermal and hygral effects are considered, the Cartesian stresses in the kth lamina are

$$\begin{Bmatrix} \sigma_x \\ \sigma_y \\ \tau_{xy} \end{Bmatrix}_k = [\bar{Q}]_k \left(\begin{Bmatrix} \varepsilon_x^0 \\ \varepsilon_y^0 \\ \gamma_{xy}^0 \end{Bmatrix} + z \begin{Bmatrix} \kappa_x \\ \kappa_y \\ \kappa_{xy} \end{Bmatrix} - \begin{Bmatrix} \alpha_x \\ \alpha_y \\ \alpha_{xy} \end{Bmatrix}_k \Delta T - \begin{Bmatrix} \beta_x \\ \beta_y \\ \beta_{xy} \end{Bmatrix}_k \bar{M} \right) \tag{6.36}$$

These components of stress can easily be transformed into principal material direction stresses using stress transformations. Similarly, one could initially transform the mid-surface strains and curvatures into material direction strains. The principal direction (on-axis) strains are related to the mid-surface strains and curvatures by the strain transformation matrix:

$$\begin{Bmatrix} \varepsilon_1 \\ \varepsilon_2 \\ \gamma_{12} \end{Bmatrix}_k = [T_k] \begin{Bmatrix} \varepsilon_x \\ \varepsilon_y \\ \gamma_{xy} \end{Bmatrix}_k = [T_\varepsilon] \left(\begin{Bmatrix} \varepsilon_x^0 \\ \varepsilon_y^0 \\ \gamma_{xy}^0 \end{Bmatrix} + z \begin{Bmatrix} \kappa_x \\ \kappa_y \\ \kappa_{xy} \end{Bmatrix} \right) \tag{6.37}$$

Corresponding to these strains are the on-axis stresses. If both thermal and hygral effects are considered, the principal material direction stresses in the kth lamina are

$$\begin{Bmatrix} \sigma_1 \\ \sigma_2 \\ \tau_{12} \end{Bmatrix}_k = [Q]_k \begin{Bmatrix} \varepsilon_1 - \alpha_1 \Delta T - \beta_1 \bar{M} \\ \varepsilon_2 - \alpha_2 \Delta T - \beta_2 \bar{M} \\ \gamma_{12} \end{Bmatrix}_k \tag{6.38}$$

These equations are used in subsequent sections to determine stress variations through a laminate, failure modes, and general mechanical behavior of laminates.

6.5.1 Analysis of symmetric laminates

For symmetric laminates, the analysis procedure is simplified since $[B]=0$, and $\{M^T\}=[M^H]=0$. If the thermal and hygral effects were not assumed to be representable by a uniform laminate temperature and average moisture content, respectively, these moments may not be zero. Since there is no bending-extension coupling, the mid-surface strains and curvatures can be obtained using a simplified form of Equation (6.35):

$$\{\varepsilon^0\} = [A']\{\hat{N}\} \quad \{\kappa\} = [D']\{\hat{M}\}$$

Symmetric laminates are the most widely used and extensively studied. Tsai [8] and Azzi and Tsai [9] used symmetric laminates to correlate experimental results with theoretical predictions from CLT. They dealt primarily with cross-ply and angle-ply laminates of varying thickness and found good correlation between experimental laminate stiffness and CLT predictions. The symmetric laminate is well suited for experimental studies since there are no residual curvatures caused by thermal moments, and the response to loads is simple to predict.

6.5.1.1 Cross-ply laminate

Consider the laminate shown in Figure 6.15, for which plane stress is assumed. The material is AS/3501 graphite/epoxy, with the material properties from Table 4.2 (reference [42] in Chapter 4). The stacking sequence is $[0/90_5]_s$, and each ply has a thickness $t_{ply}=t=0.005$ in. The laminate thickness is $h=12\,t=0.06$ in. The solution is formulated in terms of the total laminate thickness h until stress and strain are computed. Thermal effects are considered, but hygral effects are neglected. The material properties used are as follows: $E_1=20.0\times10^6$ psi, $E_2=1.30\times10^6$ psi, $G_{12}=1.03\times10^6$ psi, $\nu_{12}=0.30$, $\alpha_1=-0.17$ µin./in./°F, and $\alpha_2=15.57$ µin./in./°F. The reduced stiffness matrix is

$$[Q] = \begin{bmatrix} 20.14 & 0.392 & 0 \\ 0.392 & 1.307 & 0 \\ 0 & 0 & 1.03 \end{bmatrix} \times 10^6 \text{ psi}$$

Figure 6.15 $[0/90_5]_s$ cross-ply laminate.

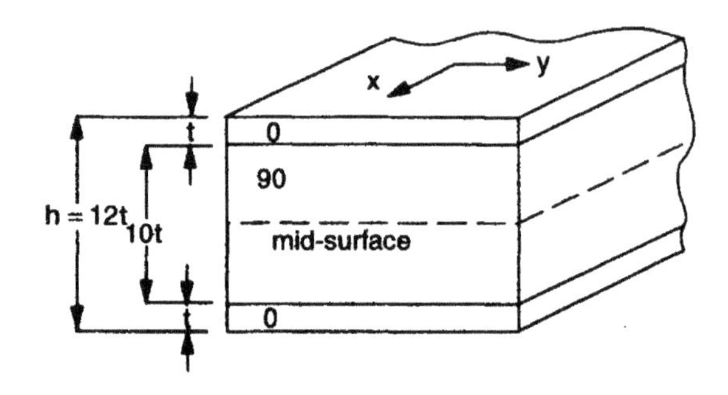

The $[\bar{Q}]$ for each lamina is

$$[\bar{Q}]_0 = [Q] \quad [\bar{Q}]_{90} = \begin{bmatrix} 1.307 & 0.392 & 0 \\ 0.392 & 20.14 & 0 \\ 0 & 0 & 1.03 \end{bmatrix} \times 10^6 \text{ psi}$$

The $[A]$ and $[D]$ matrices are formed following the procedures of Section 6.2.3.1, which are well suited for numerical work.

$$A_{ij} = \sum [\bar{Q}]_k t_k = \left\{ 2[\bar{Q}]_0 + 10[\bar{Q}]_{90} \right\} t = \begin{bmatrix} 53.35 & 4.704 & 0 \\ 4.704 & 204 & 0 \\ 0 & 0 & 12.36 \end{bmatrix} \times 10^6 t$$

$$D_{ij} = \sum [\bar{Q}]_k \left(t_k \bar{z}_k^2 + \frac{t_k^3}{12} \right) = 2[\bar{Q}]_0 \left\{ t \left(\frac{11t}{2} \right)^2 + \frac{t^3}{12} \right\} + [\bar{Q}]_{90} \frac{(10t)^3}{12}$$

$$= \begin{bmatrix} 1330.7 & 56.45 & 0 \\ 0.392 & 1757.6 & 0 \\ 0 & 0 & 117.1 \end{bmatrix} \times 10^6 t^3$$

Incorporating $h = 12\, t$, these matrices are

$$[A] = \begin{bmatrix} 4.45 & 0.392 & 0 \\ 0.392 & 17.0 & 0 \\ 0 & 0 & 1.03 \end{bmatrix} \times 10^6 h \quad [D] = \begin{bmatrix} 0.768 & 0.033 & 0 \\ 0.033 & 1.02 & 0 \\ 0 & 0 & 0.068 \end{bmatrix} \times 10^6 h^3$$

Inversion of $[A]$ and $[D]$ yields

$$[A'] = \begin{bmatrix} 0.226 & -0.005 & 0 \\ -0.005 & 0.0589 & 0 \\ 0 & 0 & 0.972 \end{bmatrix} \left(\frac{10^{-6}}{h} \right)$$

$$[D'] = \begin{bmatrix} 1.30 & -0.042 & 0 \\ -0.042 & 0.9850 & 0 \\ 0 & 0 & 11.6 \end{bmatrix} \left(\frac{10^{-6}}{h^3} \right)$$

Using Equation (6.25), the thermal loads are expressed as

$$\{N^{\mathrm{T}}\} = \sum [\bar{Q}]_k \{\alpha\}_k t_k \Delta T = \left(2[\bar{Q}]_0\{\alpha\}_0 + 10[\bar{Q}]_{90}\{\alpha\}_{90}\right) t\Delta T$$

Using α_1 and α_2 given earlier, application of Equation (3.22) yields

$$\{\alpha\}_0 = \begin{Bmatrix} \alpha_x \\ \alpha_y \\ \alpha_{xy} \end{Bmatrix}_0 = \begin{Bmatrix} -0.17 \\ 15.57 \\ 0 \end{Bmatrix} \mu\text{in./in./°F}$$

$$\{\alpha\}_{90} = \begin{Bmatrix} \alpha_x \\ \alpha_y \\ \alpha_{xy} \end{Bmatrix}_{90} = \begin{Bmatrix} 15.57 \\ -0.17 \\ 0 \end{Bmatrix} \mu\text{in./in./°F}$$

Using these $\{\alpha\}_0$ and $\{\alpha\}_{90}$ values, the thermal loads are

$$\{N^{\mathrm{T}}\} = \left(2[\bar{Q}]_0\{\alpha\}_0 + 10[\bar{Q}]_{90}\{\alpha\}_{90}\right) t\Delta T = \begin{Bmatrix} 209 \\ 68.3 \\ 0 \end{Bmatrix} t\Delta T = \begin{Bmatrix} 17.4 \\ 5.69 \\ 0 \end{Bmatrix} h\Delta T$$

The Cartesian components of mid-surface strains and curvatures are obtained by solving Equation (6.35) and are

$$\begin{Bmatrix} \varepsilon^0 \\ -- \\ k \end{Bmatrix} = \begin{bmatrix} A' & | & 0 \\ -- & -- & -- \\ 0 & | & D' \end{bmatrix} \begin{Bmatrix} \hat{N} \\ -- \\ \hat{M} \end{Bmatrix} = \begin{bmatrix} A' & | & 0 \\ -- & -- & -- \\ 0 & | & D' \end{bmatrix} \begin{Bmatrix} N + N^{\mathrm{T}} \\ -------- \\ M \end{Bmatrix}$$

Prior to addressing the problem with applied loads, it is instructive to examine the effect of residual stresses due to curing. Assuming $\{N\} = \{M\} = 0$, the curvatures are zero ($\{k\} = 0$), but mid-surface strains exist and are

$$\{\varepsilon^0\} = \begin{Bmatrix} \varepsilon_x^0 \\ \varepsilon_y^0 \\ \varepsilon_{xy}^0 \end{Bmatrix} = [A']\{N^{\mathrm{T}}\} = \begin{Bmatrix} 3.90 \\ 0.245 \\ 0 \end{Bmatrix} \times 10^{-6}\Delta T$$

Assuming the laminate is cured at 370 °F and room temperature is 70 °F, $\Delta T = 70° - 370° = -300$ °F, and the mid-surface strains are

$$\begin{Bmatrix} \varepsilon_x^0 \\ \varepsilon_y^0 \\ \gamma_{xy}^0 \end{Bmatrix} = \begin{Bmatrix} 3.90 \\ 0.245 \\ 0 \end{Bmatrix} \times 10^{-6}(-300) = \begin{Bmatrix} -1170 \\ -73.5 \\ 0 \end{Bmatrix} \mu\text{in./in.}$$

The variation of strain through the laminate is given by Equation (6.4) as $\{\varepsilon\}_k = \{\varepsilon^0\} + z\{\kappa\}$. Since $\{\kappa\} = 0$, the strain is uniform through the laminate, but the stresses are not. Stresses are determined from the general relationship between stress and strain defined in Section 6.5 as $\{\sigma\}_k = [\bar{Q}]_k\ (\{\varepsilon\}_k - \{\alpha\}_k \Delta T)$. The residual curing stresses for the 0° lamina are

$$\begin{Bmatrix} \sigma_x \\ \sigma_y \\ \tau_{xy} \end{Bmatrix}_0 = [\bar{Q}]_0 \left(\begin{Bmatrix} \varepsilon_x^0 \\ \varepsilon_y^0 \\ \gamma_{xy}^0 \end{Bmatrix} - \begin{Bmatrix} \alpha_x \\ \alpha_y \\ \alpha_{xy} \end{Bmatrix}_0 \Delta T \right)$$

$$= \begin{bmatrix} 20.14 & 0.392 & 0 \\ 0.392 & 1.307 & 0 \\ 0 & 0 & 1.03 \end{bmatrix} \times 10^6 \begin{Bmatrix} -1170 - (-300)(-0.17) \\ -73.5 - (-300)(15.57) \\ 0 \end{Bmatrix} \times 10^{-6}$$

$$= \begin{Bmatrix} -22.8 \\ 5.55 \\ 0 \end{Bmatrix} \text{ksi}$$

Similarly, for the 90° lamina

$$\begin{Bmatrix} \sigma_x \\ \sigma_y \\ \tau_{xy} \end{Bmatrix}_{90} = \begin{bmatrix} 1.307 & 0.392 & 0 \\ 0.392 & 20.14 & 0 \\ 0 & 0 & 1.03 \end{bmatrix} \times 10^6 \begin{Bmatrix} -1170 - (-300)(-0.17) \\ -73.5 - (-300)(15.57) \\ 0 \end{Bmatrix} \times 10^{-6}$$

$$= \begin{Bmatrix} 4.55 \\ -1.12 \\ 0 \end{Bmatrix} \text{ksi}$$

The cross-ply laminate is unique in that once the Cartesian components of stress are determined, it is simple to define the principal material direction (on-axis) stresses. This is because in either lamina, one of the on-axis directions coincides with the x-direction, and the other with the y-direction as illustrated in Figure 6.16.

The residual stresses developed during curing depend on both macroscopic and microscopic conditions. On the macroscopic level, the stacking sequence gives rise to stresses resulting from the variation of stiffness and thermal expansion coefficients across the ply interfaces between homogeneous layers. A microscopic effect for thermoplastic matrix composites is morphology associated with crystallinity gradients during cooling, which influences the macroscopic internal stresses. Processing conditions are closely related to macroscopic stresses [10–12]. Experimental techniques for defining residual stress in thermoplastic and thermosetting composites using techniques of layer removal [11,13–15], photoelasticity [16], and laminate failure [17,18] are often used to investigate processing-induced stresses. Detailed discussions of these methods are beyond the scope of this text.

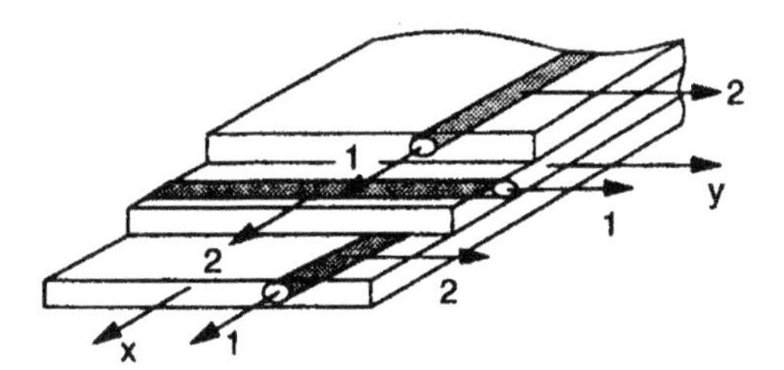

Figure 6.16 Orientation of 1–2 fiber directions with global x–y directions for a cross-ply laminate.

Consider the effect of externally applied loads. The mid-surface strains and curvatures can be determined by solving the relation

$$\begin{Bmatrix} \varepsilon^0 \\ -- \\ \kappa \end{Bmatrix} = \begin{bmatrix} A' & | & 0 \\ -- & -- & -- \\ 0 & | & D' \end{bmatrix} \begin{Bmatrix} \hat{N} \\ -- \\ \hat{M} \end{Bmatrix}$$

Assume the only load is N_x, which can be either tensile or compressive. With $\Delta T = -300$ °F,

$$\{\hat{N}\} = \begin{Bmatrix} N_x + N_x^{\mathrm{T}} \\ N_y^{\mathrm{T}} \\ 0 \end{Bmatrix} = \begin{Bmatrix} N_x - 313 \\ -102 \\ 0 \end{Bmatrix}$$

Since $\{M\} = 0$, [D′] is not required, and $\{\kappa\} = 0$. Therefore,

$$\begin{Bmatrix} \varepsilon_x^0 \\ \varepsilon_y^0 \\ \gamma_{xy}^0 \end{Bmatrix} = \begin{Bmatrix} 0.226 & -0.005 & 0 \\ -0.005 & 0.0589 & 0 \\ 0 & 0 & 0.972 \end{Bmatrix} = \begin{Bmatrix} N_x - 313 \\ -102 \\ 0 \end{Bmatrix} \left(\frac{10^{-6}}{h} \right)$$

With the laminate thickness being $h = 0.06$ in., this becomes

$$\begin{Bmatrix} \varepsilon_x^0 \\ \varepsilon_y^0 \\ \gamma_{xy}^0 \end{Bmatrix} = \begin{Bmatrix} 3.76N_x - 1171 \\ -0.086N_x - 73.5 \\ 0 \end{Bmatrix} \times 10^{-6}$$

The stress in each lamina (for $\Delta T = -300$ °F) is

$$\begin{Bmatrix} \sigma_x \\ \sigma_y \\ \tau_{xy} \end{Bmatrix}_k = [\bar{Q}]_k \left(\begin{Bmatrix} 3.76N_x - 1170 \\ -0.0867N_x - 73.5 \\ 0 \end{Bmatrix} - (-300) \begin{Bmatrix} \alpha_x \\ \alpha_y \\ \alpha_{xy} \end{Bmatrix}_k \right)$$

Using the appropriate αs for each lamina, as determined from Equation (3.22), the stresses in each are

$$\begin{Bmatrix} \sigma_x \\ \sigma_y \\ \tau_{xy} \end{Bmatrix}_0 = \begin{Bmatrix} 75.7N_x - 22{,}800 \\ 1.36N_x + 5550 \\ 0 \end{Bmatrix} \quad \begin{Bmatrix} \sigma_x \\ \sigma_y \\ \tau_{xy} \end{Bmatrix}_{90} = \begin{Bmatrix} 4.88N_x + 4550 \\ -0.271N_x - 1120 \\ 0 \end{Bmatrix}$$

From these expressions, it is a simple matter to evaluate the stresses in each lamina as a function of applied load N_x. In order to produce a tensile stress in the x-direction in the 0° lamina, the applied load N_x must overcome the thermal stress and therefore must be greater than 22,800/75.7 = 301 lb/in. Similar assessments can be made for the other stress components.

Other loading conditions produce different stress distributions in each lamina. For example, assume the $[0/90_5]_s$ laminate under consideration is subjected to four-point bending as shown in Figure 6.17. The laminate loads in the region of constant bending moment are also shown in this figure. Assuming the laminate width is 1.5 in., the bending moment used for stress analysis is $(-PL/8)/1.5$, since bending moments have dimensions of in.-lb/in. (discussed in Section 6.2.3). Assuming no axial force is applied, the laminate loads consist of only $\{N^T\}$ and the moment $M = -PL/12$ ($\{M^T\} = 0$ for symmetric laminates). Therefore,

$$\{\hat{N}\} = \begin{Bmatrix} 17.4 \\ 5.69 \\ 0 \end{Bmatrix} h\Delta T \quad \{\hat{M}\} = \begin{Bmatrix} -PL/12 \\ 0 \\ 0 \end{Bmatrix}$$

Using Equation (6.35) with the appropriate $[A']$, $[D']$, $\{\hat{N}\}$, and $\{\hat{M}\}$ results in both mid-surface strains and curvatures, which are

$$\begin{Bmatrix} \varepsilon_x^0 \\ \varepsilon_y^0 \\ \gamma_{xy}^0 \end{Bmatrix} = \begin{Bmatrix} 3.90 \\ 0.245 \\ 0 \end{Bmatrix} \times 10^{-6}\Delta T \quad \begin{Bmatrix} \kappa_x \\ \kappa_y \\ \kappa_{xy} \end{Bmatrix} = \begin{Bmatrix} -0.1083 \\ 0.0035 \\ 0 \end{Bmatrix} \left(\frac{PL}{h^3}\right) \times 10^{-6}$$

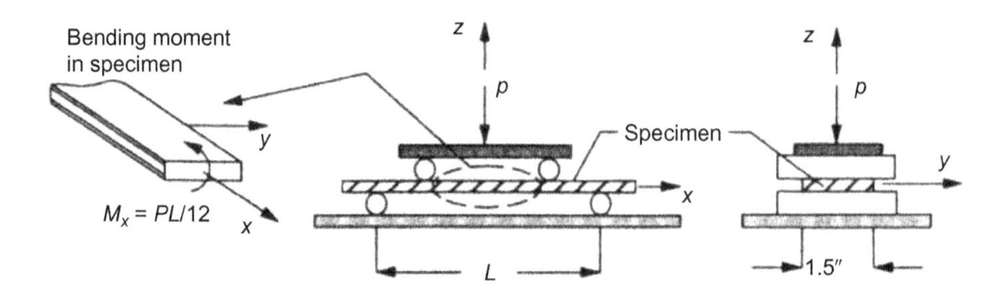

Figure 6.17 Four-point bending of a $[0/90_5]_s$ graphite/epoxy laminate.

For a total laminate thickness of $h = 0.06$ in. and $\Delta T = -300$ °F, the strain variation through the laminate is

$$\begin{Bmatrix} \varepsilon_x^0 \\ \varepsilon_y^0 \\ \gamma_{xy}^0 \end{Bmatrix} = \begin{Bmatrix} 3.90 \\ 0.245 \\ 0 \end{Bmatrix} \times 10^{-6}(\Delta T) + z \begin{Bmatrix} -0.1083 \\ 0.0035 \\ 0 \end{Bmatrix} \left(\frac{PL}{0.06^3}\right)$$
$$= \begin{Bmatrix} -1170 \\ -73.5 \\ 00 \end{Bmatrix} + z \begin{Bmatrix} -501 \\ 16 \\ 0 \end{Bmatrix} (PL) \ \mu\text{in./in.}$$

The stress variation through the lamina is

$$\begin{Bmatrix} \sigma_x \\ \sigma_y \\ \tau_{xy} \end{Bmatrix}_k = [\bar{Q}]_k \left(\begin{Bmatrix} -1170 \\ -73.5 \\ 0 \end{Bmatrix} + z(PL) \begin{Bmatrix} -501 \\ 16 \\ 0 \end{Bmatrix} - \begin{Bmatrix} \alpha_x \\ \alpha_y \\ \alpha_{xy} \end{Bmatrix}_k \Delta T \right)$$

We have already established $\Delta T = -300$ °F as well as

$$\begin{Bmatrix} \alpha_x \\ \alpha_y \\ \alpha_{xy} \end{Bmatrix}_0 = \begin{Bmatrix} -0.17 \\ 15.57 \\ 0 \end{Bmatrix} \mu\text{in./in./°F and} \begin{Bmatrix} \alpha_x \\ \alpha_y \\ \alpha_{xy} \end{Bmatrix}_{90} = \begin{Bmatrix} 15.57 \\ -0.17 \\ 0 \end{Bmatrix} \mu\text{in./in./°}F$$

Therefore, for the 0° lamina,

$$\begin{Bmatrix} \sigma_x \\ \sigma_y \\ \tau_{xy} \end{Bmatrix}_0 = \begin{Bmatrix} -24,431 \\ 5960 \\ 0 \end{Bmatrix} + z(PL) \begin{Bmatrix} -10,084 \\ -175 \\ 0 \end{Bmatrix}$$

The variable z has two ranges for the 0° lamina: $0.025 \le z \le 0.030$ and $-0.030 \le z \le -0.025$. For the 90° lamina, the stress is

$$\begin{Bmatrix} \sigma_x \\ \sigma_y \\ \tau_{xy} \end{Bmatrix}_{90} = \begin{Bmatrix} 4526 \\ -1146 \\ 0 \end{Bmatrix} + z(PL) \begin{Bmatrix} -649 \\ 126 \\ 0 \end{Bmatrix}$$

where $-0.025 \le z \le 0.025$. The maximum tensile and compressive stresses occur at the top and bottom surfaces of the beam ($z = 0.030$ for compression and $z = -0.030$ for tension) and are functions of P and L.

Altering the stacking sequence while retaining the same ply thickness for each lamina results in a laminate whose response to axial loads is unchanged, but which has a different response to bending. Consider, for example, the laminate shown in Figure 6.18.

Since $A_{ij} = \sum [\bar{Q}]_k t_k$, the $[A]$ matrix is the same as that of the $[0/90_5]_s$. The response of either laminate to axial loads is the same. The stress variation through each

Figure 6.18 $[0/90_5]_s$ graphite/epoxy cross-ply laminate.

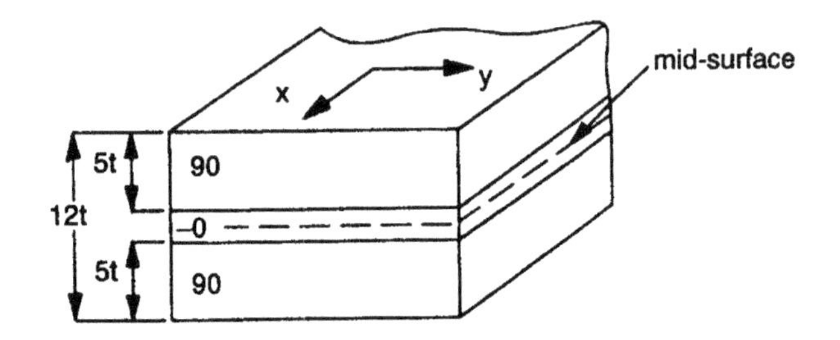

laminate due to an axial load changes as a result of the relative position of each lamina, but the stress magnitudes remain the same.

The response of each laminate to bending is different, a result of the changes in the $[D]$ matrix that occur when the stacking sequence is altered. For the $[90_5/0]_s$ laminate

$$D_{ij} = \sum [\bar{Q}]_k \left(t_k \bar{z}_k^2 + \frac{t_k^3}{12} \right) = [\bar{Q}]_0 \left(\frac{2t}{12} \right)^3 + 2[\bar{Q}]_{90} \left\{ t\frac{(7t)^2}{2} + \frac{(5t)^3}{12} \right\}$$

$$= \begin{bmatrix} 72.67 & 18.03 & 0 \\ 18.03 & 913.9 & 0 \\ 0 & 0 & 47.3 \end{bmatrix} \times 10^6 t^3$$

Using $h = 12t$ results in

$$[D] = \begin{bmatrix} 0.042 & 0.0104 & 0 \\ 0.0104 & 0.5289 & 0 \\ 0 & 0 & 0.0274 \end{bmatrix} \times 10^6 h^3$$

Comparing this to the $[D]$ for the $[0/90_5]_s$ laminate shows that its flexural response is different.

Applying the bending moment shown in Figure 6.17 and retaining the same $\{\hat{N}\}$ produces a strain variation through the laminate of

$$\begin{Bmatrix} \varepsilon_x \\ \varepsilon_y \\ \gamma_{xy} \end{Bmatrix} = \begin{Bmatrix} -1170 \\ -73.5 \\ 0 \end{Bmatrix} + z \begin{Bmatrix} -9231 \\ 181 \\ 0 \end{Bmatrix} (PL)\ \mu\text{in./in.}$$

The resulting stresses in each lamina are computed to be

$$\begin{Bmatrix} \sigma_x \\ \sigma_y \\ \tau_{xy} \end{Bmatrix}_0 = \begin{Bmatrix} -24{,}431 \\ 5960 \\ 0 \end{Bmatrix} + z(PL) \begin{Bmatrix} -185{,}841 \\ -3382 \\ 0 \end{Bmatrix}$$

$$\begin{Bmatrix} \sigma_x \\ \sigma_y \\ \tau_{xy} \end{Bmatrix}_{90} = \begin{Bmatrix} 4526 \\ -1146 \\ 0 \end{Bmatrix} + z(PL) \begin{Bmatrix} -11{,}994 \\ 27 \\ 0 \end{Bmatrix}$$

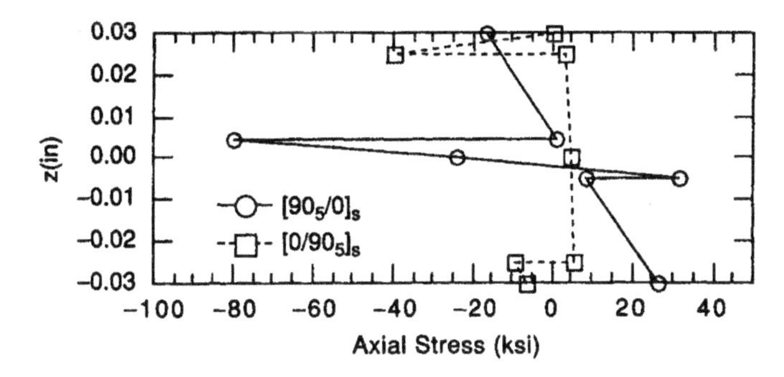

Figure 6.19 Variation of σ_x through $[0/90_5]_s$ and $[90/0]_s$ graphite/epoxy laminates.

Assuming $L = 12$ in. and $p = 5$ lb, the variation of σ_x through the thickness of each laminate is shown in Figure 6.19. The magnitudes of σ_y are not as significant as those of σ_x and are not shown. The shear stress for each laminate is zero.

The stress distribution through the laminate is not symmetric for cases of pure bending since the strains are functions of $\{N^T\}$ and $\{\hat{M}\}$. If thermal effects are ignored, the distribution is symmetric. The components of $[D]$ for the $[0/90_5]_s$ laminate are smaller than those of the $[90_5/0]_s$ laminate, meaning it is more flexible, which results in higher stresses. The 0° lamina for each laminate is oriented with its fibers in the direction of the applied moment. Placing the 0° lamina farther from the midsurface increases the overall laminate stiffness. Both laminates respond in the same manner to axial loads, but their response to flexure can be altered by the appropriate arrangement of each lamina.

These examples do not represent typical cross-ply laminate behavior. The requirement for a laminate to be classified as cross-ply is that each lamina be either 0° or 90°. The laminate itself can be either symmetric or antisymmetric.

6.5.1.2 Angle-ply laminates

Two symmetric angle-ply laminates with $[B] = 0$ and $\{M^T\} = \{M^H\} = 0$ are considered. The first is the $[\pm 45]_{2s}$ laminate referred to in Section 4.3.3 as a candidate for determining G_{12}, and the second is a $[15/-15_4]_s$ laminate.

$[\pm 45]_{2s}$ Laminate. The stacking sequence and loading condition for determination of G_{12} is shown in Figure 6.20. The test coupon geometry for this specimen is identical to that of the 90° tensile test specimen described in Section 4.3.1. Since a tensile test is involved, end tabs are required. Each lamina is assumed to have a thickness t, with a total thickness of $8t$. The stiffness matrix for each lamina is

$$[\bar{Q}]_{\pm 45} = \begin{bmatrix} \bar{Q}_{11} & \bar{Q}_{12} & \pm\bar{Q}_{16} \\ \bar{Q}_{12} & \bar{Q}_{22} & \pm\bar{Q}_{16} \\ \pm\bar{Q}_{16} & \pm\bar{Q}_{16} & \bar{Q}_{66} \end{bmatrix}$$

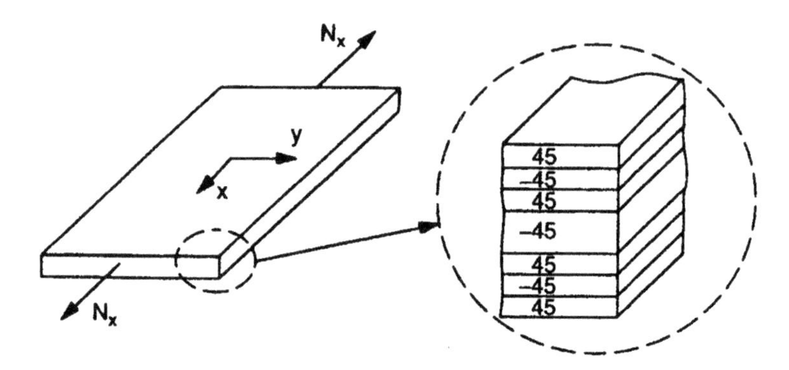

Figure 6.20 $[\pm45]_{s2}$ laminate for determination of G_{12}.

Where the $+\bar{Q}_{ij}$ and $-\bar{Q}_{ij}$ terms correspond to the +45° and −45° lamina, respectively. Because of symmetry, the +45° and −45° laminas combine to form an $[A]$ matrix in which $A_{16}=A_{26}=0$:

$$[A]=\begin{bmatrix}\bar{Q}_{11} & \bar{Q}_{12} & 0\\ \bar{Q}_{12} & \bar{Q}_{11} & 0\\ 0 & 0 & \bar{Q}_{66}\end{bmatrix}(8t)$$

Estimating G_{12} from this laminate requires strain gage data. The strain gages are applied after the laminate is cured, so they cannot be used to evaluate residual strains, and thermal effects are not considered. Hygral effects are also neglected. From the applied loads shown in Figure 6.20, the load–strain relationship for the laminate is

$$\begin{Bmatrix}N_x\\0\\0\end{Bmatrix}=\begin{bmatrix}\bar{Q}_{11} & \bar{Q}_{12} & 0\\ \bar{Q}_{12} & \bar{Q}_{22} & 0\\ 0 & 0 & \bar{Q}_{66}\end{bmatrix}(8t)\begin{Bmatrix}\varepsilon_x^0\\ \varepsilon_y^0\\ \varepsilon_{xy}^0\end{Bmatrix}$$

From this we get

$$N_x=8\left(\bar{Q}_{11}\varepsilon_x^0+\bar{Q}_{12}\varepsilon_y^0\right)t$$

$$0=8\left(\bar{Q}_{12}\varepsilon_x^0+\bar{Q}_{11}\varepsilon_y^0\right)t\Rightarrow\varepsilon_y^0=-\frac{\bar{Q}_{12}}{\bar{Q}_{11}}\varepsilon_x^0$$

$$0=8\bar{Q}_{66}\gamma_{xy}^0t\Rightarrow\gamma_{xy}^0=0$$

N_x has units of load per unit length of laminate and can be changed to an expression for σ_x by dividing through by the total laminate thickness $(8t)$. Using the ε_y^0 just given, and converting from load to stress,

$$\sigma_x=\frac{N_x}{8t}=\frac{\bar{Q}_{11}^2-\bar{Q}_{12}^2}{\bar{Q}_{11}}\varepsilon_x^0$$

The mid-surface strains are

$$\varepsilon_x^0 = \frac{N_x}{8t}\left(\frac{\bar{Q}_{11}}{\bar{Q}_{11}^2 - \bar{Q}_{12}^2}\right) \quad \varepsilon_y^0 = \frac{-N_x}{8t}\left(\frac{\bar{Q}_{12}}{\bar{Q}_{11}^2 - \bar{Q}_{12}^2}\right) \quad \gamma_{xy}^0 = 0$$

Since $\{k\} = 0$, the strain through the laminate is constant. The stress varies according to the stiffness in each ply, $[\bar{Q}]_k$. The stresses in either the +45° or −45° lamina are

$$\begin{Bmatrix} \sigma_x \\ \sigma_y \\ \tau_{xy} \end{Bmatrix}_{\pm 45} = \begin{bmatrix} \bar{Q}_{11} & \bar{Q}_{12} & \pm\bar{Q}_{16} \\ \bar{Q}_{12} & \bar{Q}_{22} & \pm\bar{Q}_{16} \\ \pm\bar{Q}_{16} & \pm\bar{Q}_{16} & \bar{Q}_{66} \end{bmatrix} \begin{Bmatrix} \bar{Q}_{11} \\ -\bar{Q}_{12} \\ 0 \end{Bmatrix} \frac{\sigma_x}{\bar{Q}_{11}^2 - \bar{Q}_{12}^2}$$

$$= \begin{Bmatrix} 1 \\ 0 \\ \pm\bar{Q}_{16}(\bar{Q}_{11} - \bar{Q}_{12}) \end{Bmatrix} \frac{\sigma_x}{\bar{Q}_{11}^2 - \bar{Q}_{12}^2}$$

The objective is to determine G_{12}; therefore, the shear stress in the principal material direction is required. The principal direction stresses in either the +45° or −45° lamina are obtained through use of the stress transformation matrix given in Equation (2.3). Since $\sigma_y = 0$, the transformation to the principal material direction is

$$\begin{Bmatrix} \sigma_1 \\ \sigma_2 \\ \tau_{12} \end{Bmatrix} = \begin{bmatrix} m^2 & 2mn \\ n^2 & -2mn \\ -mn & m^2 - n^2 \end{bmatrix} \begin{Bmatrix} \sigma_x \\ \tau_{xy} \end{Bmatrix} = \begin{Bmatrix} \frac{\sigma_x}{2} \pm \tau_{xy} \\ \frac{\sigma_x}{2} \pm \tau_{xy} \\ \mp\frac{\sigma_x}{2} \end{Bmatrix}$$

The shear stress is a function of the applied stress σ_x and is not coupled to τ_{xy}. The strains in the principal material directions are

$$\begin{Bmatrix} \varepsilon_1 \\ \varepsilon_2 \\ \gamma_{12} \end{Bmatrix} = \begin{Bmatrix} \left(\varepsilon_x^0 + \varepsilon_x^0\right) \\ \left(\varepsilon_x^0 + \varepsilon_y^0\right) \\ -2\left(\varepsilon_x^0 + \varepsilon_y^0\right) \end{Bmatrix} \left(\frac{1}{2}\right)$$

The shear stress and shear strain for this laminate are $\tau_{12} = -\sigma_x/2$ and $\gamma_{12} = -\varepsilon_x^0 + \varepsilon_y^0$. Using the relation $\tau = G\gamma$, we see

$$G_{12} = \frac{-\sigma_x/2}{\varepsilon_y^0 - \varepsilon_x^0} = \frac{-\sigma_x/2}{\left(\varepsilon_y^0/\varepsilon_x^0 - 1\right)\varepsilon_x^0} = \frac{\sigma_x/\varepsilon_x^0}{2\left(1 - \varepsilon_y^0/\varepsilon_x^0\right)}$$

The applied load can be monitored during testing, and strain gages mounted on the specimen allow for direct determination of ε_x and ε_y. The effective modulus in the

x-direction (on a +45° lamina) and Poisson's ratio in the x–y plane are related to the measured strains by $E_x = E_{+45} = \sigma_x/\varepsilon_x^0$ and $\nu_{xy} = \nu_{+45} = -\varepsilon_y^0/\varepsilon_x^0$. Using these two relations and the expression just shown, G_{12} is

$$G_{12} = \frac{E_x}{2(1+\nu_{xy})}$$

[15/−15$_4$]$_s$ Laminate. Consider the [15/−15$_4$]$_s$ angle-ply laminate in Figure 6.21 with $E_1 = 22.2 \times 10^6$ psi, $E_2 = 1.58 \times 10^6$ psi, $G_{12} = 0.81 \times 10^6$ psi, $\nu_{12} = 0.30$, $\alpha_1 = 0.011$ μin./in./°F, $\alpha_2 = 12.5$ μin./in./°F, and $t_{ply} = 0.005$ in. This example (with the exception of material properties) has been discussed by Tsai [19] and Jones [2]. Hygral effects are not considered. For this material and the ply orientations given,

$$[Q] = \begin{bmatrix} 22.35 & 0.4932 & 0 \\ 0.4932 & 1.591 & 0 \\ 0 & 0 & 0.81 \end{bmatrix} \times 10^6 \text{ psi}$$

$$[\bar{Q}]_{\pm 15} = \begin{bmatrix} 19.73 & 1.726 & \pm 4.730 \\ 1.726 & 1.749 & \pm 0.461 \\ \pm 4.730 & \pm 0.461 & 2.042 \end{bmatrix} \times 10^6 \text{ psi}$$

The terms in $[\bar{Q}]$ for each lamina have units of 10^6 psi, which are not shown in the computation of $[A]$ and $[D]$ below, but are reflected in the answer. The laminate is symmetric, so $[B] = 0$.

$$[A_{ij}] = \sum [\bar{Q}]_k t_k = \left(2[\bar{Q}]_{15} + 8[\bar{Q}]_{-15}\right)(0.005)$$
$$= \begin{bmatrix} 9.86 & 0.863 & -1.42 \\ 0.863 & 0.875 & -0.138 \\ -1.42 & -0.138 & 1.02 \end{bmatrix} \times 10^5$$

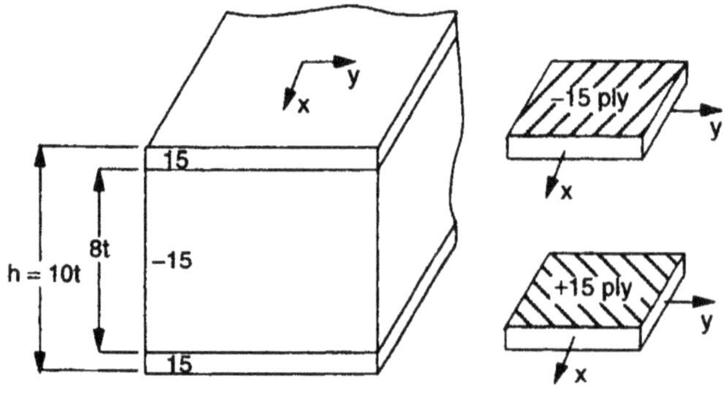

Figure 6.21 [15/−15$_4$]$_s$ symmetric angle-ply laminate.

$$[D_{ij}] = \sum [\bar{Q}]_k \left(t_k \bar{z}_k^2 + \frac{t_k^3}{12} \right)$$

$$= 2[\bar{Q}]_{15} \left\{ (0.005) \left(\frac{0.045}{2} \right)^2 + \frac{0.005^3}{12} \right\} + [\bar{Q}]_{-15} \frac{(0.040)^3}{12}$$

$$= \begin{bmatrix} 206 & 18.0 & -1.18 \\ 18.0 & 18.2 & -0.115 \\ -1.18 & -0.115 & 21.3 \end{bmatrix}$$

The thermal loads are established after the coefficients of thermal expansion for each lamina are determined. For the symmetric laminate, $\{M^T\} = 0$. The coefficients of thermal expansion for each lamina are

$$\begin{Bmatrix} \alpha_x \\ \alpha_y \\ \alpha_{xy} \end{Bmatrix}_k = \begin{bmatrix} m^2 & n^2 \\ n^2 & m^2 \\ 2mn & -2mn \end{bmatrix} \begin{Bmatrix} \alpha_1 \\ \alpha_2 \end{Bmatrix}$$

Therefore, for the 15° lamina, the αs are

$$\begin{Bmatrix} \alpha_x \\ \alpha_y \\ \alpha_{xy} \end{Bmatrix}_{15} = \begin{bmatrix} 0.933 & 0.067 \\ 0.067 & 0.933 \\ 0.25 & -0.25 \end{bmatrix} \begin{Bmatrix} 0.011 \\ 12.5 \end{Bmatrix} = \begin{Bmatrix} 0.848 \\ 11.66 \\ -6.24 \end{Bmatrix} \mu\text{in./in./°F}$$

Similarly, for the −15° lamina,

$$\begin{Bmatrix} \alpha_x \\ \alpha_y \\ \alpha_{xy} \end{Bmatrix}_{-15} = \begin{bmatrix} 0.933 & 0.067 \\ 0.067 & 0.933 \\ -0.25 & 0.25 \end{bmatrix} \begin{Bmatrix} 0.011 \\ 12.5 \end{Bmatrix} = \begin{Bmatrix} 0.848 \\ 11.66 \\ 6.24 \end{Bmatrix} \mu\text{in./in./°F}$$

Using these coefficients for each lamina, the thermal loads are

$$\{N^T\} = \sum [\bar{Q}]_k \begin{Bmatrix} \alpha_x \\ \alpha_y \\ \alpha_{xy} \end{Bmatrix}_k \Delta T(t_k)$$

$$= 2\left\{ [\bar{Q}]_{15} \begin{Bmatrix} \alpha_x \\ \alpha_y \\ \alpha_{xy} \end{Bmatrix}_{15} + 4[\bar{Q}]_{-15} \begin{Bmatrix} \alpha_x \\ \alpha_y \\ \alpha_{xy} \end{Bmatrix}_{-15} \right\} \Delta T(0.005)$$

$$= \begin{Bmatrix} 0.3656 \\ 0.9493 \\ 0.1011 \end{Bmatrix} \Delta T$$

The mid-surface strains and curvatures are found by solving

$$\begin{Bmatrix} \varepsilon^0 \\ -- \\ \kappa \end{Bmatrix} = \begin{bmatrix} A' & | & 0 \\ -- & -- & --- \\ 0 & | & D' \end{bmatrix} \begin{Bmatrix} \hat{N} \\ -- \\ \hat{M} \end{Bmatrix} = \begin{bmatrix} A' & | & 0 \\ -- & -- & --- \\ 0 & | & D' \end{bmatrix} \begin{Bmatrix} N+N^{\mathrm{T}} \\ -------- \\ M \end{Bmatrix}$$

Assuming the only load applied is N_x, the mid-surface strains are

$$\begin{Bmatrix} \varepsilon_x^0 \\ \varepsilon_y^0 \\ \gamma_{xy}^0 \end{Bmatrix} = \begin{bmatrix} 1.36 & -1.06 & 1.74 \\ -1.06 & 12.5 & 0.217 \\ 1.74 & 0.217 & 12.2 \end{bmatrix} \times 10^{-6} \left(\begin{Bmatrix} N_x \\ 0 \\ 0 \end{Bmatrix} + \begin{Bmatrix} 0.3656 \\ 0.9493 \\ 0.1011 \end{Bmatrix} \Delta T \right)$$

$$= \begin{Bmatrix} 1.36 \\ -1.06 \\ 1.74 \end{Bmatrix} \times 10^{-6} N_x + \begin{Bmatrix} -0.333 \\ 11.5 \\ 2.076 \end{Bmatrix} \times 10^{-6} \Delta T$$

Without an applied moment $\{k\} = 0$, the strain in each lamina is $\{\varepsilon^0\}$. The stress in each lamina is

$$\begin{Bmatrix} \sigma_x \\ \sigma_y \\ \gamma_{xy} \end{Bmatrix}_{15} = [\bar{Q}]_{15} \left(\begin{Bmatrix} \varepsilon_x^0 \\ \varepsilon_y^0 \\ \gamma_{xy}^0 \end{Bmatrix} - \begin{Bmatrix} \alpha_x \\ \alpha_y \\ \alpha_{xy} \end{Bmatrix}_{15} \Delta T \right)$$

$$= \begin{Bmatrix} 33.22 \\ 1.296 \\ 9.497 \end{Bmatrix} N_x + \begin{Bmatrix} 15.77 \\ 1.515 \\ 11.32 \end{Bmatrix} \Delta T$$

$$\begin{Bmatrix} \sigma_x \\ \sigma_y \\ \gamma_{xy} \end{Bmatrix}_{-15} = [\bar{Q}]_{-15} \left(\begin{Bmatrix} \varepsilon_x^0 \\ \varepsilon_y^0 \\ \gamma_{xy}^0 \end{Bmatrix} - \begin{Bmatrix} \alpha_x \\ \alpha_y \\ \alpha_{xy} \end{Bmatrix}_{-15} \Delta T \right)$$

$$= \begin{Bmatrix} 16.76 \\ -0.309 \\ -2.391 \end{Bmatrix} N_x + \begin{Bmatrix} -3.88 \\ -0.399 \\ -2.843 \end{Bmatrix} \Delta T$$

The on-axis stresses (in the principal material directions) are found by using the stress transformation equation for each lamina, resulting in

$$\begin{Bmatrix} \sigma_1 \\ \sigma_2 \\ \tau_{12} \end{Bmatrix}_{15} = [T_\sigma] \begin{Bmatrix} \sigma_x \\ \sigma_y \\ \tau_{xy} \end{Bmatrix}_{15} = \begin{Bmatrix} 36.0 \\ -1.485 \\ -0.044 \end{Bmatrix} N_x + \begin{Bmatrix} 20.68 \\ -3.39 \\ 6.11 \end{Bmatrix} \Delta T$$

$$\begin{Bmatrix} \sigma_1 \\ \sigma_2 \\ \tau_{12} \end{Bmatrix}_{-15} = [T_\sigma] \begin{Bmatrix} \sigma_x \\ \sigma_y \\ \tau_{xy} \end{Bmatrix}_{-15} = \begin{Bmatrix} 16.855 \\ -0.404 \\ 2.327 \end{Bmatrix} N_x + \begin{Bmatrix} -2.175 \\ -2.105 \\ -3.364 \end{Bmatrix} \Delta T$$

Assuming $\Delta T = -300$ °F and $N_x = 500$ lb/in., the stresses are

$$\begin{Bmatrix} \sigma_x \\ \sigma_y \\ \gamma_{xy} \end{Bmatrix}_{15} = \begin{Bmatrix} 33.22 \\ 1.296 \\ 9.497 \end{Bmatrix} (500) + \begin{Bmatrix} 15.77 \\ 1.515 \\ 11.32 \end{Bmatrix} (-300) = \begin{Bmatrix} 11{,}879 \\ 194 \\ 1353 \end{Bmatrix}$$

$$\begin{Bmatrix} \sigma_x \\ \sigma_y \\ \gamma_{xy} \end{Bmatrix}_{-15} = \begin{Bmatrix} 16.76 \\ -0.309 \\ -2.391 \end{Bmatrix} (500) + \begin{Bmatrix} -3.88 \\ -0.399 \\ -2.843 \end{Bmatrix} (-300) = \begin{Bmatrix} 9544 \\ -35 \\ -343 \end{Bmatrix}$$

$$\begin{Bmatrix} \sigma_1 \\ \sigma_2 \\ \tau_{12} \end{Bmatrix}_{15} = \begin{Bmatrix} 36.0 \\ -1.485 \\ -0.044 \end{Bmatrix} (500) + \begin{Bmatrix} -20.68 \\ -3.39 \\ 6.11 \end{Bmatrix} (-300) = \begin{Bmatrix} 11{,}796 \\ 275 \\ -1855 \end{Bmatrix}$$

$$\begin{Bmatrix} \sigma_1 \\ \sigma_2 \\ \tau_{12} \end{Bmatrix}_{-15} = \begin{Bmatrix} 16.855 \\ -0.404 \\ 2.327 \end{Bmatrix} (500) + \begin{Bmatrix} -2.175 \\ -2.105 \\ -3.364 \end{Bmatrix} (-300) = \begin{Bmatrix} 9080 \\ 430 \\ 2173 \end{Bmatrix}$$

The distribution of these stresses through the laminate is shown in Figure 6.22.

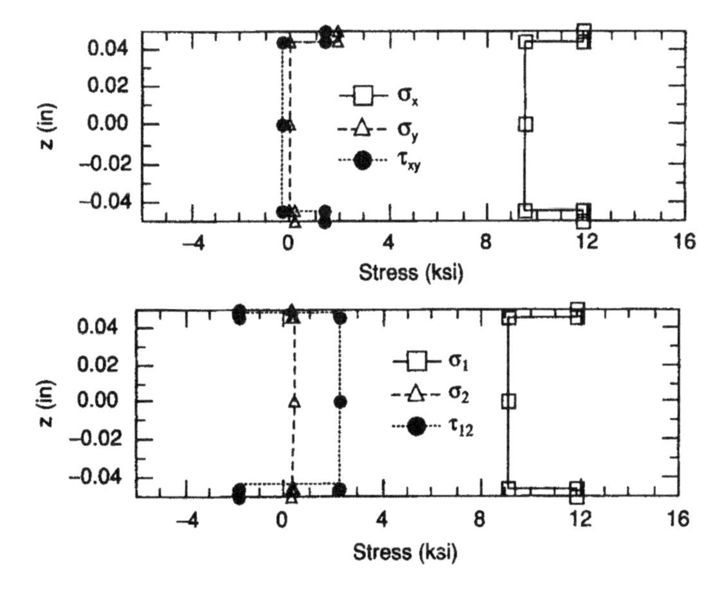

Figure 6.22 Stress variation in a $[15/-15_4]_s$ laminate.

6.5.2 Antisymmetric laminates

The analysis procedure for antisymmetric laminates is identical to that for symmetric laminates, except that $[B] \neq 0$. The loss of symmetry complicates analysis because $\{M^T\} \neq 0$ and $\{M^H\} \neq 0$. Extension-bending coupling exists, and strains are not constant through the laminate.

For antisymmetric laminates, residual stresses must be considered. Residual stresses occur because of differences in elastic moduli and thermal expansion coefficients parallel and transverse to the fibers. When lamina is not symmetrically disposed with respect to the laminate mid-surface, the in-plane residual stresses result in out-of-plane warping, which comes from coupling of the bending and stretching deformations. Residual curvatures resulting from post-cure cool-down are not accurately determined by CLT, as demonstrated by Hyer [20–22] for thermal effects and Harper [23] for moisture effects. As demonstrated by examples in the following section, CLT predicts a saddle shape at room temperature for a [90/0] laminate. In reality, this laminate produces a cylindrical shape at room temperature (often two stable cylindrical shapes are observed for this laminate, since snap through is possible), as illustrated in Figure 6.23. This type of unbalanced laminate has been used to evaluate residual stresses by Narin and Zoller [24,25], who found correlations to within 5% of predicted curvatures using measurements of postcure curvature.

In predicting the cured shape for antisymmetric laminates, Hyer [21] assumed a displacement field in the z-direction of $w(x, y) = (ax^2 + by^2)/2$. This application of kinematic assumptions regarding mid-surface strains results in nonlinear displacement fields for U_0 and V_0. As a result, the basic definitions of mid-surface strain and curvature fundamental to CLT are no longer valid. Although it is important to predict residual curvatures for antisymmetric laminates, a nonlinear theory is not attractive because of difficulties in obtaining solutions. The techniques of CLT remain applicable to a large class of problems.

6.5.2.1 Cross-ply laminate

Consider the glass/epoxy cross-ply laminate in Figure 6.24, with $E_1 = 5.6 \times 10^6$ psi, $E_2 = 1.2 \times 10^6$ psi, $G_{12} = 0.6 \times 10^6$ psi, $\nu_{12} = 0.26$, $\alpha_1 = 4.77$ μin./in./°F, and $\alpha_1 = 12.24$ μin./in./°F. This example examines the influence of ply thickness for each lamina on curing strains and curvatures predicted by CLT. The thickness of the 0°

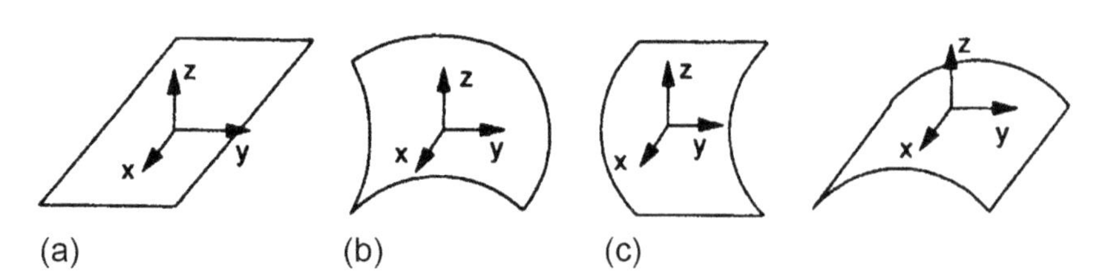

Figure 6.23 [0/90] laminate shapes (a) at elevated curing temperatures; (b) saddle shape, as predicted by CLT; and (c) two stable cylindrical shapes that actually exist.

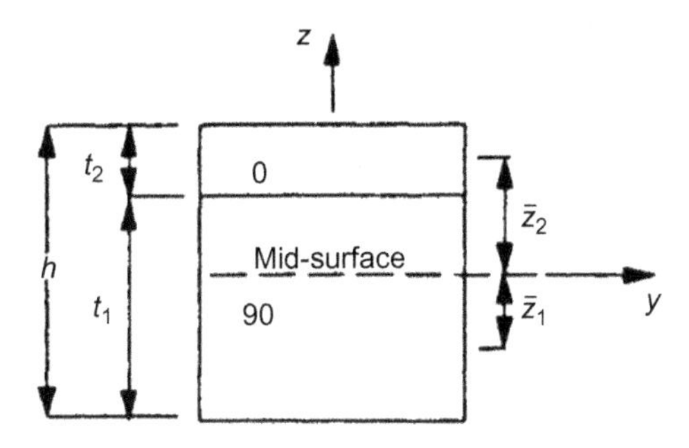

Figure 6.24 Antisymmetric cross-ply laminate.

lamina is t_2. and the 90° lamina has a thickness of t_1. The total laminate thickness is $h = t_1 + t_2 = 0.30$ in., and the centroid of each lamina with respect to the mid-surface is at $\bar{z}_1 = -[(t_1 + t_2)/2 - t_1/2] = -t_2/2$ and $\bar{z}_2 = [(t_1 + t_2)/2 - t_2/2] = t_1/2$.

The [A], [B], and [D] matrices are written in terms of t_1 and t_2 as

$$[A] = [\bar{Q}]_{90}t_1 + [\bar{Q}]_0 t_2 = \left(\begin{bmatrix} 1.218 & 0.317 & 0 \\ 0.317 & 5.682 & 0 \\ 0 & 0 & 0.60 \end{bmatrix} t_1 + \begin{bmatrix} 5.682 & 0.317 & 0 \\ 0.317 & 1.218 & 0 \\ 0 & 0 & 0.60 \end{bmatrix} t_2 \right) \times 10^6$$

$$[B] = [\bar{Q}]_{90}t_1\bar{z}_1 + [\bar{Q}]_0 t_2\bar{z}_2$$
$$= \left(\begin{bmatrix} 1.218 & 0.317 & 0 \\ 0.317 & 5.682 & 0 \\ 0 & 0 & 0.60 \end{bmatrix} t_1\left(\frac{-t_1}{2}\right) + \begin{bmatrix} 5.682 & 0.317 & 0 \\ 0.317 & 1.218 & 0 \\ 0 & 0 & 0.60 \end{bmatrix} t_2\left(\frac{t_1}{2}\right) \right) \times 10^6$$

$$[D] = [\bar{Q}]_{90}\left(t_1\bar{z}_1^2 + \left(\frac{t_1^3}{12}\right)\right) + [\bar{Q}]_0\left(t_2\bar{z}_{12}^2 + \left(\frac{t_2^3}{12}\right)\right)$$
$$= \left(\begin{bmatrix} 1.218 & 0.317 & 0 \\ 0.317 & 5.682 & 0 \\ 0 & 0 & 0.60 \end{bmatrix} \left(t_1\left(\frac{-t_2}{2}\right)^2 + \frac{t_1^3}{12}\right) + \begin{bmatrix} 5.682 & 0.317 & 0 \\ 0.317 & 1.218 & 0 \\ 0 & 0 & 0.60 \end{bmatrix} \left(t_2\left(\frac{t_1}{2}\right)^2 + \frac{t_2^3}{12}\right) \right) \times 10^6$$

Similarly, $\{N^T\}$ and $\{M^T\}$ are

$$\{N^T\} = \left([\bar{Q}]_{90}\{\alpha\}_{90}t_1 + [\bar{Q}]_0\{\alpha\}_0 t_2\right)\Delta T$$
$$= \begin{Bmatrix} 16.42t_1 + 29.28t_2 \\ 29.28t_1 + 16.42t_1 \\ 0 \end{Bmatrix} \Delta T$$

$$\{M^T\} = [\bar{Q}]_{90}\{\alpha\}_{90}t_1\bar{z}_1 + [\bar{Q}]_0\{\alpha\}_0 t_2\bar{z}_2$$
$$= \begin{Bmatrix} 7.281t_1t_2 \\ -7.281t_1t_2 \\ 0 \end{Bmatrix} \Delta T$$

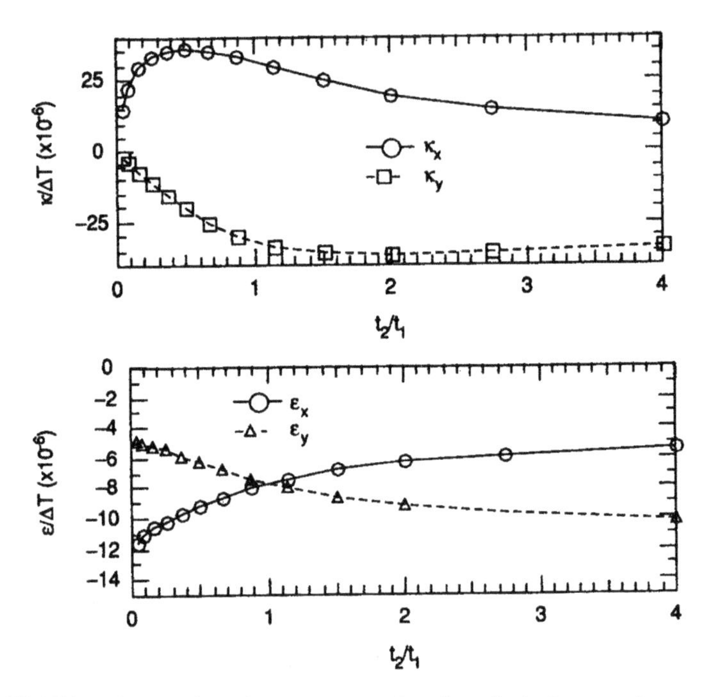

Figure 6.25 Mid-surface strain and curvature as a function of t_2/t_1 for an antisymmetric cross-ply laminate.

The mid-surface curing strains and curvatures are found by solving Equation (6.35) with $\{\hat{N}\}=\{N^{\mathrm{T}}\}$ and $\{\bar{M}\}=\{M^{\mathrm{T}}\}$. Assume the temperature difference is $\Delta T=-1$. The variation of mid-surface strain and curvature as a function of t_2/t_1 (or t_0/t_{90}) is shown in Figure 6.25. As indicated here, the saddle shape predicted by CLT is evident from the fact that one curvature is positive and the other negative. As the ratio of t_2/t_1 becomes larger, both curvatures and strains approach a limiting value, and there is a reduced coupling effect due to thermal expansion coefficient mismatch between adjacent lamina.

The antisymmetric cross-ply laminate should experience only k_x and k_y curvatures. A slight variation in fiber angle for one of the lamina can result in an x–y curvature and a warped shape after curing.

6.5.2.2 Angle-ply laminate

Consider the $[-15_4/15]$ angle-ply laminate in Figure 6.26 with an axial load of 1000 lb applied. Assuming the specimen is 2 in wide, $N_x=500$ lb/in. Thermal effects are considered with $\Delta T=-300$ °F. For this material, $E_1=22.2\times10^6$ psi, $E_2=1.58\times10^6$ psi, $G_{12}=0.81\times10^6$ psi, $\nu_{12}=0.30$, $\alpha_1=0.011$ μin./in./°F, $\alpha_2=12.5$ μin./in./°F, and $t=0.005$ in. Using these properties,

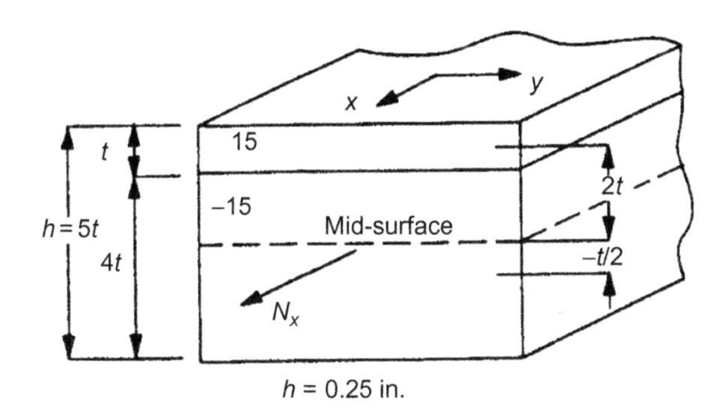

Figure 6.26 $[-15_4/15]$ angle-ply laminate.

$$[Q] = \begin{bmatrix} 22.35 & 0.4932 & 0 \\ 0.4932 & 1.591 & 0 \\ 0 & 0 & 0.81 \end{bmatrix} \times 10^6 \text{ psi}$$

Following the analysis procedures previously outlined,

$$[\bar{Q}]_{\pm 15} = \begin{bmatrix} 19.73 & 1.726 & \pm 4.730 \\ 1.726 & 1.749 & \pm 0.461 \\ \pm 4.730 & \pm 0.461 & 2.042 \end{bmatrix} \times 10^6 \text{ psi}$$

$$[A_{ij}] = \sum [\bar{Q}]_k t_k = \left([\bar{Q}]_{15} + 4[\bar{Q}]_{-15}\right)(0.005)$$
$$= \begin{bmatrix} 4.93 & 0.432 & -0.709 \\ 0.432 & 0.437 & -0.069 \\ -0.709 & -0.069 & 0.511 \end{bmatrix} \times 10^5$$

$$[B_{ij}] = \sum [\bar{Q}]_k t_k \bar{z}_k = [\bar{Q}]_{15}(2t) + [\bar{Q}]_{-15}(4t)(-t/2)$$
$$= \begin{bmatrix} 0 & 0 & 473 \\ 0 & 0 & 46.1 \\ 473 & 46.1 & 0 \end{bmatrix}$$

$$[D_{ij}] = \sum [\bar{Q}]_k \left(t_k \bar{z}_k^2 + \frac{t_k^3}{12}\right) = [\bar{Q}]_{15}\left\{(0.005)(0.01)^2 + \frac{0.005^3}{12}\right\}$$
$$+ [\bar{Q}]_{-15}\left\{(0.02)(-0.0025)^2 + \frac{0.02^3}{12}\right\} = \begin{bmatrix} 25.69 & 2.247 & -1.333 \\ 2.247 & 18.2 & -0.130 \\ -1.333 & -0.130 & 2.659 \end{bmatrix}$$

From these we obtain

$$[A'] = \begin{bmatrix} 1.589 & -2.249 & 0.439 \\ -2.249 & 25.02 & 0.054 \\ 0.439 & 0.054 & 13.37 \end{bmatrix} \times 10^{-6}$$

$$[B'] = \begin{bmatrix} -1.56 & -0.193 & -7.28 \\ -1.93 & -0.0238 & -0.90 \\ -7.28 & -0.90 & -15.4 \end{bmatrix} \times 10^{-4}$$

$$[D'] = \begin{bmatrix} 5.86 & -3.97 & 5.55 \\ -3.97 & 48.1 & 0.685 \\ 555 & 0.685 & 53.5 \end{bmatrix} \times 10^{-2}$$

The coefficients of thermal expansion for this material and these ply orientations has been previously determined to be

$$\begin{Bmatrix} \alpha_x \\ \alpha_y \\ \alpha_{xy} \end{Bmatrix}_{\pm 15} = \begin{Bmatrix} 0.848 \\ 11.66 \\ \mp 6.24 \end{Bmatrix} \ \mu\text{in./in./°F}$$

Noting that $[\bar{Q}]$ has a magnitude of 10^6 and $\{\alpha\}$ a magnitude of 10^{-6}, the thermal loads and moments are

$$\{N^{\mathrm{T}}\} = \sum [\bar{Q}]_k \begin{Bmatrix} \alpha_x \\ \alpha_y \\ \alpha_{xy} \end{Bmatrix}_k \Delta T(t_k) = \left\{ [\bar{Q}]_{15} \begin{Bmatrix} \alpha_x \\ \alpha_y \\ \alpha_{xy} \end{Bmatrix}_{15} + 4[\bar{Q}]_{-15} \begin{Bmatrix} \alpha_x \\ \alpha_y \\ \alpha_{xy} \end{Bmatrix}_{-15} \right\}$$

$$(-300)(0.005) = \begin{Bmatrix} -55.05 \\ -142.35 \\ -15.09 \end{Bmatrix}$$

$$\{M^{\mathrm{T}}\} = \sum [\bar{Q}]_k \begin{Bmatrix} \alpha_x \\ \alpha_y \\ \alpha_{xy} \end{Bmatrix}_k \Delta T(t_k \bar{z}_k) = \left\{ [\bar{Q}]_{15} \begin{Bmatrix} \alpha_x \\ \alpha_y \\ \alpha_{xy} \end{Bmatrix}_{15} (0.010) + 4[\bar{Q}]_{-15} \begin{Bmatrix} \alpha_x \\ \alpha_y \\ \alpha_{xy} \end{Bmatrix}_{-15} \right.$$

$$\left. (-0.0025) \right\} (-300)(0.005) = \begin{Bmatrix} 0 \\ 0 \\ 0.101 \end{Bmatrix}.$$

Incorporating the axial load $N_x = 500$ into the expressions for thermal loads and moments results in

$$\{\hat{N}\} = \begin{Bmatrix} 444.95 \\ -142.35 \\ -15.09 \end{Bmatrix} \quad \{\hat{M}\} = \begin{Bmatrix} 0 \\ 0 \\ 0.101 \end{Bmatrix}$$

Solving for the mid-surface strains and curvatures,

$$\{\varepsilon^0\} = [A']\{\hat{N}\} + [B']\{\hat{M}\} = \begin{Bmatrix} 946.5 \\ -4563.1 \\ -169.6 \end{Bmatrix} \times 10^{-6}$$

$$\{k\} = [B']^{\mathrm{T}}\{\hat{N}\} + [D']\{\hat{M}\} = \begin{Bmatrix} -500.7 \\ -834.9 \\ -2338.4 \end{Bmatrix} \times 10^{-4}$$

The strain variation through the laminate is

$$\begin{Bmatrix} \varepsilon_x \\ \varepsilon_y \\ \gamma_{xy} \end{Bmatrix} = \begin{Bmatrix} \varepsilon_x^0 \\ \varepsilon_y^0 \\ \gamma_{xy}^0 \end{Bmatrix} + z\begin{Bmatrix} \kappa_x \\ \kappa_y \\ \kappa_{xy} \end{Bmatrix} = \left(\begin{Bmatrix} 946.5 \\ -4563.1 \\ -169.6 \end{Bmatrix} + z\begin{Bmatrix} -50,070 \\ -83,490 \\ -233,840 \end{Bmatrix}\right) \times 10^{-6}$$

where $-0.0125 \leq z \leq 0.0125$. The stress distribution through the laminate is linear and is defined by

$$\begin{Bmatrix} \sigma_x \\ \sigma_y \\ \gamma_{xy} \end{Bmatrix}_k = [\bar{Q}]_k\left(\left(\begin{Bmatrix} 946.5 \\ -4563.1 \\ -169.6 \end{Bmatrix} + z\begin{Bmatrix} -50,070 \\ -83,490 \\ -233,840 \end{Bmatrix}\right) \times 10^{-6} - \begin{Bmatrix} \alpha_x \\ \alpha_y \\ \alpha_{xy} \end{Bmatrix}_k \Delta T\right)$$

The stresses at the interface of the $+15°$ and $-15°$ lamina ($z = 0.0075$) are dual-valued because of the change in $[\bar{Q}]$ at that location and are similar to the effects illustrated in Figure 6.3. The stress at the interface must be computed using both $[\bar{Q}]_{+15}$ and $[\bar{Q}]_{-15}$ with $z = 0.0075$. The strain at the interface is

$$\begin{Bmatrix} \varepsilon_x \\ \varepsilon_y \\ \gamma_{xy} \end{Bmatrix} = \left(\begin{Bmatrix} 946.5 \\ -4563.1 \\ -169.6 \end{Bmatrix} + (0.0075)\begin{Bmatrix} -50,070 \\ -83,490 \\ -233,840 \end{Bmatrix}\right) \times 10^{-6}$$
$$= \begin{Bmatrix} 570.9 \\ -5189.3 \\ -1923.4 \end{Bmatrix} \times 10^{-6}$$

The stresses in each lamina are

$$\begin{Bmatrix} \sigma_x \\ \sigma_y \\ \tau_{xy} \end{Bmatrix}_{+15} = [\bar{Q}]_{+15}\left(\begin{Bmatrix} 570.9 \\ -5189.3 \\ -1923.4 \end{Bmatrix} - \begin{Bmatrix} 0.848 \\ 11.66 \\ -6.24 \end{Bmatrix}(-300)\right) = \begin{Bmatrix} -4.59 \\ -3.28 \\ -4.63 \end{Bmatrix} \text{ ksi}$$

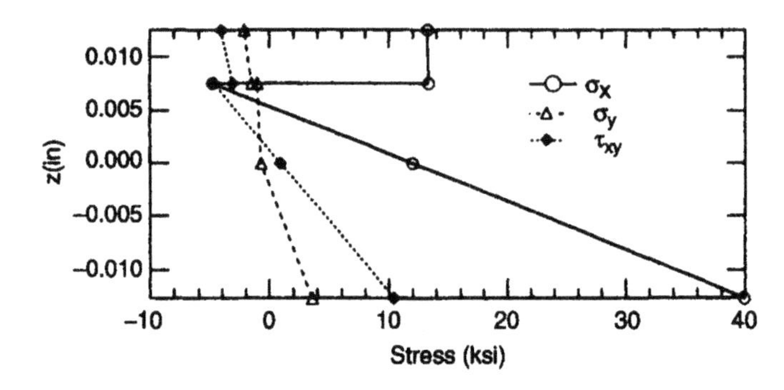

Figure 6.27 Stress variation through a $[-15_4/15]$ angle-ply laminate.

$$\begin{Bmatrix} \sigma_x \\ \sigma_y \\ \tau_{xy} \end{Bmatrix}_{-15} = [\bar{Q}]_{-15} \left(\begin{Bmatrix} 570.9 \\ -5189.3 \\ -1923.4 \end{Bmatrix} - \begin{Bmatrix} 0.848 \\ 11.66 \\ -6.24 \end{Bmatrix} (-300) \right) = \begin{Bmatrix} 13.61 \\ -1.51 \\ -3.23 \end{Bmatrix} \text{ksi}$$

The variation of each in-plane stress component through the laminate is shown in Figure 6.27.

6.5.3 Nonsymmetric laminates

For completely nonsymmetric laminates, the $[A]$, $[B]$, and $[D]$ matrices are generally fully populated, and for cases in which thermal effects are considered, $[M^T] \neq 0$. Consider the [90/45/0/−45] laminate in Figure 6.28. The elastic constants, loading condition, and temperature difference are $E_1 = 26.25 \times 10^6$ psi, $E_2 = 1.50 \times 10^6$ psi, $G_{12} = 1.04 \times 10^6$ psi, $\nu_{12} = 0.28$, $\alpha_1 = 2.0$ μin./in./°F, $\alpha_2 = 15.0$ μin./in./°F, $\Delta T = -300$ °F, $t = 0.005$ in., $\{M\} = 0$, and $N_x = 500$ lb/in. Hygral effects are not considered.

Figure 6.28 [90/45/0/−45] laminate.

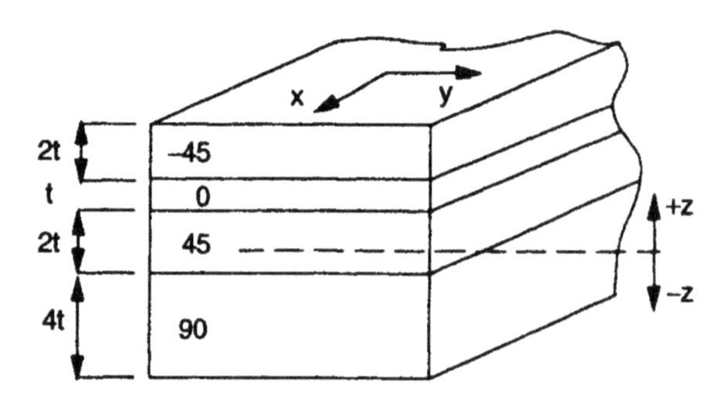

For this material, we compute

$$[Q] = [\bar{Q}]_0 = \begin{bmatrix} 26.37 & 0.422 & 0 \\ 0.422 & 1.507 & 0 \\ 0 & 0 & 1.04 \end{bmatrix} \times 10^6$$

$$[\bar{Q}]_{90} = \begin{bmatrix} 1.507 & 0.422 & 0 \\ 0.422 & 26.37 & 0 \\ 0 & 0 & 1.04 \end{bmatrix} \times 10^6$$

$$[\bar{Q}]_{\pm 45} = \begin{bmatrix} 8.22 & 6.140 & \pm 6.215 \\ 6.140 & 8.22 & \pm 6.215 \\ \pm 6.215 & \pm 6.215 & 6.758 \end{bmatrix} \times 10^6$$

Following standard procedures $[A]$, $[B]$, and $[D]$ are

$$[A] = [\bar{Q}]_{90}(4t) + [\bar{Q}]_{45}(2t) + [\bar{Q}]_0(t) + [\bar{Q}]_{45}(2t)$$

$$= \begin{bmatrix} 0.326 & 0.133 & 0 \\ 0.133 & 0.699 & 0 \\ 0 & 0 & 0.161 \end{bmatrix} \times 10^6$$

$$[B] = [\bar{Q}]_{90}(4t)(-2.5t) + [\bar{Q}]_{45}(2t)(0.5t) + [\bar{Q}]_0(t)(2t) + [\bar{Q}]_{-45}(2t)(3.5t)$$

$$= \begin{bmatrix} 2.59 & 1.14 & -0.932 \\ 1.14 & -4.87 & -0.932 \\ -0.932 & -0.932 & 1.14 \end{bmatrix} \times 10^3$$

$$[D] = [\bar{Q}]_{90}\left((4t)(-2.5t)^2 + \frac{(4t)^3}{12}\right) + [\bar{Q}]_{45}\left((2t)(0.5t)^2 + \frac{(2t)^3}{12}\right)$$

$$+ [\bar{Q}]_0\left((t)(2t)^2 + \frac{(t)^3}{12}\right) + [\bar{Q}]_{-45}\left((2t)(3.5t)^2 + \frac{(2t)^3}{12}\right)$$

$$= \begin{bmatrix} 46.2 & 22.0 & -18.6 \\ 22.0 & 128.0 & -18.6 \\ -18.6 & -18.6 & 26.7 \end{bmatrix}$$

In order to establish the mid-surface strains and curvatures, $[A']$, $[B']$, and $[D']$ are required and are

$$[A'] = \begin{bmatrix} 6.56 & -0.990 & -2.48 \\ -0.990 & 2.91 & -0.541 \\ -2.48 & -0.541 & 10.2 \end{bmatrix} \times 10^{-6}$$

$$[B'] = \begin{bmatrix} -3.64 & -0.501 & 0.114 \\ -0.272 & 1.46 & 1.73 \\ 1.99 & -0.191 & -4.16 \end{bmatrix} \times 10^{-4}$$

$$[D'] = \begin{bmatrix} 5.35 & -0.372 & 1.26 \\ -0.372 & 1.62 & 1.29 \\ 1.26 & 1.29 & 7.95 \end{bmatrix} \times 10^{-2}$$

The thermal loads and moments are determined once the coefficients of thermal expansion for each lamina are established from Equation (3.22) to be

$$\{a\}_0 = \begin{Bmatrix} 2 \\ 15 \\ 0 \end{Bmatrix} \mu\text{in./in. °F} \quad \{a\}_{90} = \begin{Bmatrix} 15 \\ 2 \\ 0 \end{Bmatrix} \mu\text{in./in. °F}$$

$$\{a\}_{\pm 45} = \begin{Bmatrix} 8.5 \\ 8.5 \\ \mp 13 \end{Bmatrix} \mu\text{in./in. °F}$$

Thermal loads and moments are determined from Equations (6.25) and (6.26) to be

$$[N^{\mathrm{T}}] = [\bar{Q}]_{90}(0.02)\{\alpha\}_{90} + [\bar{Q}]_{45}(0.01)\{\alpha\}_{45} + [\bar{Q}]_0(0.005)\{\alpha\}_0 + [\bar{Q}]_{-45}(0.01)\{\alpha\}_{-45} = \begin{Bmatrix} -477 \\ -637 \\ 0 \end{Bmatrix}$$

$$[M^{\mathrm{T}}] = [\bar{Q}]_{90}(0.02)(-0.0125)\{\alpha\}_{90} + [\bar{Q}]_{45}(0.01)(0.0025)\{\alpha\}_{45} + [\bar{Q}]_0(0.005)(0.01)\{\alpha\}_0 + [\bar{Q}]_{-45}(0.01)(0.0175)\{\alpha\}_{-45} = \begin{Bmatrix} -1.6 \\ 1.6 \\ 0 \end{Bmatrix}$$

Combining the thermal loads and moments with the applied loads yields

$$\{\hat{N}\} = \begin{Bmatrix} 23 \\ -637 \\ 0 \end{Bmatrix} \quad \{\hat{M}\} = \begin{Bmatrix} -1.6 \\ 1.6 \\ 0.801 \end{Bmatrix}$$

The mid-surface strains and curvatures are

$$\{\varepsilon^{\circ}\} = [A']\{\hat{N}\} + [B']\{\hat{M}\} = \left\{ \begin{array}{c} 1292 \\ -1461 \\ -395 \end{array} \right\} \times 10^{-6}$$

$$\{k\} = [B']\{\hat{N}\} + [D']\{\hat{M}\} = \left\{ \begin{array}{c} -727 \\ -518 \\ -457 \end{array} \right\} \times 10^{-4}$$

The variation of strain and stress through the laminate is

$$\left\{ \begin{array}{c} \varepsilon_x \\ \varepsilon_y \\ \gamma_{xy} \end{array} \right\} = \left\{ \begin{array}{c} \varepsilon_x^0 \\ \varepsilon_{yx}^0 \\ \gamma_{xy}^0 \end{array} \right\} + z \left\{ \begin{array}{c} \kappa_x \\ \kappa_y \\ \kappa_{xy} \end{array} \right\} = \left\{ \begin{array}{c} 1292 \\ -1461 \\ -395 \end{array} \right\} \times 10^{-6} + z \left\{ \begin{array}{c} -0.0727 \\ -0.0518 \\ -0.0457 \end{array} \right\}$$

$$\left\{ \begin{array}{c} \sigma_x \\ \sigma_y \\ \tau_{xy} \end{array} \right\}_k = [\bar{Q}]_k \left(\left\{ \begin{array}{c} 0.001292 - 0.0727z \\ -0.001461 - 0.0518z \\ -0.000395 - 0.0457z \end{array} \right\} - \left\{ \begin{array}{c} \alpha_x \\ \alpha_y \\ \alpha_{xy} \end{array} \right\}_k \Delta T \right)$$

In this expression, $[\bar{Q}]_k$, $\{\alpha\}_k$, and z vary from lamina to lamina, causing linear variations of stress. For example, in the 0° lamina, the stress is

$$\left\{ \begin{array}{c} \sigma_x \\ \sigma_y \\ \tau_{xy} \end{array} \right\}_0 = \left[\begin{array}{ccc} 26.37 & 0.42 & 0 \\ 0.42 & 1.51 & 0 \\ 0 & 0 & 1.04 \end{array} \right] \times 10^6 \left(\left\{ \begin{array}{c} 0.001292 - 0.0727z \\ -0.001461 - 0.0518z \\ -0.000395 - 0.0457z \end{array} \right\} - \left\{ \begin{array}{c} 2 \\ 15 \\ 0 \end{array} \right\} \times 10^{-6}(-300) \right)$$

For the 0° lamina $0.0075 \leq z \leq 0.0125$ in., and for these values of z,

$$\left\{ \begin{array}{c} \sigma_x \\ \sigma_y \\ \tau_{xy} \end{array} \right\}_{z=0.0075} = \left\{ \begin{array}{c} 36.7 \\ 4.576 \\ -0.77 \end{array} \right\} \text{ksi} \quad \left\{ \begin{array}{c} \sigma_x \\ \sigma_y \\ \tau_{xy} \end{array} \right\}_{z=0.0125} = \left\{ \begin{array}{c} 27 \\ 4.02 \\ -1.01 \end{array} \right\} \text{ksi}$$

After application of the same procedure to each lamina, the variation of Cartesian stress components through the laminate is defined and shown in Figure 6.29. Using stress transformations, the principal material direction stress components can be established. Note that the stresses are dual-valued at lamina interfaces.

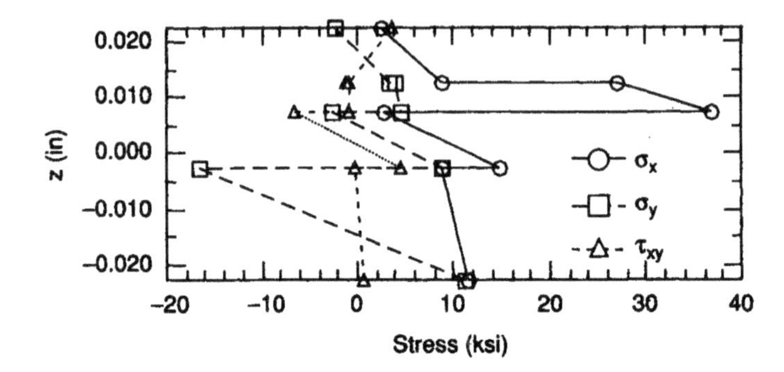

Figure 6.29 Stress distribution through a [90/45/0/−45] laminate.

6.6 Laminate failure analysis

In many cases, stress analysis needs to be supplemented by considering laminate failure. One approach to estimating laminate failure is through experimentally defining the failure loads for specific laminates. The large number of possible laminate configurations makes this a time-consuming and expensive procedure. Numerous failure theories for lamina were presented in Chapter 5, which form the basis of laminate failure analysis presented herein. Laminate failure analysis involves two initial phases:

1. Establish the stress distribution through the laminate, recalling that the principal material direction stresses are required for the failure theories presented in Chapter 5.
2. Apply an appropriate failure theory to each lamina.

The predicted failure of a lamina does not imply total laminate failure. Some laminates can function and carry load past the point at which first ply failure occurs. The Tsai–Hill failure criterion is used to present laminate failure analysis.

6.6.1 Cross-ply laminate

The symmetric cross-ply laminate in Section 6.5.1.1 is used to discuss failure analysis. For the $[0/90_5]_s$ laminate in Figure 6.30, the total thickness is $h = 0.060$ in. Because of symmetry, $[B] = 0$, and there is no curvature due to curing. The only applied load is N_x,

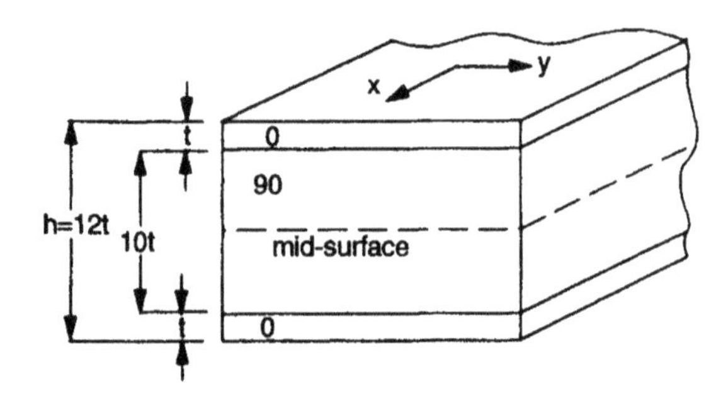

Figure 6.30 $[0/90_5]_s$ cross-ply laminate for failure analysis.

which is unspecified for discussion purposes. The mid-surface strains were determined in Section 6.5.1.1 and are

$$\begin{Bmatrix} \varepsilon_x^0 \\ \varepsilon_y^0 \\ \gamma_{xy}^0 \end{Bmatrix} = \begin{Bmatrix} 3.767 \\ -0.0867 \\ 0 \end{Bmatrix} N_x + \begin{Bmatrix} 3.90 \\ 0.245 \\ 0 \end{Bmatrix} \Delta T$$

The Cartesian components of stress in each lamina are

$$\begin{Bmatrix} \sigma_x \\ \sigma_y \\ \tau_{xy} \end{Bmatrix} = \left[\bar{Q}\right]_k \left(\begin{Bmatrix} 3.767 \\ -0.0867 \\ 0 \end{Bmatrix} N_x + \left(\begin{Bmatrix} 3.90 \\ 0.245 \\ 0 \end{Bmatrix} - \begin{Bmatrix} \alpha_x \\ \alpha_y \\ \alpha_{xy} \end{Bmatrix}_k \right) \Delta T \right)$$

Since the lamina has either 0° or 90° fiber orientations, the principal material direction stresses are easy to determine. The on-axis stress components are required for failure analysis, and by accounting for $\left[\bar{Q}\right]$ and $\{\alpha\}$ for each lamina (refer to Section 6.5.1.1), they are

$$\begin{Bmatrix} \sigma_1 \\ \sigma_2 \\ \tau_{12} \end{Bmatrix}_0 = \begin{Bmatrix} \sigma_x \\ \sigma_y \\ \tau_{xy} \end{Bmatrix}_0 = \begin{Bmatrix} 75.7 \\ 1.36 \\ 0 \end{Bmatrix} N_x + \begin{Bmatrix} 76 \\ -18.5 \\ 0 \end{Bmatrix} \Delta T$$

$$\begin{Bmatrix} \sigma_1 \\ \sigma_2 \\ \tau_{12} \end{Bmatrix}_{90} = \begin{Bmatrix} \sigma_y \\ \sigma_x \\ \tau_{xy} \end{Bmatrix}_{90} = \begin{Bmatrix} -0.271 \\ 4.88 \\ 0 \end{Bmatrix} N_x + \begin{Bmatrix} 3.73 \\ -14.83 \\ 0 \end{Bmatrix} \Delta T$$

The failure strengths are $X = X' = 210$ ksi, $Y = 7.5$ ksi, $Y' = 29.9$ ksi, and $S = 13.5$ ksi. Initially, thermal effects are neglected ($\Delta T = 0$). Stresses are due only to the unknown applied load N_x. The general form of the Tsai–Hill failure theory is

$$\left(\frac{\sigma_1}{X}\right)^2 - \left(\frac{\sigma_1 \sigma_2}{X^2}\right) + \left(\frac{\sigma_2}{Y}\right)^2 + \left(\frac{\tau_{12}}{S}\right)^2 = 1$$

This can be written as

$$\sigma_1^2 - \sigma_1 \sigma_2 + \left(\frac{X}{Y}\right)^2 \sigma_2^2 + \left(\frac{X}{S}\right)^2 \tau_{12}^2 = X^2$$

Since $X = X'$, the sign of σ_1 is not considered. The sign of σ_2 must be considered since $Y \neq Y'$. Substituting X, Y, and S into the preceding equation yields two forms of the governing failure equation:

$$\sigma_2 > 0 \quad \sigma_1^2 - \sigma_1 \sigma_2 + 784\sigma_2^2 + 242\tau_{12}^2 = \left(210 \times 10^3\right)^2$$

$$\sigma_2 < 0 \quad \sigma_1^2 - \sigma_1 \sigma_2 + 49.3\sigma_2^2 + 242\tau_{12}^2 = \left(210 \times 10^3\right)^2$$

Assuming $N_x > 0$, the 0° lamina experiences a tensile σ_2. The failure criterion for this lamina is

$$(75.7N_x)^2 - (75.7)(1.36)N_x^2 + 784(1.36N_x)^2 = \left(210 \times 10^2\right)^2$$

Solving yields $N_x = 2.567$ kip/in. For the 90° lamina σ_2 is tensile, and

$$(-0.271N_x)^2 - (-0.271)(4.88)N_x^2 + 784(4.88N_x)^2 = \left(210 \times 10^3\right)^2$$

The solution to this equation is $N_x = 1.537$ kip/in. The 90° lamina is predicted to fail first, which is not surprising since it is primarily supported by a matrix in the direction of load application. The failure of this lamina does not mean the entire laminate has failed. After the 90° lamina fails, the 0° lamina can sustain load. This feature of the cross-ply laminate makes it a useful experimental test specimen, and symmetric cross-ply laminates are used extensively for the study of matrix cracking and damage.

6.6.1.1 Post-first-ply-failure analysis

The failed lamina contains a series of matrix cracks as depicted in Figure 6.31. These cracks form perpendicular to the applied load (parallel to the fibers) and result in a loss of transverse stiffness in the failed ply. The failed ply is assumed to have stiffness only in the fiber direction, which is completely uncoupled from the transverse extensional stiffness.

Matrix cracks are a frequently observed and extensively studied mode of matrix damage. The number of cracks (termed crack density) increases with load (or number of cycles in fatigue) until saturation is reached. Saturation is referred to as the characteristic damage state (CDS). Macroscopic damage modes, such as delamination, do not usually appear until the CDS is reached. The study of matrix cracks has advanced the understanding of damage and failure mechanisms in laminated composites [26–32] and is beyond the scope of this text.

In order to completely assess the effects of first ply failure, we must analyze the problem again, as discussed by Tsai [19]. The analysis is complicated by now having two materials to consider: the original material (0° lamina) and the failed material (90° lamina). The original material remains unchanged with $[Q] = [\bar{Q}]_0$. The failed ply is

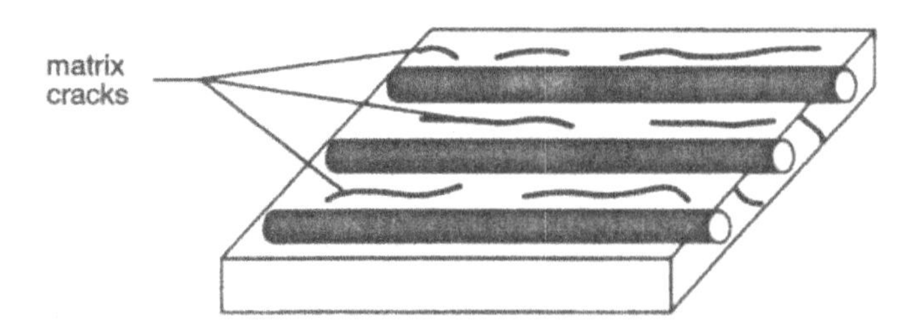

Figure 6.31 Matrix cracks in the failed lamina of a $[0/90_5]_s$ laminate.

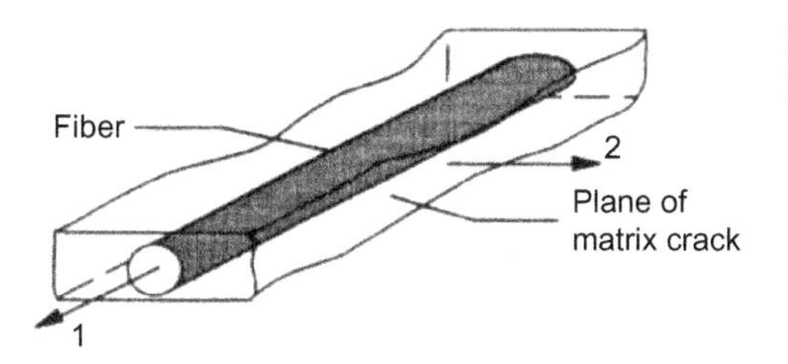

Figure 6.32 Isolated segment of failed lamina.

characterized by degrading the 90° lamina. Degradation of the failed ply applies only to the matrix, which is assumed to have failed. Although some damage may have been sustained by the fibers, they are assumed to behave as they originally did. A schematic of an isolated segment of the failed lamina is shown in Figure 6.32.

Although the matrix has failed, it is assumed that it has not separated from the fiber. Therefore, in the fiber direction, the stresses and strains can be related by Q_{11}. No loads can be supported by or transferred across the cracks. Therefore, there is no coupling between axial and transverse strains, and $Q_{12}=0$ for the failed ply. Similarly, shear stress is assumed to be eliminated since there is no available path for shear transfer across the cracks, and $Q_{66}=0$. Since cracks form parallel to the fibers, no transverse normal loads can be supported, and $Q_{22}=0$. Based on these arguments, the stiffness matrix for the *failed ply only* is

$$[Q]=\begin{bmatrix} Q_{11} & 0 & 0 \\ 0 & 0 & 0 \\ 0 & 0 & 0 \end{bmatrix}=\begin{bmatrix} 20.14 & 0 & 0 \\ 0 & 0 & 0 \\ 0 & 0 & 0 \end{bmatrix}\times 10^6$$

For the failed lamina, the $[\bar{Q}]_{90}$ matrix is

$$[\bar{Q}]_{90}=\begin{bmatrix} 0 & 0 & 0 \\ 0 & 20.14 & 0 \\ 0 & 0 & 0 \end{bmatrix}\times 10^6$$

This results in new $[A]$ and $[A']$ matrices. The new $[A]$ matrix is formed exactly as it originally was, with the degraded $[\bar{Q}]_{90}$ replacing the original $[\bar{Q}]_{90}$. The new $[A]$ matrix is

$$\begin{aligned}
[A_{ij}] &= \sum [\bar{Q}]_k t_k = \left\{2[\bar{Q}]_0 + 10[\bar{Q}]_{90}\right\} t \\
&= \left\{2\begin{bmatrix} 20.14 & 0.392 & 0 \\ 0.392 & 1.307 & 0 \\ 0 & 0 & 1.03 \end{bmatrix}\times 10^6 + 10\begin{bmatrix} 0 & 0 & 0 \\ 0 & 20.14 & 0 \\ 0 & 0 & 0 \end{bmatrix}\times 10^6\right\} t \\
&= \begin{bmatrix} 40.24 & 0.784 & 0 \\ 0.784 & 204.01 & 0 \\ 0 & 0 & 2.06 \end{bmatrix}\times 10^6 t
\end{aligned}$$

Using $h = 12t$, the original and degraded $[A]$ matrices can be compared:

Original $[A]$

$$[A] = \begin{bmatrix} 4.45 & 0.392 & 0 \\ 0.392 & 17.0 & 0 \\ 0 & 0 & 1.03 \end{bmatrix} \times 10^6 h$$

Degraded $[A]$

$$[A] = \begin{bmatrix} 3.353 & 0.0653 & 0 \\ 0.0653 & 17.0 & 0 \\ 0 & 0 & 0.1717 \end{bmatrix} \times 10^6 h$$

The only component not affected by the degraded lamina is A_{22}, which is consistent with the assumption that in the 90° lamina the fibers can sustain load. From the two matrices shown, it is apparent that the overall stiffness in the x-direction (direction of the applied load) has been reduced. The change in $[A]$ means that the mid-surface strains have changed. Using $h = 0.06$ in., the strains in the failed lamina are

$$\begin{Bmatrix} \varepsilon_x^0 \\ \varepsilon_y^0 \\ \gamma_{xy}^0 \end{Bmatrix} = [A'] \begin{Bmatrix} N_x \\ 0 \\ 0 \end{Bmatrix} = \begin{Bmatrix} 4.97 \\ -0.019 \\ 0 \end{Bmatrix} \times 10^{-6} N_x$$

In the 0° lamina, the stresses are

$$\begin{Bmatrix} \sigma_x \\ \sigma_y \\ \tau_{xy} \end{Bmatrix}_0 = [\bar{Q}]_0 \begin{Bmatrix} \varepsilon_x^0 \\ \varepsilon_y^0 \\ \varepsilon_{xy}^0 \end{Bmatrix} = \begin{bmatrix} 20.14 & 0.392 & 0 \\ 0.392 & 1.307 & 0 \\ 0 & 0 & 1.03 \end{bmatrix} \begin{Bmatrix} 4.97 \\ -0.019 \\ 0 \end{Bmatrix} N_x$$

$$= \begin{Bmatrix} 100 \\ 192 \\ 0 \end{Bmatrix} N_x$$

In the 90° lamina, the degraded $[\bar{Q}]_{90}$ is used and the stresses are

$$\begin{Bmatrix} \sigma_x \\ \sigma_y \\ \tau_{xy} \end{Bmatrix}_{90} = [\bar{Q}]_{90} \begin{Bmatrix} \varepsilon_x^0 \\ \varepsilon_y^0 \\ \gamma_{xy}^0 \end{Bmatrix} = \begin{bmatrix} 0 & 0 & 0 \\ 0 & 20.14 & 0 \\ 0 & 0 & 0 \end{bmatrix} \begin{Bmatrix} 4.97 \\ -0.019 \\ 0 \end{Bmatrix} N_x = \begin{Bmatrix} 0 \\ -0.385 \\ 0 \end{Bmatrix} N_x$$

The 0° lamina now supports a greater stress in both the x- and y-directions, since the portion of the applied load originally supported by the 90° lamina has been transferred to the 0° lamina. The 90° lamina will only support a load in the y-direction. Since the 90° lamina has failed, the failure criterion is applied to the 0° lamina, yielding

$$(100N_x)^2 - (100N_x)(1.92N_x) + 784(1.92N_x)^2 = (210 \times 10^3)^2$$

Solving this equation results in $N_x = 1.857$ kip/in. Although higher than the first ply failure load ($N_x = 1.537$ kip/in.), this load is not significantly higher, nor does it represent the catastrophic failure load for the laminate. Failure of the 0° lamina begins with matrix cracks parallel to the fiber, as with the 90° lamina. In order to define the load at which total laminate failure occurs, the 0° lamina is also degraded. The new $[\bar{Q}]$ for the failed 0° lamina is

$$[\bar{Q}]_0 = [\bar{Q}]_0 = \begin{bmatrix} 20.14 & 0 & 0 \\ 0 & 0 & 0 \\ 0 & 0 & 0 \end{bmatrix} \times 10^6$$

Reformulation of the extensional stiffness matrix using the degraded properties for both the 0° and 90° lamina results in

$$A_{ij} = \sum [\bar{Q}]_k t_k = \left\{ 2[\bar{Q}]_0 + 10[\bar{Q}]_{90} \right\} 0.005)$$
$$= \left\{ 2 \begin{bmatrix} 20.14 & 0 & 0 \\ 0 & 0 & 0 \\ 0 & 0 & 0 \end{bmatrix} \times 10^6 + 10 \begin{bmatrix} 0 & 0 & 0 \\ 0 & 20.14 & 0 \\ 0 & 0 & 0 \end{bmatrix} \times 10^6 \right\} 0.005$$
$$= \begin{bmatrix} 0.2014 & 0 & 0 \\ 0 & 1.007 & 0 \\ 0 & 0 & 0 \end{bmatrix} \times 10^6$$

This results in new mid-surface strains, which are

$$\begin{Bmatrix} \varepsilon_x^0 \\ \varepsilon_y^0 \\ \gamma_{xy}^0 \end{Bmatrix} = [A'] \begin{Bmatrix} N_x \\ 0 \\ 0 \end{Bmatrix} = \begin{Bmatrix} 4.97 \\ 0 \\ 0 \end{Bmatrix} \times 10^{-6} N_x$$

The resulting stresses in each lamina are

$$\begin{Bmatrix} \sigma_x \\ \sigma_y \\ \tau_{xy} \end{Bmatrix}_0 = [\bar{Q}]_0 \begin{Bmatrix} \varepsilon_x^0 \\ \varepsilon_y^0 \\ \gamma_{xy}^0 \end{Bmatrix} = \begin{bmatrix} 20.14 & 0 & 0 \\ 0 & 0 & 0 \\ 0 & 0 & 0 \end{bmatrix} \begin{Bmatrix} 4.97 \\ 0 \\ 0 \end{Bmatrix} N_x = \begin{Bmatrix} 100 \\ 0 \\ 0 \end{Bmatrix} N_x$$
$$\begin{Bmatrix} \sigma_x \\ \sigma_y \\ \tau_{xy} \end{Bmatrix}_{90} = [\bar{Q}]_{90} \begin{Bmatrix} \varepsilon_x^0 \\ \varepsilon_y^0 \\ \gamma_{xy}^0 \end{Bmatrix} = \begin{bmatrix} 0 & 0 & 0 \\ 0 & 20.14 & 0 \\ 0 & 0 & 0 \end{bmatrix} \begin{Bmatrix} 4.97 \\ 0 \\ 0 \end{Bmatrix} N_x = \begin{Bmatrix} 0 \\ 0 \\ 0 \end{Bmatrix} N_x$$

This represents the stresses in each lamina when only the fibers are capable of supporting the load. The maximum load at which the laminate fails is estimated by considering the load at which the first ply failed ($N_x = 1.537$ kip/in.). At this load, the stresses in the unfailed 0° lamina, from Section 6.6.1, are

$$\begin{Bmatrix} \sigma_x \\ \sigma_y \\ \tau_{xy} \end{Bmatrix}_0 = \begin{Bmatrix} \sigma_1 \\ \sigma_2 \\ \tau_{12} \end{Bmatrix}_0 = \begin{Bmatrix} 75.7 \\ 1.36 \\ 0 \end{Bmatrix} N_x = \begin{Bmatrix} 75.7 \\ 1.36 \\ 0 \end{Bmatrix} (1.537) = \begin{Bmatrix} 116 \\ 2.1 \\ 0 \end{Bmatrix} \text{ksi}$$

The maximum failure load in the fiber direction is $X = 210$ ksi. The unfailed 0° fibers can be stressed an additional 94 ksi. For the failure to be complete, the stress in the outer lamina has been shown to be $100N_x$. The additional load the laminate can support above the initial failure load ($N_x = 1.537$ kip/in.) is found by equating the change in stress, $\Delta\sigma$, from the initial to final failure (94 ksi) with the stress at final failure ($100N_x$):

$$\Delta\sigma = 94 = 100N_x \Rightarrow N_x = 0.94$$

The maximum load the laminate can support is therefore

$$N_x = 1.537 + 0.94 = 2.477 \text{ kip/in.}$$

Assuming the laminate has a width of 0.50 in., a plot of σ_x vs ε_x for this laminate is shown in Figure 6.33 with experimental data for a similar material with material properties slightly different from those of the laminate in Section 6.5.1.1. The theoretical and experimental data show similar trends, but the experimental data predicts first ply failure (the point where the slope of the stress–strain curve changes) at a lower level. The bilinear response of this laminate is a result of the change in laminate stiffness after first ply failure, as indicated by the difference between the original and degraded extensional stiffness matrices. The bilinear response is typical of cross-ply laminates. Stiffness loss is an indication of matrix cracks and is used to identify the onset of matrix cracking in a test specimen.

The theoretical curve in Figure 6.33 does not include the effect of curing stresses. In Section 6.5.1.1, the stresses in the principal material directions (with $\Delta T = -300$ °F) for each lamina were determined to be

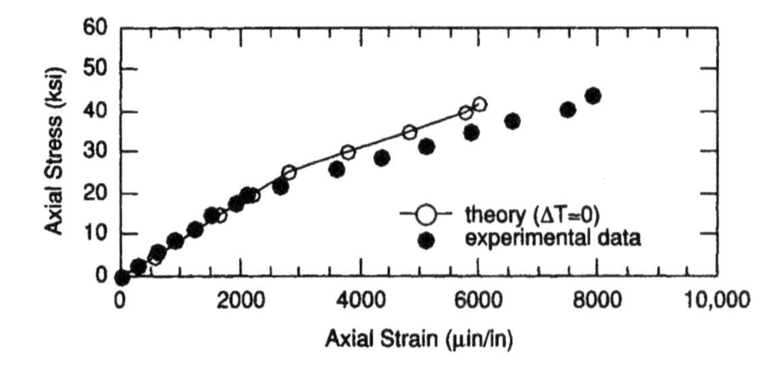

Figure 6.33 Stress–strain curve for a cross-ply laminate.

$$\begin{Bmatrix} \sigma_x \\ \sigma_y \\ \tau_{xy} \end{Bmatrix}_0 = \begin{Bmatrix} \sigma_1 \\ \sigma_2 \\ \tau_{12} \end{Bmatrix}_0 = \begin{Bmatrix} 75.7N_x - 22,800 \\ 1.36N_x + 5550 \\ 0 \end{Bmatrix}$$

$$\begin{Bmatrix} \sigma_x \\ \sigma_y \\ \tau_{xy} \end{Bmatrix}_{90} = \begin{Bmatrix} \sigma_2 \\ \sigma_2 \\ \tau_{12} \end{Bmatrix}_{90} = \begin{Bmatrix} 4.88N_x + 4550 \\ -0.271N_x - 1120 \\ 0 \end{Bmatrix}$$

Application of the Tsai–Hill failure theory to the 90° lamina results in

$$(-0.271N_x - 1120)^2 - (-0.271N_x - 1120)(4.88N_x + 4550) + 784(4.88N_x + 4550)^2 \\ = (210 \times 10^3)^2$$

Solving for N_x, we find the load for first ply failure is $N_x = 841$ kip/in. Post-first-ply-failure analysis is identical to the case in which temperature was neglected. The primary exception is that we must compute a new $[N^T]$ using the degraded $[\bar{Q}]_{90}$ resulting in

$$\{N^T\} = 2(0.005)\left(\begin{bmatrix} 20.14 & 0.393 & 0 \\ 0.393 & 1.31 & 0 \\ 0 & 0 & 2.06 \end{bmatrix}\begin{Bmatrix} -0.17 \\ 15.57 \\ 0 \end{Bmatrix}\right.$$

$$\left. +5\begin{bmatrix} 0 & 0 & 0 \\ 0 & 20.14 & 0 \\ 0 & 0 & 0 \end{bmatrix}\begin{Bmatrix} 15.57 \\ -0.17 \\ 0 \end{Bmatrix}\right)(-300) = \begin{Bmatrix} -8.086 \\ -9.633 \\ 0 \end{Bmatrix}$$

The mid-surface strains for the degraded laminate are

$$\begin{Bmatrix} \varepsilon_x^0 \\ \varepsilon_y^0 \\ \gamma_{xy}^0 \end{Bmatrix} = [A']\{\hat{N}\} = \begin{bmatrix} 4.97 & -0.019 & 0 \\ -0.019 & 0.982 & 0 \\ 0 & 0 & 97.1 \end{bmatrix} \times 10^{-6}\begin{Bmatrix} N_x - 8.086 \\ -9.633 \\ 0 \end{Bmatrix}$$

$$= \begin{Bmatrix} 4.97N_x - 40 \\ -0.019N_x - 9.206 \\ 0 \end{Bmatrix} \times 10^{-6}$$

The stresses in the laminate are now defined by

$$\begin{Bmatrix} \sigma_x \\ \sigma_y \\ \tau_{xy} \end{Bmatrix}_k = [\bar{Q}]_k\left(\begin{Bmatrix} 4.97N_x - 40 \\ -0.019N_x - 9.206 \\ 0 \end{Bmatrix} - \begin{Bmatrix} \alpha_x \\ \alpha_y \\ \alpha_{xy} \end{Bmatrix}(-300)\right)$$

The stress in each lamina is therefore

$$\begin{Bmatrix}\sigma_1\\ \sigma_2\\ \tau_{12}\end{Bmatrix}_0=\begin{Bmatrix}\sigma_x\\ \sigma_y\\ \tau_{xy}\end{Bmatrix}_0=\begin{bmatrix}20.14 & 0.392 & 0\\ 0.392 & 1.307 & 0\\ 0 & 0 & 1.03\end{bmatrix}$$
$$\left(\begin{Bmatrix}4.97N_x-40\\ -0.019N_x-9.206\\ 0\end{Bmatrix}-\begin{Bmatrix}-0.17\\ 15.57\\ 0\end{Bmatrix}(-300)\right)=\begin{Bmatrix}100\\ 1.92\\ 0\end{Bmatrix}N_x+\begin{Bmatrix}22\\ 6081\\ 0\end{Bmatrix}$$

$$\begin{Bmatrix}\sigma_2\\ \sigma_1\\ \tau_{12}\end{Bmatrix}_{90}=\begin{Bmatrix}\sigma_x\\ \sigma_y\\ \tau_{xy}\end{Bmatrix}_{90}=\begin{bmatrix}0 & 0 & 0\\ 0 & 20.14 & 0\\ 0 & 0 & 0\end{bmatrix}$$
$$\left(\begin{Bmatrix}4.97N_x-40\\ -0.019N_x-9.206\\ 0\end{Bmatrix}-\begin{Bmatrix}15.57\\ -0.17\\ 0\end{Bmatrix}(-300)\right)=\begin{Bmatrix}0\\ 0.385\\ 0\end{Bmatrix}N_x+\begin{Bmatrix}0\\ -1212\\ 0\end{Bmatrix}$$

Comparing these stresses to those for the unfailed laminate reveals that in the 0° lamina thermal coupling in the x-direction is reduced, while in the y-direction it is increased after first ply failure. The laminate can be further degraded as it was for the case without temperature effects. A plot of the results for this case is presented in Figure 6.34, along with experimental results and those from the case with $\Delta T=0$.

6.6.2 Angle-ply laminate

Consider the $[15/-15_4]_s$ angle-ply laminate from Section 6.5.1.2, for which $t_{ply}=0.005$ in. and the total laminate thickness is $h=0.50$ in. Assume this laminate is subjected to an axial load N_x as shown in Figure 6.35.

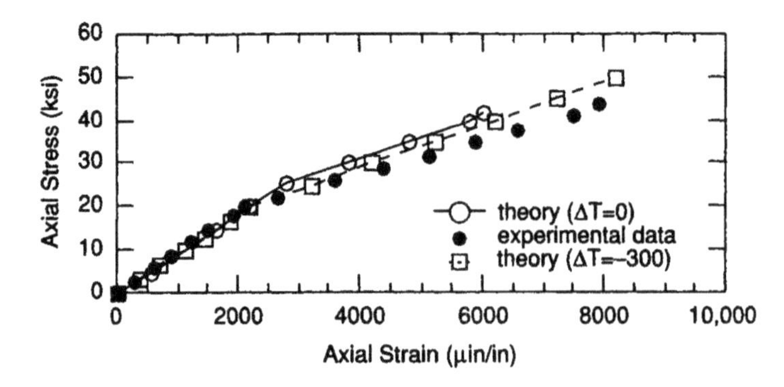

Figure 6.34 Stress–strain curve for a cross-ply laminate with $\Delta T\neq 0$.

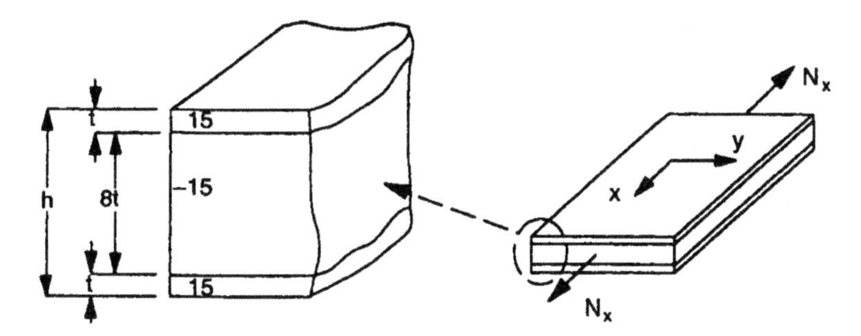

Figure 6.35 $[15/-15_4]_s$ symmetric angle-ply laminate with N_x applied.

In Section 6.5.1.2, the stresses in the principal material direction for an arbitrary axial load N_x were determined in terms of ΔT. Keeping N_x as a variable and setting $\Delta T = -300$ °F, we determine

$$\begin{Bmatrix} \sigma_1 \\ \sigma_2 \\ \tau_{12} \end{Bmatrix}_{15} = \begin{Bmatrix} 36 \\ -1.485 \\ -0.044 \end{Bmatrix} N_x + \begin{Bmatrix} -6204 \\ 1017 \\ -1833 \end{Bmatrix}$$

$$\begin{Bmatrix} \sigma_1 \\ \sigma_2 \\ \tau_{12} \end{Bmatrix}_{-15} = \begin{Bmatrix} 16.855 \\ -0.404 \\ -2.327 \end{Bmatrix} N_x + \begin{Bmatrix} -652.5 \\ 631.5 \\ 1009.2 \end{Bmatrix}$$

The failure strengths for this material are $X = 100$ ksi, $X' = 110$ ksi, $Y = 4$ ksi, $Y' = 13.9$ ksi, and $S = 9$ ksi. The direction of N_x in Figure 6.35 implies $\sigma_1 > 0$ for each lamina. Therefore, two possible conditions are considered, one for $\sigma_2 > 0$ and the other for $\sigma_2 < 0$. The Tsai–Hill failure theory is written as

$$\sigma_1^2 - \sigma_1\sigma_2 + \left(\frac{X}{Y}\right)^2 \sigma_2^2 + \left(\frac{X}{S}\right)^2 \tau_{12}^2 = X^2$$

Assuming σ_2 can be either tensile or compressive, we have two possible failure conditions to consider:

$$\text{For } \sigma_2 > 0: \sigma_1^2 - \sigma_1\sigma_2 + 625\sigma_2^2 + 123.5\tau_{12}^2 = 1 \times 10^{10}$$
$$\text{For } \sigma_2 > 0: \sigma_1^2 - \sigma_1\sigma_2 + 51.75\sigma_2^2 + 123.5\tau_{12}^2 = 1 \times 10^{10}$$

Considering the +15° lamina, the failure criterion for a tensile σ_2 is

$$(36N_x - 6204)^2 - (36N_x - 6204)(-1.485N_x - 445.5) + 625(-1.485N_x - 445.5)^2$$
$$+ 123.5(-0.044N_x + 13.2)^2 = 1 \times 10^{10}$$

This reduces to $N_x^2 - 477.8N_x - 3.618 \times 10^6 = 0$. The roots of the quadratic are $N_x = 2156, -1678$. Since the applied load is tensile, the negative root is eliminated.

This lamina is then evaluated assuming a compressive σ_2. Similarly, the $-15°$ lamina is evaluated for both tensile and compressive σ_2. The results are summarized in the following table.

Failure load N_x (lb/in.)		
Lamina	**Tensile σ_2**	**Compressive σ_2**
+15°	2156	2799
−15°	3019	3052

From this table, it is evident that the +15° lamina fails first. The stresses in each lamina at the failure load of $N_x = 2156$ are

$$\begin{Bmatrix} \sigma_1 \\ \sigma_2 \\ \tau_{12} \end{Bmatrix}_{15} = \begin{Bmatrix} 71.41 \\ -2.18 \\ -1.93 \end{Bmatrix} \text{ksi} \quad \begin{Bmatrix} \sigma_1 \\ \sigma_2 \\ \tau_{12} \end{Bmatrix}_{-15} = \begin{Bmatrix} 36.99 \\ -0.239 \\ 6.03 \end{Bmatrix} \text{ksi}$$

The failed lamina is now degraded so that the $[Q]$ and $[\bar{Q}]$ for the +15° lamina are

$$[Q]_{+15} = \begin{bmatrix} 22.35 & 0 & 0 \\ 0 & 0 & 0 \\ 0 & 0 & 0 \end{bmatrix} \times 10^6 \quad [\bar{Q}]_{+15} = \begin{bmatrix} 19.46 & 1.397 & 5.213 \\ 1.396 & 0.100 & 0.374 \\ 5.213 & 0.374 & 1.397 \end{bmatrix} \times 10^6$$

The $[A]$ matrix is now

$$[A] = 0.005(2)\left([\bar{Q}]_{+15} + 4[\bar{Q}]_{-15}\right) = \begin{bmatrix} 0.973 & 0.0824 & -0.137 \\ 0.0824 & 0.709 & -0.0145 \\ -0.137 & -0.0145 & 0.0957 \end{bmatrix} \times 10^6$$

Using the coefficients of thermal expansion from Section 6.5.1.2, the thermal loads with $\Delta T = -300$ °F are

$$\{N^T\} = \sum [\bar{Q}]_k \begin{Bmatrix} \sigma_x \\ \sigma_y \\ \tau_{xy} \end{Bmatrix}_k \Delta T(t_k) = \begin{Bmatrix} -117.2 \\ -230.4 \\ -36.4 \end{Bmatrix}$$

Combining $[N^T]$ with the applied load N_x, the mid-surface strains are

$$\begin{Bmatrix} \varepsilon_x^0 \\ \varepsilon_y^0 \\ \varepsilon_{xy}^0 \end{Bmatrix} = [A']\{\hat{N}\} = [A']\left(\begin{Bmatrix} N_x \\ 0 \\ 0 \end{Bmatrix} + \begin{Bmatrix} -117.2 \\ -230.4 \\ -36.4 \end{Bmatrix}\right)$$
$$= \begin{Bmatrix} 1.37 \\ -1.23 \\ 1.77 \end{Bmatrix} N_x + \begin{Bmatrix} 379.5 \\ -3265.5 \\ -828.5 \end{Bmatrix} \mu\text{in./in.}$$

Since there is no curvature, the strains are uniform through the laminate. The stresses in the unfailed $-15°$ lamina are

$$\begin{Bmatrix}\sigma_x\\ \sigma_y\\ \sigma_{xy}\end{Bmatrix}_{-15} = [\bar{Q}]_{-15}\left(\begin{Bmatrix}\varepsilon_x^0\\ \varepsilon_y^0\\ \gamma_{xy}^0\end{Bmatrix} - \begin{Bmatrix}\alpha_x\\ \alpha_y\\ \alpha_{xy}\end{Bmatrix}_{-15}\Delta T\right) = \begin{Bmatrix}16.165\\ -0.604\\ -2.299\end{Bmatrix}N_x + \begin{Bmatrix}-22.996\\ -12,719\\ 3460\end{Bmatrix}$$

The stresses in the principal material direction are

$$\begin{Bmatrix}\sigma_1\\ \sigma_2\\ \sigma_{12}\end{Bmatrix}_{-15} = [T_\sigma]\begin{Bmatrix}\sigma_x\\ \sigma_y\\ \sigma_{xy}\end{Bmatrix}_{-15} = \begin{Bmatrix}15.616\\ -0.057\\ -6.394\end{Bmatrix}N_x + \begin{Bmatrix}-23,171\\ -12,540\\ -2142\end{Bmatrix}$$

The σ_2 stress in the unfailed $-15°$ lamina is compressive, so failure is predicted using

$$(16.61N_x - 23,717)^2 - (15.616N_x - 23,717)(-0.057N_x - 12,540) + 51.75(-0.57N_x - 12,540)^2 + 123.5(6.39N_x - 2142)^2 = 1\times10^{10}$$

This reduces to $N_x^2 - 725N_x - 1.982\times10^5 = 0$. Solving this quadratic results in $N_x = 937$, which is less than the failure load for first ply failure. Therefore, the $-15°$ lamina fails immediately after the $+15°$ lamina. The stress–strain curve for the angle-ply laminate does not have a knee as that for the cross-ply laminate did.

6.6.3 Moisture effects

Consider the $[0/90_5]_s$ cross-ply laminate of Section 6.6.1 as shown in Figure 6.36. Both thermal and hygral effects are considered. Since the laminate is symmetric, $\{M^T\} = 0$ and $\{M^H\} = 0$. In Section 6.5.1.1, we established expressions for $[A']$, $\{N\}$, and $\{N^T\}$. With $h = 0.060$, these arc

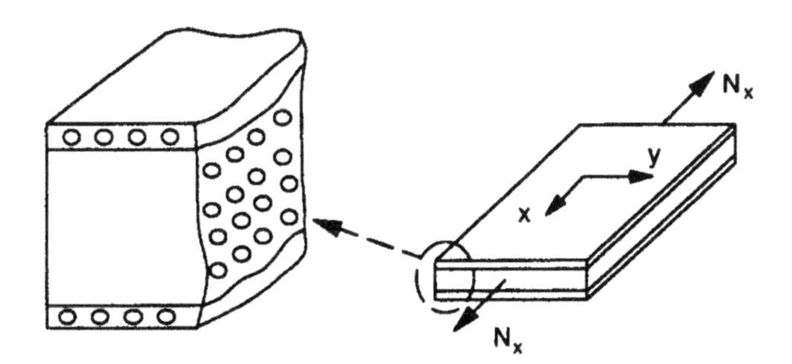

Figure 6.36 Cross-ply laminate involving thermal and hygral effects.

$$[A'] = \begin{bmatrix} 3.77 & -0.087 & 0 \\ -0.087 & 0.982 & 0 \\ 0 & 0 & 8.10 \end{bmatrix} \times 10^{-6} \quad \{N\} = \begin{Bmatrix} N_x \\ 0 \\ 0 \end{Bmatrix}$$

$$\{N^{\mathrm{T}}\} = \begin{Bmatrix} 1.044 \\ 0.341 \\ 0 \end{Bmatrix} \Delta T$$

The coefficients of hygral expansion for this material are $\beta_1 = 0.0$ and $\beta_2 = 0.44$. The moisture coefficients for each lamina are defined from Equation (3.31) to be

$$\{\beta\}_0 = \begin{Bmatrix} \beta_x \\ \beta_y \\ \beta_{xy} \end{Bmatrix}_0 \begin{Bmatrix} 0 \\ 0.44 \\ 0 \end{Bmatrix} \quad \{\beta\}_{90} = \begin{Bmatrix} \beta_x \\ \beta_y \\ \beta_{xy} \end{Bmatrix}_{90} = \begin{Bmatrix} 0.44 \\ 0 \\ 0 \end{Bmatrix}$$

The hygral loads are determined using Equation (6.29):

$$\{N^{\mathrm{H}}\} = 2(0.006)\left([\bar{Q}]_0 \{\beta\}_0 + 5 [\bar{Q}]_{90} \{\beta\}_{90} \right) \bar{M}$$
$$= \begin{Bmatrix} 0.0366 \\ 0.01724 \\ 0 \end{Bmatrix} \times 10^6 \bar{M}$$

The total laminate load $\{\hat{N}\}$ is

$$\{\hat{N}\} = \begin{Bmatrix} N_x + 1.044\Delta T + 36{,}660\bar{M} \\ 0.341\Delta T + 17{,}240\bar{M} \\ 0 \end{Bmatrix}$$

The mid-surface strains are

$$\{\varepsilon^{\circ}\} = [A']\{\hat{N}\} = \begin{Bmatrix} 3.77N_x + 3.906\Delta T + 136{,}662\bar{M} \\ -0.087N_x + 0.244\Delta T + 13{,}750\bar{M} \\ 0 \end{Bmatrix} \times 10^6$$

As previously demonstrated, the 90° lamina fails first; therefore, we focus on that lamina. The Cartesian components of stress in the 90° lamina are defined by $\{\sigma\}_{90} = [\bar{Q}]_{90}\{\varepsilon^0\} - \{\alpha\}_{90}\Delta T - \{\beta\}_{90}\bar{M}$:

$$\begin{Bmatrix} \sigma_x \\ \sigma_y \\ \tau_{xy} \end{Bmatrix}_{90} = \begin{Bmatrix} \sigma_2 \\ \sigma_1 \\ \tau_{12} \end{Bmatrix}_{90} = \begin{Bmatrix} 4.893 \\ -0.274 \\ 0 \end{Bmatrix} N_x + \begin{Bmatrix} -15.09 \\ 3.76 \\ 0 \end{Bmatrix} \Delta T + \begin{Bmatrix} -392{,}206 \\ 102{,}868 \\ 0 \end{Bmatrix} \bar{M}$$

Assume $\Delta T = -300\ °F$, and after 1 h of exposure to a humid environment, $\bar{M} = 0.00573$. These conditions result in

$$\sigma_1 = -0.274N_x - 538 \quad \sigma_2 = 4.893N_x + 2240$$

Recalling $X = X' = 210$ ksi, $Y = 7.5$ ksi, $Y' = 29.9$ ksi, and $S = 13.5$ ksi, the failure criterion can be written as

$$(-0.274N_x - 538)^2 - (-0.274N_x - 538)(4.893N_x + 2240) + \left(\frac{210}{7.5}\right)^2 (4.893N_x + 2240)^2 = \left(210 \times 10^3\right)^2$$

This reduces to $N_x^2 + 92N_x - 2.25 \times 10^6 = 0$. Solving this equation results in a predicted failure load of $N_x \approx 1420$, which is larger than the case without moisture. Moisture causes swelling, which reduces the strain and subsequent stress in the lamina. Increasing the average moisture content increases the predicted failure load. The increased failure load with exposure to moisture is misleading. As moisture content increases, the strength of the lamina decreases as a function of exposure time and $\bar{M}$. The strength parameters used in this example did not consider the degradation of strength with moisture content. The analysis presented may be applicable to short-time exposure, but not long-time exposure. Detailed discussions of the effects of moisture can be found various references [33–35].

6.7 In-plane laminate strength analysis

Laminate strength analysis focuses on the entire laminate and uses the concepts of strength ratios from Section 5.5.2.1 and presented by Tsai and Hahn [36]. The discussions presented herein do not include thermal or hygral effects. Their inclusion would alter the magnitudes of stress components, while the analysis procedures remain the same. The Tsai–Wu theory failure theory is used in the form $\left[F_{11}\sigma_1^2 + 2F_{12}\sigma_1\sigma_2 + F_{22}\sigma_2^2 + F_{66}\tau_{12}^2\right]R^2 + \left[F_1\sigma_1 + F_2\sigma_2\right]R = 1$, where R is the strength ratio. For a multidirectional laminate, the in-plane strength consists of multiple strength ratios (R and R'). These pairs of strength ratios are a function of lamina orientation and applied load, and one pair will exist for each ply of the laminate. When the analysis is completed, a failure ellipse for the entire laminate is constructed.

The cross-ply laminate in Figure 6.37 is used to illustrate the concepts behind in-plane strength analysis. For this problem, the thicknesses of the 0° and 90° laminas are varied by controlling the number of plies in each lamina.

The number of plies in each lamina is defined by N_0 (number of 0° plies) and N_{90} (number of 90° plies). Each ply is assumed to have the same thickness ($t = 0.005$ in.), and the total laminate thickness is defined by $h = (N_0 + N_{90})t$. The material properties for this laminate are $E_1 = 22.2 \times 10^6$ psi, $E_2 = 1.58 \times 10^6$ psi, $G_{12} = 0.81 \times 10^6$ psi, $\nu_{12} = 0.30$, $X = 100$ ksi, $X' = 110$ ksi, $Y = 4$ ksi, $Y' = 13.9$ ksi, and $S = 9$ ksi. Assuming $F^*_{12} = -1/2$, the strength parameters are $F_{11} = 9.09 \times 10^{-11}$, $F_{12} = -6.392 \times 10^{-10}$,

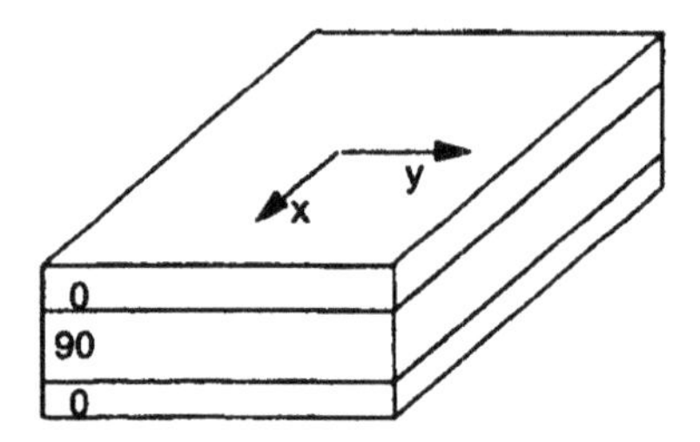

Figure 6.37 Symmetric cross-ply laminate for strength analysis.

$F_{22}=1.799\times 10^{-8}$, $F_{66}=1.235\times 10^{-8}$, $F_1=9.1\times 10^{-7}$, and $F_2=1.78\times 10^{-4}$. The stiffness matrix for each lamina is

$$[\bar{Q}]_0=\begin{bmatrix} 22.32 & 0.477 & 0 \\ 0.477 & 1.59 & 0 \\ 0 & 0 & 1.62 \end{bmatrix}\times 10^6 \quad [\bar{Q}]_{90}=\begin{bmatrix} 1.59 & 0.477 & 0 \\ 0.477 & 22.34 & 0 \\ 0 & 0 & 1.62 \end{bmatrix}\times 10^6$$

The extensional stiffness matrix is defined by $[A]=\left([\bar{Q}]_0 N_0+[\bar{Q}]_{90} N_{90}\right)t$. Bending is not considered, so $[D]$ is not required, and since the laminate is symmetric, $[B]=0$. The mid-surface strains are determined in terms of an applied load divided by the total laminate thickness h, which can be regarded as a normal stress applied to the laminate in a specified direction. The first laminate considered is a $[0/90]_s$ laminate. For this case, we have $N_0=N_{90}=2$ and $h=4t$. In order to define the failure surface for this laminate, a series of unit load vectors are applied. For example, the initial unit load vector is

$$\{\hat{N}\}=\begin{Bmatrix} N_x \\ 0 \\ 0 \end{Bmatrix}=\begin{Bmatrix} 1 \\ 0 \\ 0 \end{Bmatrix}$$

The $[A]$ and $[A']$ matrices are determined in terms of the total laminate thickness h to be

$$[A]=\begin{bmatrix} 11.965 & 0.477 & 0 \\ 0.477 & 11.965 & 0 \\ 0 & 0 & 1.62 \end{bmatrix} h\times 10^6$$

$$[A']=\begin{bmatrix} 0.0837 & -0.0033 & 0 \\ -0.0033 & 0.0837 & 0 \\ 0 & 0 & 0.6173 \end{bmatrix}\left(\frac{10^{-6}}{h}\right)$$

The mid-surface strains are

$$\{\varepsilon^0\}=[A']\begin{Bmatrix} N_x \\ 0 \\ 0 \end{Bmatrix}=\begin{Bmatrix} 0.0837 \\ -0.0033 \\ 0 \end{Bmatrix}\left(\frac{N_x}{h}\right)\times 10^{-6}$$

The stress in the principal material direction for each lamina is

$$\{\sigma\}_0 = \begin{Bmatrix} \sigma_1 \\ \sigma_2 \\ \tau_{12} \end{Bmatrix}_0 = [\bar{Q}]_0\{\varepsilon^0\} = \begin{Bmatrix} 1.866 \\ 0.0346 \\ 0 \end{Bmatrix}\left(\frac{N_x}{h}\right)$$

$$\{\sigma\}_{90} = \begin{Bmatrix} \sigma_1 \\ \sigma_2 \\ \tau_{12} \end{Bmatrix}_{90} = [\bar{Q}]_{90}\{\varepsilon^0\} = \begin{Bmatrix} 0.1313 \\ -0.0347 \\ 0 \end{Bmatrix}\left(\frac{N_x}{h}\right)$$

where N_x/h is the applied stress in the x-direction. The failure load for each lamina is identified using strength ratios. For the 0° lamina, we initially define $R = N_x/h$, so the stresses can be expressed as $\sigma_1 = 1.866R$, $\sigma_2 = 0.0346R$, and $\tau_{12} = 0$. Failure for this laminate is predicted from

$$9.09 \times 10^{-11}(1.866R)^2 + 2(-6.392 \times 10^{-10})(1.866R)(0.0346R) + 1.799 \times 10^{-8}(0.0346R)^2 + 9.1 \times 10^{-7}(1.866R) + 1.78 \times 10^{-4}(0.0346R) = 1$$

This reduces to $R^2 + 3.075 \times 10^4 R - 3.914 \times 10^9 = 0$

The roots of this quadratic are $R = 49.05$, -79.82 ksi. In a similar manner, the 90° lamina can be evaluated. For the 90° lamina, the failure criterion is $R^2 + 7.44 \times 10^4 R - 3.193 \times 10^9 = 0$. The roots of this equation are $R = 30.5$, -104.9 ksi. These roots identify the axial load at which failure of each ply is predicted. In addition, other unit load vectors could be used to predict lamina failure. For example,

$$\{\hat{N}\} = \begin{Bmatrix} 0 \\ N_y \\ 0 \end{Bmatrix} = \begin{Bmatrix} 0 \\ 1 \\ 0 \end{Bmatrix} \quad \{\hat{N}\} = \begin{Bmatrix} N_x \\ N_y \\ 0 \end{Bmatrix} = \begin{Bmatrix} 1 \\ 1 \\ 0 \end{Bmatrix}$$

$$\{\hat{N}\} = \begin{Bmatrix} N_x \\ -N_y \\ 0 \end{Bmatrix} = \begin{Bmatrix} 1 \\ -1 \\ 0 \end{Bmatrix}$$

This analysis can be repeated for other load vector combinations, resulting in different sets of R for each case. The results are then plotted for each lamina to define the boundaries of the failure envelope for specific unit load vectors.

An alternative approach is to define the principal material direction stresses in terms of arbitrary applied loads N_x, N_y, and N_{xy}. For example, assume the laminate in Figure 6.38, where $N_{xy} = 0$.

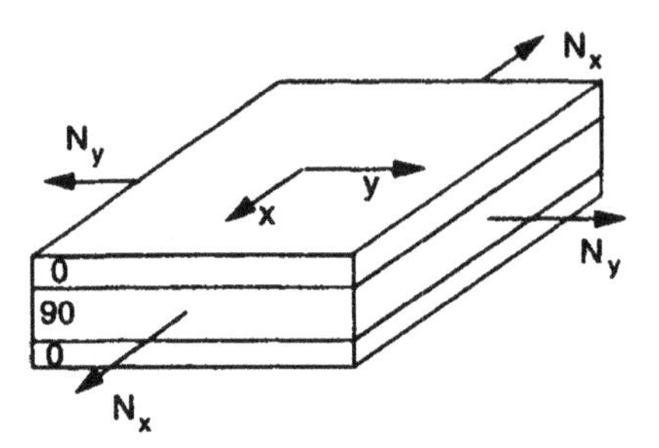

Figure 6.38 Laminate subjected to N_x and N_y.

The mid-surface strains for this laminate and the stresses in each lamina are defined by

$$\begin{Bmatrix} \varepsilon_x^0 \\ \varepsilon_y^0 \\ \gamma_{xy}^0 \end{Bmatrix} = [A'] \begin{Bmatrix} N_x \\ N_y \\ 0 \end{Bmatrix} = \begin{Bmatrix} 0.0837N_x - 0.0033N_y \\ -0.0033N_x + 0.0837N_y \\ 0 \end{Bmatrix} \left(\frac{10^{-6}}{h}\right)$$

$$\begin{Bmatrix} \sigma_1 \\ \sigma_2 \\ \tau_{12} \end{Bmatrix}_k = [\bar{Q}]_k \begin{Bmatrix} 0.0837N_x - 0.0033N_y \\ -0.0033N_x + 0.0837N_y \\ 0 \end{Bmatrix} \left(\frac{10^{-6}}{h}\right)$$

The stresses in the 1–2 plane can be written in terms of the stresses in the x–y plane, which are defined as $\sigma_x = N_x/h$ and $\sigma_y = N_y/h$. For the 0° lamina, we have

$$\begin{Bmatrix} \sigma_1 \\ \sigma_2 \\ \tau_{12} \end{Bmatrix} = \begin{bmatrix} 22.32 & 0.477 & 0 \\ 0.477 & 1.59 & 0 \\ 0 & 0 & 1.62 \end{bmatrix} \times 10^6 \begin{Bmatrix} 0.0837N_x - 0.0033N_y \\ -0.0033N_x + 0.0837N_y \\ 0 \end{Bmatrix} \left(\frac{10^{-6}}{h}\right)$$

$$= \begin{Bmatrix} 1.866N_x - 0.0337N_y \\ 0.0347N_x + 0.1315N_y \\ 0 \end{Bmatrix} \left(\frac{1}{h}\right) = \begin{pmatrix} 1.866\sigma_x - 0.0337\sigma_y \\ 0.0347\sigma_x + 0.1315\sigma_y \\ 0 \end{pmatrix}$$

The Tsai–Wu failure theory for this lamina will be

$$F_{11}(1.866\sigma_x - 0.0337\sigma_y)^2 + 2F_{12}(1.866\sigma_x - 0.0337\sigma_y)(0.0347\sigma_x + 0.13157\sigma_y)$$
$$+ F_{22}(\ldots) + \ldots + F_2(0.0347\sigma_x + 0.13157\sigma_y) = 1$$

Expanding and using the appropriate strength parameters results in

$$2.57 \times 10^{-10}\sigma_x^2 - 1.63 \times 10^{-10}\sigma_x\sigma_y + 3.17 \times 10^{-10}\sigma_y^2 + 7.77 \times 10^{-6}\sigma_x$$
$$+ 2.32 \times 10^{-5}\sigma_y = 1$$

This form of the failure criteria defines the failure ellipse in terms of σ_x and σ_y for the 0° lamina. A similar expression can be generated for the 90° lamina. This form of the failure criteria may be more useful for laminates other than the cross-ply where σ_1 and σ_2 do not coincide with σ_x and σ_y. The failure ellipses for the 0° and 90° lamina are shown in Figures 6.39 and 6.40, respectively. Included in these figures are points corresponding to failures associated with various unit vectors using strength ratios. The effect of including shear stress would be similar to that discussed in Example 5.5.

A composite failure envelope for the laminate is formed by combining the results of Figures 6.39 and 6.40, as shown in Figure 6.41. The intersection of the two failure ellipses defines the safe region for the laminate.

A similar analysis can be performed for various combinations of ply orientations and unit stress vectors (or σ_x and σ_y components). This type of analysis is not limited to cross-ply laminates.

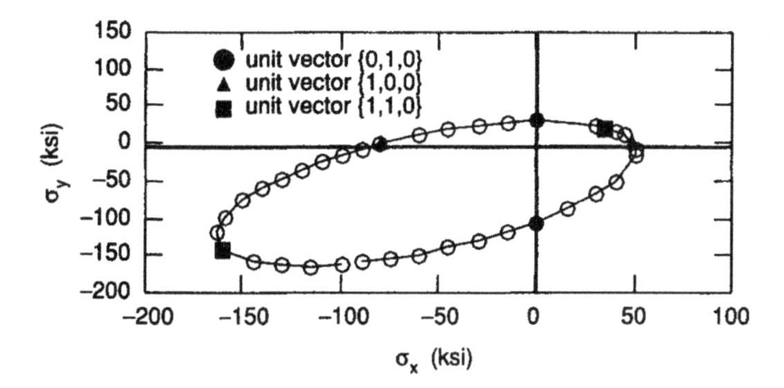

Figure 6.39 Failure ellipse for 0° lamina in a $[0/90]_s$ laminate.

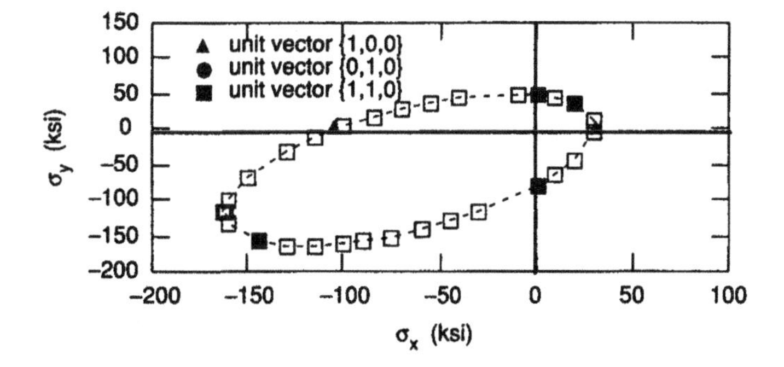

Figure 6.40 Failure ellipse for 0° lamina in a $[0/90]_s$ laminate.

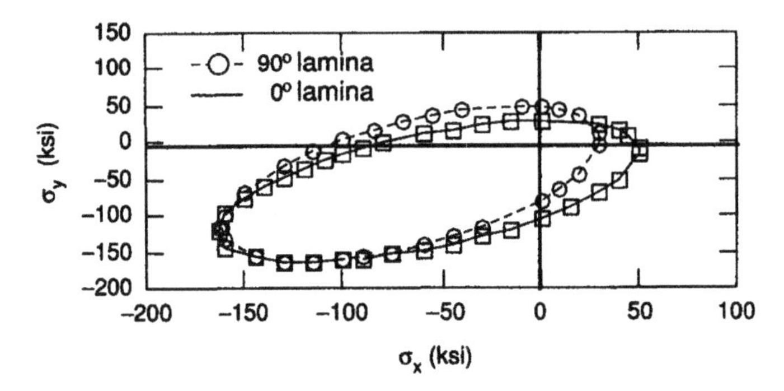

Figure 6.41 Combined failure ellipses for $[0/90]_s$ laminate.

6.8 Invariant forms of [*A*], [*B*], [*D*]

Laminate design based on strength can be supplemented with considerations of stiffness. By controlling parameters such as material and stacking sequence, the [A], [B], and [D] matrices can be tailored to meet certain design requirements. Each component of these matrices can be expressed in compact form as

$$(A_{ij}, B_{ij}, D_{ij}) = \int \bar{Q}_{ij}(1, z, z^2)\mathrm{d}z$$

In this expression, $\bar{Q}_{ij}$ is constant for a given lamina, but may vary through the laminate, as a function of fiber orientation. Each component of [A] is associated with $\int [\bar{Q}_{ij}](1)\mathrm{d}z$, whereas the components of [B] and [D] are associated with $\int [\bar{Q}_{ij}](z)\mathrm{d}z$ and $\int [\bar{Q}_{ij}](z^2)\mathrm{d}z$, respectively.

The invariant form of $\bar{Q}_{ij}$ given by Equation (3.12) is used to define [A], [B], and [D], as presented by Tsai and Pagano [37]. From Chapter 3, the invariant forms of $\bar{Q}_{ij}$ are

$$\begin{Bmatrix} \bar{Q}_{11} \\ \bar{Q}_{22} \\ \bar{Q}_{12} \\ \bar{Q}_{66} \\ \bar{Q}_{16} \\ \bar{Q}_{26} \end{Bmatrix} = \begin{bmatrix} U_1 & \cos 2\theta & \cos 4\theta \\ U_1 & -\cos 2\theta & \cos 4\theta \\ U_4 & 0 & -\cos 4\theta \\ U_5 & 0 & -\cos 4\theta \\ 0 & \sin 2\theta/2 & \sin 4\theta \\ 0 & \sin 2\theta/2 & -\sin 4\theta \end{bmatrix} \begin{Bmatrix} 1 \\ U_2 \\ U_3 \end{Bmatrix}$$

The terms U_1 to U_5 are defined by Equation (3.11). Consider the terms A_{11}, B_{11}, and D_{11}, which can be written as

$$(A_{11}, B_{11}, D_{11}) = \int \bar{Q}_{11}(1, z, z^2)\mathrm{d}z$$

Substituting the expression for $\bar{Q}_{11}$ from the invariant form results in

$$(A_{11}, B_{11}, D_{11}) = \int \left[U_1 \left(1, z, z^2\right) + U_2 \cos 2\theta \left(1, z, z^2\right) + U_3 \cos 4\theta \left(1, z, z^2\right) \right] \mathrm{d}z$$

If each lamina is made from the same material, U_1, U_2, and U_3 can be brought outside the integral. Since the limits of integration are $-h/2$ and $h/2$, and since the first term does not depend on θ, this expression can be written as

$$(A_{11}, B_{11}, D_{11}) = U_1 \left(h, 0, \frac{h^3}{12} \right) + U_2 \int_{-h/2}^{h/2} \cos 2\theta \left(1, z, z^2\right) \mathrm{d}z$$
$$+ U_3 \int_{-h/2}^{h/2} \cos 4\theta \left(1, z, z^2\right) \mathrm{d}z$$

where h is the total laminate thickness. Since all $\bar{Q}_{ij}$ terms can be expressed as invariants, each A_{ij}, B_{ij}, and D_{ij} term contains a combination of integrals in 0 and $(1, z, z^2)\mathrm{d}z$. As a result, it is convenient to define additional parameters:

$$\begin{aligned}
V_{0(A,B,D)} &= \left(h, 0, \frac{h^3}{12} \right) \\
V_{1(A,B,D)} &= \int \cos 2\theta \left(1, z, z^2\right) \mathrm{d}z \quad V_{2(A,B,D)} = \int \sin 2\theta \left(1, z, z^2\right) \mathrm{d}z \\
V_{3(A,B,D)} &= \int \cos 4\theta \left(1, z, z^2\right) \mathrm{d}z \quad V_{4(A,B,D)} = \int \sin 4\theta \left(1, z, z^2\right) \mathrm{d}z
\end{aligned} \tag{6.39}$$

For an N-layered laminate, these integrals can be simplified to

$$\begin{aligned}
V_{iA} &= \sum_{k-1}^{N} W_k (Z_{k+1} - Z_k) \\
V_{iB} &= \frac{1}{2} \sum_{k-1}^{N} W_k \left(Z_{k+1}^2 - Z_k^2\right) \\
V_{iD} &= \frac{1}{3} \sum_{k-1}^{N} W_k \left(Z_{k+1}^3 - Z_k^3\right)
\end{aligned} \tag{6.40}$$

where $i = 1$ to 4, and the subscripts A, B, and D refer to the corresponding laminate stiffness matrices. In addition, we define

$$W_k = \begin{cases} \cos 2\theta_k, & \text{for } i = 1 \\ \sin 2\theta_k, & \text{for } i = 2 \\ \cos 4\theta_k, & \text{for } i = 3 \\ \sin 4\theta_k, & \text{for } i = 4 \end{cases} \tag{6.41}$$

The angle θ_k is the fiber orientation for the kth lamina. Using this notation, the components of $[A]$, $[B]$, and $[D]$ are written as

$$\begin{Bmatrix} (A_{11}, B_{11}, D_{11}) \\ (A_{22}, B_{22}, D_{22}) \\ (A_{12}, B_{12}, D_{12}) \\ (A_{66}, B_{66}, D_{66}) \\ (A_{16}, B_{16}, D_{16}) \\ (A_{26}, B_{26}, D_{26}) \end{Bmatrix} = \begin{bmatrix} U_1 & U_2 & 0 & U_3 & 0 \\ U_1 & -U_2 & 0 & U_3 & 0 \\ U_4 & 0 & 0 & -U_3 & 0 \\ U_5 & 0 & 0 & -U_3 & 0 \\ 0 & 0 & -U_2/2 & 0 & -U_3 \\ 0 & 0 & -U_2/2 & 0 & U_3 \end{bmatrix} \begin{Bmatrix} V_{0(A,B,D)} \\ V_{1(A,B,D)} \\ V_{2(A,B,D)} \\ V_{3(A,B,D)} \\ V_{4(A,B,D)} \end{Bmatrix} \quad (6.42)$$

This form of $[A]$, $[B]$, and $[D]$ can be used to identify useful parameters such as optimum fiber orientation. For example, consider a general laminate as shown in Figure 6.42. In the x, y coordinate system shown, this laminate will respond to applied loads according to

$$\begin{Bmatrix} \hat{N} \\ \hat{M} \end{Bmatrix} = \begin{bmatrix} A & B \\ B & D \end{bmatrix} \begin{Bmatrix} \varepsilon^\circ \\ k \end{Bmatrix}$$

Instead of using the x, y coordinate system, it may be better to examine the laminate response in an x', y' coordinate system.

Rotating the laminate through some angle Φ, a set of transformed $[A]$, $[B]$, and $[D]$ matrices is developed. The transformed laminate stiffness matrices are designated as $[\bar{A}]$, $[\bar{B}]$, and $[\bar{C}]$, which is analogous to the transformation from $[Q]$ to $[\bar{Q}]$. The orientation of fibers in each lamina in the x–y coordinate system is 0, and in the x'–y' coordinate system, it is θ'. These orientations are related to the angle Φ through the relation $\theta' = \theta - \Phi$, as illustrated in Figure 6.43. The transformed angle θ' replaces θ in the invariant representation of $[\bar{Q}]$.

Trigonometric identities for sine and cosine of the angles $n(\theta - \Phi)$ are needed to explicitly define $[\bar{A}]$, $[\bar{B}]$, and $[\bar{D}]$. For example, the general form of A_{11} expressed using the invariants of Equation (6.41) is $A_{11} = U_1 V_{0A} + U_2 V_{1A} + U_3 V_{3A}$. Substituting for V_{0A}, V_{1A}, and V_{3A} from Equation (6.39), and recalling that θ' is used in the transformed system,

$$\bar{A}_{11} = U_1 h + U_2 \int \cos 2\theta' \mathrm{d}z + U_3 \int \cos 4\theta' \mathrm{d}z$$

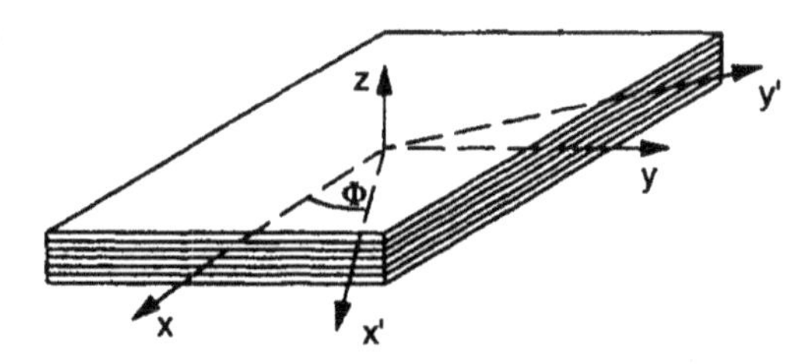

Figure 6.42 General laminate forms of [A], [B], and [D].

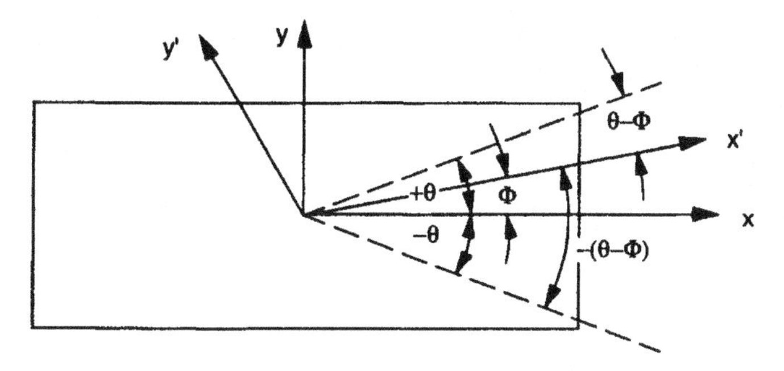

Figure 6.43 Definition of θ for [A], [B], and [D].

Trig identities for $2\theta' = 2(\theta - \Phi)$ and $4\theta' = 4(\theta - \Phi)$ result in a transformed $\bar{A}_{11}$ in terms of 0 (the lamina orientation in the x–y system) and Φ (the orientation angle of the entire laminate with respect to the X axis). Therefore,

$$\bar{A}_{11} = U_1 h + U_2 \int (\cos 2\theta \cos 2\Phi + \sin 2\theta \sin 2\Phi) dz + U_3 \int (\cos 4\theta \cos 4\Phi + \sin 4\theta \sin 4\Phi) dz$$

Using the definitions of $\int \cos m\theta \, dz$ from Equation (6.39),

$$\bar{A}_{11} = U_1 h + U_2 [V_{1A} \cos 2\Phi + V_{2A} \sin 2\Phi] + U_3 [V_{3A} \cos 4\Phi + V_{4A} \sin 4\Phi]$$

In a similar manner, the entire $[\bar{A}]$ matrix is defined as

$$\begin{Bmatrix} \bar{A}_{11} \\ \bar{A}_{22} \\ \bar{A}_{12} \\ \bar{A}_{66} \\ \bar{A}_{16} \\ \bar{A}_{26} \end{Bmatrix} = \begin{bmatrix} U_1 V_{0A} & U_2 V_{1A} & U_2 V_{2A} & U_3 V_{3A} & U_3 V_{4A} \\ U_1 V_{0A} & -U_2 V_{1A} & -U_2 V_{2A} & U_3 V_{3A} & U_3 V_{4A} \\ U_4 V_{0A} & 0 & 0 & -U_3 V_{3A} & -U_3 V_{4A} \\ U_5 V_{0A} & 0 & 0 & -U_3 V_{3A} & -U_3 V_{4A} \\ 0 & U_2 V_{2A}/2 & -U_2 V_{1A}/2 & U_3 V_{4A} & -U_3 V_{3A} \\ 0 & U_2 V_{2A}/2 & -U_2 V_{1A}/2 & -U_3 V_{4A} & U_3 V_{3A} \end{bmatrix} \begin{Bmatrix} 1 \\ \cos 2\Phi \\ \sin 2\Phi \\ \cos 4\Phi \\ \sin 4\Phi \end{Bmatrix} \tag{6.43}$$

The $[\bar{B}]$ and $[\bar{D}]$ matrices have the same form as $[\bar{A}]$, except the V_{iA} terms are replaced with V_{iB} and V_{iD} as defined by Equation (6.39). The explicit representations of $[\bar{B}]$ and $[\bar{D}]$ are

$$\begin{Bmatrix} \bar{B}_{11} \\ \bar{B}_{22} \\ \bar{B}_{12} \\ \bar{B}_{66} \\ \bar{B}_{16} \\ \bar{B}_{26} \end{Bmatrix} = \begin{bmatrix} U_1 V_{0B} & U_2 V_{1B} & U_2 V_{2B} & U_3 V_{3B} & U_3 V_{4B} \\ U_1 V_{0B} & -U_2 V_{1B} & -U_2 V_{2B} & U_3 V_{3B} & U_3 V_{4B} \\ U_4 V_{0B} & 0 & 0 & -U_3 V_{3B} & -U_3 V_{4B} \\ U_5 V_{0B} & 0 & 0 & -U_3 V_{3B} & -U_3 V_{4B} \\ 0 & U_2 V_{2B}/2 & -U_2 V_{1B}/2 & U_3 V_{4B} & -U_3 V_{3B} \\ 0 & U_2 V_{2B}/2 & -U_2 V_{1B}/2 & -U_3 V_{4B} & U_3 V_{3B} \end{bmatrix} \begin{Bmatrix} 1 \\ \cos 2\Phi \\ \sin 2\Phi \\ \cos 4\Phi \\ \sin 4\Phi \end{Bmatrix} \tag{6.44}$$

$$\begin{Bmatrix} \bar{D}_{11} \\ \bar{D}_{22} \\ \bar{D}_{12} \\ \bar{D}_{66} \\ \bar{D}_{16} \\ \bar{D}_{26} \end{Bmatrix} = \begin{bmatrix} U_1V_{0D} & U_2V_{1D} & U_2V_{2D} & U_3V_{3D} & U_3V_{4D} \\ U_1V_{0D} & -U_2V_{1D} & -U_2V_{2D} & U_3V_{3D} & U_3V_{4D} \\ U_4V_{0D} & 0 & 0 & -U_3V_{3D} & -U_3V_{4D} \\ U_5V_{0D} & 0 & 0 & -U_3V_{3D} & -U_3V_{4D} \\ 0 & U_2V_{2D}/2 & -U_2V_{1D}/2 & U_3V_{4D} & -U_3V_{3D} \\ 0 & U_2V_{2D}/2 & -U_2V_{1D}/2 & -U_3V_{4D} & U_3V_{3D} \end{bmatrix} \begin{Bmatrix} 1 \\ \cos 2\Phi \\ \sin 2\Phi \\ \cos 4\Phi \\ \sin 4\Phi \end{Bmatrix} \tag{6.45}$$

Jones [2] presents a complete discussion of special cases of $[\bar{A}]$, $[\bar{B}]$, and $[\bar{D}]$. Two special cases are considered herein that involve

$$\int_{-z}^{z} (\text{odd function})dz = 0 \quad \text{and} \quad \int_{-z}^{z} (\text{even function})dz = \text{finite}$$

Consider first the antisymmetric laminate in Figure 6.44a. The $+\alpha$ and $-\alpha$ fiber orientations influence the integrals of $V_{i(A,B,D)}$ in Equation (6.39). The odd and even integrands for the antisymmetric laminate are

$$\begin{aligned} &\text{Odd}: \cos 2\theta(z),\ \cos 4\theta(z),\ \sin 2\theta(1,z^2),\ \sin 4\theta(1,z^2) \\ &\text{Even}: \cos 2\theta(1,z^2),\ \cos 4\theta(1,z^2),\ \sin 2\theta(z),\ \sin 4\theta(z) \end{aligned}$$

These result in $V_{2A}=V_{4A}=V_{1B}=V_{3B}=V_{2D}=V_{4D}=0$, which in turn lead to $\bar{A}_{16}=\bar{A}_{26}=\bar{B}_{11}=\bar{B}_{12}=\bar{B}_{22}=\bar{D}_{16}=\bar{D}_{26}=0$.

Next, we consider the symmetric laminate in Figure 6.44b for which odd and even integrands of $V_{i(A,B,D)}$ are

$$\begin{aligned} &\text{Odd}: \cos 2\theta(z),\ \cos 4\theta(z),\ \sin 2\theta(z),\ \sin 4\theta(z) \\ &\text{Even}: \cos 2\theta(1,z^2),\ \cos 4\theta(1,z^2),\ \sin 2\theta(1,z^2),\ \sin 4\theta(1,z^2) \end{aligned}$$

This results in $[\bar{B}]=0$.

The usefulness of $[\bar{A}]$, $[\bar{B}]$, and $[\bar{D}]$ is illustrated by considering the $[-15/15_4]_s$ angle-ply laminate in Figure 6.45. The material properties are $E_1=26.25\times10^6$ psi, $E_2=1.50\times10^6$ psi, $G_{12}=1.04\times10^6$ psi, and $\nu_{12}=0.28$. The applied load is assumed to be N_x, and the total laminate thickness is $h=0.05$ in. (each ply is 0.005 in) The stiffness matrix and invariants are

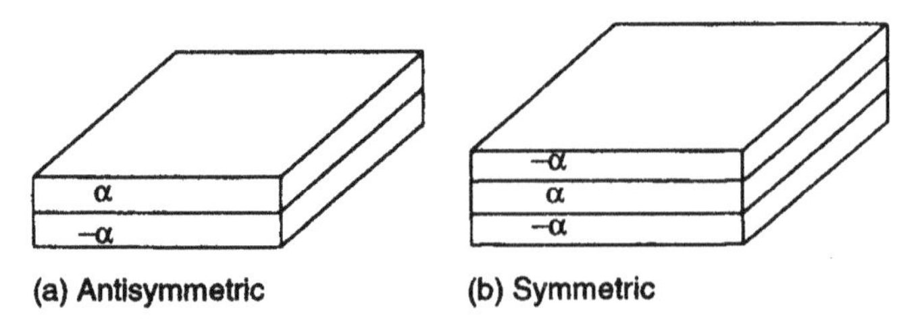

Figure 6.44 (a) Antisymmetric and (b) symmetric laminates.

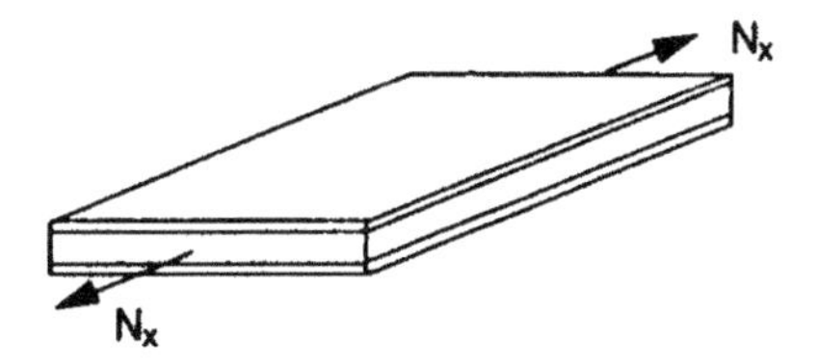

Figure 6.45 $[-15/15_4]_s$ angle-ply laminate.

$$[Q]=\begin{bmatrix}26.36 & 0.422 & 0\\ 0.422 & 1.507 & 0\\ 0 & 0 & 1.04\end{bmatrix}\times 10^6 \quad \begin{matrix}U_1=11.07\times 10^6 & U_2=12.43\times 10^6\\ U_3=1.297\times 10^6 & U_4=3.270\times 10^6\\ U_5=3.890\times 10^6 & \end{matrix}$$

Since the laminate is symmetric and only an axial load is applied, $[\bar{B}]=0$ and $[\bar{D}]$ is not needed. The V_{iA} terms are determined from Equations (6.39) and (6.40) to be

$$\begin{aligned}
V_{0A}&=10t=h=0.050\\
V_{1A}&=\cos\,2\theta_k\;(2_{k+1}-z_k)=2[\cos(-30)+4\,\cos(30)]t=0.0433\\
V_{2A}&=\sin\,2\theta_k\;(z_{k+1}-z_k)=2[\sin(-30)+4\,\sin(30)]t=0.0150\\
V_{3A}&=\cos\,4\theta_k\;(z_{k+1}-z_k)=2[\cos(-60)+4\,\cos(60)]t=0.0250\\
V_{4A}&=\sin\,4\theta_k\;(z_{k+1}-z_k)=2[\sin(-60)+4\,\sin(60)]t=0.02598
\end{aligned}$$

Incorporating these into the expression for $[\bar{A}]$ results in

$$\begin{Bmatrix}\bar{A}_{11}\\ \bar{A}_{22}\\ \bar{A}_{12}\\ \bar{A}_{66}\\ \bar{A}_{16}\\ \bar{A}_{26}\end{Bmatrix}=\begin{bmatrix}0.554 & 0.533 & 0.187 & 0.032 & 0.034\\ 0.554 & -0.533 & -0.187 & 0.032 & 0.034\\ 0.164 & 0 & 0 & -0.032 & -0.034\\ 0.195 & 0 & 0 & -0.032 & -0.034\\ 0 & 0.094 & -0.267 & 0.034 & -0.033\\ 0 & 0.094 & 0.267 & -0.034 & 0.033\end{bmatrix}\begin{Bmatrix}1\\ \cos\,2\Phi\\ \sin\,2\Phi\\ \cos\,4\Phi\\ \sin\,4\Phi\end{Bmatrix}$$

Assuming that it is desired to have $\bar{A}_{16}=\bar{A}_{26}$. From the preceding relation,

$$\begin{aligned}&0.094\cos 2\Phi-0.267\sin 2\Phi+0.034\cos 4\Phi-0.033\sin 4\Phi\\ &\qquad=0.094\cos 2\Phi-0.267\sin 2\Phi-0.034\cos 4\Phi+0.033\sin 4\Phi\end{aligned}$$

Solving this expression results in $\tan 4\Phi=1.037$, from which $\Phi=11.51°$. This means that in order to get the desired $\bar{A}_{16}=\bar{A}_{26}$, the entire laminate must be oriented at 11.51° to the x-axis. An alternative view is that instead of a $[-15/15_4]_s$ laminate rotated through 11.51°, one could achieve this response by using a $[-3.49/26.51_4]_s$ laminate. In a similar manner, we may wish to have $\bar{A}_{11}=\bar{A}_{22}$. Following the same procedures as before results in $\Phi=-35.4°$.

6.9 Analysis of hybrid laminates

A *hybrid laminate* is one in which two or more fiber/matrix systems are combined to form new, sometimes superior composite material systems. For example, a laminate with both glass and graphite (or other fiber combinations) can be constructed. A hybrid offers certain advantages over conventional laminates. A hybrid made by using different lamina materials (glass/epoxy and carbon/epoxy, etc.) is called a *laminar hybrid*. Combining different materials in a single lamina produces an *interlaminar* or *intraply hybrid*. Both types of hybrid laminates are shown in Figure 6.46.

One reason for using hybrid laminates is economics. The cost of manufacturing can be reduced by mixing less expensive fibers (glass) with more expensive fibers (graphite). For example, a mixture of 20% (by volume) graphite fibers with glass fibers can produce a composite with 75% of the strength and stiffness, and 30% of the cost, of an all-graphite composite. Applications and physical properties of some intraply hybrids can be found in survey papers [38–42]. A problem associated with understanding the behavior of intraply hybrids is the *hybrid effect*, first noted by Hayashi [43]. It is related to an inability to accurately model certain mechanical properties to using rule-of-mixture approaches and has been investigated to a degree [44–49]. A problem with intraply hybrid composites is the significant scatter in ultimate strength data.

Conventional CLT analysis procedures can be applied to hybrid laminates, provided the constitutive relationship for the material is known. In forming the $[A]$, $[B]$, and $[D]$ matrices for a laminar hybrid, each lamina may have a different $[Q]$ and $[\bar{Q}]$. Intraply hybrids can be treated as if they were conventional orthotropic lamina since $[Q]$ is the same for each lamina. For the purpose of illustration, two material combinations are used for laminar hybrids and combined to form an intraply hybrid. A simple rule of mixtures approximation is used to determine the modulus of the intraply hybrid, assuming a 50% mixture of graphite/epoxy and glass/epoxy. The material properties used the following examples that are presented in Table 6.1.

The analysis of hybrid laminates using CLT is illustrated by considering the [0/90] laminate in Figure 6.47. In addition to the laminate shown in this figure, a [0/90] intraply laminate with the same dimensions is also considered. Both laminates are assumed to be subjected to a normal force of $N_x = 500$ lb/in., with temperature effects included.

The $[Q]$ for each material is

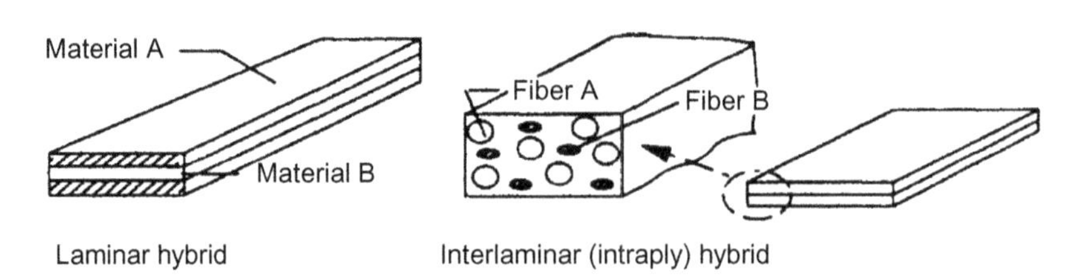

Figure 6.46 Laminar and intralaminar hybrid laminates.

Table 6.1 **Material properties for hybrid laminate examples**

Property	Glass/epoxy	Graphite/epoxy	Intraply hybrid
E_1 (Msi)	5.5	26.3	15.9
E_2 (Msi)	1.20	1.50	1.35
G_{12} (Msi)	0.60	1.00	0.80
ν_{12}	0.26	0.28	0.27
α_1 (μin./in./°F)	3.5	2.0	2.75
α_2 (μin./in./°F)	11.4	15.0	13.2

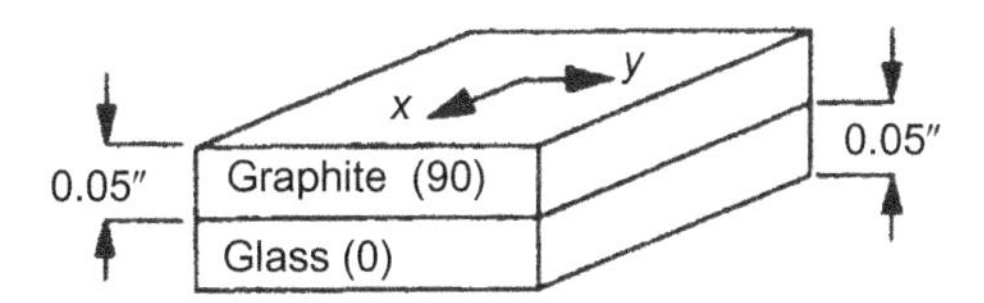

Figure 6.47 [0/90] hybrid laminate.

$$[Q]_{\text{glass}} = \begin{bmatrix} 5.603 & 0.318 & 0 \\ 0.318 & 1.223 & 0 \\ 0 & 0 & 0.60 \end{bmatrix} \times 10^6$$

$$[Q]_{\text{graphite}} = \begin{bmatrix} 26.42 & 0.422 & 0 \\ 0.422 & 1.507 & 0 \\ 0 & 0 & 1.00 \end{bmatrix} \times 10^6$$

$$[Q]_{\text{intraply}} = \begin{bmatrix} 16.00 & 0.367 & 0 \\ 0.367 & 1.358 & 0 \\ 0 & 0 & 0.80 \end{bmatrix} \times 10^6$$

The $[A]$, $[B]$, and $[D]$ matrices for the laminar hybrid are formed using standard CLT procedures:

$$[A] = \sum [\bar{Q}]_k t_k = \left([\bar{Q}]_{\text{glass}} + [\bar{Q}]_{\text{graphite}} \right) (0.05)$$
$$= \begin{bmatrix} 0.3555 & 0.037 & 0 \\ 0.037 & 1.384 & 0 \\ 0 & 0 & 0.08 \end{bmatrix} \times 10^6$$

$$[B] = \sum [\bar{Q}]_k t_k \bar{z}_k = \left([\bar{Q}]_{\text{glass}}(-0.025) + [\bar{Q}]_{\text{graphite}}(0.025) \right) (0.05)$$
$$= \begin{bmatrix} -5.12 & 0.130 & 0 \\ 0.130 & 31.49 & 0 \\ 0 & 0 & 0.50 \end{bmatrix} \times 10^3$$

$$[D]=\sum[\bar{Q}]_k\left(t_k\bar{z}_k^2+\frac{t_k^3}{12}\right)=\left([\bar{Q}]_{\text{glass}}+[\bar{Q}]_{\text{graphite}}\right)\left(0.05(0.025)^2+\frac{(0.05)^2}{12}\right)$$

$$=\begin{bmatrix}296.3 & 30.8 & 0\\ 30.8 & 1151.8 & 0\\ 0 & 0 & 66.67\end{bmatrix}$$

The thermal loads and moments are determined in a similar manner:

$$\{N^{\text{T}}\}=\sum[\bar{Q}]_k\ \{\alpha\}_k t_k \Delta T=\left([\bar{Q}]_{\text{glass}}\{\alpha\}_{\text{glass}}+[\bar{Q}]_{\text{graphite}}\{\alpha\}_{\text{graphite}}\right)(0.05)\,\Delta T$$

$$=\begin{Bmatrix}2.334\\ 3.711\\ 0\end{Bmatrix}\Delta T$$

$$\{M^{\text{T}}\}=\sum[\bar{Q}]_k\ \{\alpha\}_k\bar{z}_k t_k\Delta T=\left([\bar{Q}]_{\text{graphite}}\{\alpha\}_{\text{graphite}}-[\bar{Q}]_{\text{glass}}[\alpha]_{\text{glass}}\right)(0.025)(0.05)\,\Delta T$$

$$=\begin{Bmatrix}0.267\\ 54.99\\ 0\end{Bmatrix}\times 10^{-3}\Delta T$$

Assuming $\Delta T=-300$ °F results in thermal loads and moments of

$$\{N^{\text{T}}\}=\begin{Bmatrix}-700.2\\ -1113.3\\ 0\end{Bmatrix}\quad \{M^{\text{T}}\}=\begin{Bmatrix}-0.081\\ -19.497\\ 0\end{Bmatrix}$$

Recalling that for this laminate we apply $N_x=500$,

$$\{\hat{N}\}=\begin{Bmatrix}-200.2\\ -1113.3\\ 0\end{Bmatrix}\quad \{\hat{M}\}=\begin{Bmatrix}-0.081\\ -19.497\\ 0\end{Bmatrix}$$

The resulting mid-surface strains and curvatures are determined from

$$\{\varepsilon^0\}=[A']\{\hat{N}\}+[B']\{\hat{M}\}=\begin{Bmatrix}-629\\ -1096\\ 0\end{Bmatrix}\times 10^{-6}$$

$$\{k\}=[B']\{\hat{N}\}+[D']\{\hat{M}\}=\begin{Bmatrix}-1202\\ 1337\\ 0\end{Bmatrix}\times 10^{-5}$$

The stresses in each lamina are defined by

$$\begin{Bmatrix} \sigma_x \\ \sigma_y \\ \tau_{xy} \end{Bmatrix} = [\bar{Q}]_k \left(\begin{Bmatrix} -629 \\ -1096 \\ 0 \end{Bmatrix} \times 10^{-6} + z \begin{Bmatrix} -1202 \\ 1337 \\ 0 \end{Bmatrix} \times 10^{-5} - \begin{Bmatrix} \alpha_x \\ \alpha_y \\ \alpha_{xy} \end{Bmatrix}_k \Delta T \right)$$

For the 0° lamina, $-0.05 \le z \le 0$ and

$$\begin{Bmatrix} \sigma_x \\ \sigma_y \\ \tau_{xy} \end{Bmatrix}_0 = \begin{Bmatrix} 3097 \\ 2976 \\ 0 \end{Bmatrix} + z \begin{Bmatrix} -63,096 \\ 12,529 \\ 0 \end{Bmatrix}$$

For the 90° lamina, $0 \le z \le 0.050$ and

$$\begin{Bmatrix} \sigma_x \\ \sigma_y \\ \tau_{xy} \end{Bmatrix}_{90} = \begin{Bmatrix} 5624 \\ -11,471 \\ 0 \end{Bmatrix} + z \begin{Bmatrix} -12,472 \\ 348,163 \\ 0 \end{Bmatrix}$$

The intraply laminate is analyzed exactly like conventional laminated composites. Detailed derivations of $[A]$, $[B]$, $[D]$, $[N^{\mathrm{T}}]$, etc. are omitted. Assume that a [0/90] intraply laminate with the dimensions shown in Figure 6.47 is subjected to an axial load of $N_x = 500$ lb/in. and $\Delta T = -300$ °F. The laminate load and moment are

$$\{\hat{N}\} = \begin{Bmatrix} -516 \\ -1016 \\ 0 \end{Bmatrix} \quad \{\hat{M}\} = \begin{Bmatrix} 11.21 \\ -11.21 \\ 0 \end{Bmatrix}$$

These result in mid-surface strains and curvatures of

$$\{\varepsilon^0\} = [A']\{\hat{N}\} + [B']\{\hat{M}\} = \begin{Bmatrix} -473 \\ -1766 \\ 0 \end{Bmatrix} \times 10^{-6}$$

$$\{\kappa\} = [B']\{\hat{N}\} + [D']\{\hat{M}\} = \begin{Bmatrix} 240 \\ 2900 \\ 0 \end{Bmatrix} \times 10^{-5}$$

The strain through the laminate is

$$\begin{Bmatrix} \varepsilon_x \\ \varepsilon_y \\ \gamma_{xy} \end{Bmatrix} = \begin{Bmatrix} \varepsilon_x^0 \\ \varepsilon_y^0 \\ \gamma_{xy}^0 \end{Bmatrix} + z \begin{Bmatrix} k_x \\ k_y \\ k_{xy} \end{Bmatrix} = \begin{Bmatrix} -473 \\ -1766 \\ 0 \end{Bmatrix} \times 10^{-6} + z \begin{Bmatrix} 240 \\ 2900 \\ 0 \end{Bmatrix} \times 10^{-5}$$

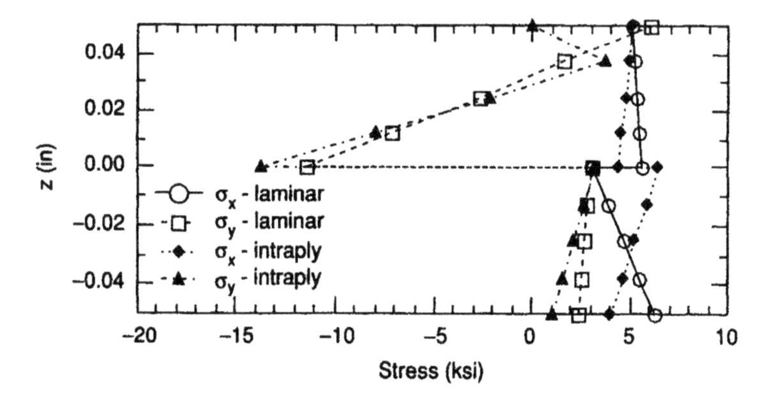

Figure 6.48 Normal stresses in [0/90] laminar and intraply hybrids.

Using this strain, the stresses in each lamina with $\Delta T = -300$ °F are determined to be

$$\begin{Bmatrix} \sigma_x \\ \sigma_y \\ \tau_{xy} \end{Bmatrix}_0 = [\bar{Q}]_0 \left(\begin{Bmatrix} \varepsilon_x^0 \\ \varepsilon_y^0 \\ \gamma_{xy}^0 \end{Bmatrix} + z \begin{Bmatrix} k_x \\ k_y \\ k_{xy} \end{Bmatrix} - \begin{Bmatrix} \alpha_x \\ \alpha_y \\ \alpha_{xy} \end{Bmatrix}_0 \Delta T \right)$$
$$= \begin{Bmatrix} 6434 \\ 3109 \\ 0 \end{Bmatrix} + z \begin{Bmatrix} 49,019 \\ 40,806 \\ 0 \end{Bmatrix}$$

$$\begin{Bmatrix} \sigma_x \\ \sigma_y \\ \tau_{xy} \end{Bmatrix}_{90} = [\bar{Q}]_{90} \left(\begin{Bmatrix} \varepsilon_x^0 \\ \varepsilon_y^0 \\ \gamma_{xy}^0 \end{Bmatrix} + z \begin{Bmatrix} k_x \\ k_y \\ k_{xy} \end{Bmatrix} - \begin{Bmatrix} \alpha_x \\ \alpha_y \\ \alpha_{xy} \end{Bmatrix}_{90} \Delta T \right)$$
$$= \begin{Bmatrix} 4390 \\ -13,767 \\ 0 \end{Bmatrix} + z \begin{Bmatrix} 13,902 \\ 464,591 \\ 0 \end{Bmatrix}$$

The stress distributions through both the laminar and intraply hybrids are shown in Figure 6.48. The shear stress is zero for both laminates because $\gamma_{xy}^0 = 0$ and $k_{xy} = 0$. A small variation in fiber orientation in either the 0° or 90° lamina can cause a slight mid-surface shear strain and curvature k_{xy}. The effect these have on the stress in each laminate depends on the thickness of each lamina.

6.10 Short fiber composites

Short fiber composites (SFCs) cannot generally be analyzed using CLT techniques. The response of an SFC may or may not be orthotropic. SFCs generally consist of chopped fibers or whiskers dispersed in a matrix. Random fiber orientation and discontinuity through the matrix makes it difficult to model response characteristics using CLT. Analysis of SFCs generally requires experimental evaluation as presented in ASTM STP 772 [50]. Establishing a predictive capability for SFCs requires appropriate modeling of the mechanism of stress transfer between fiber and matrix. Subsequently, modulus and strength predictions can be made. These topics are briefly presented and follow the discussions of Agarwal and Broutman [51]. In some special cases, CLT techniques can be applied to SFCs as described by Halpin and Pagano [52] and Halpin, Jerine, and Whitney [53].

6.10.1 Stress transfer and modulus predictions

In a composite, loads are not applied directly to the fibers, but rather to the matrix. The loads experienced by the matrix are then transferred to the fibers through the fiber ends. For composites consisting of long continuous fibers, the effects of load transfer at the fiber ends can be neglected. For short discontinuous fiber composites, end effects have a significant effect on the behavior of the composite. The shear-lag analysis presented by Rosen [54] is a modification of the analysis developed by Dow [55] and is one of the most widely used techniques for assessing stress transfer. In order to evaluate the stress distribution along a length of fiber, consider the model shown in Figure 6.49.

The normal stress in the composite (σ_c) comprises a matrix stress (σ_m) and a fiber stress (σ_f). The normal stress in the composite is assumed to be transferred to the fiber by a shear stress (τ). Assuming the fiber has a cylindrical shape with a radius r and length L, the stresses are expressed in terms of forces by satisfying the condition of equilibrium:

$$\sum F = 0 = \sigma_f(\pi r^2) + (2\pi r)\tau(\mathrm{d}z) - (\pi r^2)(\sigma_f + \mathrm{d}\sigma_f)$$

The equation of equilibrium can be manipulated to yield a relation between σ_f and τ at the fiber–matrix interface as the differential equation $\mathrm{d}\sigma_f/\mathrm{d}z = 2\tau/r$. Integrating over an arbitrary length of fiber results in

$$\sigma_f = \sigma_{f0} + \frac{2}{r}\int_0^z \tau\,\mathrm{d}z$$

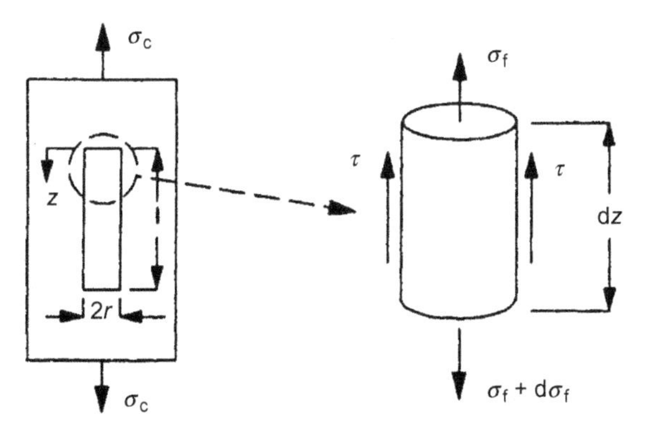

Figure 6.49 Equilibrium model for a short fiber.

where σ_{f0} is the stress at the fiber end. At the fiber end, matrix yielding (adjacent to the fiber) or separation of the matrix from the fiber may occur because of large stress concentrations. For simplicity, it is assumed that σ_{f0} is negligible, resulting in

$$\sigma_f = \frac{2}{r}\int_0^z \tau \, dz \tag{6.46}$$

The solution of this equation requires knowledge of the shear stress distribution along the fiber. Assumptions regarding the distribution of τ are required. A common assumption is that the matrix surrounding the fiber is perfectly plastic so that the interfacial shear stress along the fiber is constant. It is assumed to have a value equal to the matrix yield stress ($\tau = \tau_y$) which results in

$$\sigma_f = \frac{2}{r}\int_0^z \tau_y \, dz = \frac{2\tau_y z}{r} \tag{6.47}$$

The mechanism for generating normal stress in the fiber is through shear transfer. The normal stress at each fiber end is zero and the maximum normal stress occurs at the mid-length of the fiber ($z = L/2$); therefore,

$$(\sigma_f)_{max} = \frac{\tau_y L}{r} \tag{6.48}$$

The magnitude of $(\sigma_f)_{max}$ is limited. Assuming the fiber and matrix have not separated, simple rule-of-mixtures approximations are applied. The normal strain in the composite (ε_c) must equal both the fiber (ε_f) and matrix (ε_m) strains so that $\varepsilon_c = \varepsilon_f = \varepsilon_m$. For the uniaxial state of stress in Figure 6.49, the stress and strain in the fiber and composite are related by $\sigma_c = E_c \varepsilon_c$ and $\sigma_f = E_f \varepsilon_f$. In order to maintain the condition $\varepsilon_c = \varepsilon_f = \varepsilon_m$, it is a simple matter to show that $(\sigma_f)_{max} = (E_f/E_c)\sigma_c$. The elastic modulus of the composite can be approximated using the rule-of-mixtures relation given by Equation (3.36).

Transfer of shear into normal stress does not necessarily occur over the entire length of the fiber. The length of fiber over which load is transferred is called the load-transfer length, designated as L_t and shown in Figure 6.50. The minimum load

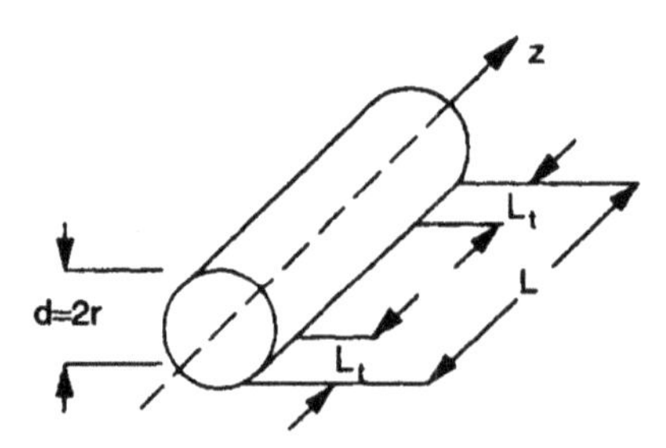

Figure 6.50 Load-transfer length.

transfer length in which the maximum normal stress in the fiber can be achieved is obtained from Equation (6.48) by noting that the fiber diameter is $d=2r$ and letting $L=L_t$, resulting in

$$\frac{L_t}{d}=\frac{(\sigma_f)_{max}}{2\tau_y} \tag{6.49}$$

Equation (6.49) can be expressed in terms of the elastic moduli of the fiber and composite, since $(\sigma_f)_{max}=(E_f/E_c)\sigma_c$:

$$\frac{L_t}{d}=\frac{E_f\sigma_c}{2E_c\tau_y} \tag{6.50}$$

The distribution of stress along the fiber depends on assumptions of perfectly plastic or elastic–plastic matrix material. The problem of interfacial stresses has been investigated using various techniques [51,56–64]. The distribution of stresses as a function of fiber length, from Agarwal and Broutman [51], is presented in Figure 6.51.

A simple model can be obtained using the Halpin–Tsai Equations (3.47–3.50). Assume an aligned SFC, as shown in Figure 6.52, as opposed to random fiber orientations. The L and T designations in Figure 6.52 refer to the longitudinal and transverse directions of the laminated sheet, respectively.

The parameter ξ in the Halpin–Tsai equations is assumed to have two distinct values, one for the longitudinal (L) and one for the transverse (T) directions. As discussed in Halpin [65], it is assumed that $\xi=2L/d$ for the longitudinal direction and $\xi=2$ for the transverse direction. The moduli estimated from the Halpin–Tsai equations are

$$\frac{E_L}{E_m}=\frac{1+(2L/d)\eta_L v_f}{1-\eta_L v_f} \quad \frac{E_T}{E_m}=\frac{1+2\eta_T v_f}{1-\eta_T v_f} \tag{6.51}$$

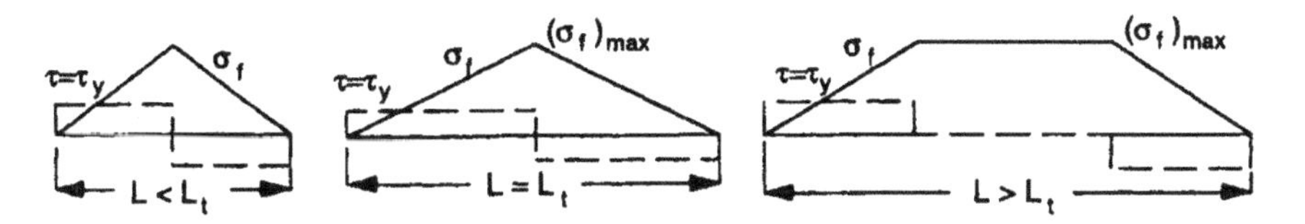

Figure 6.51 Fiber and shear stress variation vs. fiber length.

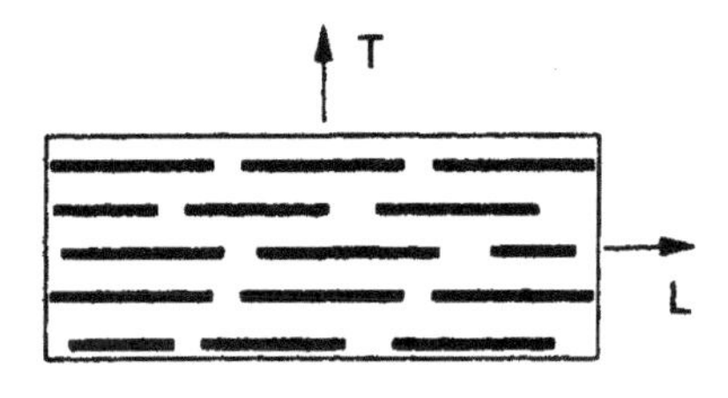

Figure 6.52 Aligned short fiber composites.

where

$$\eta_L = \frac{(E_f/E_m) - 1}{(E_f/E_m) + 2L/d} \quad \eta_T = \frac{(E_f/E_m) - 1}{(E_f/E_m) + 2} \tag{6.52}$$

These approximations are for an aligned fiber composite. For a random fiber composite, as discussed in Sandors [50], the elastic modulus is approximated from

$$E_{random} = \frac{3}{8}E_L + \frac{5}{8}E_T \tag{6.53}$$

where E_L and E_T are approximated from Equation (6.51).

The stress in the composite can also be estimated using rule-of-mixtures approximations. Assuming the same perfectly plastic matrix that was used to define Equations (6.47)–(6.50), an estimate of σ_c can be made. The variation of normal stress in the fiber (σ_f) depends on the fiber length. We define the average normal fiber stress as $\overline{\sigma}_f = (1/L)\int_0^L \sigma_f \, dx$. This is determined from the area under the σ_f curve in Figure 6.51:

$$\begin{aligned} \overline{\sigma}_f &= \frac{1}{2}(\sigma_f)_{max} = \frac{\tau_y L}{d} && \text{for } L = L_t \\ \overline{\sigma}_f &= (\sigma_f)_{max} = \left(1 - \frac{L_t}{2L}\right) && \text{for } L > L_t \end{aligned} \tag{6.54}$$

For very SFCs ($L = L_T$), the shear failure stress is for the matrix, not the fiber. Using a simple rule-of-mixtures approximation, the composite stress is $\sigma_c = \sigma_f \upsilon_f + \sigma_m \upsilon_m$. The ultimate strength of the composite depends on fiber length, diameter, and the ultimate strength of each constituent material (σ_{fU}, σ_{mU}). For a composite with very short fibers, the matrix properties dominate the solution, whereas for composites with fiber lengths greater than L_t, the ultimate fiber strength dominates the solution. The ultimate stress in the composite for either case is

$$\begin{aligned} L \le L_t \quad \sigma_{cU} &= \frac{\tau_y L}{d}\upsilon_f + \sigma_{mU}\upsilon_m \\ L > L_t \quad \sigma_{cU} &= \sigma_{fU}\left(1 - \frac{L_t}{2L}\right)\upsilon_f + \sigma_{mU}\upsilon_m \end{aligned} \tag{6.55}$$

Another form of SFC is a ribbon-reinforced, or tape-reinforced, composite as shown in Figure 6.53. It has a higher strength and stiffness in the longitudinal and transverse directions than conventional SFCs. Approximations to the elastic moduli for this type of composite can be made using the Halpin–Tsai equations. Because of the geometry of each ribbon, the volume fraction of fibers (v_f) used in the Halpin–Tsai equations is replaced by the volume fraction of ribbon (v_f). Using Figure 6.53, it can be shown that v_f is approximated by

$$\upsilon_r = \frac{1}{2[1 - (a/W_r)][1 + (t_m/t_r)]} \tag{6.56}$$

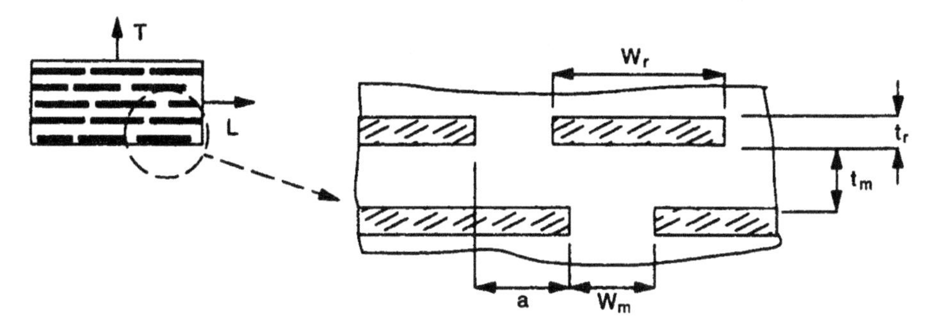

Figure 6.53 Ribbon composite schematic.

The Halpin–Tsai approximations to the longitudinal and transverse moduli for a ribbon composite are

$$E_L = E_r v_r + E_m v_m$$
$$\frac{E_T}{E_m} = \frac{1 + 2(W_r/t_r)\eta_r v_r}{1 - \eta_r v_r} \tag{6.57}$$

where

$$\eta_r = \frac{(E_r/E_m) - 1}{(E_r/E_m) + 2(W_r/t_r)} \tag{6.58}$$

The elastic modulus of the ribbon (E_r) must be experimentally determined.

6.10.2 Laminate approximation

Halpin and Pagano [52] developed an approximate method for estimating the stiffness of SFCs based on approximations from laminate analysis. They assumed the material exists in sheet form as shown in Figure 6.54. Their procedure is based on estimating the mechanical behavior of short fiber laminates using micromechanical approaches presented in Halpin [65]. Based on micromechanics, modulus estimates are a function of the ratio of fiber length to diameter (*L*/*d*) and volume fractions of fiber and matrix.

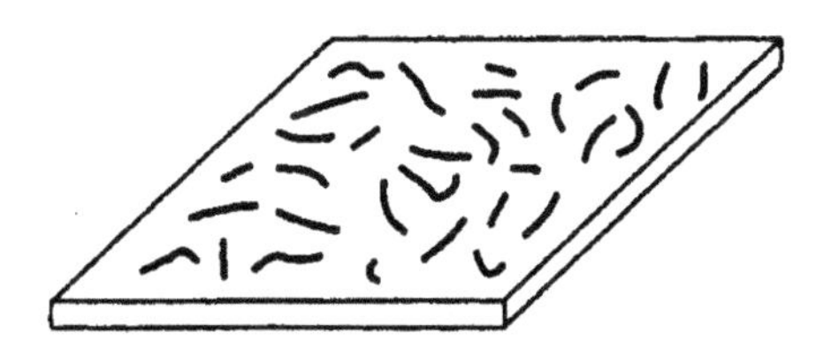

Figure 6.54 Short fiber composite sheet.

The Halpin–Tsai equations were used in Halpin [65] to establish the different parameters in terms of (L/d). It was found that only E_{11} is sensitive to (L/d).

Provided the sheet thickness is much less than the average fiber length, the reinforcement can be considered a random 2D array of fibers. These in turn are assumed to be quasi-isotropic, which allows one to express the principal material direction elastic properties as

$$E_1 = E_2 = \bar{E} \quad \nu_{12} = \nu_{21} = \bar{\nu} \quad G_{12} = \bar{G} = \frac{\bar{E}}{2(1+\bar{\nu})} \tag{6.59}$$

Halpin and Pagano found it convenient to use invariant forms of U_1 and U_5: $U_1 = (3Q_{11} + 3Q_{22} + 2Q_{12} + 4Q_{66})/8$ and $U_5 = (Q_{11} + Q_{22} - 2Q_{12} + 4Q_{66})/8$. From CLT, $Q_{11} = E_1/(1 - \nu_{12}\nu_{21})$, $Q_{22} = E_2/(1 - \nu_{12}\nu_{21})$, $Q_{12} = \nu_{12}Q_{22} = \nu_{21}Q_{11}$, and $Q_{66} = G_{12}$. The estimates of the moduli in these expressions come from the predictions in Halpin [65], in which Equations (6.51) and (6.52) are used to define E_1 and E_2. Substitution of the appropriate moduli into the invariant forms of U_1 and U_2 allows Equation (6.59) to be expressed as

$$\bar{G} = U_5 \quad \bar{E} = \frac{4U_5(U_1 - U_5)}{U_1} \quad \bar{\nu} = \frac{(U_1 - 2U_5)}{U_1} \tag{6.60}$$

Halpin and Pagano [52] showed a reasonable approximation to E using this procedure.

6.10.3 Laminate analogy

The laminate analogy for SFCs was developed by Halpin et al. [53]. Although a typical SFC may appear to consist of random fiber orientations (Figure 6.55a), it is assumed that there is actually a fiber bias (Figure 6.55b).

In the biased fiber orientation, not all fibers are considered to have a preferred direction of orientation. The percentage of fibers orientated at some angle θ can be expressed as $f(\theta)/h$, where h is the thickness of the composite and the angle θ may

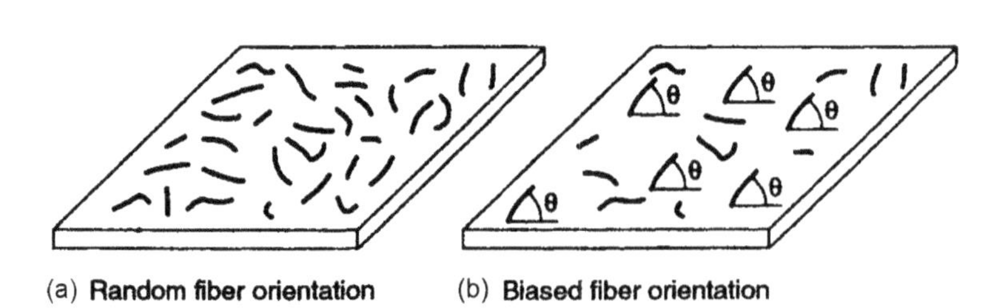

Figure 6.55 Short fiber composites (a) with random fiber orientations, and (b) with biased fiber orientations.

be positive or negative. The percentage of fibers oriented at some angle is estimated by an experimentally determined angular distribution function,

$$\int_0^{\infty} f(\theta)\,\mathrm{d}\theta = 1.0$$

Accounting for fiber orientation variability in this manner allows each component of the extensional stiffness matrix to be defined as

$$[A_{ij}] = \sum_{k=1}^{N} \frac{f(\theta_k)}{h} A'_{ij}(\theta_k) \tag{6.61}$$

The $A'_{ij}(\theta_k)$ terms are determined from the general material behavior for an SFC as defined by the procedures of Halpin [65]. For example, assume that the procedures in Halpin [65] are used to establish the following $[Q]$ matrix for an SFC:

$$[Q] = \begin{bmatrix} 2.5 & 0.35 & 0 \\ 0.35 & 1.2 & 0 \\ 0 & 0 & 0.75 \end{bmatrix} \times 10^6$$

Next, assume that the SFC (with a total thickness t) defined by this $[Q]$ matrix is examined, and the fiber bias recorded in terms of a percent of total fibers at some angle θ to a reference axis (assumed to be the x-axis). Assume that the following information has been collected:

Fiber orientation θ (degrees)	**Percent fibers with θ orientation**
5	35%
15	25%
30	20%
45	15%
60	5%

For each angle, the $A'_{ij}(\theta_k)$ has to be established. Using CLT procedures results in $A'_{ij}(\theta_k) = t[\bar{Q}_{ij}]$. Consequently,

$$A_{ij} = \frac{1}{t}\left\{0.35[\bar{Q}_{ij}]_5 + 0.25[\bar{Q}_{ij}]_{15} + 0.20[\bar{Q}_{ij}]_{30} + 0.15[\bar{Q}_{ij}]_{45} + 0.05[\bar{Q}_{ij}]_{60}\right.$$

Substituting the appropriate numerical values yields

$$[A] = \begin{bmatrix} 2.41 & 0.35 & 0.18 \\ 0.35 & 1.48 & 0.18 \\ 0.18 & 0.18 & 0.75 \end{bmatrix} \times 10^6$$

This approximation for predicting the behavior of an SFC is based on the laminate analogy. Since the procedure is based on a single-ply laminate, there is no bending–extension coupling ($[B]=0$). An approximation to the bending stiffness can be made by analogy to CLT, resulting in

$$[D_{ij}] = [\bar{Q}_{ij}]\frac{t^3}{12}$$

SFCs are not generally treated as laminates. The procedures presented here are only a demonstration of approximations that can be made in order to use CLT in dealing with estimates of the elastic modulus for SFCs.

6.11 Delamination and edge effects

Delamination can result from numerous causes, including improper curing, impact damage, stacking sequence, vibrations, etc. The intent of this section is to illustrate the fundamental principles behind delamination by describing the early investigations of delamination and edge effects [66–72], which were critical in understanding and assessing the phenomenon. CLT does not account for interlaminar stresses such as σ_z, τ_{zx}, and τ_{zy}. In addition, the shear stress τ_{xy} is predicted to exist in locations where it cannot possibly exist. Consider a symmetric four lamina angle-ply laminate of width $2b$ and thickness $4h$ with an axial load N_x applied. From CLT, only the $[A]$ matrix is required to obtain the mid-surface strains. It can be shown that the shear stress τ_{xy} can be expressed in terms of the components of $[Q]$, the extensional stiffness matrix $[A]$, fiber orientation, and applied load as

$$\tau_{xy} = \{mn[(Q_{11}-Q_{12})(m^2-An^2)+(Q_{12}-Q_{22})(n^2-Am^2)]$$
$$-2mn(m^2-n^2)Q_{66}(1+A)\}BN_x$$

In general, $A=A_{12}/A_{22}\neq 0$ and $B=A_{22}/(A_{11}A_{22}-A_{12}^2)\neq 0$, since $Q_{11}\neq Q_{12}\neq Q_{22}\neq Q_{66}\neq 0$. Therefore, the only possible conditions for $\tau_{xy}=0$ are for $\theta=0°$ and $\theta=90°$. Therefore, for fiber orientation other than 0° and 90°, a τ_{xy} is predicted to exist. This is not physically possible at the top and bottom of the laminate ($y=\pm b$) for the loading condition given. In somewhat general terms, we can make observations;

1. Each lamina within the laminate has a different extensional stiffness due to fiber orientations.
2. Under an applied load, the laminas tend to slide over each other because of the different stiffness of each lamina.
3. The lamina are elastically connected.
4. Due to (3), reactive shear stresses are developed within each lamina (on their top and bottom surfaces).

In general, the results from CLT are accurate as long as the plane stress assumption is valid. This is a good assumption for interior regions of the laminate where geometric discontinuities (such as free edges) do not exist. The existence of a boundary layer near the free edge indicates that a three-dimensional state of stress exists at the free edge. The boundary layer is a region wherein stress transfer between adjacent lamina is accomplished through the action of interlaminar stresses as described by Pipes and Pagano [66] and illustrated in Figure 6.56. The width of the boundary layer depends on the elastic properties of the lamina, the fiber orientation, and the laminate geometry. A simple rule of thumb is that the boundary layer for hard-polymer-matrix composites is approximately equal to the laminate thickness.

The primary consequences of the laminate boundary layer are delamination-induced failures initiating within the boundary layer region and distortions of surface strain and deformations due to the presence of interlaminar shear components. Failures initiating in the boundary layer region of a finite width test specimen may not yield the true strength of the laminate. The free body diagram of Figure 6.56 is useful in understanding the physical mechanism of shear transfer between layers. The fact that τ_{xy} must vanish on a free edge means that the couple caused by τ_{xy} acting along the other edge of the free body diagram must be reacted. The only possible reacting

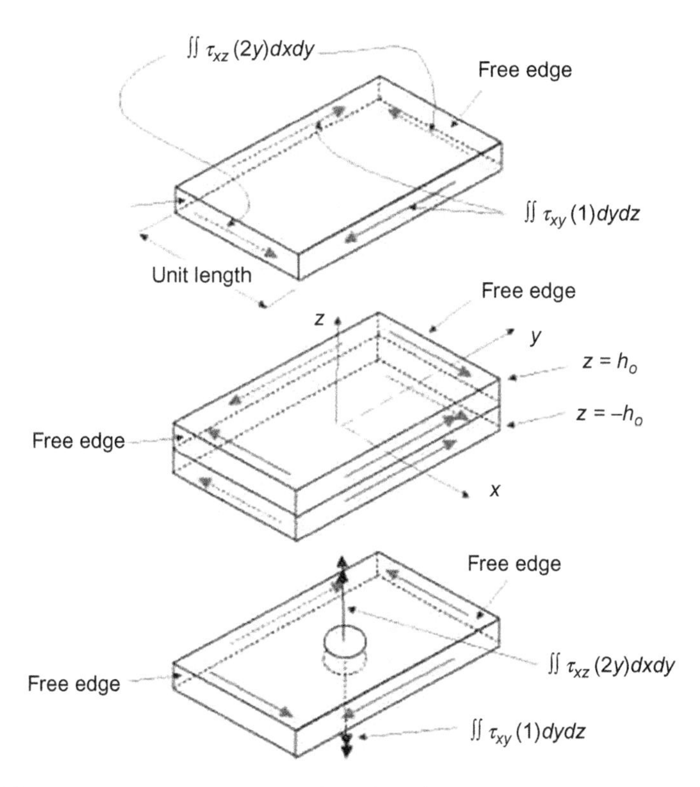

Figure 6.56 Interlaminar shear stress transfer (after Pipes and Pagano [66]).

couple to satisfy moment equilibrium is caused by τ_{xz} acting on part of the lower face of the layers at the interface with the adjacent layer. The mechanisms for failure of an angle-ply laminates subjected to an axial load N_x are termed a first-mode mechanism. Solutions for the mode of failure vary in complexity from elasticity solutions down to approximate solutions.

A traditional elasticity solution for the interlaminar stresses in a laminated composite subjected to uniform axial extension was developed by Pipes and Pagano [66]. A similar development was forwarded by Puppo and Evenson [67] for generalized plane stress. The solution in Ref. [66] produced the set of coupled second-order partial differential equations given below:

$$\bar{Q}_{66}U_{,yy}+\bar{Q}_{55}U_{,zz}+\bar{Q}_{26}V_{,yy}+\bar{Q}_{45}V_{,zz}+\left(\bar{Q}_{36}+\bar{Q}_{45}\right)W_{,yz}=0$$
$$\bar{Q}_{26}U_{,yy}+\bar{Q}_{45}U_{,zz}+\bar{Q}_{22}V_{,yy}+\bar{Q}_{44}V_{,zz}+\left(\bar{Q}_{23}+\bar{Q}_{44}\right)W_{,yz}=0$$
$$\left(\bar{Q}_{45}+\bar{Q}_{36}\right)U_{,yz}+\left(\bar{Q}_{44}+\bar{Q}_{23}\right)V_{,yz}+\bar{Q}_{44}W_{,yy}+\bar{Q}_{33}W_{,zz}=0$$

In these equations, U, V, and W are the prescribed displacement fields in the x, y, and z directions. These equations do not admit a closed-form solution. Therefore, an approximate numerical technique was required to solve them. Using a finite difference approximation and a model of 1/4 of the plate, the solution for a $[45/-45]_s$ high modulus graphite/epoxy is shown in Figure 6.57.

The stress components σ_y, τ_{yz}, and σ_z are quite small. However, as $y/b \rightarrow 1$, the τ_{xz} stress component becomes large. The actual distribution of τ_{xz} over the

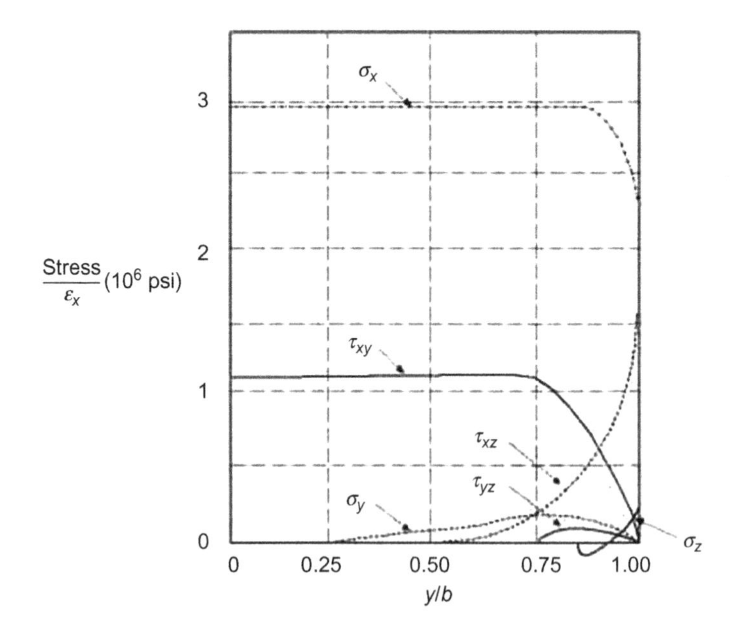

Figure 6.57 Stress variation of interlaminar stresses (after Pipes and Pagano [66]).

cross-section varies with distance from the laminate mid-surface. In addition to the $[45/-45]_s$ laminate discussed above, other four layer laminates with $[+\theta/-\theta]_s$ stacking sequences were also investigated. They found that $\tau_{xz} \approx \tau_{xz-\max}$ at $\theta = 35°$.

It was hypothesized that although τ_{xz} is the largest component of stress at the free boundary, it is the σ_z component that is responsible for generating delamination resulting in failure. Figure 6.58 shows a free body diagram for part of the top lamina in an eight-layer $[\pm 15/\pm 45]_s$ laminate subjected to an applied load in the x-direction. Recalling Figure 6.56, the mechanism for interlaminar stress transfer involves a balance of moments and stress. Two observations can be made from Figure 6.58:

1. Since τ_{xy} is zero at the free edge and finite elsewhere, the moment it produces must be balanced by the moment caused by the interlaminar shear stress τ_{xz}.
2. The moment due to σ_z must be balanced by the moment due to σ_y.

Based on observations of the elasticity solution for σ_z, a distribution of σ_z as shown in Figure 6.59 was hypothesized [68]. From the free body diagram of Figure 6.58, it was

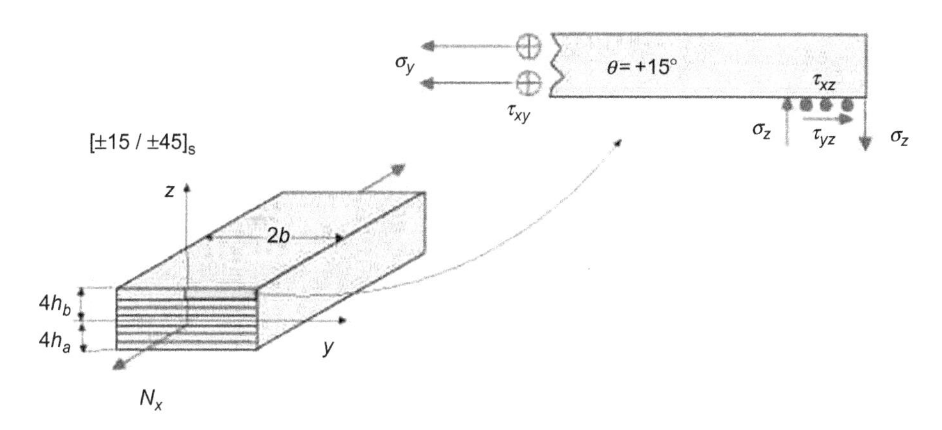

Figure 6.58 Free body diagram of top lamina in a $[\pm 15/\pm 45]_s$ laminate (after Pipes and Pagano [66]).

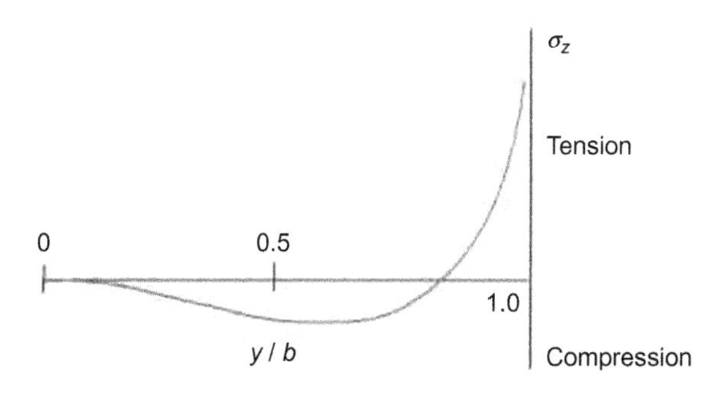

Figure 6.59 Hypothesized distribution of σ_z (after Pagano and Pipes [68]).

observed that a tensile σ_y in the $+15°$ layer implies a tensile σ_z at the free edge, with the converse holding for a compressive σ_y. The tensile σ_y in the $+15°$ layer is consistent with CLT.

It was also reasoned that if a compressive σ_y could be generated then a compressive σ_z at the free edge would result, and the laminate would be less likely to delaminate. They examined $[\pm15/\pm45]_s$ and $[15/\pm45/15]_s$ stacking sequences and observed that the σ_z stress distribution through the laminate was largest for the $[\pm15/\pm45]_s$ laminate. Therefore, the $[15/\pm45/15]_s$ sequence should provide a greater strength and lesser tendency to delaminate. They similarly reasoned that a stacking sequence of $[\pm45/\pm15]_s$ would lead to compressive stresses that are mirror images of the tensile stresses in the $[\pm15/\pm45]_s$ laminate and would be stronger. The interlaminar shear stress in the two cases is essentially the same. Thus, they concluded that σ_z may be the key to a successful laminate design that would resist delamination. Therefore, in order to reduce the possibility of delamination, the normal stress σ_z at the free edge should be small (optimally it should be compressive). A series of experimental tests were conducted on six ply symmetric laminates and the theoretical values of σ_z were determined at three points. The results are shown in Table 6.2. In general, it was observed that if the outside plies (θ_1) were 0°, the specimen failed at a higher load.

Table 6.2 Theoretical normal stress σ_z at free edge of a laminate (after Pagano and Pipes [68])

Stress location	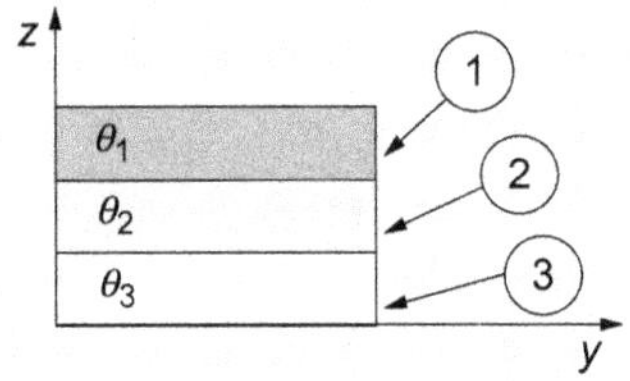 z, θ_1, θ_2, θ_3, 1, 2, 3, y			
Specimen	**Point 1**	**Point 2**	**Point 3**	
$[0/\pm15]_s$	−4.7	−11.8	−14.1	Preferred sequence
$[15/0/-15]_s$	2.4	2.4	0	
$[\pm15/0]_s$	2.4	9.4	14.1	
$[0/\pm30]_s$	−18.6	−46.5	−55.8	Preferred sequence
$[30/0/-30]_s$	9.3	9.3	0	
$[\pm30/0]_s$	9.3	37.2	55.8	
$[0/\pm45]_s$	−8.8	−22.0	−26.4	Preferred sequence
$[45/0/-45]_s$	4.4	4.4	0	
$[\pm45/0]_s$	4.4	17.6	26.4	
$[0/\pm60]_s$	−0.6	−1.5	−1.8	Preferred sequence
$[60/0/-60]_s$	0.3	0.3	0	
$[\pm60/0]_s$	0.3	1.2	1.8	

As seen in Table 6.2, this corresponds to $\sigma_z < 0$. The general concepts regarding interlaminar stresses may be summarized by some observations.

(i) Plane stress CLT is recovered along the central plane $y = 0$, provided $b/h >> 2$.
(ii) The force and moment which are statically equivalent to various interlaminar stresses in planes of $z =$ constant can be determined by $\sum F$ and $\sum M$.
(iii) The interlaminar stresses are confined to a narrow region of dimension comparable to the laminate thickness, adjacent to the free edges $y = \pm b$.

There is no closed form solution to the problem of interlaminar shear stress. Numerical techniques (finite element and finite difference) give good approximations. The preceding work and discussions may be found in Refs. [66–69]. In summary, there are three classes of interlaminar stress problems:

- $[\pm\theta]$ laminates exhibit only shear coupling (no Poisson mismatch between layers), so τ_{xz} is the only nonzero interlaminar stress.
- [0/90] laminates exhibit only a Poisson mismatch between layers (no shear coupling), so τ_{yz} and σ_z are the only nonzero interlaminar stress.
- Combinations of the above ($[\pm\theta_1/\pm\theta_2]$) laminates exhibit both shear coupling and a Poisson mismatch between layers, so they have τ_{xz}, τ_{yz}, and σ_z as the only nonzero interlaminar stress.

The hypothesis that σ_z is the primary driving force behind delamination seems to be correct from observed failures. In static analysis, Pagano and Pipes [70] developed a means for assessing σ_z. Their analysis contains no description of the influence of material and geometric parameters on the state of interlaminar stress. They considered the case of uniaxial tension on a symmetric laminate with a thickness of $2h$ and a width of $2b$. Based on the stress distribution in Figure 6.59, the model in Figure 6.60 was formulated. In this model, the normal stress σ_z is assumed to act only over the region of boundary layer effect ($y/2h$).

Taking moments about the mid-point between the tensile and compressive forces gives a relationship between $\sigma_{m'}$ and σ_m ($\sigma_{m'} = \sigma_m/5$). The moment resulting from these forces is the integral of the normal stress σ_y over the lamina thickness and can be related to σ_m. The laminate investigated is symmetric with equal thickness

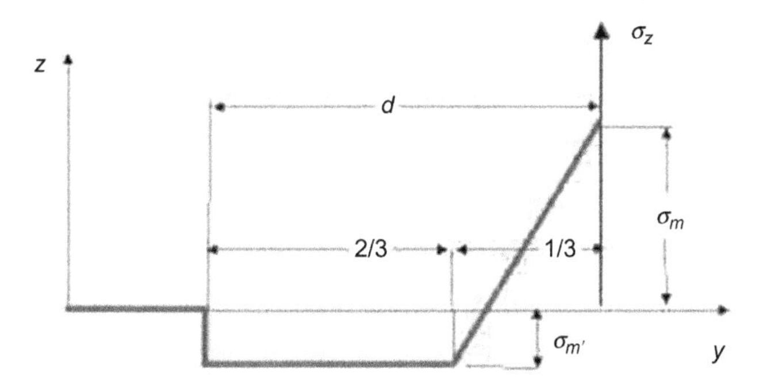

Figure 6.60 Model of σ_z distribution near free edge of a laminate (after Pagano and Pipes [70]).

layers of $+\theta$ and $-\theta$. This combination of angle plies is designated as a unit defined by a new form of $[\bar{Q}]$, defined as $[\bar{Q}^*] = \frac{1}{2}[\bar{Q}(+\theta) + \bar{Q}(-\theta)]$. Using this, an extensional matrix $[A]$ is defined based on $[\bar{Q}^*]$. The normal stress σ_y is then related to the thickness of each unit. An effective Poisson's ratio for an angle-ply unit measuring normal strain in the y-direction under an axial load in the x-direction, assuming the applied load in the x-direction is σ_o. The y-direction stresses in each unit are related to the thickness of each unit, and subsequently the normal stress, σ_m, in the z-direction. By maximizing the σ_m function (a compressive stress), the critical height ratio of each unit can be determined. Following this procedure, a symmetric laminate may be designed to reduce the possibility of delamination due to σ_y. To do this, one must keep in mind that σ_y should be compressive (refer to Figure 6.58).

Pagano [71] developed a complex procedure for a symmetric angle-ply laminate subjected to an axial strain ε_x in order to refine the previous solution. His solution introduced two new parameters to the CLT plate equations. These parameters may be considered the first moments of τ_{xz} and τ_{yz}. They are basically the weighted average of the moment resulting from the shear stresses τ_{xz} and τ_{yz} and were expressed as $(R_x, R_y) = \int_{-h/2}^{h/2} (\tau_{xz}, \tau_{yz}) z \, dz$. These were incorporated into the CLT load–strain relationship with the additional terms R_x and R_y accounted for. With the additional two terms, correction factors were needed, as identified in Ref. [72]. The end result is a set of six equations containing elements of the $[A]$, $[B]$, and $[D]$ matrices and derivatives of the assumed displacement fields ($u = \varepsilon x + U(y) + z\Psi(y)$, $v = V(y) + z\Omega(y)$, and $w = W(y) + z\Phi(y)$). Applying appropriate boundary conditions, the general solution of the equations requires both complementary and particular solutions, making it a complex procedure. Additional contributions to the edge delamination phenomenon for a variety of conditions have been made since these initial investigations. The preceding is intended as a guide to understanding the problem.

6.12 Problems

The material properties for the following problems can be found in Tables 4.2 and 4.3.

6.1 Compute $[A]$, $[B]$, and $[D]$ for a T300/5208 (reference [42] in Chapter 4) laminate with each of the following stacking sequences, and discuss your results. Assume that the thickness of each ply is 0.005 in.
(A) [0/60/−60], (B) $[0/60/-60]_s$, and (C) $[60/15/-30/-75]_s$

6.2 Determine a laminate stacking arrangement different from those in Problem 6.1 that results in a quasi-isotropic laminate for an E-glass/epoxy. Assume each ply has a thickness of 0.010 in. Verify the stacking arrangements by computing $[A]$, $[B]$, and $[D]$.

6.3 Assuming that each layer of a woven fiber laminate can be modeled as a cross-ply laminate, defined as either a [0/90] or [90/0], it can be shown that the $[A]$ matrix of the laminate is quasi-isotropic with $Q_{11} = Q_{22}$. Assume an n layer laminate with a total thickness $t = nt_{\text{ply}}$ has been constructed of quasi-isotropic lamina. Determine expressions for $[A]$, $[B]$, and $[D]$ in terms each component of $[Q]$ and nt_{ply}.

6.4 A [0/90] laminate with identical ply thicknesses of 0.005 in. is to be constructed. Determine the elements of the extension-bending coupling matrix $[B]$ knowing that for the material used

$$[Q]=\begin{bmatrix} 5.68 & 0.317 & 0 \\ 0.317 & 1.22 & 0 \\ 0 & 0 & 0.60 \end{bmatrix}\times 10^6$$

6.5 For the stacking arrangement given, plot the variation of σ_x, σ_y, and τ_{xy} through the thickness of an AS/3501 (reference [42] in Chapter 4) laminate subjected to the loading condition shown. Assume each ply has a thickness of 0.005 in., and neglect thermal and hygral effects.

(A) $[0/45/-45]_s$, (B) $[45/0/-45]_s$, and (C) $[30/45/-45/0]_s$

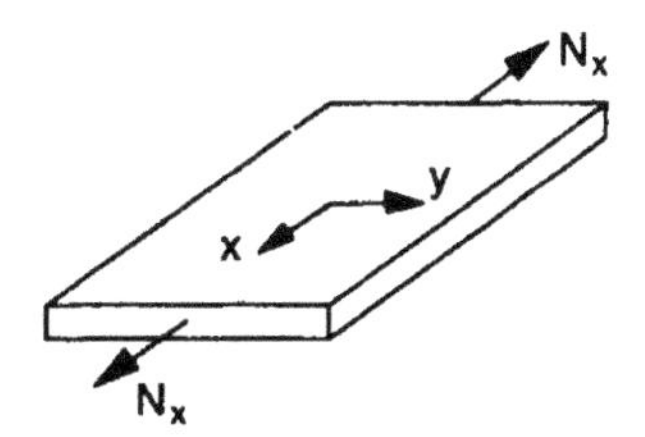

6.6 Plot σ_x, σ_y, and τ_{xy} through a $[0/90_5/45_2]_s$ T300/5208 (reference [42] in Chapter 4) laminate for the loading conditions shown. Neglect thermal and hygral effects, and assume each ply has a thickness of 0.005 in.

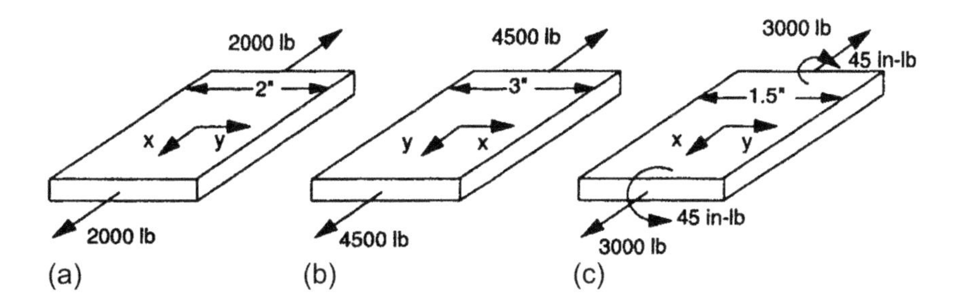

6.7 Predict first ply failure for an S-glass/XP-251 laminate subjected to a normal tensile load N_x lb/in. Use the Tsai–Hill theory, and neglect thermal and hygral effects. Assume that each ply has a thickness of 0.010 in. The laminate stacking sequence is

(A) $[0/90_4/45/-45]_s$ and (B) $[90/0_4/-45/45]_s$

6.8 Work Problem 6.7 (A) or (B) using AS/3501 (reference [42] in Chapter 4), assuming it is loaded with only a shear force of $-N_{xy}$ lb/in., and $t_{ply}=0.005$ in.

6.9 Use the Tsai–Hill failure theory to predict the load N at which first ply failure of the [0/45/0/–45] T300/5208 (reference [42] in Chapter 4) laminate occurs. Neglect thermal and hygral effects and assume that $t_{ply} = 0.005$ in.

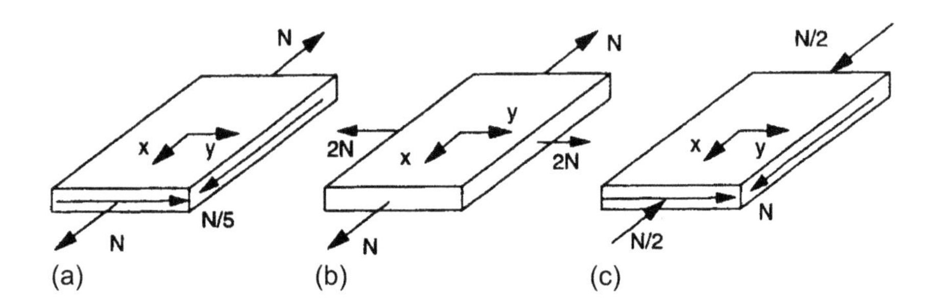

6.10 A $[0/90]_s$ laminate is constructed with the temperature gradient between the top (T_0) and bottom (T_1) lamina taken into account. The thermal loads and moments are developed based on the thermal strain in each lamina being defined as $\{\varepsilon^T\}_k = \{\alpha\}_k T(z)$, where $T(z) = C_1 + C_2 z$ and $C_1 = (T_0 + T_1)/2$. All lamina has the same thickness (0.005 in.). The coefficients of thermal expansion are $\alpha_1 = 2 \times 10^{-6}$ in./in./°F and $\alpha_2 = 15 \times 10^{-6}$ in./in./°F. After formulating the problem, integrating through the thickness to define $[A]$, $[B]$, $[D]$, $\{N\}^T$, and $\{M\}^T$, we know $[B] = [B'] = 0$ and

$$[A'] = \begin{bmatrix} 3.59 & -0.109 & 0 \\ -0.109 & 3.59 & 0 \\ 0 & 0 & 48 \end{bmatrix} \times 10^{-6} \qquad [D'] = \begin{bmatrix} 64.6 & -5.9 & 0 \\ -5.9 & 326 & 0 \\ 0 & 0 & 1440 \end{bmatrix} \times 10^{-3}$$

$$\{N\}^T = \begin{Bmatrix} 0.422 \\ 0.422 \\ 0 \end{Bmatrix} (T_0 + T_1) \qquad \{M\}^T = \begin{Bmatrix} 9.10 \\ 4.60 \\ 0 \end{Bmatrix} (T_0 - T_1) \times 10^{-6}$$

Determine the Cartesian components of strain (in terms of T_0 and T_1) at $z = +0.01$.

6.11 Determine the residual stresses σ_x, σ_y, and τ_{xy} through the thickness of the following laminates made AS/3501 (reference [42] in Chapter 4, with $t_{ply} = 0.005$). Assume $\Delta T = -300$ °F.

(A) $[0/90_2/45]_s$, (B) $[90/0_2/45]_s$, and (C) $[45/-45/0/90_4/45]_s$

6.12 An E-glass/epoxy ($t_{ply} = 0.010$) laminate is attached to two rigid walls while in its stress-free state. Stress concentrations between the wall and the laminate and hygral effects are neglected. Determine the temperature drop ΔT required to produce first ply failure using the Tsai–Hill failure theory if the laminate stacking sequence is

(A) $[0/90_2/45]_s$ and (B) $[45/-45/0/90_4/45]_s$

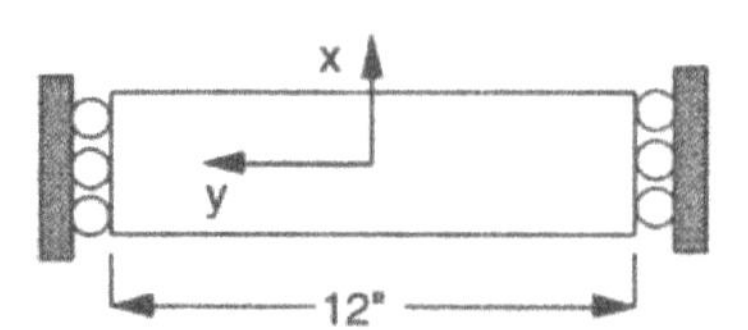

6.13 Assume a T300/5208 (reference [42] in Chapter 4) graphite/epoxy laminate ($t_{ply}=0.005$) is cured so that $\Delta T=-300$ °F. Determine the residual deformations (mid-surface strains and curvatures) for $\bar{M}=0.005$ and $\bar{M}=0.0125$ if the laminate is defined by
(A) $[0/90_2]$ and (B) $[0/90_2/45/-45]$

6.14 Assume the laminate of Problem 6.13 (A) or (B) is subjected to a load N_x. Determine the N_x sufficient to produce first ply failure for $\bar{M}=0.005$, or the lamina in which failure is predicted as a result of curing strains and curvatures. Use the Tsai–Hill theory.

6.15 Assume the laminate of Problem 6.13 (A) or (B) is subjected to the loading shown. Determine the load N_x sufficient to produce first ply failure for $\bar{M}=0.001$, or the lamina in which failure is predicted as a result of curing strains and curvatures. Use the Tsai–Hill theory.

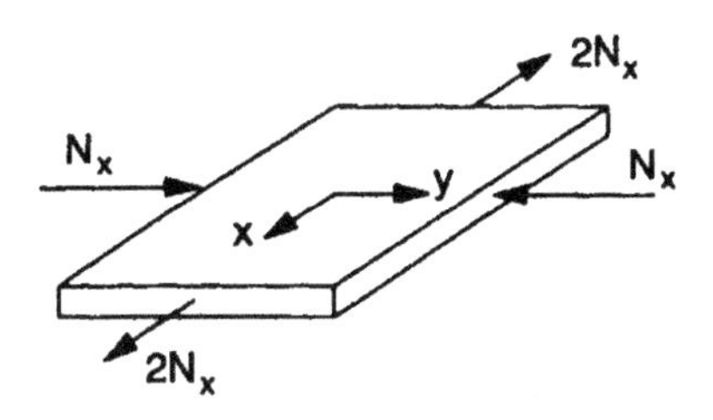

6.16 Determine the load N_x that can be supported by the laminate shown prior to catastrophic failure. The laminate is AS/3501 (reference [42] in Chapter 4, with $t_{ply}=0.005$). The laminate was cured at a temperature of 370 °F and is tested at 70 °F. Use the Tsai–Hill theory and neglect hygral effects.
(A) $[90_4/0]_s$ and (B) $[0/45/-45/90_2]_s$

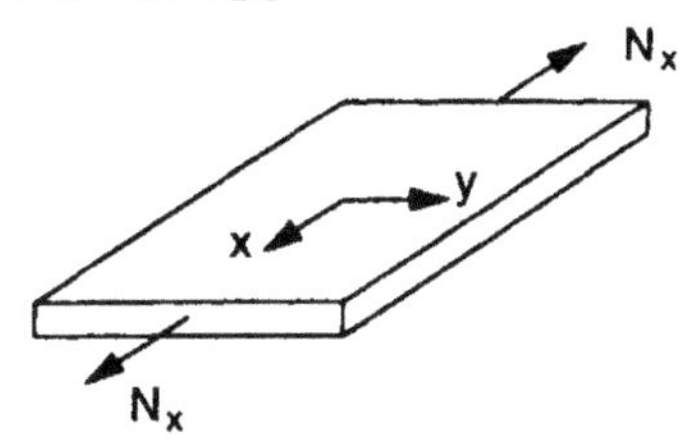

6.17 An AS/3501 (reference [42] in Chapter 4) laminate is to be used as a cantilever beam. A rigid attachment is fixed to the free end of the beam. The beam must be able to support the loads shown and experience no failure at a planc 15 in. from the free end of the beam. Each ply of laminate is 0.005 in. thick. Find the load P that results in first ply failure on plane A for the following stacking arrangement. Assume $\Delta T=-250$ °F and neglect hygral effects.
(A) $[0/90_4]_s$ and (B) $[30/0/90_2/45]_s$

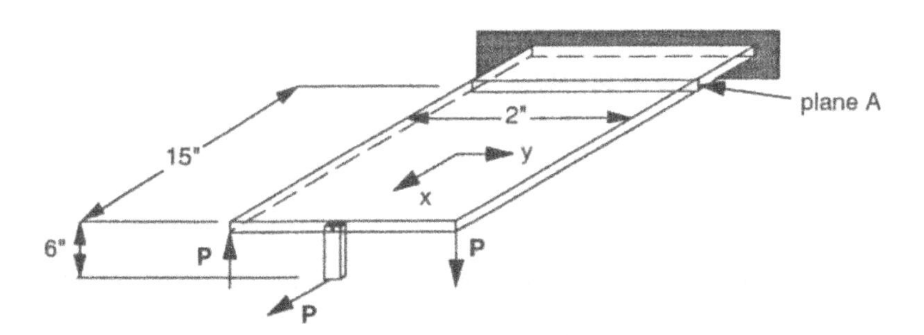

6.18 A $[90/30/-45/0]_s$ laminate is to be used in a design that requires $\bar{A}_{11}=\bar{A}_{22}$. Determine the angle Φ at which the laminate must be oriented in order to meet the design requirement. Neglect thermal and hygral effects. Assume $t_{ply}=0.010$ in. The material is
(A) E-glass/epoxy and (B) B(4)/5505 (reference [42] in Chapter 4)

6.19 The $[90/30/-45/0]_s$ laminate of Problem 6.18 is required to respond so that $\bar{D}_{11}=\bar{D}_{12}$. Determine the angle Φ at which the laminate must be oriented in order to achieve this requirement. Neglect thermal and hygral effects. Assume that $t_{ply}=0.010$ in. and the material is
(A) E-glass/epoxy and (B) AS/3501 (reference [42] in Chapter 4)

6.20 T300/5208 graphite/epoxy (material 2, reference [42] in Chapter 4) and Scotchply type 1002 glass/epoxy (material 1) are used to construct a laminar hybrid. The thickness of a graphite/epoxy lamina is 0.005 in. and the thickness of a glass/epoxy lamina is 0.010 in. Laminate stacking sequences are designated as a combination of angles and subscripted material numbers. Numbers that are not subscripted indicate the number of lamina at a particular angle. For example, $[0_1/2(90_2)/45_1]_s$ means that material 1 (glass/epoxy) is oriented at 0°, then two (2) 90° plies of material 2 (graphite/epoxy), followed by one ply of material 1 at 45°. The subscript "s" designates a symmetric laminate. Plot the distribution of σ_x, σ_y, and τ_{xy} through the laminate defined below if it is subjected to an applied axial load of $N_x=250$ lb/in. and $\Delta T=-300$ °F. Neglect hygral effects.
(A) $[45_1/45_2/90_1/90_2]_s$ and (B) $[30_1/45_2/0_1/90_2]_s$

In Problems 6.21 and 6.22, an intraply hybrid made from materials 1 and 2 above is also considered. For the intraply hybrid, a simple rule-of-mixtures approximation is used to define material properties, as presented in the following table.

Property	Percent AS/3501		
	30%	**50%**	**70%**
E_1	11.8×10^6	15.9×10^6	20.1×10^6
E_2	1.29×10^6	1.35×10^6	1.40×10^6
G_{12}	0.732×10^6	0.80×10^6	0.91×10^6
ν_{12}	0.266	0.27	0.274
α_1	3.34×10^{-6}	2.75×10^{-6}	1.51×10^{-6}
α_2	12.31×10^{-6}	13.2×10^{-6}	12.4×10^{-6}

Assume that the thickness of each intraply lamina is 0.010 in.

6.21 Assume that the graphite/epoxy constituent of a $[0/90_2/45_2]_s$ intraply hybrid has volume fractions of 30% and 70%. Plot the residual stress distribution (in the x–y plane) through the laminate assuming $\Delta T=-300$ °F. Neglect hygral effects.

6.22 An intraply hybrid with 50% AS/3501 is to be used as an interface layer between glass/epoxy and graphite/epoxy lamina. Each lamina has a thickness of 0.10 in. Plot the stress distribution through the laminate assuming $\Delta T=-300$ °F and that it is subjected to an applied load $N_x=2000$ lb/in. Neglect hygral effects.
(A) [0/45/0] (B) [0/90/0]

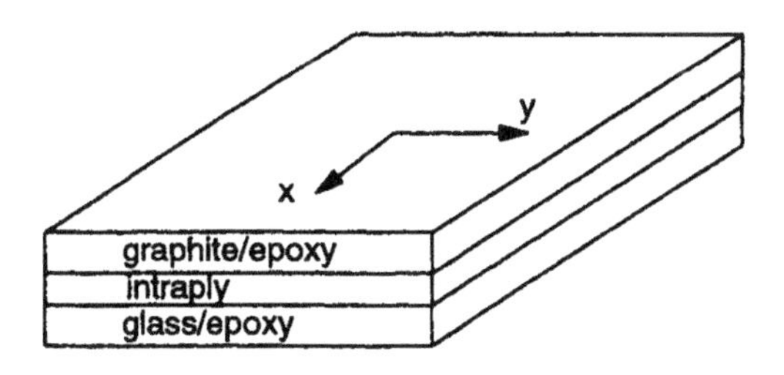

6.23 Assume a randomly oriented SFC is made from a combination of E-glass fiber and a polyester matrix with a 30% volume fraction of fibers. The relevant properties for each constituent are $E_f = 10.5 \times 10^6$ psi, $E_m = 0.50 \times 10^6$ psi, $d = 400\ \mu$ in., and $L = 0.125$ in. Estimate the longitudinal and transverse moduli as well as E_{random}.

6.24 Assume the fiber and matrix properties of Problem 6.23 are applicable to an SFC with a variable volume fraction of fibers and $L/d = 1000$. Plot E_{random} vs. v_f for $0.25 \leq v_f \leq 0.75$.

6.25 For the section of ribbon-reinforced composite shown, determine E_L and E_T, assuming $a = 0.300$, $E_r = 12.5 \times 10^6$ psi, $W_r = 0.375$, $t_r = 0.125$, $E_m = 0.50 \times 10^6$ psi, $W_m = 0.250$, and $t_m = 0.350$.

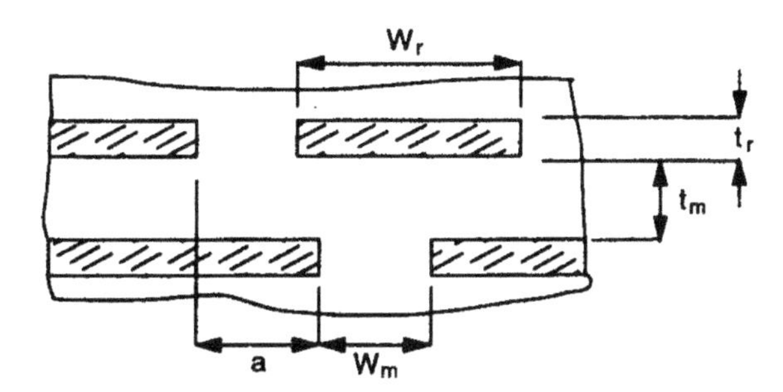

6.26 Use the material properties and volume fraction of fibers in Problem 6.23 to determine $E_1(E_L)$ and $E_2(E_T)$. Assume $v_{12} = 0.25$ and $G_{12} = 0.75 \times 10^6$ psi. Assume that an SFC has a biased fiber orientation as shown in the following table. Determine $[A]$ for a laminate with a total thickness of 0.50 in.

θ (°)	%
0	10
10	15
30	40
60	15
90	20

References

[1] Ashton JE, Halpin JC, Petit PH. Primer on composite materials. Westport, CT: Technomic; 1969.
[2] Jones RM. Mechanics of composite materials. New York: Hemisphere Publishing; 1975.
[3] Vinson JR, Sierakowski RL. The behavior of structures composed of composite materials. Dordrecht, The Netherlands: Martinus Nijhoff Publishers; 1986.
[4] Pister KS, Dong SB. Elastic bending of layered plates. J Eng Mech Div ASCE 1959;85:1–10.
[5] Reisner E, Stavsky Y. Bending and stretching of certain types of heterogeneous aelotropic elastic plates. J Appl Mech 1961;28:402–8.
[6] Ugural AC. Stresses in plates and shells. New York: McGraw-Hill; 1981.
[7] Reisner E. The effect of transverse shear deformation on the bending of elastic plates. J Appl Mech 1945;12:A66–79.
[8] Tsai SW. Structural behavior of composite materials, NASA CR-71; 1964.
[9] Azzi VD, Tsai SW. Elastic moduli of laminated anisotropic composites. Exp Mech 1965;5:177–85.
[10] Mason J-AE, Copeland SD, Seferis JC. Intrinsic process characterization and scale up of advanced thermoplastic structures. In: SPE ANTEC '88 conference proceedings, Atlanta, GA; 1988.
[11] Chapman TJ, Gillespie JW, Mason J-AE, Serefis JC, Pipes RB. Thermal skin/core residual stresses induced during process cooling of thermoplastic matrix composites. In: Proceedings of the American Society of Composites, third technical conference. Westport, CT: Technomic; 1988. p. 449–58.
[12] Chapman TJ, Gillespie JW, Pipes RB, Mason J-AE, Serefis JC. Prediction of process-induced residual stresses in thermoplastic composites. J Compos Mater 1990;24:616–28.
[13] Jeronimidis G, Parkyn AT. Residual stresses in carbon fibre–thermoplastic matrix laminates. J Compos Mater 1988;22:401–15.
[14] Manechy CE, Miyanao Y, Shimbo M, Woo TC. Residual-stress analysis of an epoxy plate subjected to rapid cooling on both surfaces. Exp Mech 1986;26:306–12.
[15] Manson J-AE, Serefis JC. Internal stress determination by process simulated laminates. In: SPE ANTEC '87 conference proceedings, Los Angeles, CA; 1987.
[16] Narin JA, Zoller P. Matrix solidification and the resulting residual thermal stresses in composites. J Mater Sci 1985;20:355–67.
[17] Kim RY, Hahn HT. Effect of curing stresses on the first-ply-failure in composite laminates. J Compos Mater 1979;13:2–16.
[18] Whitney JM, Daniel IM, Pipes RB. Experimental mechanics of fiber reinforced composite materials, SESA Monograph No. 4. Westport, CT: Technomic; 1982.
[19] Tsai SW. Strength characteristics of composite materials, NASA CR-224; April 1965.
[20] Hyer MW. Some observations on the cured shape of thin unsymmetric laminates. J Compos Mater 1981;15:175–94.
[21] Hyer MW. Calculations of the room-temperature shapes of unsymmetric laminates. J Compos Mater 1981;15:296–310.
[22] Hyer MW. The room-temperature shape of four layer unsymmetrical cross-ply laminates. J Compos Mater 1982;16:318–40.
[23] Harper BD. The effect of moisture induced swelling upon the shapes of antisymmetric cross-ply laminates. J Compos Mater 1987;21:36–48.
[24] Narin JA, Zoller P. Residual thermal stresses in semicrystalline thermoplastic matrix composites. In: Proc. 5th Int. conference of composite materials, San Diego, CA; 1985.

[25] Narin JA, Zoller P. Residual stresses in amorphous and semicrystalline thermoplastic matrix composites. Toughened composites, ASTM STP-937. Philadelphia: ASTM; 1987.
[26] Master JE, Reifsnider KL. An investigation of cumulative damage development in quasi-isotropic graphite/epoxy laminates. Damage in composite materials, ASTM STP 775. Philadelphia: ASTM; 1982, pp. 40–62.
[27] Highsmith AL, Reifsnider KL. Stiffness reduction mechanisms in composite laminates. Damage in composite materials, ASTM STP 775. Philadelphia: ASTM; 1982, pp. 103–117.
[28] Jamison RD, Reifsnider KL. "Advanced fatigue damage development in graphite/epoxy laminates," AFWAL-TR-82–3103, Air Force Wright Aeronautical Laboratories; December 1982.
[29] Bailey JE, Curtis PT, Parvizi A. On transverse cracking and longitudinal splitting behavior of glass and carbon fiber reinforced epoxy cross-ply laminates and the effect of poisson and thermally generated strains. Proc Royal Soc London A 1979;366:599–623.
[30] Harrison RP, Bader MG. Damage development in CFRP laminates under monotonic and cyclic stressing. Fiber Sci Technol 1983;18:163–80.
[31] Jamison RD. The role of microdamage in tensile failure of graphite/epoxy laminates. Fiber Sci Technol 1985;25:83–99.
[32] Sun CT, Jen KC. On the effect of matrix cracks on laminate strength. J Reinf Plast 1987;6:208–22.
[33] Whitney JM, Ashton JE. Effects of environment on the elastic response of layered composite plates. AIAA J 1970;9:1708–12.
[34] Pipes RB, Vinson JR, Chow TW. On the hygrothermal response of laminated composite systems. J Compos Mater 1976;10:129–36.
[35] Springer GS, editor. Environmental effects on composite materials. Westport, CT: Technomic; 1981.
[36] Tsai SW, Hahn HT. Introduction to composite materials. Westport, CT: Technomic; 1980.
[37] Tsai SW, Pagano NJ. Invariant properties of composite materials. In: Tsai SW, Halpin JC, Pagano NJ, editors. Composite materials workshop. Westport, CT: Technomic; 1968.
[38] Lovell DR. Hybrid laminates of glass/carbon fibers—1. Reinf Plast 1978;22:216–25.
[39] Lovell DR. Hybrid laminates of glass/carbon fibers—2. Reinf Plast 1978;22:252–61.
[40] Summerscales J, Short D. Carbon fiber and glass fiber hybrid reinforced plastics. Composites 1978;9:157–64.
[41] Short D, Summerscales J. Hybrids—a review, part 1, techniques, design and construction. Composites 1979;10:215–25.
[42] Short D, Summerscales J. Hybrids—a review, part 2. Physical properties. Composites 1980;11:33–40.
[43] Hayashi T. Development on new material properties by hybrid composition. Fukugo Zairyo (Compos Mater) 1972;1:18–26.
[44] Phillips LN. The hybrid effect—does it exist? Composites 1976;7:7–14.
[45] Zwben C. Tensile strength of hybrid composites. J Mater Sci 1977;12:1325–33.
[46] Marom G, Fisher S, Tuler FR, Wagner HD. Hybrid effects in composites: conditions for positive or negative effects versus rule of mixtures behavior. J Mater Sci 1978;13:1419–25.
[47] Chamis CC, Lark RF, Sinclair JH. Mechanical property characterization of intraply hybrid composites. In: Test methods and design allowables for fibrous composites, ASTM STP 734. Philadelphia: ASTM; 1981. p. 261–9.
[48] Manders PW, Bader MG. The strength of hybrid glass/carbon fiber composites, part 1. Failure strain enhancement and failure mode. J Mater Sci 1981;16:2233–9.

[49] Fariborz SJ, Yang CL, Harlow DG. The tensile behavior of intraply hybrid composites I: model and simulation. J Compos Mater 1985;19:334–54.
[50] Sandors BA, editor. Short fiber reinforced composite materials—ASTM STP 772. Philadelphia: ASTM; 1982.
[51] Agarwal BD, Broutman LJ. Analysis and performance of fiber composites. New York: John Wiley and Sons; 1980.
[52] Halpin JC, Pagano NJ. The laminate approximation of randomly oriented fibrous composites. J Compos Mater 1969;3:720–4.
[53] Halpin JC, Jerine K, Whitney JM. The laminate analogy for 2 and 3 dimensional composite materials. J Compos Mater 1971;5:36–49.
[54] Rosen BW. Mechanics of composite strengthening. Fiber composite materials. Metals Park, OH: American Society for Metals; 1964 [chapter 3].
[55] Dow NF. Study of stresses near a discontinuity in a filament reinforced composite sheet, General Electric Company report No. TISR63SD61; August 1963.
[56] Carrara AS, McGarry FJ. Matrix and interface stresses in a discontinuous fiber composite model. J Compos Mater 1968;2:222–43.
[57] MacLaughlin TF, Barker RM. Effect of modulus ratio on stress near a discontinuous fiber. Exp Mech 1972;12:178–83.
[58] Owen DR, Lyness JF. Investigations of bond failure in fibre reinforced materials by the finite element methods. Fibre Sci Technol 1972;5:129–41.
[59] Broutman LJ, Agarwal BD. A theoretical study of the effect of an interfacial layer on the properties of composites. Polym Eng Sci 1974;14:581–8.
[60] Lin TH, Salinas D, Ito YM. Elastic-plastic analysis of unidirectional composites. J Compos Mater 1972;6:48–60.
[61] Agarwal BD, Lifshitz JM, Broutman LJ. Elastic-plastic finite element analysis of short fiber composites. Fibre Sci Technol 1974;7:45–62.
[62] Agarwal BD. Micromechanics analysis of composite materials using finite element methods [Ph.D. Thesis], Chicago, IL: Illinois Institute of Technology; May 1972.
[63] Agarwal BD, Bansal RK. Plastic analysis of fibre interactions in discontinuous fibre composites. Fibre Sci Technol 1977;10:281–97.
[64] Agarwal BD, Bansal RK. Effects of an interfacial layer on the properties of fibrous composites: a theoretical analysis. Fibre Sci Technol 1979;12:149–58.
[65] Halpin JC. Strength and expansion estimates for oriented short fiber composites. J Compos Mater 1969;3:732–4.
[66] Pipes RB, Pagano NJ. Interlaminar stresses in composite laminates under uniform axial extension. J Compos Mater 1970;4:538–48.
[67] Puppo AH, Evenson HA. Interlaminar shear in laminated composites under generalized plane stress. J Compos Mater 1970;4:204–20.
[68] Pagano NJ, Pipes RB. The influence of stacking sequence on laminate strength. J Compos Mater 1971;5:50–7.
[69] Pagano NJ, Pipes RB. Some observations on the interlaminar strength of composite laminates. Int J Mech Sci 1973;15:679–86.
[70] Pipes RB, Pagano NJ. Interlaminar stresses in composite laminates—an approximate elasticity solution. J Appl Mech Sept. 1974;41:668–72.
[71] Pagano NJ. On the calculation of interlaminar normal stresses in composite laminates. J Compos Mater 1974;8:65–81.
[72] Whitney JM, Sun CT. A higher order theory for extensional motion of laminated composites. J Sound Vib 1973;30:85–97.

Laminated composite beam analysis

7

7.1 Introduction

Beams, rods, bars, plates, and shells are the most common components of composite structures. A complete analysis of a composite structure involves the assessment of both stress and deflection. Deflection analysis for beams, rods, bars, plates, and shells can be developed from classical lamination theory. The complexity of the development depends on the fundamental assumptions made regarding displacement fields and admissible allowed loading conditions. Both simple and advanced beam theories are presented in this chapter. In addition, the effect of shear deformation is considered. The development of deflection relationships for beams, rods, bars, plates, and shells are found in various references, including Refs. [1–3].

7.2 Equations of equilibrium for beams, rods, and columns

Beams, rods, columns, and beam-columns are among the most commonly used structural components. The designation for each is based on the dimensions and loads to which it is subjected. These components can be defined in terms of the general dimensions for a prismatic beam, rod, or column as shown in Figure 7.1. Each component has a width b, height h, and length L, where $b/L << 1$ and $h/L << 1$.

In addition to the dimensions for classifying each component, loading conditions are also relevant. The loading conditions pertaining to each component are described below and illustrated in Figure 7.2 along with the special case of the beam-column.

- *Beam*: A beam is a component subjected to loads applied lateral to the longitudinal axis (x-axis of Figure 7.1). Lateral loads typically occur in the x–z plane and are in the z-direction. They can be concentrated or distributed loads on the top, bottom, or both surfaces. Loads are applied in such a manner that bending (curvature in the x–z plane) results.
- *Rod*: A rod is a component subjected to an applied tensile load in the axial direction (along the x-axis).
- *Column*: A column is a component subjected to an applied compressive load in the axial direction (along the x-axis). The basic difference between a rod and a column is that the latter can experience instability (buckling) at applied load levels below those causing failure due to stress.
- *Beam-column*: A beam-column is a special case in which the structural component is subjected to both lateral loads (like a beam) and compressive loads (like a column).

A generalized state of stress for the x–y and x–z planes is shown in Figure 7.3. The lengths of each side of the elements shown are dx, dy, and dz. A similar state of stress exists on the y–z plane. The partial derivatives of stress components account for the

Laminar Composites. http://dx.doi.org/10.1016/B978-0-12-802400-3.00007-6

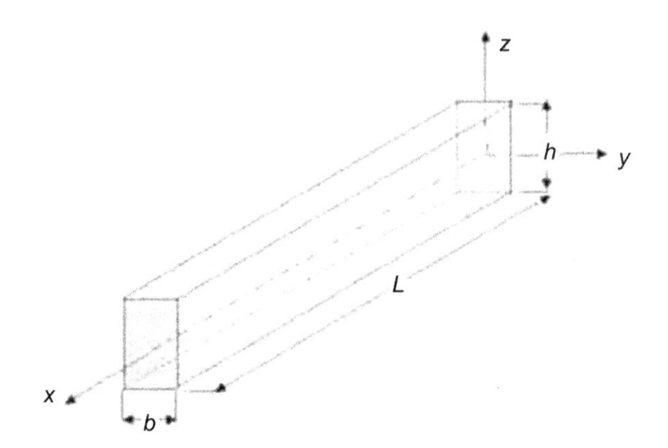

Figure 7.1 Schematic of beam, rod, and column dimensions.

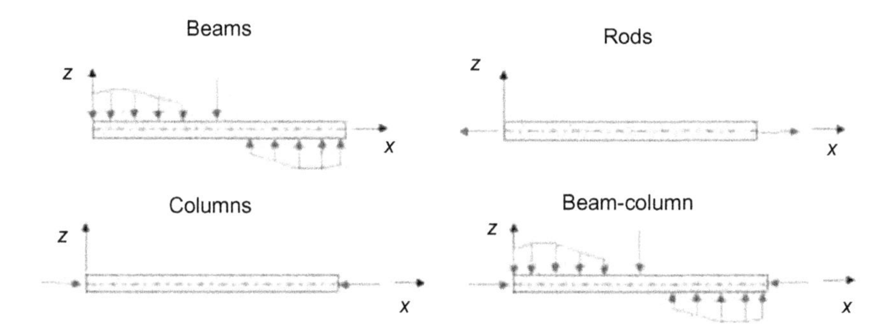

Figure 7.2 Loading conditions for beams, rods, columns, and beam-columns.

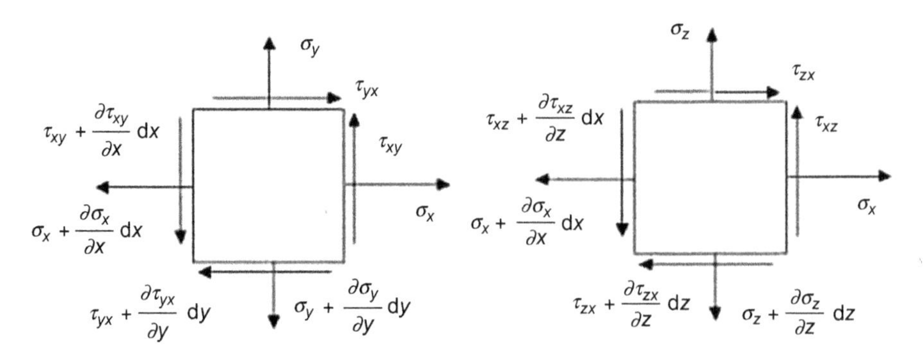

Figure 7.3 Generalized state of stress for the x–y and x–z planes.

possible variation of stress through the representative volume element. A balance of forces in the x, y, and z directions results in the three equations of equilibrium given in Equation (7.1).

$$
\begin{aligned}
&\frac{\partial \sigma_x}{\partial x}+\frac{\partial \tau_{xy}}{\partial y}+\frac{\partial \tau_{xz}}{\partial z}+F_x=0\\
&\frac{\partial \tau_{xy}}{\partial x}+\frac{\partial \sigma_y}{\partial y}+\frac{\partial \tau_{yz}}{\partial z}+F_y=0\\
&\frac{\partial \tau_{xz}}{\partial x}+\frac{\partial \tau_{yz}}{\partial y}+\frac{\partial \sigma_z}{\partial z}+F_z=0
\end{aligned}
\tag{7.1}
$$

The F_x, F_y, and F_z terms are called body forces and represent a force per unit volume (such as gravity or magnetic forces). A complete derivation of Equation (7.1) can be found in many texts, including Ref. [4]. The relationship between applied force and stress as well as applied moment and stress through a laminate is given by Equations (6.8) and (6.9). These two equations are somewhat modified and presented as Equations (7.2) and (7.3).

$$
\begin{Bmatrix} N_x \\ N_y \\ N_{xy} \\ Q_x \\ Q_y \end{Bmatrix} = \sum_{k=1}^{N} \int_{z_{k-1}}^{z_k} \begin{Bmatrix} \sigma_x \\ \sigma_y \\ \tau_{xy} \\ \tau_{xz} \\ \tau_{yz} \end{Bmatrix}_k \mathrm{d}z \tag{7.2}
$$

$$
\begin{Bmatrix} M_x \\ M_y \\ M_{xy} \end{Bmatrix} = \sum_{k=1}^{N} \int_{z_{k-1}}^{z_k} z \begin{Bmatrix} \sigma_x \\ \sigma_y \\ \tau_{xy} \end{Bmatrix}_k \mathrm{d}z \tag{7.3}
$$

In these equations, the subscript "k" refers to the kth lamina, with z_{k-1} and z_k defining the boundaries of each lamina as depicted in Figure 6.6.

The first equation of (7.1) applies to equilibrium in the x-direction. Assuming that body forces (F_x, F_y, F_z) are neglected, this equation can be integrated through the laminate thickness, or treated as a summation of the integral over each lamina. Using the latter procedure for an N lamina laminate gives

$$
\sum_{k=1}^{N} \int_{z_{k-1}}^{z_k} \left\{ \left(\frac{\partial \sigma_x}{\partial x}\right)_k + \left(\frac{\partial \tau_{xy}}{\partial y}\right)_k + \left(\frac{\partial \tau_{xz}}{\partial z}\right)_k \right\} \mathrm{d}z = 0
$$

This expression can be rearranged and represented as

$$
\frac{\partial}{\partial x}\left[\sum \int (\sigma_x)_k \mathrm{d}z\right] + \frac{\partial}{\partial y}\left[\sum \int (\tau_{xy})_k \mathrm{d}z\right] + \frac{\partial}{\partial z}\left[\sum \int (\tau_{xz})_k \mathrm{d}z\right] = 0
$$

Each summation has limits of $k = 1$ to N, and each integral has limits of z_{k-1} to z_k. Referring to Equation (7.2), the first two terms of the equation above can be replaced by $\partial N_x/\partial x$ and $\partial N_{xy}/\partial y$, respectively. The third term of this equation becomes

$$\sum_{k=1}^{N} (\tau_{xz})_k \Big|_{z_{k-1}}^{z_k}$$

The τ_{xz} component of stress is the interlaminar shear stress, and is canceled at adjacent lamina interfaces for all but the extreme values of z_k, namely, at $z = \pm h/2$. Therefore, the final term is expressed as an applied shear stress at either or both extreme surfaces of the laminate, and is represented as

$$\sum_{k=1}^{N} (\tau_{xz})_k \Big|_{z_{k-1}}^{z_k} = \tau_{xz}(h/2) - \tau_{xz}(-h/2) = \tau_{1x} - \tau_{2x}$$

Thus, the first equation of (7.1) therefore becomes

$$\frac{\partial N_x}{\partial x} + \frac{\partial N_{xy}}{\partial y} + (\tau_{1x} - \tau_{2x}) = 0 \tag{7.4}$$

Neglecting body forces and following similar procedures for the remaining two equations of (7.1) can be written as

$$\frac{\partial N_{xy}}{\partial x} + \frac{\partial N_y}{\partial y} + (\tau_{1y} - \tau_{2y}) = 0 \tag{7.5}$$

$$\frac{\partial Q_x}{\partial x} + \frac{\partial Q_y}{\partial y} + (p_1 - p_2) = 0 \tag{7.6}$$

In each equation, τ_{ix}, τ_{iy}, and p_i are applied stresses on either the top or the bottom surface of the laminate as illustrated in Figure 7.4 for the top surface. The subscript 2 (τ_{2x}, etc.) designates the bottom surface.

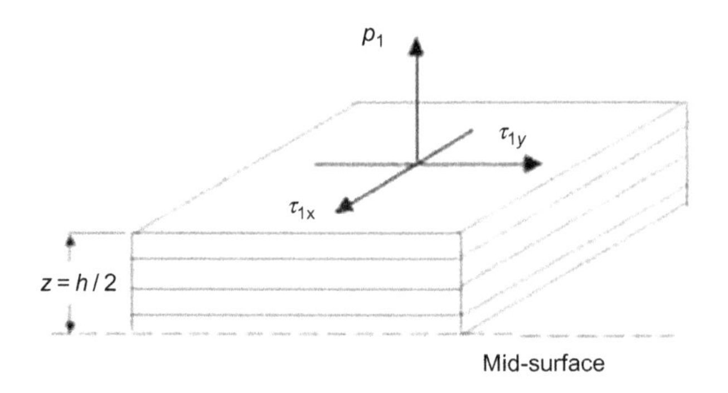

Figure 7.4 Applied surface stresses τ_{1x}, τ_{1y}, and p_1.

Two equations are required for a balance of moments about the x and y axes. These are developed based on the force equilibrium conditions in the x and y directions from Equation (7.1). The required expressions, neglecting body forces, for the x and y directions are

$$x\text{-direction}: \int_{-h/2}^{h/2} z\left\{\frac{\partial \sigma_x}{\partial x}+\frac{\partial \tau_{xy}}{\partial y}+\frac{\partial \tau_{xz}}{\partial z}\right\}\mathrm{d}z=0$$

$$y\text{-direction}: \int_{-h/2}^{h/2} z\left\{\frac{\partial \tau_{xy}}{\partial x}+\frac{\partial \sigma_y}{\partial y}+\frac{\partial \tau_{yz}}{\partial z}\right\}\mathrm{d}z=0$$

The integral over the entire laminate in the x-direction can be rearranged as a summation of integrals over each lamina so that

$$\frac{\partial}{\partial x}\left[\sum\int(\sigma_x)_k z\mathrm{d}z\right]+\frac{\partial}{\partial y}\left[\sum\int(\tau_{xy})_k z\mathrm{d}z\right]+\frac{\partial}{\partial z}\left[\sum\int(\tau_{xz})_k z\mathrm{d}z\right]=0$$

where the limits on Σ are from $k=1$ to N, and the limits of the integral are z_{k-1} to z_k. Incorporating Equation (7.3) for M_x and M_{xy} into this expression results in

$$\frac{\partial M_x}{\partial x}+\frac{\partial M_{xy}}{\partial y}+\sum_{k=1}^{N}\int_{z_{k-1}}^{z_k}\left(\frac{\partial \tau_{xz}}{\partial z}\right)_k z\ \mathrm{d}z=0$$

The last term in this equation is integrated by parts by allowing the substitutions: $u=z$, $\mathrm{d}u=\mathrm{d}z$, $v=(\tau_{xz})_k$, and $\mathrm{d}v=(\partial\tau_{xz}/\partial z)_k\mathrm{d}z$. Therefore

$$\sum_{k=1}^{N}\int_{z_{k-1}}^{z_k}\left(\frac{\partial \tau_{xz}}{\partial z}\right)_k z\ \mathrm{d}z=\sum_{k=1}^{N}\left\{z(\tau_{xz})_k\Big|_{z_{k-1}}^{z_k}-\int_{z_{k-1}}^{z_k}(\tau_{xz})_k\mathrm{d}z\right\}$$

The first term in this expression is the interlaminar shear stresses, which cancel at adjacent lamina interfaces, and may be present only on the top and bottom surfaces of the laminate, which results in

$$\sum_{k=1}^{N} z(\tau_{xz})_k\Big|_{z_{k-1}}^{z_k}=\frac{h}{2}(\tau_{1x}-\tau_{2x})$$

where τ_{1x} and τ_{2x} were previously defined. The second equation is Q_x from Equation (7.2). For the x-direction, the moment equation of equilibrium is

$$\frac{\partial M_x}{\partial x}+\frac{\partial M_{xy}}{\partial y}-Q_x+\frac{h}{2}(\tau_{1x}-\tau_{2x})=0 \tag{7.7}$$

Following a similar procedure for the y-direction results in

$$\frac{\partial M_{xy}}{\partial x} + \frac{\partial M_y}{\partial y} - Q_y + \frac{h}{2}\left(\tau_{1y} - \tau_{2y}\right) = 0 \tag{7.8}$$

The equilibrium equations in (7.4)–(7.8) are used to develop relations between load and deformation for specific problems. Although these equations represent equilibrium conditions for plates, the proper assumptions made with regard to particular beam configurations allow them to be implemented into beam analysis. The governing equation for laminate analysis relating load and strain to laminate stacking sequence are given by Equations (6.15) and (6.20). In these equations, each strain component is represented in terms of displacements in the x, y, and z directions. In Chapter 6, it was assumed that a series expansion could be used to represent displacements in x, y, and z such that

$$U = \sum_{i=0}^{\infty} z^i \Phi(x, y) \quad V = \sum_{i=0}^{\infty} z^i \Psi(x, y) \quad W = \sum_{i=0}^{\infty} z^i \Theta(x, y)$$

Assuming small displacements, only the first term of each expression in x and y are retained, yielding assumed displacement fields of $U = U_o(x, y) + z\Phi(x, y)$, $V = V_o(x, y) + z\Psi(x, y)$, and $W = W_o(x, y)$. The definitions of strain in Chapter 2 are used to express each strain component in terms of displacements. In order to streamline the derivation, a shorthand notation for partial differentiation is used. Instead of using $\partial U/\partial x$ to represent partial differentiation of U with respect to x, the notation $U,_x$ is used. In this representation, it is understood that the "," replaces the "∂", and the subscript "x" represents the parameter with respect to which the partial derivative is being taken. Multiple subscripts to the right of the "," indicate the number of partial derivatives to be taken. For example, $\partial^3 U/\partial x^3$ is replaced by $U,_{xxx}$. Using this notation, the strains are

$$\varepsilon_x = U,_x = U_{o,x} + z\Phi,_x$$

$$\varepsilon_y = V,_y = V_{o,y} + z\Psi,_y$$

$$\gamma_{xy} = U,_y + V,_x = \left(U_{o,y} + V_{o,x}\right) + z\left(\Phi,_y + \Psi,_x\right)$$

$$\gamma_{xz} = U,_z + W,_x = W_{o,x} + \Phi$$

$$\gamma_{yz} = V,_z + W,_y = W_{o,y} + \Psi$$

Each displacement is a function of x and y. Using the definitions of $\{\varepsilon^o\}$ and $\{\kappa\}$ from Equations (6.2) and (6.3) results in

$$\begin{Bmatrix} \varepsilon_x^o \\ \varepsilon_y^o \\ \gamma_{xy}^o \end{Bmatrix} = \begin{Bmatrix} U_{o,x} \\ V_{o,y} \\ U_{o,y} + V_{o,x} \end{Bmatrix} \quad \begin{Bmatrix} \kappa_x \\ \kappa_y \\ \kappa_{xy} \end{Bmatrix} = \begin{Bmatrix} \Phi,_x \\ \Psi,_y \\ \Phi,_y + \Psi,_x \end{Bmatrix}$$

The remaining two shear strains are expressed as $\gamma_{xz} = W_{o,x} + \Phi$ and $\gamma_{yz} = W_{o,y} + \Psi$, where Φ and Ψ are functions of x and y which are defined according to the problem being considered. These expressions for strain can be substituted into Equations (6.15) and (6.20) to define load–displacement relations in terms of the components of $[A]$, $[B]$, and $[D]$. Then, using the equilibrium Equations (7.4)–(7.8), the equilibrium conditions in terms of displacement fields are developed.

7.3 Elementary beam analysis

Elementary laminated beam analysis follows closely the procedures developed for beam analysis in introductory strength of materials. The simplified analysis of beams is initially presented for problems without thermal and hygral effects. All displacements are assumed to be small, and the following assumptions are made.

1. Loading is in the x–z plane only.
2. Since $b/L << 1$ (refer to Figure 7.1), Poisson's effect in the y-direction is ignored (classical beam assumption from introductory strength of materials).
3. The governing equations do not contain terms dependent on displacements in the y-direction.

As a result of these assumptions, Equations (6.19) and (6.20) become

$$\begin{Bmatrix} N_x \\ M_x \end{Bmatrix} = \begin{bmatrix} A_{11} & B_{11} \\ B_{11} & D_{11} \end{bmatrix} \begin{Bmatrix} \varepsilon_x^o \\ \kappa_x \end{Bmatrix}$$

Based on assumptions (2) and (3), this simplified analysis pertains only to symmetric laminates. If $[B] \neq 0$, out-of-plane deformations can exist and the simple beam approach cannot be used with any assurance of an accurate solution. Since $[B] = 0$, the governing equations are

$$N_x = A_{11}\varepsilon_x^o \quad M_x = D_{11}\kappa_x \quad Q_x = A_{55}\gamma_{xz}$$

We further assume that $\gamma_{xz} = 0$, but Q_x can exist. In general, $\gamma_{xz} \neq 0$ for laminated composites. This affects lateral deformations, natural frequencies of vibration, and buckling loads.

Eliminating the out-of-plane deformations by setting $N_y = N_{xy} = M_y = M_{xy} = Q_y = \tau_{1y} = \tau_{2y} = 0$ reduces the number of governing equations to three, which are given below

$$\begin{aligned} &\frac{dN_x}{dx} + \tau_{1x} - \tau_{2x} = 0 \\ &\frac{dM_x}{dx} - Q_x + \frac{h}{2}(\tau_{1x} - \tau_{2x}) = 0 \\ &\frac{dQ_x}{dx} + p_1 - p_2 = 0 \end{aligned} \tag{7.9}$$

For simplicity, it is assumed that no shear is applied to the top or bottom surfaces ($\tau_{1x} = \tau_{2x} = 0$). In addition, the applied surface loads in the z-direction (p_1 and p_2) are expressed in a more convenient form as $p(x) = p_1 - p_2$. Thus, the simplified beam equations are

$$\frac{dN_x}{dx} = 0 \quad \frac{dM_x}{dx} - Q_x = 0 \quad \frac{dQ_x}{dx} + p(x) = 0$$

The loads N_x, M_x, and Q_x come directly from the equations of classical lamination theory, and are represented as load or moment per unit width in the y-direction. Since variations of load in the y-direction are not allowed, all relevant equations of equilibrium can be multiplied by the beam width b. Using this procedure results in definitions of axial load (P), bending moment (M), transverse shear (V), and load distribution ($q(x)$), which is more familiar to conventional beam deflection analysis. They are

$$P = N_x b \quad M = M_x b \quad V = Q_x b \quad q(x) = p(x) b$$

The axial load and bending moment can be expressed in terms of partial derivatives by replacing ε_x^o with $U_{o,x}$ and κ_x with $\Phi_{,x}$. Since it is assumed that $\gamma_{xz} = 0$, the expression for Φ becomes

$$\gamma_{xz} = W_{o,x} + \Phi = 0 \Rightarrow \Phi = -W_{o,x}$$

Therefore, $\kappa_x = \Phi_{,x} = -W_{o,xx}$. Since there is no variation of deformation in the y-direction, the axial load and bending moment are

$$P = N_x b = bA_{11}\varepsilon_x^o \Rightarrow P = bA_{11}\frac{dU_o}{dx}$$

$$M = M_x b = bD_{11}\kappa_x \Rightarrow M = -bD_{11}\frac{d^2W_o}{dx^2}$$

Incorporating these into the equations of equilibrium yields

$$\frac{dP}{dx} = 0 \quad \frac{dV}{dx} + q(x) = 0 \quad \frac{dM}{dx} - V = 0 \tag{7.10}$$

Since $dP/dx = 0$, $P =$ constant, and

$$\frac{dU_o}{dx} = \frac{P}{bA_{11}}$$

The displacement in the x-direction for a beam (rod) subjected to a tensile axial load P is defined by integrating this equation to get

$$U_o = \frac{P}{bA_{11}}x + C_1 \tag{7.11}$$

where C_1 is a constant of integration determined from boundary conditions. The stress in each lamina is determined from

$$(\sigma_x)_k = \left(\bar{Q}_{11}\right)_k \varepsilon_x^o = \left(\bar{Q}_{11}\right)_k U_{o,x} \tag{7.12}$$

This expression is not valid for buckling, which requires a more refined analysis. It is valid for compressive loads, provided that buckling is not a potential problem.

The remaining two equations for $\mathrm{d}V/\mathrm{d}x$ and $\mathrm{d}M/\mathrm{d}x = 0$ are combined to evaluate bending effects, starting with the shear force–bending moment and moment–displacement relations

$$\frac{\mathrm{d}M}{\mathrm{d}x} - V = 0 \quad M = -bD_{11}\frac{\mathrm{d}^2 W_o}{\mathrm{d}x^2}$$

These expressions are combined to yield

$$\frac{\mathrm{d}}{\mathrm{d}x}\left(-bD_{11}\frac{\mathrm{d}^2 W_o}{\mathrm{d}x^2}\right) = V$$

Therefore, $V = -bD_{11}\dfrac{\mathrm{d}^3 W_o}{\mathrm{d}x^3}$. In addition, from Equation (7.10), $\mathrm{d}V/\mathrm{d}x + q(x) = 0$. Substituting the expression above for V into this equation results in

$$bD_{11}\frac{\mathrm{d}^4 W_o}{\mathrm{d}x^4} = q(x) \tag{7.13}$$

Equation (7.13) is analogous to the differential equation describing load distribution and mid-plane deformation from introductory strength of materials. For an isotropic material, the bD_{11} term is the flexural rigidity EI. In a similar manner, the bA_{11} term of Equation (7.11) is EA for an isotropic material.

The stress in any lamina is determined by first solving Equation (7.13) for W_o and subsequently using $\kappa_x = -W_{o,xx}$, which results in

$$(\sigma_x)_k = z\left(\bar{Q}_{11}\right)_k \kappa_x = -z\left(\bar{Q}_{11}\right)_k \left(W_{o,xx}\right) \tag{7.14}$$

In situations involving axial loads and flexure, the normal stress at any location in the laminate can be expressed as

$$(\sigma_x)_k = \left(\bar{Q}_{11}\right)_k \left(\frac{\mathrm{d}U_o}{\mathrm{d}x} - z\frac{\mathrm{d}^2 W_o}{\mathrm{d}x^2}\right) \tag{7.15}$$

The mid-surface displacement of laminated composite beams is obtained in a manner identical to that for elastic isotropic beams. The primary difference between isotropic and laminated beam solutions is that the former yields the displacement of the neutral bending axis, while the latter defines the displacement of the mid-surface. These two reference axes may not be the same, depending on the cross-sectional shape of the beam.

The procedures for solving the differential equations describing displacement are presented in detail in many texts, such as Refs. [4,5]. Two options are generally available, depending on the beam loading conditions. One approach is to solve Equation (7.13) directly, while the other involves the solution of the second-order differential equation $M = -bD_{11}\left(\mathrm{d}^2W_o/\mathrm{d}x^2\right)$. In either case, it is generally convenient to use singularity (or Macaulay) functions to define either $M(x)$ with a single equation, or to define $q(x)$. Regardless of the approach, the boundary conditions at supports and/or the free end of the beam are required to define constants of integration. These boundary conditions are the same as those learned in introductory strength of materials courses.

Example 7.1 Consider the cantilever beam in Figure E7.1. Assume the beam has uniform material properties throughout, with $E_1 = 25 \times 10^6$ psi, $E_2 = 1.5 \times 10^6$ psi, $G_{12} = 0.75 \times 10^6$ psi, and $\nu_{12} = 0.28$. In addition, the beam is assumed to have a total thickness of 1.0 in., with each lamina being 0.20 in., and the width is $b = 1.0$in., an applied load of $q(x) = 1.5$ lb/in. and an overall length of $L = 12$ in. Several laminate stacking sequences are to be explored to study the effect of lamina location within the laminate on deflection. These stacking sequences are:

Case 1: $[0]_5$	Case 2: $[90]_5$	Case 3: [0/45/90/45/0]
Case 4: [90/45/0/45/90]	Case 5: [90/0/45/0/90]	Case 6: [0/90/45/90/0]

The uniform load distribution makes direct integration of Equation (7.13) in an obvious method of solution. Defining a moment equation valid for the entire beam is also acceptable and requires solving a second-order differential equation as opposed to the fourth-order equation of Equation (7.13).

Since $q(x) = -q_o$, Equation (7.13) is written as $\mathrm{d}^4W_o/\mathrm{d}x^4 = -q_o/bD_{11}$. After integrating four times

$$W_o(x) = \frac{-q_o}{bD_{11}}x^4 + \frac{1}{6}C_1x^3 + \frac{1}{2}C_2x^2 + C_3x + C_4$$

The constants of integration are evaluated from the boundary conditions at the fixed end $(x=0)$ and the free end $(x=L)$. At the fixed end, the deflection and slope are $W_o(0) = 0 \rightarrow C_4 = 0$ and $\mathrm{d}W_o(0)/\mathrm{d}x = 0 \rightarrow C_3 = 0$. At the free end of the beam, the shear force and bending moment are

$$V(L) = \frac{\mathrm{d}^3W_o(L)}{\mathrm{d}x^3} = 0 = \frac{-q_o}{bD_{11}}L + C_1 \rightarrow C_1 = \frac{q_oL}{bD_{11}}$$

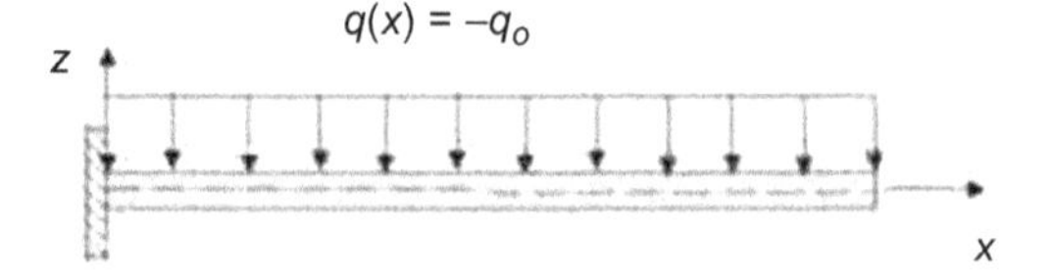

Figure E7.1 Cantilever beam with a uniform load.

$$M(L) = \frac{d^2W_o(L)}{dx^2} = 0 = \frac{-q_o}{bD_{11}}L^2 + C_1L + C_2 \rightarrow C_2 = -\frac{q_oL^2}{2bD_{11}}$$

Substituting the constants of integration into the deflection equation for $W_o(x)$ yields the mid-surface displacements as a function of x for the cantilevered beam under consideration

$$W_o(x) = -\frac{q_o}{24bD_{11}}\left(x^4 - 4x^3L + 6x^2L^2\right)$$

The magnitude of deflection, moment, and stress depends on the material and the stacking sequence. Displacements are primarily controlled by the applied load, D_{11} (as a result of $\bar{Q}_{11}$), and x. For the material specified, $(\bar{Q}_{11})_0 = 25.12 \times 10^6$, $(\bar{Q}_{11})_{45} = 7.617 \times 10^6$, and $(\bar{Q}_{11})_{90} = 1.507 \times 10^6$. The $[0]_5$ laminate is the stiffest and has the least mid-surface deflection, while the $[90]_5$ laminate experiences the greatest deflection. Cases 3 and 6 with the 0° lamina at the top and bottom surfaces deflect less than cases 4 and 5 with the 90° at the top and bottom surfaces. This is not surprising if one considers the formulation of D_{11}, which can be approximated by $\sum(\bar{Q}_{ij})z^3$. A stiffer lamina at larger distances from the mid-surface (cases 3 and 6) creates a larger D_{11} and smaller deflection than a less stiff lamina at the same distance from the mid-surface (cases 4 and 5). The numerical value of D_{11} and free end deflection for each case are presented in the table below.

Case	**1**	**2**	**3**	**4**	**5**	**6**
D_{11}	2.09×10^6	0.126×10^6	1.77×10^6	0.242×10^6	0.539×10^6	1.674×10^6
W_o (in.)	−0.00186	−0.03086	−0.00219	−0.01607	−0.00721	−0.00232

In order to determine stresses throughout the laminate, we need to determine the curvature. The curvature as a function of x can be established from $W_o(x)$ as

$$\kappa_x = \frac{\partial^2 W_o(x)}{\partial x^2} = \frac{q_o}{24bD_{11}}\left(12x^2 - 24xL + 12L^2\right) = \frac{q_o}{2bD_{11}}\left(x^2 - 2xL + L^2\right)$$

Subsequently, the normal stress at any location in the laminate is

$$(\sigma_x)_k = z\left(\bar{Q}_{11}\right)_k(\kappa_x) = \frac{z\left(\bar{Q}_{11}\right)_k q_o}{2bD_{11}}\left(x^2 - 2xL + L^2\right)$$

The stresses are largest at the fixed end. For the dimensions given, these stresses are

$$(\sigma_x)_k = \frac{108z\left(\bar{Q}_{11}\right)_k}{D_{11}}$$

Example 7.2 The second example uses singularity functions as presented in introductory strength of materials courses to determine the deflection of a beam. The beam and its corresponding free-body diagram for defining $M(x)$ are shown in Figure E7.2. Solving a second-order differential equation is sometimes more convenient in solving beam problems. The distributed load must be modeled as a continuous function of spanwise location and is therefore multivalued. For the region of the beam bounded by $0<x<L$, it has a value of $-q_o$, and for the region $L<x<2L$, its value is $+q_o$.

The resulting moment equation is $M(x)=(q_o/2)\langle x\rangle^2-(q_o/2)\langle x-L\rangle^2+F\langle x-3L/2\rangle-R_A\langle x\rangle$, where R_A is the support reaction at end A of the beam and is determined from equations of static equilibrium to be $R_A=(3q_oL+F)/4$. The reaction at B is not required for analysis, but is easily determined to be $R_B=(q_oL+3F)/4$. The governing moment equation is

$$M(x)=-bD_{11}\frac{\mathrm{d}^2W_o(x)}{\mathrm{d}x^2}=\frac{q_o}{2}\langle x\rangle^2-\frac{q_o}{2}\langle x-L\rangle^2+F\left\langle x-\frac{3}{2}L\right\rangle-R_A\langle x\rangle$$

Integrating the bending moment equation twice results in

$$-bD_{11}W_o(x)=\frac{q_o}{24}\langle x\rangle^4-\frac{q_o}{24}\langle x-L\rangle^4+\frac{F}{6}\left\langle x-\frac{3}{2}L\right\rangle^3-\frac{R_A}{6}\langle x\rangle^3+C_1x+C_2$$

where C_1 and C_2 are constants of integration. Using the boundary conditions $W_o(0)=W_o(2L)=0$ and the reaction R_A results in $C_1=(6q_oL+5F)L^2/32$ and $C_2=0$. The beam deflection as a function of x position is now written as

$$W_o(x)=-\frac{1}{bD_{11}}\left\{\frac{q_o}{24}\langle x\rangle^4-\frac{q_o}{24}\langle x-L\rangle^4+\frac{F}{6}\left\langle x-\frac{3}{2}L\right\rangle^3-\frac{3q_oL+F}{24}\langle x\rangle^3+\frac{(6q_oL+5F)L^2}{32}x\right\}$$

The magnitude of the deflection at any location x obviously depends on q_o, F, L, b, and most importantly, D_{11}. As with the previous example, case 1 has the least deflection and case 2 the most deflection, with all other cases having deflections between these two limits.

The analysis of laminated composite beams using the simplified theory is identical to the analysis for isotropic beams, with minor exceptions. The flexural rigidity for isotropic beams (EI) is replaced by bD_{11} for laminated composite beams, and the

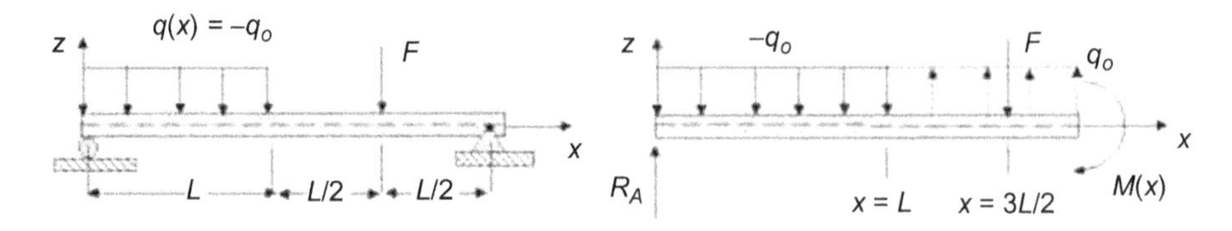

Figure E7.2 Simply supported beam and corresponding free-body diagram for defining M(x).

resulting stresses are functions of individual ply stiffness as well as position relative to the beam's mid-surface. The variation of normal stresses through this beam can be determined from Equation (7.15).

7.4 Advanced beam theory

The simplified beam analysis procedures in the previous section are an oversimplification of beam deflections. The procedures for simplified analysis are good approximations for initial design considerations, but are generally not applicable for a thorough beam deflection and stress analysis. Those procedures are only applicable to constant width beams. Advanced beam theory incorporates most of the beam assumptions of the simplified theory ($h/L \ll 1$, and each ply is assumed to be "perfectly" bonded to adjacent plies so that no interlaminar slipping occurs). The mid-surface x-axis is the reference axis for defining deflections, and the positive sense of applied loads (following Ref. [2]) is shown in Figure 7.5. In this figure, $m_x(x)$ and $p_x(x)$ are distributed moments and normal forces along the beam. These terms include both those forces and moments resulting from directly applied loads and those resulting from τ_{1x} and τ_{2x} as defined by Equation (7.9).

Thermal and hygral effects are included in advanced beam theory. Coupling of torsion with extension and flexure is not permitted; therefore, only symmetric laminates are considered. The Bernoulli–Euler hypothesis (plane sections before deformation remain plane after deformation) is the kinematic assumption approximating deformations.

Both static and dynamic effects are considered. Beam displacements are defined as functions of position and time. The only admissible loads (with the exception of thermal and hygral loads) are in the x–z plane, namely, N_x, M_x, and Q_x. Although no forces or moments are applied directly in the y-direction, the applied loads can create strains in the y-direction. These strains result from Poisson's ratio in the x–y plane (ν_{xy}) in each lamina. Although typically small, they can affect the state of stress in a lamina. The inclusion of out-of-plane deformations and strains results in a more complex set of governing equations, which are similar to the plate equations for classical lamination theory presented in various texts, including Refs. [1–3]. These strains are generally neglected, and displacements in the y-direction are not allowed. Therefore, the only permissible displacement fields are $U(x,z,t) = U_0(x,z,t) + z\Phi(x,t)$ and $W(x,z,t) = W_0(x,t)$. Inclusion of more terms in each assumed displacement field

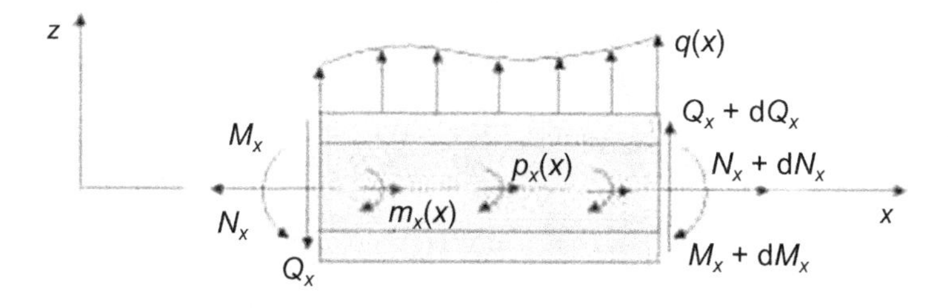

Figure 7.5 Beam loading for the advanced beam theory.

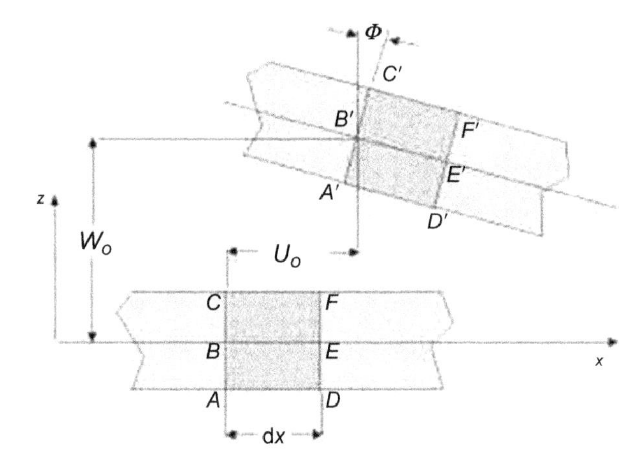

Figure 7.6 Kinematics of beam deflections.

results in a more complicated set of governing equations, however, the procedure for establishing these equations remains the same.

The x, z, and t notations are neglected with the understanding that each displacement field is a function of both position and time. The geometric relations between U_0, W_o, Φ, and z are shown in Figure 7.6. The region defined by $ACFD$ deforms into region $A'C'F'D'$. After deformation, line BE (laminate mid-surface) moves up and to the right to $B'E'$. In addition, a rotation Φ takes place. As in the previous beam theory $\gamma_{yz}=U_{o,z}+W_{o,x}=0$, leading to the relationship $\Phi=-W_{o,x}$. With $\varepsilon_x=U_{o,x}$, the expression for Φ given above results in

$$\varepsilon_x=U_{o,x}+z\Phi_{,x}=U_{o,x}-zW_{o,xx}=\varepsilon_x^o+z\kappa_x$$

From the assumption $\gamma_{yz}=0$, the previously defined relationship between Φ and W_o resulted. In general, this is not a valid assumption, and $\Phi\neq-W_{o,x}$. This means that $A'B'C'$ is not perpendicular to the deformed centerline of the beam, as illustrated in Figure 7.7, where the rotational difference $\Phi-W_{o,x}$ denotes a rotation of elements along the centerline, and is discussed in a subsequent section.

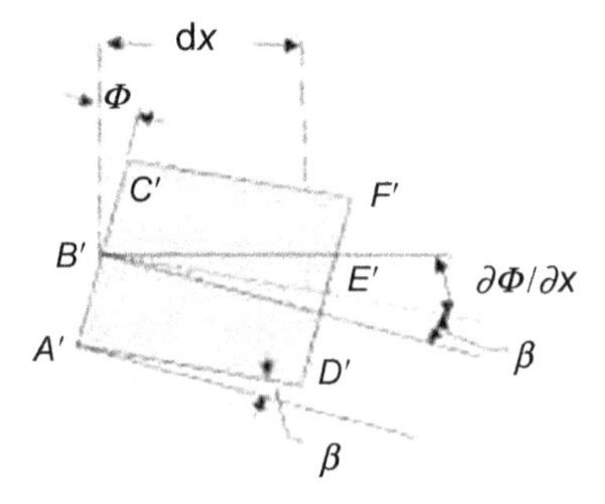

Figure 7.7 Beam deformations for which $\Phi\neq-W_{o,x}$.

The starting point for the load–strain relationship in Chapter 6 was the relationship $\{\sigma\} = [Q]\{\varepsilon\}$. Since beam analysis deals only with loads in the x–z plane (N_x, M_x, and Q_x), a more compact stress–strain relationship can be identified. The only stresses considered are σ_x and τ_{xz}. The treatment of variable thickness beams requires a formulation different from classical lamination theory. The definition of $[A]$, $[B]$, and $[D]$ used in the simple beam theory applies to uniform width beams. The normal stress in the kth layer of the laminate is expressed as

$$(\sigma_x)_k = (E_x)_k\{\varepsilon_x\}_k \tag{7.16}$$

where $(E_x)_k$ is the effective elastic modulus in the x-direction for the kth lamina. It is defined by the elastic constants and fiber orientation of the lamina in a manner consistent with Equation (3.19) as

$$\frac{1}{(E_x)_k} = \frac{\cos^4\theta_k}{(E_1)_k} + \left(\frac{1}{G_{12}} - \frac{2\nu_{12}}{E_1}\right)_k \sin^2\theta_k \cos^2\theta_k + \frac{\sin^4\theta_k}{(E_2)_k} \tag{7.17}$$

The pertinent equations for load–stress relationships for a beam are

$$N_x = \int_{-h/2}^{h/2} \sigma_x \mathrm{d}A \quad M_x = \int_{-h/2}^{h/2} z\sigma_x \mathrm{d}A \quad Q_x = \int_{-h/2}^{h/2} \tau_{xz} \mathrm{d}A$$

The integral from $-h/2$ to $+h/2$ can be replaced by the summation over the N layers of the integral over each ply thickness as done in developing Equation (6.31). The stress in each lamina depends on $(E_x)_k$ and the strain in each lamina, $(\varepsilon_x)_k$. Including thermal and hygral effects results in

$$(\varepsilon_x)_k = \varepsilon_x^o + z\kappa_x - (\alpha_x)_k \Delta T - (\beta_x)_k \bar{M} \tag{7.18}$$

where ΔT and $\bar{M}$ are the temperature difference and average moisture content. The $(\alpha_x)_k$ and $(\beta_x)_k$ terms for each layer are defined by Equations (3.23) and (3.31), respectively.

The differential area in the load–stress relationships above is expressed as $\mathrm{d}A = b_k \mathrm{d}z$. This form of representing $\mathrm{d}A$ accounts for variable width beams such as I-beams. Substituting Equation (7.18) into Equation (7.16) and integrating over the laminate thickness to satisfy equilibrium of forces in the x-direction results in

$$N_x = \int_{-h/2}^{h/2} (E_x b)_k \{\varepsilon_x^o + z\kappa_x - (\alpha_x)_k \Delta T - (\beta_x)_k \bar{M}\} \mathrm{d}z$$

$$= \sum_{k=1}^{N} \int_{h_{k-1}}^{h_k} (E_x b)_k \{\varepsilon_x^o + z\kappa_x - (\alpha_x)_k \Delta T - (\beta_x)_k \bar{M}\} \mathrm{d}z$$

Following procedures identical to those presented in Chapter 6 for classical lamination theory results in the following expression for N_x.

$$N_x = A_x\varepsilon_x^o + B_x\kappa_x - N_x^{\mathrm{T}} - N_x^{\mathrm{H}} \tag{7.19}$$

Application of identical procedures to M_x produces

$$M_x = B_x\varepsilon_x^o + D_x\kappa_x - M_x^{\mathrm{T}} - M_x^{\mathrm{H}} \tag{7.20}$$

The expression for Q_x is represented in a slightly different manner as

$$Q_x = \sum_{k=1}^{N} \int_{h_{k-1}}^{h_k} (\tau_{xz})_k b_k \mathrm{d}z \tag{7.21}$$

The forces and moments have units of lb (N) and in. lb (N m), respectively. This is consistent with units of forces and moments for isotropic beams and is different than those used in CLT.

The A_x, B_x, and D_x terms in Equations (7.19) and (7.20) are expressed in a manner similar to their counterparts $[A]$, $[B]$, and $[D]$ from classical lamination theory, and are

$$\begin{aligned} A_x &= \sum_{k=1}^{N} (bE_x)_k t_k \\ B_x &= \sum_{k=1}^{N} (bE_x)_k t_k \bar{z}_k \\ D_x &= \sum_{k=1}^{N} (bE_x)_k \left(t_k \bar{z}_k^2 + \frac{t_k^3}{12} \right) \end{aligned} \tag{7.22}$$

Similarly, N_x^{T}, M_x^{T}, N_x^{H}, and M_x^{H} are analogous to Equations (6.25), (6.26), (6.29), and (6.30) and are expressed as

$$\begin{aligned} N_x^{\mathrm{T}} &= \sum_{k=1}^{N} (bE_x\alpha_x)_k t_k \Delta T \quad M_x^{\mathrm{T}} = \sum_{k=1}^{N} (bE_x\alpha_x)_k t_k \bar{z}_k \Delta T \\ N_x^{\mathrm{H}} &= \sum_{k=1}^{N} (bE_x\beta_x)_k t_k \bar{M} \quad M_x^{\mathrm{H}} = \sum_{k=1}^{N} (bE_x\beta_x)_k t_k \bar{z}_k \bar{M} \end{aligned} \tag{7.23}$$

The t_k and $\bar{z}_k$ terms are identical to those in Chapter 6 representing the thickness and centroid location relative to the laminate mid-surface for the kth lamina.

The equations of equilibrium including dynamic effects are developed as an extension of the three equations in Equation (7.9). Including dynamic effects is a bridge from classical static, to dynamic analysis for beams, plates, and shells. Application of Newton's second law in the x and z directions result in a set of governing equations.

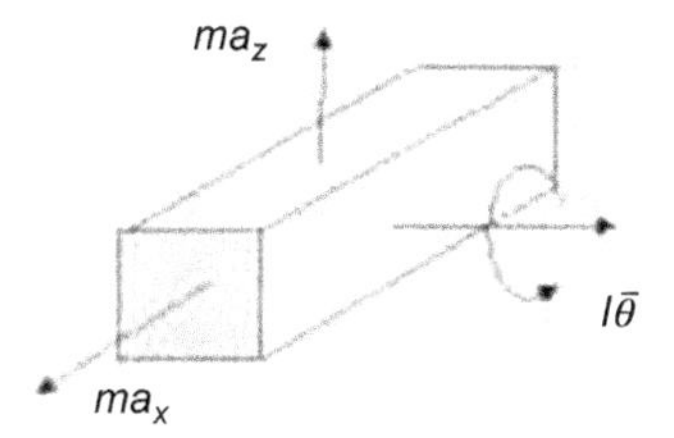

Figure 7.8 Mass acceleration diagram for a laminated composite beam.

Using applied loads and moments from Figure 7.5 as a free-body diagram, and equating the unbalanced forces and moments to the associated mass–acceleration diagram of Figure 7.8, the equations of motion can be developed.

Applying Newton's second law to forces in the x-direction results in

$$\sum F_x = ma_x = m\ddot{U} = m(\ddot{U}_o + z\ddot{\Phi})$$

Recalling that $p_x(x) = \tau_{1x} - \tau_{2x}$ and using the first equation from Equation (7.9) results in

$$N_{x,x} + p_x(x) = \bar{\rho}_1 \ddot{U}_o + \bar{\rho}_2 \ddot{\Phi} \tag{7.24}$$

In the z-direction the equation of motion is

$$\sum F_z = ma_z = m\ddot{W} = m\ddot{W}_o$$

Similarly, recalling that $q(x) = p_1 - p_2$ and using Equation (7.9) results in

$$Q_{x,x} + q(x) = \bar{\rho}_1 \ddot{W}_o \tag{7.25}$$

The corresponding equation of motion for moments is

$$\sum M_x = I\ddot{\Theta}$$

Incorporating the moment expression from Equation (7.9) yields

$$M_{x,x} - Q_x + m_x(x) = \bar{\rho}_2 \ddot{U}_o + \bar{\rho}_3 \ddot{\Phi} \tag{7.26}$$

In these equations, $\bar{\rho}_i$ is the average mass density with units of lb s^2/in.^{4}, to account for the possibility of variable density lamina throughout the laminate thickness. Each $\bar{\rho}_i$ is associated with a specific inertia.

$$\text{Longitudinal inertia}: \ \bar{\rho}_1 = \sum_{k=1}^{N} \int_{h_{k-1}}^{h_k} (b\rho)_k \mathrm{d}z = \sum_{k=1}^{N} (b\rho t)_k \tag{7.27}$$

$$\text{Extension-rotation coupling inertia}: \ \overline{\rho}_2 = \sum_{k=1}^{N} \int_{h_{k-1}}^{h_k} (b\rho)_k z \ \mathrm{d}z = \sum_{k=1}^{N} (b\rho)_k \bar{z}_k t_k \tag{7.28}$$

$$\text{Rotational inertia}: \ \overline{\rho}_3 = \sum_{k=1}^{N} \int_{h_{k-1}}^{h_k} (b\rho)_k z^2 \mathrm{d}z = \sum_{k=1}^{N} (b\rho)_k \bar{z}_k^2 \left(t_k + \frac{t_k^3}{12} \right) \tag{7.29}$$

Solving Equation (7.26) for Q_x, substituting into Equation (7.25) and using the expression for M_x from Equation (7.20) results in

$$B_x \varepsilon^o_{x,xx} + D_x \kappa_{x,xx} + m_x(x)_{,x} + q(x) - M^{\mathrm{T}}_{x,xx} - M^{\mathrm{H}}_{x,xx} = \overline{\rho}_1 \ddot{W}_o + \overline{\rho}_2 \ddot{U}_{o,x} + \overline{\rho}_3 \ddot{\Phi}_{,x}$$

Similarly, substituting Equation (7.19) into Equation (7.24) produces

$$A_x \varepsilon^o_{x,x} + B_x \kappa_{x,x} + p_x(x) - N^{\mathrm{T}}_{x,x} - N^{\mathrm{H}}_{x,x} = \overline{\rho}_1 \ddot{U}_o + \overline{\rho}_2 \ddot{\Phi}$$

Substitution of the definitions for ε^o_x, κ_x, and Φ in terms of displacements $\left(\varepsilon^o_x = U_{o,x},\ \kappa_x = -W_{o,xx},\ \Phi = -W_{o,x}\right)$ into these equations results in the governing equations of motion

$$A_x U_{0,xx} + B_x W_{o,xxx} + p_x(x) - N^{\mathrm{T}}_{x,x} - N^{\mathrm{H}}_{x,x} = \overline{\rho}_1 \ddot{U}_o + \overline{\rho}_2 \ddot{\Phi} \tag{7.30}$$

$$B_x U_{o,xxx} - D_x W_{o,xxxx} + m_x(x)_{,x} + q(x) - M^{\mathrm{T}}_{x,xx} - M^{\mathrm{H}}_{x,xx} = \overline{\rho}_1 \ddot{W}_o + \overline{\rho}_2 \ddot{U}_{o,x} + \overline{\rho}_3 \ddot{W}_{o,xx} \tag{7.31}$$

Equations (7.30) and (7.31) pertain to the generalized conditions in which dynamic effects are considered. Although dynamic effects are an important consideration in many instances, their use is considered in a subsequent section. For now, only static analysis is considered, and $\ddot{U}_o = \ddot{U}_{o,x} = \ddot{W}_o = \ddot{W}_{o,x} = \ddot{W}_{o,xx} = 0$. Thus, Equation (7.30) can be solved for $U_{o,xx}$. Taking the partial derivative of the resulting equation with respect to x and substituting it into Equation (7.31) yields

$$\left(\frac{B_x^2 - A_x D_x}{A_x} \right) W_{0,xxxx} + \frac{B_x}{A_x} \left[N^{\mathrm{T}}_{x,xx} + N^{\mathrm{H}}_{x,xx} - p_{x,x}(x) \right] + \left[q(x) + m_x(x)_{,x} - M^{\mathrm{T}}_{x,xx} - M^{\mathrm{H}}_{x,xx} \right] = 0$$

This expression and the corresponding one for $U_{o,xx}$ are presented in a simplified manner by introducing the definitions

$$\bar{A}_x \equiv \left(\frac{A_x D_x - B_x^2}{D_x} \right) \quad \bar{B}_x \equiv \left(\frac{A_x D_x - B_x^2}{B_x} \right) \quad \bar{D}_x \equiv \left(\frac{A_x D_x - B_x^2}{A_x} \right) \tag{7.32}$$

Using Equation (7.32), the partial differential equations relating mid-surface displacements to loads are written as

$$U_{o,xxx}=\frac{1}{\bar{B}_x}\left[q(x)+m_x(x)_{,x}-M^{\mathrm{T}}_{x,xx}-M^{\mathrm{H}}_{x,xx}\right]-\frac{1}{\bar{A}_x}\left[p_x(x)_{,x}-N^{\mathrm{T}}_{x,xx}-N^{\mathrm{H}}_{x,xx}\right] \tag{7.33}$$

$$W_{o,xxxx}=\frac{1}{\bar{D}_x}\left[q(x)+m_x(x)_{,x}-M^{\mathrm{T}}_{x,xx}-M^{\mathrm{H}}_{x,xx}\right]-\frac{1}{\bar{B}_x}\left[p_x(x)_{,x}-N^{\mathrm{T}}_{x,xx}-N^{\mathrm{H}}_{x,xx}\right] \tag{7.34}$$

These equations for displacement and deflection of the mid-surface can be solved, provided sufficient boundary conditions are established. A total of three boundary conditions for Equation (7.33) and four for Equation (7.34) are required.

The variation of stress through the beam can be determined once the displacements are established. The normal stress distribution is determined from Equation (7.16) with the appropriate use of Equations (7.17) and (7.18) in which ε^o_x and κ_x are defined in terms of U_o and W_o. The resulting expression for normal stress in the kth lamina is

$$(\sigma_x)_k=(E_x)_k\left[U_{0,x}-zW_{o,xx}-(\alpha_x)_k\Delta T-(\beta_x)_k\bar{M}\right] \tag{7.35}$$

Defining the transverse shear stress $(\tau_{xz})_k$ is not as straight forward. The Bernoulli–Euler hypothesis regarding negligible transverse shear stress corresponds to the fact that τ_{xz} cannot be determined from a simple stress–strain relationship. This component of stress is determined from techniques developed in strength of materials. The transverse shear stress is related to bending moment by

$$\tau=\left(\frac{\mathrm{d}M}{\mathrm{d}x}\right)\left(\frac{1}{Ib}\right)\int z\,\mathrm{d}A$$

where I is the second area moment of inertia and b is the beam width at a particular location. The normal stress (σ_x) is related to the bending moment, inertia, and z position by $\sigma_x=Mz/I$. Next, we replace $\mathrm{d}(Mz/I)/\mathrm{d}x$ with $\mathrm{d}\sigma_x/\mathrm{d}x$. The definition of I for a laminated composite beam has not been explicitly forwarded. It is incorporated in the load–strain relationship (and subsequently stress) through $[A]$, $[B]$, and $[D]$ for classical lamination theory, and through A_x, B_x, and D_x for beam analysis. The variation of transverse shear stress through the beam is expressed in terms of the width of the kth lamina as

$$(\tau_{xz})_k=\frac{1}{b_k}\int_{h_o}^{z}(\sigma_{x,x})\mathrm{d}A \tag{7.36}$$

where b_k is the beam width at the z location for which τ_{xz} is computed. The lower and upper limits of integration are the outside laminate surface (h_o) and the z plane at which τ_{xz} is required, as shown in Figure 7.9. The differential area is defined

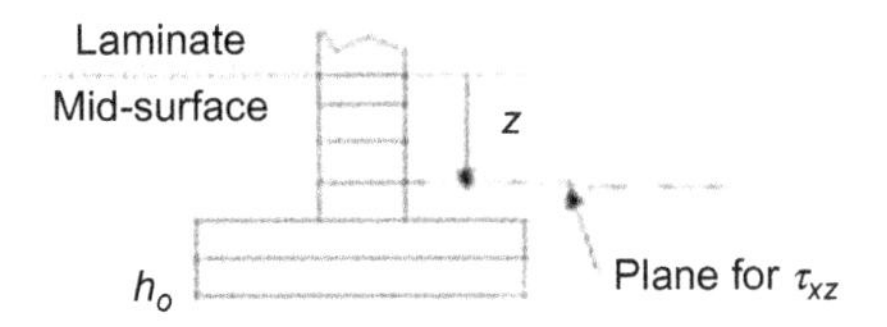

Figure 7.9 Definition of integration limits for Equation (7.36).

as $dA = b_k dz$, but is left as dA in Equation (7.36) until further evaluation of τ_{xz} is required.

The mid-surface elongation and deflection of laminated composite beams as defined by Equations (7.33) and (7.34) can be solved by integration and the use of appropriate boundary conditions to define constants of integration. Equations (7.33) and (7.34) are written in terms of $\bar{A}_x$, $\bar{B}_x$, and $\bar{D}_x$, each of which is defined by combinations of A_x, B_x, and D_x. An initial assumption for the advanced beam theory was that only symmetric laminates be considered. This assumption results in $B_x = 0$ and is analogous to $[B] = 0$ in classical lamination theory. As discussed in Chapter 6, this also leads to $M_x^T = M_x^H = 0$. The M_x^T and M_x^H terms of Equations (7.33) and (7.34) are included to account for the possibility of thermal and hygral gradients through and along the beam after curing. Hygral effects are not considered in any of the following examples. Their inclusion would be handled in a manner identical to that for the thermal case, with the exception that $\alpha_x \Delta T$ for the thermal case be replaced by $\beta_x \bar{M}$ for the hygral case.

Example 7.3 The beam in Figure E7.3-1 is assumed to be a symmetric laminate. Thermal and hygral effects are neglected. The direction of the uniformly distributed load coupled with neglecting thermal and hygral effects results in the simplified form of Equations (7.33) and (7.34) shown below:

$$U_{o,xxx} = \frac{-q_o}{\bar{B}_x} \quad W_{o,xxxx} = \frac{-q_o}{\bar{D}_x}$$

Integrating $U_{o,xxx}$ three times and $W_{o,xxxx}$ four times results in

$$W_o(x) = \frac{-q_o}{24\bar{D}_x}x^4 + \frac{C_1}{6}x^3 + \frac{C_2}{2}x^2 + C_3 x + C_4$$

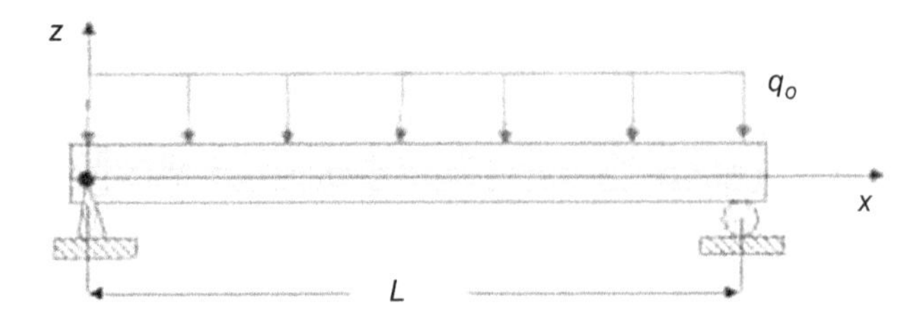

Figure E7.3-1 Simply supported beam with a uniformly distributed load.

$$U_o(x) = \frac{-q_o}{6\bar{B}_x}x^3 + \frac{C_5}{2}x^2 + C_6x + C_7$$

The constants of integration are determined from the following boundary conditions:

$$W_o(0) = W_o(L) = 0 \quad M_x(0) = M_x(L) = 0 \quad N_x(0) = N_x(L) = 0 \quad U_o(0) = 0$$

Due to the roller at $x = L$, $U_o(L) \neq 0$. The boundary conditions for $M_x(x)$ and $N_x(x)$ require the use of two additional relationships, which are obtained from Equations (7.19) and (7.20) without the thermal and hygral effects.

$$N_x = A_x\varepsilon_x^o + B_x\kappa_x = A_xU_{o,x} - B_xW_{o,xx}$$

$$M_x = B_x\varepsilon_x^o + D_x\kappa_x = B_xU_{o,x} - D_xW_{o,xx}$$

The proper use of these equations requires appropriate differentiation of the expressions for $W_o(x)$ and $U_o(x)$ given above in terms of integration constants. From the two boundary conditions $W_o(0) = 0$ and $U_o(0) = 0$, the constants C_4 and C_7 are both zero. Using $N_x(0) = 0$ and $M_x(0) = 0$ in accordance with the two equations above results in

$$N_x(0) = 0 = A_xC_6 - B_xC_2 \quad M_x(0) = 0 = B_xC_6 - D_xC_2$$

Simultaneous solution of these two equations results in $C_2 = C_6 = 0$. The boundary conditions of $N_x(L) = M_x(L) = 0$ result in two additional equations

$$N_x(L) = 0 = A_x\left(\frac{-q_oL^2}{2\bar{B}_x} + C_5L\right) - B_x\left(\frac{-q_oL^2}{2\bar{D}_x} + C_1L\right)$$

$$M_x(L) = 0 = B_x\left(\frac{-q_oL^2}{2\bar{B}_x} + C_5L\right) - D_x\left(\frac{-q_oL^2}{2\bar{D}_x} + C_1L\right)$$

Solving these equations yields $C_1 = q_oL/2\bar{D}_x$ and $C_5 = q_oL/2\bar{B}_x$. Using the integration constants already determined and the final boundary condition of $W_o(L) = 0$ yields $C_3 = -q_oL^3/24\bar{D}_x$. The general equations defining extension and deflection of the mid-surface of this beam can now be defined explicitly as

$$U_o(x) = \frac{q_oL^3}{24\bar{B}_x}\left[-4\left(\frac{x}{L}\right)^3 + 6\left(\frac{x}{L}\right)^2\right]$$

$$W_o(x) = \frac{q_oL^4}{24\bar{D}_x}\left[-\left(\frac{x}{L}\right)^4 + 2\left(\frac{x}{L}\right)^3 - \left(\frac{x}{L}\right)\right]$$

The normal stress at any point within the laminate can be defined from Equations (7.16) and (7.18) in which $\varepsilon_x^0 = U_{o,x}$ and $\kappa_x = -W_{o,xx}$, and is

$$(\sigma_x)_k = \left(\frac{q_o (E_x)_k L^2}{2}\right)\left[\left(\frac{x}{L}\right) - \left(\frac{x}{L}\right)^2\right]\left(\frac{1}{\bar{B}_x} - \frac{z}{\bar{D}_x}\right)$$

The transverse shear stress in the laminate is defined from Equation (7.36) as

$$(\tau_{xz})_k = \frac{q_o L}{2b_k}\left(1 - 2\frac{x}{L}\right)\int_{h_o}^{z} (E_x)\left(\frac{1}{\bar{B}_x} - \frac{z}{\bar{D}_x}\right) \mathrm{d}A$$

The integral between h_o and z can be approximated as a summation of integrals over discrete regions (or individual lamina), similar to the integral relationships developed in Chapter 6. The differential area is $\mathrm{d}A = b_k \mathrm{d}z$, and E_x is replaced by $(E_x)_k$. The transverse shear stress in the kth lamina is therefore expressed as

$$(\tau_{xz})_k = \frac{q_o L}{2b_k}\left(1 - 2\frac{x}{L}\right)\sum_{k=1}^{n}\int_{h_{k-1}}^{h_k} (E_x b)_k \left(\frac{1}{\bar{B}_x} - \frac{z}{\bar{D}_x}\right) \mathrm{d}z$$

Where the limits of the summation are from the first lamina (the outside surface) to the plane defined by z. It should be noted that on the outside surface of the laminate ($z = +h/2$), $\tau_{xz} = 0$ in the absence of applied shear stresses at these surfaces. Evaluation of the integral can be carried out between the limits indicated, with the end result being expressions similar to those used to define A_x and B_x in Equation (7.22). Therefore, the shear stress in the kth lamina is

$$(\tau_{xz})_k = \frac{q_o L}{2b_k}\left(1 - 2\frac{x}{L}\right)\sum_{k=1}^{n} (E_x b)_k \left(\frac{t_k}{\bar{B}_x} - \frac{t_k \bar{z}_k}{\bar{D}_x}\right)$$

Explicit definitions of U_o, W_o, σ_x, and τ_{xz} for this beam depend upon material, fiber orientations within each lamina, and the beam's cross-sectional geometry. For the purpose of illustration, an E-Glass/Epoxy ($E_1 = 7.57 \times 10^6$ psi, $E_2 = 2.03 \times 10^6$ psi, $G_{12} = 1.25 \times 10^6$ psi, and $\nu_{12} = 0.21$) material is considered. The laminate is assumed to have a stacking arrangement of $[0/\pm 45/90]_s$. Two cross sections are considered: one with a constant width through the beam thickness, and the other with a variable width as shown in Figure E7.3-2. Each lamina has the same thickness ($t_k = 0.10$ in.).

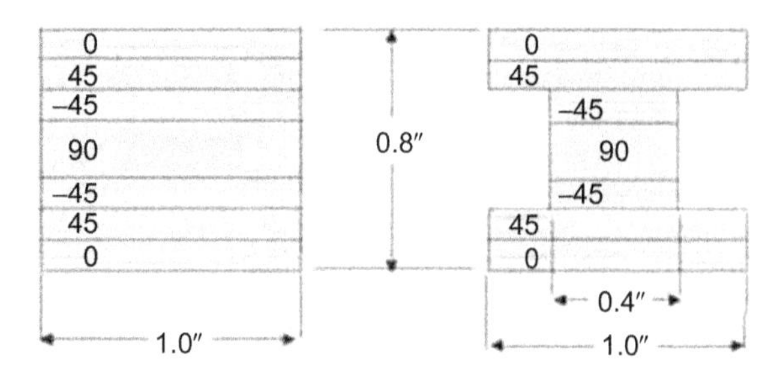

Figure E7.3-2 Simply supported composite beam cross sections.

Using Equation (7.17), $(E_x)_k$ for each lamina is $(E_x)_0 = 7.57\times10^6$ psi, $(E_x)_{90} = 2.03\times10^6$ psi, and $(E_x)_{45} = 2.92\times10^6$ psi. Due to material and geometric symmetry of each cross-section, $B_x = 0$ for both beam sections. The A_x and D_x terms must be evaluated for each beam. For the constant width beam, they are

$$A_x = 2\left[(E_x)_0 + 2(E_x)_{45} + (E_x)_{90}\right](1.0)(0.1) = 3.09\times10^6\,\text{psi}$$

$$D_x = 2\left[(E_x)_0\left(0.1(0.35)^2 + \frac{(0.1)^3}{12}\right) + (E_x)_{45}\left(0.1(0.25)^2 + \frac{(0.1)^3}{12}\right)\right.$$
$$\left. + (E_x)_{-45}\left(0.1(0.15)^2 + \frac{(0.1)^3}{12}\right) + (E_x)_{90}\left(0.1(0.05)^2 + \frac{(0.1)^3}{12}\right)\right]$$
$$(1.0) = 0.23862\times10^6\,\text{psi}$$

Based on these values of A_x, B_x, and D_x, the constant width beam has $A_x = \bar{A}_x = 3.09\times10^6$ psi, $\bar{B}_x = \infty$, and $D_x = \bar{D}_x = 0.23862\times10^6$ psi. For the variable width beam, the following parameters are determined

$$\bar{A}_x = A_x = 2\left[(E_x)_0(1.0) + (E_x)_{45}(1.0) + (E_x)_{-45}(0.4) + (E_x)_{90}(0.4)\right](0.1)$$
$$= 2.49\times10^6\,\text{psi}$$

$$\bar{B}_x = \infty$$

$$\bar{D}_x = D_x = 2\left[(E_x)_0\left(0.1(0.35)^2 + \frac{(0.1)^3}{12}\right)(1.0) + (E_x)_{45}\left(0.1(0.25)^2 + \frac{(0.1)^3}{12}\right)(1.0)\right.$$
$$\left. + (E_x)_{-45}\left(0.1(0.15)^2 + \frac{(0.1)^3}{12}\right)(0.4) + (E_x)_{90}\left(0.1(0.05)^2 + \frac{(0.1)^3}{12}\right)(0.4)\right]$$
$$= 0.2296\times10^6\,\text{psi}$$

Since $\bar{B}_x = \infty$ for both cases, $U_o(x)$ and the normal stress and shear stress equations are easier to manipulate than if $\bar{B}_x$ were finite. The mid-surface deflection for each cross section is virtually identical (compare the $\bar{D}_x$ for each case). Similarly, the variation of normal stress through each beam's thickness is virtually identical.

The variation of transverse shear stress through each laminated composite beam is a function of both spanwise location and position within the beam, referenced from the mid-surface. Since each laminate is symmetric with $B_x = 0$, the transverse shear stress equation at $x = 0$ reduces to

$$(\tau_{xz})_k = \frac{q_o L}{2b_k}\sum_{k=1}^{n}\frac{(E_x b t \bar{z})_k}{\bar{D}_x}$$

For each lamina, E_x, t, and $\bar{z}$ are identical for both cross sections. The major difference between the two is the beam width (b) for the −45° and 90° laminas. This causes a

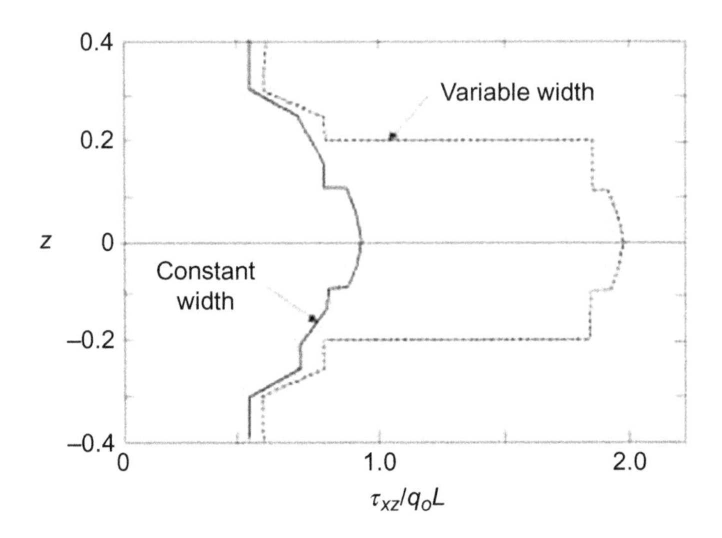

Figure E7.3-3 Variation of τ_{xz}/q_oL through each cross section at $x=0$.

large jump in the shear stress in these laminas. The approximate variation of shear stress, τ_{xz}/q_oL, through beam is illustrated in Figure E7.3-3. The effect of ply modulus (E_x) influences both beam sections at $z=\pm0.3$ more than at other interfaces. This is a result of the larger difference in modulus between the $0°\left((E_x)_0=7.57\times10^6\text{psi}\right)$ and $+45°\left((E_x)_{45}=2.92\times10^6\text{psi}\right)$ laminas than between the $-45°\left((E_x)_{-45}=2.92\times10^6\text{psi}\right)$ and $90°\left((E_x)_{90}=2.03\times10^6\text{psi}\right)$ laminas. Aside from the modulus-induced effect, the stress distribution through the uniform width beam is the same as the parabolic shear stress distribution in uniform width isotropic beams. The effect of the reduced beam width between $-0.2\leq z\leq0.2$ produces a large jump in shear stress at the interface of the $+45°$ and $-45°$ laminas. This distribution in τ_{xz} is also similar in shape to that which would exist in an isotropic beam of identical geometry.

Example 7.4 Consider the simply supported beam in Figure E7.4-1. The only applied loads are assumed to be end loads N_1, N_2 and a distributed axial load $p_x(x)$. Thermal and hygral effects are neglected. From the applied loads, Equations (7.33) and (7.34) reduce to

$$U_{o,xxx}=-\frac{p_x(x)_{,x}}{\bar{A}_x}\quad W_{o,xxxx}=-\frac{p_x(x)_{,x}}{\bar{B}_x}$$

Figure E7.4-1 Axially loaded simply supported beam.

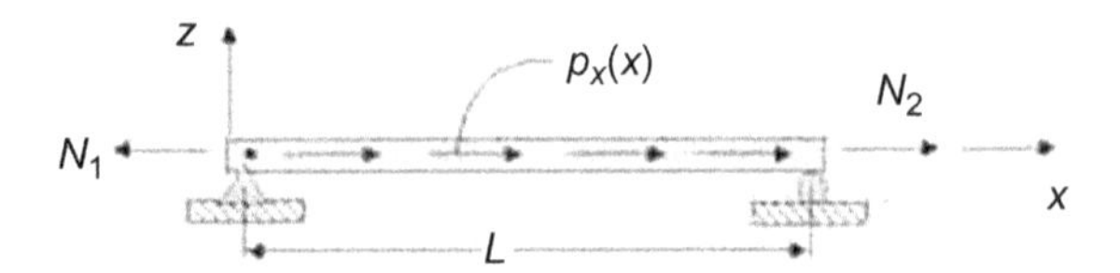

These equations are easily solved by integration. In order to provide for a cleaner appearing solution, the following definition is introduced

$$P(x)=\iiint p_x(x)\mathrm{d}x\mathrm{d}x\mathrm{d}x$$

Using this definition, integration of each equation results in

$$U_o(x)=-\frac{P(x)_{,x}}{\bar{A}_x}+\frac{C_1}{2}x^2+C_2x+C_3$$

$$W_o(x)=-\frac{P(x)_{,x}}{\bar{B}_x}+\frac{C_4}{6}x^3+\frac{C_5}{2}x^2+C_6x+C_7$$

The boundary conditions needed for defining the constants of integration are

$$W_o(0)=W_o(L)=M_x(0)=M_x(L)=U_o(0)=0,N_x(0)=N_1,N_x(L)=N_2,\text{ and } U_o(L)\neq 0$$

Definitions of $N_x(x)$ and $M_x(x)$ in terms of U_o and W_o are required. The existence of $P(x)$ and $P(x)_{,x}$ in the governing equations do not allow for a simple substitution of $U_o(0)=0$ and $W_o(0)=0$, and a direct evaluation of C_3 and C_7. Solving for integration constants can sometimes be simplified (depending upon the form of $P(x)$) by using the relationship

$$N_1=N_2+\int_0^L p_x(x)\mathrm{d}x=N_2+P(x)_{,xx}$$

The boundary conditions of $U_o(0)=0$ and $W_o(0)=0$ result in $C_3=P(0)_{,x}/\bar{A}_x$ and $C_7=P(0)_{,x}/\bar{B}_x$.

The boundary condition $W_o(L)=0$ yields

$$C_4\frac{L^3}{6}+C_5\frac{L^2}{2}+C_6+C_7=\frac{P(L)}{\bar{B}_x}$$

The moment equation $M_x=B_xU_{o,x}-D_xW_{o,xx}$ is expressed as

$$M_x=B_x\left(C_2+C_1x-\frac{P(x)_{,xx}}{\bar{A}_x}\right)-D_x\left(C_5+C_4x-\frac{P(x)_{,xx}}{\bar{B}_x}\right)$$

Using Equation (7.32), it can easily be shown that $D_x/\bar{B}_x-B_x/\bar{A}_x=0$, which results in

$$M_x=B_x(C_2+C_1x)-D_x(C_5+C_4x)$$

Substituting the appropriate boundary conditions results in

$$M_x(0)=0=B_xC_2-D_xC_5 \quad M_x(L)=0=B_x(C_2+C_1L)-D_x(C_5+C_4L)$$

The equation for normal force $N_x=A_xU_{o,x}-B_xW_{o,xx}$ can be written as

$$N_x=A_x\left(C_2+C_1x-\frac{P(x)_{,xx}}{\bar{A}_x}\right)-B_x\left(C_5+C_4x-\frac{P(x)_{,xx}}{\bar{B}_x}\right)$$

From Equation (7.32), it can be shown that $B_x/\bar{B}_x-A_x/\bar{A}_x=-1$. Using this and the boundary conditions into the relationship for N_x above results in

$$N_x(0)=N_1=A_xC_2-B_xC_5-P(0)_{,xx}$$

$$N_x(L)=N_2=A_x(C_2+C_1L)-B_x(C_5+C_4L)-P(L)_{,xx}$$

The final boundary condition of $W_o(L)=0$ gives

$$-\frac{P(L)}{\bar{B}_x}+C_4\frac{L^3}{6}+C_5\frac{L^2}{2}+C_6L+C_7=0$$

Solving these seven equations for the integration constants provides

$$\begin{aligned}
&C_1=\frac{1}{L\bar{A}_x}\left[N_2-N_1+P(L)_{,xx}-P(0)_{,xx}\right] \quad C_2=\frac{1}{\bar{A}_x}\left[P(0)_{,xx}+N_1\right] \quad C_3=\frac{P(0)_{,x}}{\bar{A}_x}\\
&C_4=\frac{1}{L\bar{B}_x}\left[N_2-N_1+P(L)_{,xx}-P(0)_{,xx}\right] \quad C_5=\frac{1}{\bar{B}_x}\left[P(0)_{,xx}+N_1\right]\\
&C_6=\frac{1}{L\bar{B}_x}\left\{[P(L)-P(0)]-\frac{L^2}{6}\left[2N_1+N_2+P(L)_{,xx}+2P(0)_{,xx}\right]\right\} \quad C_7=\frac{P(0)}{\bar{B}_x}
\end{aligned}$$

The explicit form of $U_o(x)$ and $W_o(x)$ can now be written as

$$U_o(x)=\frac{1}{\bar{A}_x}\left\{-P(x)_{,x}+\frac{x^2}{2L}\left[N_2-N_1+P(L)_{,xx}-P(0)_{,xx}\right]+x\left[P(0)_{,xx}+N_1\right]+P(0)_{,x}\right\}$$

$$\begin{aligned}
W_o(x)=\frac{1}{\bar{B}_x}\Bigg\{&-P(x)_{,x}+\frac{x^3}{6L}\left[N_2-N_1+P(L)_{,xx}-P(0)_{,xx}\right]+\frac{x^2}{2}\left[P(0)_{,xx}+N_1\right]\\
&+\frac{x}{L}[P(L)-P(0)]-\frac{L^2}{6}\left[2N_1+N_2+P(L)_{,xx}+2P(0)_{,xx}\right]+P(0)\Bigg\}
\end{aligned}$$

The normal stress and transverse shear stress are evaluated from Equations (7.35) and (7.36). The transverse shear stress for an axially loaded beam is generally zero. An exact solution to this problem requires an initial definition of N_1, N_2, and $p_x(x)$. For the purpose of illustration, it is assumed that $N_2=0$, $N_1\neq 0$, and

$p_x(x) = p_o = \text{constant}$. Since $p_x(x)$ is constant, $N_1 = \int_0^L p_o dx = p_o L$. Similarly, $P(x)_{,x} = P(x)_{,xx} = 0$. Therefore, the equations defining $U_o(x)$ and $W_o(x)$ reduce to

$$U_o(x) = \frac{p_o L}{\bar{A}_x}\left[x - \frac{x^2}{2L}\right] = \frac{p_o L^2}{\bar{A}_x}\left[\frac{x}{L} - \frac{1}{2}\left(\frac{x}{L}\right)^2\right]$$

$$W_o(x) = \frac{p_o}{\bar{B}_x}\left[-\frac{x^3}{6} + \frac{x^2}{2L} - \frac{xL^2}{3}\right] = \frac{p_o L^3}{6\bar{B}_x}\left[-\left(\frac{x}{L}\right)^3 + 3\left(\frac{x}{L}\right)^2 - 2\left(\frac{x}{L}\right)\right]$$

The normal stress at any point within the beam is expressed as

$$(\sigma_x)_k = (E_x)_k(U_{o,x} - zW_{o,xx}) = p_o(E_x)_k(L-x)\left(\frac{1}{\bar{A}_x} - \frac{z}{\bar{B}_x}\right)$$
$$= p_o L(E_x)_k\left(1 - \frac{x}{L}\right)\left(\frac{1}{\bar{A}_x} - \frac{z}{\bar{B}_x}\right)$$

For both of the beam sections (constant and variable thickness) considered in Example 7.3, $W_o(x) = 0$, since $\bar{B}_x = \infty$. The normal stress can be represented as $(\sigma_x)_k = p_o L(E_x)_k/\bar{A}_x(1 - x/L)$ and is largest at $x = 0$, where it is represented simply as $(\sigma_x)_k = p_o L(E_x)_k/\bar{A}_x$. A representation of the axial displacement ($U_o(x)/p_o L^2$) is shown in Figure E7.4-2 for both beam cross sections considered in Example 7.3. The distribution of normal stress through each laminate is not shown.

Example 7.5 This example considers only thermal and hygral effects. The simply supported beam in Figure E7.5-1 is analyzed in terms of thermal and hygral effects. The temperatures on the top and bottom surfaces of the beam are assumed to be

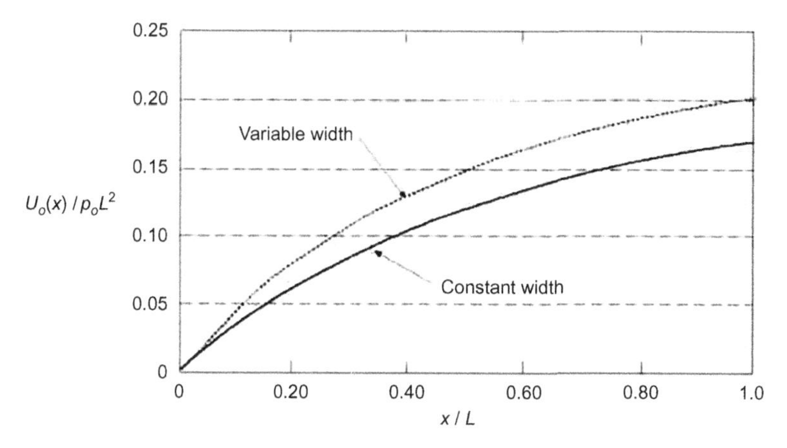

Figure E7.4-2 $U_o(x)/p_o L^2$ for simply supported laminated beam.

Figure E7.5-1 Simply supported beam with thermal and hygral effects considered.

different. It is also assumed that no mechanical loads are applied to the beam. For this case, Equations (7.33) and (7.34) reduce to

$$U_{o,xxx} = \frac{1}{\bar{B}_x}\left[m_x(x)_{,x} - M^{\mathrm{T}}_{x,xx} - M^{\mathrm{H}}_{x,xx}\right] + \frac{1}{\bar{A}_x}\left[N^{\mathrm{T}}_{x,xx} + N^{\mathrm{H}}_{x,xx}\right]$$

$$W_{o,xxxx} = \frac{1}{\bar{D}_x}\left[m_x(x)_{,x} - M^{\mathrm{T}}_{x,xx} - M^{\mathrm{H}}_{x,xx}\right] + \frac{1}{\bar{B}_x}\left[N^{\mathrm{T}}_{x,xx} + N^{\mathrm{H}}_{x,xx}\right]$$

Applied loads are assumed to be zero, but the $m_x(x)$ term is retained to account for hygro-thermal moments which may result from various temperature or moisture profiles present in the beam. Different temperatures on the top and bottom of the beam can induce the $m_x(x)$ term. As in the previous example, the analysis can be streamlined by introducing the following definitions:

$$N_T(x) = \int\int N_x^T \mathrm{d}x\mathrm{d}x \quad N_{\mathrm{H}}(x) = \int\int N_x^H \mathrm{d}x\mathrm{d}x \quad M_{\mathrm{T}}(x) = \int\int M_x^{\mathrm{T}} \mathrm{d}x\mathrm{d}x \quad M_{\mathrm{H}}(x) = \int\int M_x^{\mathrm{H}} \mathrm{d}x\mathrm{d}x$$

$$m'(x) = \int\int\int m_x(x)\mathrm{d}x\mathrm{d}x\mathrm{d}x$$

Incorporating these definitions into the integration of the governing equations yields

$$U_o(x) = \frac{1}{\bar{A}_x}\left[N_{\mathrm{T}}(x)_{,x} + N_{\mathrm{H}}(x)_{,x}\right] - \frac{1}{\bar{B}_x}\left[M_{\mathrm{T}}(x)_{,x} + M_{\mathrm{H}}(x)_{,x} - m'(x)_{,x}\right] + \frac{C_1}{2}x^2 + C_2x + C_3$$

$$W_o(x) = \frac{1}{\bar{B}_x}\left[N_{\mathrm{T}}(x) + N_{\mathrm{H}}(x)\right] - \frac{1}{\bar{D}_x}\left[M_{\mathrm{T}}(x) + M_{\mathrm{H}}(x) - m'(x)\right] + \frac{C_4}{6}x^3 + \frac{C_5}{2}x^2 + C_6x + C_7$$

The boundary conditions for displacement in this problem are $U_o(0) = W_o(0) = W_o(L) = 0$, which result in

$$U_o(0) = 0 \Rightarrow C_3 = \frac{1}{\bar{B}_x}\left[M_{\mathrm{T}}(0)_{,x} + M_{\mathrm{H}}(0)_{,x} - m'(0)_{,x}\right] - \frac{1}{\bar{A}_x}\left[N_{\mathrm{T}}(0)_{,x} + N_{\mathrm{H}}(0)_{,x}\right] \quad \text{(a)}$$

$$W_o(0)=0\Rightarrow C_7=\frac{1}{\bar{D}_x}[M_{\mathrm{T}}(0)+M_{\mathrm{H}}(0)-m'(0)]-\frac{1}{\bar{B}_x}[N_{\mathrm{T}}(0)+N_{\mathrm{H}}(0)] \tag{b}$$

$$W_o(L)=0\Rightarrow\frac{1}{\bar{B}_x}[N_{\mathrm{T}}(L)+N_{\mathrm{H}}(L)]-\frac{1}{\bar{D}_x}[M_{\mathrm{T}}(L)+M_{\mathrm{H}}(L)-m'(L)]+\frac{C_4}{6}L^3$$

$$+\frac{C_5}{2}L^2+C_6L+C_7=0$$

Using the established relationship for C_7, this becomes

$$\frac{C_4}{6}L^3+\frac{C_5}{2}L^2+C_6L=\frac{1}{\bar{D}_x}[M_{\mathrm{T}}(L)-M_{\mathrm{T}}(0)+M_{\mathrm{H}}(L)-M_{\mathrm{H}}(0)-m'(L)+m'(0)]$$
$$+\frac{1}{\bar{B}_x}[N_{\mathrm{T}}(0)-N_{\mathrm{T}}(L)+N_{\mathrm{H}}(0)-N_{\mathrm{H}}(L)] \tag{c}$$

The boundary conditions for N_x and M_x are similar to those in previous examples, except for the inclusion of thermal and hygral effects. They are expressed as

$$N_x(x)=A_xU_{o,x}-B_xW_{o,xx}-N_x^{\mathrm{T}}(x)-N_x^{\mathrm{H}}(x)$$
$$=A_x\left\{\frac{1}{\bar{A}_x}\left[N_{\mathrm{T}}(x)_{,xx}+N_{\mathrm{H}}(x)_{,xx}\right]-\frac{1}{\bar{B}_x}\left[M_{\mathrm{T}}(x)_{,xx}+M_{\mathrm{H}}(x)_{,xx}-m'(x)_{,xx}\right]+C_1x+C_2\right\}$$
$$-B_x\left\{\frac{1}{\bar{B}_x}\left[N_{\mathrm{T}}(x)_{,xx}+N_{\mathrm{H}}(x)_{,xx}\right]-\frac{1}{\bar{D}_x}\left[M_{\mathrm{T}}(x)_{,xx}+M_{\mathrm{H}}(x)_{,xx}-m'(x)_{,xx}\right]+C_4x+C_5\right\}$$
$$-N_x^{\mathrm{T}}(x)-N_x^{\mathrm{H}}(x)$$
$$=\left(\frac{A_x}{\bar{A}_x}-\frac{B_x}{\bar{B}_x}\right)\left[N_{\mathrm{T}}(x)_{,xx}+N_{\mathrm{H}}(x)_{,xx}\right]+\left(\frac{B_x}{\bar{D}_x}-\frac{A_x}{\bar{B}_x}\right)\left[M_{\mathrm{T}}(x)_{,xx}+M_{\mathrm{H}}(x)_{,xx}-m'(x)_{,xx}\right]$$
$$+A_x(C_1x+C_2)-B_x(C_4x+C_5)-N_x^{\mathrm{T}}(x)-N_x^{\mathrm{H}}(x)$$

Using Equation (7.32), $(A_x/\bar{A}_x-B_x/\bar{B}_x)=1$ and $(B_x/\bar{D}_x-A_x/\bar{B}_x)=0$. This reduces the equation for $N_x(x)$ to

$$N_x(x)=\left[N_{\mathrm{T}}(x)_{,xx}+N_{\mathrm{H}}(x)_{,xx}\right]+A_x(C_1x+C_2)-B_x(C_4x+C_5)-N_x^{\mathrm{T}}(x)-N_x^{\mathrm{H}}(x)$$

The equation for $M_x(x)$ is developed in a similar manner. After appropriate algebraic manipulation and the use of Equation (7.32), this equation is expressed as

$$M_x=B_xU_{o,x}-D_xW_{o,xx}-M_x^{\mathrm{T}}(x)-M_x^{\mathrm{H}}(x)$$
$$=\left[M_{\mathrm{T}}(x)_{,xx}+M_{\mathrm{H}}(x)_{,xx}-m'(x)_{,xx}\right]+B_x(C_1x+C_2)-D_x(C_4x+C_5)-M_x^{\mathrm{T}}(x)-M_x^{\mathrm{H}}(x)$$

Substitution of the boundary conditions $N_x(0)=M_x(0)=0$ into these equations results in

$$A_xC_2-B_xC_5=N_x^{\mathrm{T}}(0)+N_x^{\mathrm{H}}(0)-N_{\mathrm{T}}(0)_{,xx}-N_{\mathrm{H}}(0)_{,xx} \tag{d}$$

$$B_xC_2-D_xC_5=M_x^{\mathrm{T}}(0)+M_x^{\mathrm{H}}(0)+M_{\mathrm{T}}(0)_{,xx}+M_{\mathrm{H}}(0)_{,xx}-m'(0)_{,xx} \tag{e}$$

In a similar manner, the boundary conditions of $N_x(L)=M_x(L)=0$ result in

$$A_x(C_1L+C_2)-B_x(C_4L+C_5)=N_x^{\mathrm{T}}(L)+N_x^{\mathrm{H}}(L)+N_{\mathrm{T}}(L)_{,xx}+N_{\mathrm{H}}(L)_{,xx} \tag{f}$$

$$B_x(C_1L+C_2)-D_x(C_4L+C_5)=M_x^{\mathrm{T}}(L)+M_x^{\mathrm{H}}(L)+M_{\mathrm{T}}(L)_{,xx}+M_{\mathrm{H}}(L)_{,xx}-m(L)_{,xx} \tag{g}$$

Solving Equations (a)–(g) simultaneously results in the following expressions for the constants of integration.

$$C_1=\frac{1}{\bar{A}_xL}\left[N_x^{\mathrm{T}}(L)-N_x^{\mathrm{T}}(0)+N_x^{\mathrm{H}}(L)-N_x^{\mathrm{H}}(0)+N_{\mathrm{T}}(0)_{,xx}-N_{\mathrm{T}}(L)_{,xx}+N_{\mathrm{H}}(0)_{,xx}-N_{\mathrm{H}}(L)_{,xx}\right]$$
$$-\frac{1}{\bar{B}_xL}\left[M_x^{\mathrm{T}}(L)-M_x^{\mathrm{T}}(0)+M_x^{\mathrm{H}}(L)-M_x^{\mathrm{H}}(0)+M_{\mathrm{T}}(0)_{,xx}-M_{\mathrm{T}}(L)_{,xx}+M_{\mathrm{H}}(0)_{,xx}-M_{\mathrm{H}}(L)_{,xx}\right.$$
$$\left.+m'(L)_{,xx}-m'(0)_{,xx}\right]$$

$$C_2=\frac{1}{\bar{A}_x}\left[N_x^{\mathrm{T}}(0)+N_x^{\mathrm{H}}(0)-N_{\mathrm{T}}(0)_{,xx}-N_{\mathrm{H}}(0)_{,xx}\right]$$
$$-\frac{1}{\bar{B}_x}\left[M_x^{\mathrm{T}}(0)+M_x^{\mathrm{H}}(0)-M_{\mathrm{T}}(0)_{,xx}-M_{\mathrm{H}}(0)_{,xx}-m'(0)_{,xx}\right]$$

$$C_3=\frac{1}{\bar{B}_x}\left[M_{\mathrm{T}}(0)_{,x}+M_{\mathrm{H}}(0)_{,x}-m'(0)_{,x}\right]-\frac{1}{\bar{A}_x}\left[N_{\mathrm{T}}(0)_{,x}+N_{\mathrm{H}}(0)_{,x}\right]$$

$$C_4=\frac{1}{\bar{B}_xL}\left[N_x^{\mathrm{T}}(L)-N_x^{\mathrm{T}}(0)+N_x^{\mathrm{H}}(L)-N_x^{\mathrm{H}}(0)+N_{\mathrm{T}}(0)_{,xx}-N_{\mathrm{T}}(L)_{,xx}+N_{\mathrm{H}}(0)_{,xx}-N_{\mathrm{H}}(L)_{,xx}\right]$$
$$-\frac{1}{\bar{D}_xL}\left[M_x^{\mathrm{T}}(L)-M_x^{\mathrm{T}}(0)+M_x^{\mathrm{H}}(L)-M_x^{\mathrm{H}}(0)+M_{\mathrm{T}}(0)_{,xx}-M_{\mathrm{T}}(L)_{,xx}+M_{\mathrm{H}}(0)_{,xx}-M_{\mathrm{H}}(L)_{,xx}\right.$$
$$\left.+m'(0)_{,xx}-m'(L)_{,xx}\right]$$

$$C_5=\frac{1}{\bar{B}_x}\left[N_x^{\mathrm{T}}(0)+N_x^{\mathrm{H}}(0)-N_{\mathrm{T}}(0)_{,xx}-N_{\mathrm{H}}(0)_{,xx}\right]$$
$$-\frac{1}{\bar{D}_x}\left[M_x^{\mathrm{T}}(0)+M_x^{\mathrm{H}}(0)-M_{\mathrm{T}}(0)_{,xx}-M_{\mathrm{H}}(0)_{,xx}+m'(0)_{,xx}\right]$$

$$C_6 = \frac{1}{\bar{B}_x}\left\{\frac{1}{L}[N_T(0) - N_T(L) + N_H(0) - N_H(L)] + \frac{L}{6}\left[N_x^T(L) - 4N_x^T(0) + N_x^H(L) - 4N_x^H(0)\right.\right.$$
$$\left.\left. + 4N_T(0)_{,xx} - N_T(L)_{,xx} + 4N_H(0)_{,xx} - N_H(L)_{,xx}\right]\right\}$$
$$+ \frac{1}{\bar{D}_x}\left\{\frac{1}{L}[M_T(0) - M_T(L) + M_H(0) - M_H(L) + m'(0) - m'(L)] + \frac{L}{6}\left[4M_x^T(0) - M_x^T(L) + 4M_x^H(0)\right.\right.$$
$$\left.\left. - M_x^H(L) + M_T(L)_{,xx} - 4M_T(0)_{,xx} + M_H(L)_{,xx} - 4M_H(0)_{,xx} + 4m'(0)_{,xx} - m'(L)_{,xx}\right]\right\}$$

$$C_7 = \frac{1}{\bar{D}_x}[M_T(0) + M_H(0) - m'(0)] - \frac{1}{\bar{B}_x}[N_T(0) + N_H(0)]$$

Substituting these into the equations for $U_o(x)$ and $W_o(x)$ results in a set of expressions which are evaluated once $N_x^T(x)$, $N_x^H(x)$, $M_x^T(x)$, $M_x^H(x)$, and $m_x(x)$ are defined, and their appropriate derivatives evaluated at the boundaries $x = 0$ and $x = L$.

To demonstrate numerical calculations, we assume the beam is the same constant width beam explored in Example 7.3. The solution for a problem involving a variable width beam parallels that of the uniform width beam. The primary difference is the numerical value of $\bar{A}_x$, $\bar{B}_x$, and $\bar{D}_x$ as well as $N_x^T(x)$, $N_x^H(x)$, $M_x^T(x)$, and $M_x^H(x)$. For the uniform width beam, it has already been established that $\bar{A}_x = 3.09 \times 10^6$ psi, $\bar{B}_x = \infty$, and $\bar{D}_x = 0.23862 \times 10^6$ psi. The coefficients of thermal and hygral expansions for this material are $\alpha_1 = 3.5\,\mu$in./in./°F, $\alpha_2 = 11.4\,\mu$in./in./°F, $\beta_1 = 0$, and $\beta_2 = 0.44$. Recalling that $\alpha_x = m^2\alpha_1 + n^2\alpha_2$ and $\beta_x = m^2\beta_1 + n^2\beta_2$ (where $m = \cos\theta$ and $n = \sin\theta$), the coefficients of thermal and hygral expansions for each lamina are

$$(\alpha_x)_0 = 3.5 \times 10^{-6} \quad (\alpha_x)_{\pm 45} = 7.45 \times 10^{-6} \quad (\alpha_x)_{90} = 11.4 \times 10^{-6}$$
$$(\beta_x)_0 = 0 \quad (\beta_x)_{\pm 45} = 0.22 \quad (B_x)_{90} = 0.44$$

Equation (7.23) allows for the explicit evaluation of $N_x^T(x)$, $N_x^H(x)$, $M_x^T(x)$, and $M_x^H(x)$. Due to symmetry, $M_x^T(x) = M_x^H(x) = 0$. Recalling that $(E_x)_0 = 7.57 \times 10^6$ psi, $(E_x)_{90} = 2.03 \times 10^6$ psi, and $(E_x)_{45} = 2.92 \times 10^6$ psi, the remaining two terms are

$$N_x^T = \sum_{k=1}^{N}(bE_x t\alpha_x)_k \Delta T = 2(1.0)(0.1)[(7.5 \times 10^6)(3.5 \times 10^{-6})$$
$$+ (2.92 \times 10^6)(7.45 \times 10^{-6})(2)$$
$$+ (2.03 \times 10^6)(11.4 \times 10^{-6})]\Delta T = 18.58\Delta T$$

$$N_x^H = \sum_{k=1}^{N}(bE_x t\beta_x)_k \Delta T = 2(1.0)(0.1)[(7.5 \times 10^6)(0) + (2.92 \times 10^6)(0.22)(2)$$
$$+ (2.03 \times 10^6)(0.44)]\bar{M} = 0.4356 \times 10^6\bar{M}$$

The beam deformation and deflections depend upon evaluating $N_T(x)$, $N_H(x)$, $M_T(x)$, $M_H(x)$, and $m'(x)$ as defined by the integrals $N_T(x)=\int\int N_x^T \mathrm{d}x\mathrm{d}x$, $N_H(x)=\int\int N_x^H \mathrm{d}x\mathrm{d}x$, $M_T(x)=\int\int M_x^T \mathrm{d}x\mathrm{d}x$, $M_H(x)=\int\int M_x^H \mathrm{d}x\mathrm{d}x$, and $m'(x)=\int\int\int m_x(x)\mathrm{d}x\mathrm{d}x\mathrm{d}x$. Each of these parameters depends on the thermal and hygral gradients in the beam. Assuming thermal and hygral equilibrium exist throughout the beam thickness (i.e., no gradients in the z-direction), $m_x(x)=0$. Otherwise, $m_x(x)$ must be established as a function of span, and is based upon N_x^T and N_x^H through the beam thickness. In other words, these could be functions of z, as well as x. This type of variation in moisture content and temperature complicates the solution. The spanwise variation of temperature and moisture defines $N_T(x)$ and $N_H(x)$.

For simplicity, we assume that the moisture content is constant along the beam and that equilibrium in moisture content through the beam thickness has been reached. Therefore, with $\bar{M}=$ constant along the span of the beam we get

$$N_H(x)=\int\int N_x^H(x)\mathrm{d}x\mathrm{d}x=\int\int\left(0.4356\times10^6\bar{M}\right)\mathrm{d}x\mathrm{d}x=0.4356\times10^6\bar{M}x^2$$

No constants of integration appear in this expression since they are accounted for by $N_H(0)$ and $N_H(L)$, terms in C_1 through C_7. Assuming thermal equilibrium through the beam thickness exists $(T_1=T_2)$, and a uniform temperature distribution along the span of the beam results in

$$N_T(x)=\int\int N_x^T(x)\mathrm{d}x\mathrm{d}x=\int\int(18.58\Delta T)\mathrm{d}x\mathrm{d}x=18.58\Delta Tx^2$$

Using the expressions for $N_H(x)$ and $N_T(x)$, evaluating the constants C_1 through C_7, and substituting them into the expressions for $U_o(x)$ and $W_o(x)$ with appropriate numerical values of $\bar{A}_x$, $\bar{B}_x$, and $\bar{D}_x$ results in

$$U_o(x)=\frac{x}{\bar{A}_x}[18.63\Delta T+435{,}600\bar{M}]=\left(3.236\times10^{-7}\right)[18.63\Delta T+435{,}600\bar{M}]x$$
$$W_o(x)=0$$

If the temperature varied linearly with x $\left(N_x^T=18.58\Delta Tx\right)$, we would have

$$N_T(x)=\int\int(18.58\Delta Tx)\mathrm{d}x\mathrm{d}x=3.105\Delta Tx^3$$

Evaluating C_1 through C_7 for this case (with $N_H(x)=0.4356\times10^6\bar{M}x^2$) yields

$$U_o(x)=\frac{x}{\bar{A}_x}[9.315x\Delta T+435{,}600\bar{M}]=\left(3.236\times10^{-7}\right)\left[9.315x^2\Delta T+435{,}600\bar{M}x\right]$$
$$W_o(x)=0$$

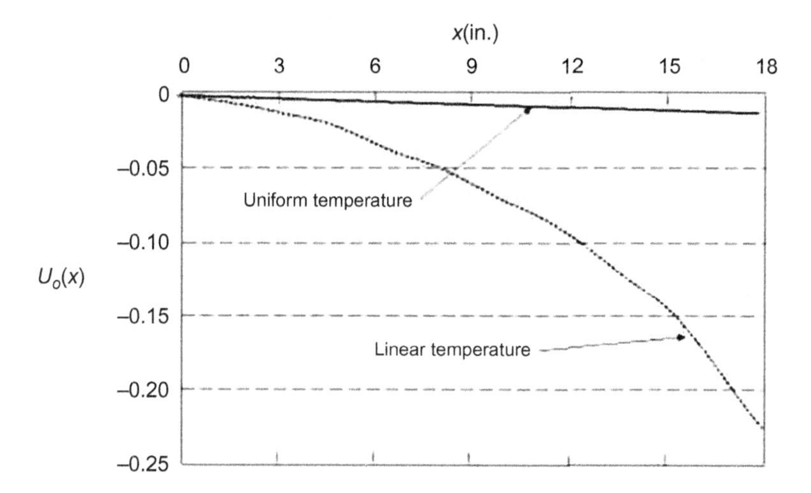

Figure E7.5-2 $U_o(x)$ for uniform and linear temperature variations.

Assuming a beam length of $x = 18''$, a moisture content of $\bar{M} = 5\% = 0.005$, and a temperature variation of $\Delta T = -250\,°\text{F}$, the displacement $U_o(x)$ for the uniform and linear temperature cases above become

$$(U_o(x))_{\text{uniform}} = (3.236 \times 10^{-7})[18.63(-250) + 435,600(0.005)]x$$
$$= x[-0.001507 + 0.0007048] = -0.000802x$$
$$(U_o(x))_{\text{linear}} = (3.236 \times 10^{-7})[9.315x^2(-250) + 435,600(0.005)x]$$
$$= -0.000753x^2 + 0.000705x$$

A plot of $U_o(x)$ for both of these cases is presented in Figure E7.5-2.

Similar results can be obtained for a variety of temperature and/or moisture profiles along the span of the beam. The degree of complexity of the resulting equations for displacement depends upon the temperature and moisture profiles, both through the beam thickness and along the span of the beam.

7.5 Superposition

The method of superposition is applied to composite beams in the same manner described in introductory strength of materials. General expressions for $U_o(x)$ and $W_o(x)$ for selected loads and beam support conditions are presented in Table 7.1. These solutions do not include thermal or hygral effects, distributed moments or axial forces ($m_x(x)$ and $p_x(x)$ from Figure 7.5). In most cases, beams with symmetric ply orientations are used, resulting in $\bar{B}_x = \infty$, which leads to $U_o(x) = 0$. Expressions for $U_o(x)$

Table 7.1 Beam deflection equations for selected loading and support conditions

Case	Beam loading and support	Deflection equations
1	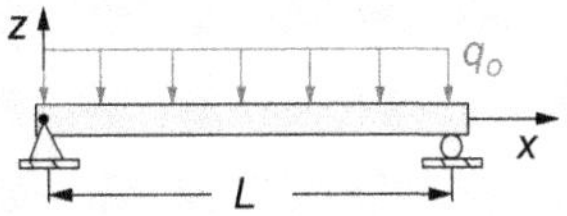 z, q_o, x, L	$U_o(x)=\frac{q_oL^3}{24\bar{B}_x}\left\{6\left(\frac{x}{L}\right)^2-4\left(\frac{x}{L}\right)^3\right\}$ $W_o(x)=\frac{q_oL^4}{24\bar{D}_x}\left\{-\left(\frac{x}{L}\right)^4+2\left(\frac{x}{L}\right)^3-\left(\frac{x}{L}\right)\right\}$
2	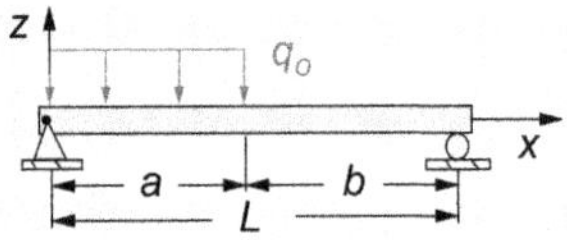 z, q_o, x, a, b, L	$U_o(x)=\frac{q_o}{12\bar{B}_x}\left\{2\langle x-a\rangle^3-2x^3+3ax^2\left(\frac{a-2L}{L}\right)\right\}$ $W_o(x)=\frac{q_o}{24\bar{D}_x}\left\{\langle x-a\rangle^4-x^4+2a(a-2L)\frac{x^3}{L}+(4a^2L+8L^3-8aL^2-a^3)\frac{ax}{L}\right\}$
3	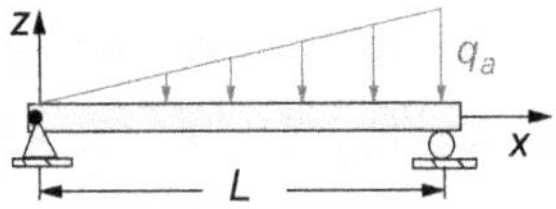 z, q_a, x, L	$U_o(x)=\frac{q_ox^2}{24\bar{B}_x}\left\{2L-\frac{x^2}{L}\right\}$ $W_o(x)=\frac{q_o}{120\bar{D}_x}\left\{5Lx^3-4xL^3-\frac{x^5}{L}\right\}$
4	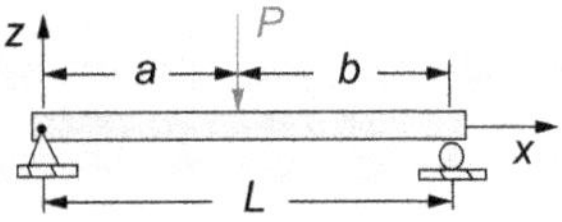 z, P, x, a, b, L	$U_o(x)=\frac{P}{2\bar{B}_x}\left\{x^2\left(\frac{L-a}{L}\right)-\langle x-a\rangle^2\right\}$ $W_o(x)=\frac{P}{6\bar{D}_x}\left\{x^3\left(\frac{L-a}{L}\right)-\langle x-a\rangle^3+\left(\frac{a(a-L)(2L-a)}{L}\right)x\right\}$
5	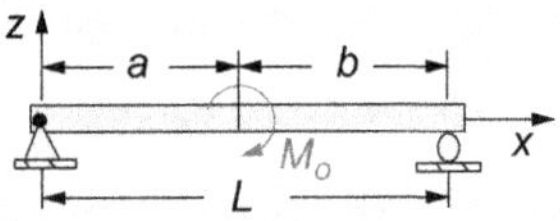 z, M_o, x, a, b, L	$U_o(x)=\frac{M_o}{\bar{B}_x}\left\{\langle x-a\rangle-\frac{x^2}{2L}\right\}$ $W_o(x)=\frac{M_o}{6\bar{D}_x}\left\{3\langle x-a\rangle^2-\frac{x^2}{L}+6\left(\frac{6aL-3a^2-2L^2}{L}\right)x\right\}$
6	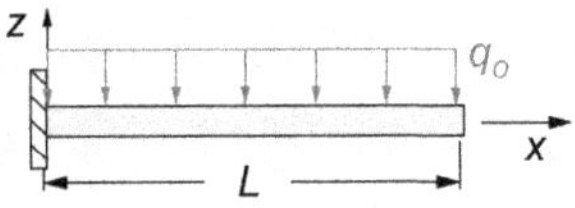 z, q_o, x, L	$U_o(x)=\frac{q_ox}{\bar{B}_x}\{3L(x-L)-x^2\}$ $W_o(x)=\frac{q_ox^2}{24\bar{D}_x}\{L(4x-6L)-x^2\}$
7	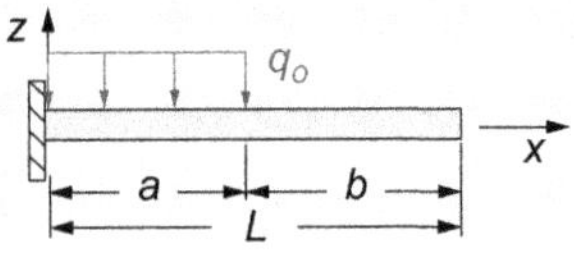 z, q_o, x, a, b, L	$U_o(x)=\frac{q_o}{6\bar{B}_x}\left\{\langle x-a\rangle^3-x^3+3a(x+a-L)x\right\}$ $W_o(x)=\frac{q_o}{24\bar{D}_x}\left\{\langle x-a\rangle^4-x^4+ax^2(4x-6L+6a)\right\}$
8	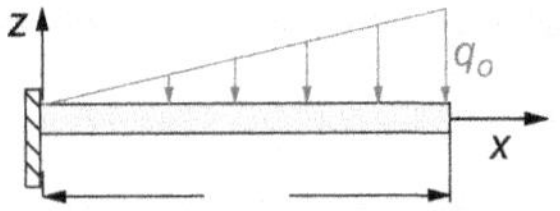 z, q_o, x	$U_o(x)=\frac{q_ox}{24\bar{B}_x}\left\{L(6x-8L)-\frac{x^3}{L}\right\}$ $W_o(x)=\frac{q_ox^2}{120\bar{D}_x}\left\{10L(x-2L)-\frac{x^3}{L}\right\}$
9	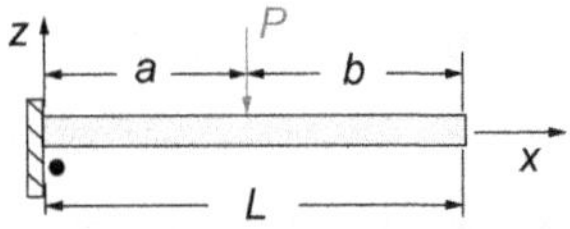 z, P, x, a, b, L	$U_o(x)=\frac{P}{2\bar{B}_x}\left\{x^2-2ax-\langle x-a\rangle^2\right\}$ $W_o(x)=\frac{P}{6\bar{D}_x}\left\{x^3-3ax^2-\langle x-a\rangle^3\right\}$
10	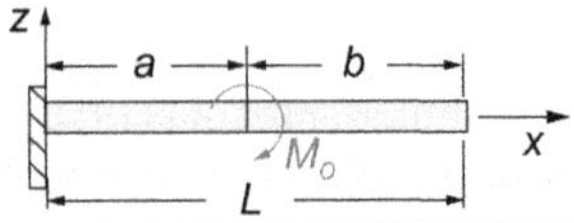 z, M_o, x, a, b, L	$U_o(x)=\frac{M_o}{\bar{B}_x(L-a)}\{(L-a)\langle x-a\rangle-x^2+ax\}$ $W_o(x)=\frac{M_o}{6\bar{D}_x(L-a)}\left\{3(L-a)\langle x-a\rangle^2-x^3+3ax^2\right\}$

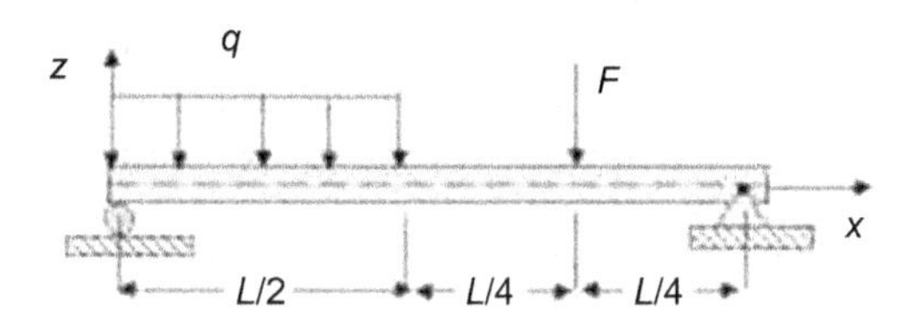

Figure 7.10 Simply supported beam with distributed and concentrated loads.

are included in Table 7.1 for special cases in which nonsymmetric laminates are used, but the degree of thermal and hygral coupling are insignificant (small M_x^{T} and M_x^{H}). One should note the similarity between the expressions for $W_o(x)$ in Table 7.1 and those in conventional strength of materials texts for identical loading conditions. The results for nonsymmetric beams should not be considered accurate since N_x^{T}, $M_x^{\mathrm{T}}, N_x^{\mathrm{H}}$, and M_x^{H} have been omitted. In several cases, the expressions are given in terms of singularity functions (expressed as $\langle x-a\rangle^n$) defined and discussed in various references, including Refs. [5,6].

As an illustration of the use of superposition, consider the beam in Figure 7.10. The appropriate expressions for $U_o(x)$ and $W_o(x)$ are obtained by combining the solutions from case 2 of Table 7.1 (with $a=L/2$), and case 4 (with $a=3L/4$). The expressions for $U_o(x)$ and $W_o(x)$ for each case are given below.

$$(U_o(x))_{\text{case 2}} = (q_o/12\bar{B}_x)\left\{2\langle x-L/2\rangle^3 - 2x^3 + 3(L/2)x^2(-3/2)\right\}$$

$$(U_o(x))_{\text{case 4}} = (P/2\bar{B}_x)\left\{x^2(1/2) - \langle x-L/2\rangle^2\right\}$$

$$\begin{aligned}(W_o(x))_{\text{case 2}} = (q_0/24\bar{D}_x)\Big\{&\langle x-L/2\rangle^4 - x^4 + 2(L/2)(-3/2)x^3 \\ &+ \left(4(L/2)^2(L) + 8L^3 - 8(L/2)L^2 - (L/2)^3\right)(x/2)\}\end{aligned}$$

$$(W_o(x))_{\text{case 4}} = (q_0/6\bar{D}_x)\left\{x^3(1/2) - \langle x-L/2\rangle^3 - (3L^2/8)x\right\}$$

Combining the two cases and simplifying results in

$$U_o(x) = \frac{q_o}{12\bar{B}_x}\left\{2\left\langle x-\frac{L}{2}\right\rangle^3 - 2x^3 - \frac{9}{4}x^2L\right\} + \frac{P}{2\bar{B}_x}\left\{\frac{x^2}{4} - \left\langle x-\frac{3L}{4}\right\rangle^2\right\}$$

$$W_o(x) = \frac{q_o}{24\bar{D}_x}\left\{\left\langle x-\frac{L}{2}\right\rangle^3 - x^4 - \frac{3}{2}x^3L + \frac{39}{16}xL^3\right\} + \frac{P}{6\bar{D}_x}\left\{\frac{x^3}{4} - \left\langle x-\frac{3L}{4}\right\rangle^2 - \frac{15}{64}xL^2\right\}$$

The methods of superposition can also be used to determine reactions for statically indeterminate problems in the same manner learned in undergraduate strength of materials.

7.6 Beams with shear deformation

In classical lamination theory, as well as the previously defined beam problems, shear deformation was assumed to be negligible. This assumption of $\gamma_{xz} = 0$ resulted in $\kappa_x = -W_{o,xx}$. In the following discussions, shear deformation is not restricted to be zero. Removing the restriction of $\gamma_{xz} = 0$ is often referred to as shear deformation theory (SDT). The assumed displacement fields are as before $U = U_o + z\Phi$ and $W = W_o$. The resulting strain–displacement relations are $\varepsilon_x = \partial U/\partial x = U_{o,x} + z\Phi_{,x}$ and $\gamma_{xz} = \partial U/\partial z + \partial W/\partial x = \Phi + W_{o,x}$.

The equations of equilibrium presented in Equations (7.24) through (7.26) remain valid, as do the definitions of $\overline{p}_1$, $\overline{p}_2$, and $\overline{p}_3$ given in Equations (7.27)–(7.29), respectively. The dynamic equations for beams including shear deformation can be derived as in Section 7.4. The primary focus of this section, however, is on the effects of shear deformation. Therefore, only the static case is considered. The definitions of N_x and M_x in terms of lb (N) or in. lb (N m) units are

$$(N_x, M_x) = \int_{-h/2}^{h/2} (1, z)\sigma_x \mathrm{d}A = \sum_{k=1}^{N} \int_{-h_{k-1}}^{h_k} b_k(1, z)(\sigma_x)_k \mathrm{d}z$$

The shear force Q_x is

$$Q_x = \int_{-h/2}^{h/2} \tau_{xz} \mathrm{d}A = \sum_{k=1}^{N} \int_{-h_{k-1}}^{h_k} (b\tau_{xz})_k \mathrm{d}z$$

The load–strain relations are

$$N_x = A_x \varepsilon_x^o + B_x \kappa_x \quad M_x = B_x \varepsilon_x^o + D_x \kappa_x \quad Q_x = KA_{55}\gamma_{xz}$$

where $\varepsilon_x^0 = U_{o,x}$, $\kappa_x = \Phi_{,x}$, and $\gamma_{xz} = \Phi + W_{o,x}$. Substituting these into the load–strain relationships and incorporating thermal and hygral effects as before results in

$$\begin{aligned} N_x &= A_x U_{o,x} + B_x \Phi_{,x} - N_x^{\mathrm{T}} - N_x^{\mathrm{H}} \\ M_x &= B_x U_{o,x} + D_x \Phi_{,x} - M_x^{\mathrm{T}} - M_x^{\mathrm{H}} \\ Q_x &= KA_{55}\left(\Phi + W_{0,x}\right) \end{aligned}$$

where K is a shear correction factor, which depends upon the cross-sectional shape of the beam. Numerical values of K can be found in many advanced strength of materials texts, including Refs. [7,8].

The static equations of equilibrium (obtained by setting the right-hand side of Equation (7.24) through Equation (7.26) equal to zero) are

$$N_{x,x} + p_x(x) = 0 \quad M_{x,x} - Q_x + m_x(x) = 0 \quad Q_{x,x} + q(x) = 0$$

Taking the partial derivatives of N_x, M_x, and Q_x and substituting them into these expressions results in the following equations of static equilibrium for an arbitrary width beam, including shear effects

$$\begin{aligned} &A_x U_{o,xx} + B_x \Phi_{,xx} - N^{\mathrm{T}}_{x,x} - N^{\mathrm{H}}_{x,x} + p_x(x) = 0 \\ &B_x U_{o,xx} + D_x \Phi_{,xx} - M^{\mathrm{T}}_{x,x} - M^{\mathrm{H}}_{x,x} - Q_x + m_x(x) = 0 \\ &KA_{55}\left(\Phi_{,x} + W_{o,xx}\right) + q(x) = 0 \end{aligned} \tag{7.37}$$

Solving the first equation of (7.37) for $U_{o,xx}$ and substituting into the second equation, along with the expression for Q_x results in

$$\bar{D}_x \Phi_{,xx} - KA_{55}\left(\Phi + W_{o,x}\right) - \frac{B_x}{A_x} p_x(x) + m_x(x) + \frac{B_x}{A_x}\left(N^{\mathrm{T}}_{x,x} + N^{\mathrm{H}}_{x,x}\right) - \left(M^{\mathrm{T}}_{x,x} + M^{\mathrm{H}}_{x,x}\right) = 0$$

where $\bar{D}_x$ is defined by Equation (7.32). Taking the partial derivative of this expression with respect to x and substituting the third equation of (7.37), an expression for $\Phi_{,xxx}$ is developed, and is

$$\Phi_{,xxx} = -\frac{q(x)}{\bar{D}_x} + \frac{p_x(x)_{,x}}{\bar{B}_x} - \frac{m(x)_{,x}}{\bar{D}_x} - \frac{N^{\mathrm{T}}_{x,xx} + N^{\mathrm{H}}_{x,xx}}{\bar{B}_x} + \frac{M^{\mathrm{T}}_{x,xx} + M^{\mathrm{H}}_{x,xx}}{\bar{D}_x} \tag{7.38}$$

The third equation of (7.37) can be written as

$$\Phi_{,xxx} = -\frac{q(x)_{,xx}}{KA_{55}} - W_{o,xxxx}$$

Substituting this into Equation (7.38) results in

$$W_{0,xxxx} = \frac{q(x)}{\bar{D}_x} - \frac{q(x)_{,xx}}{KA_{55}} - \frac{p_x(x)_{,x}}{\bar{B}_x} + \frac{m(x)_{,x}}{\bar{D}_x} + \frac{N^{\mathrm{T}}_{x,xx} + N^{\mathrm{H}}_{x,xx}}{\bar{B}_x} - \frac{M^{\mathrm{T}}_{x,xx} + M^{\mathrm{H}}_{x,xx}}{\bar{D}_x} \tag{7.39}$$

Similarly, the first equation of (7.37) can be differentiated with respect to x and Equation (7.38) can be substituted into the resulting expression, yielding

$$U_{0,xxx} = -\frac{p_x(x)_{,x}}{\bar{A}_x} + \frac{q(x)}{\bar{B}_x} + \frac{m_x(x)_{,x}}{\bar{B}_x} + \frac{N^{\mathrm{T}}_{x,xx} + N^{\mathrm{H}}_{x,xx}}{\bar{A}_x} - \frac{M^{\mathrm{T}}_{x,xx} + M^{\mathrm{H}}_{x,xx}}{\bar{B}_x} \tag{7.40}$$

Equations (7.39) and (7.40) are identical to (7.33) and (7.34) with the exception of the additional $q(x)_{,xx}/KA_{55}$ term in Equation (7.39). The solution procedure for deflections of a beam including shear effects are similar to those in Section 7.4, except that an additional equation for Φ exists. In order to illustrate the solution procedure, the

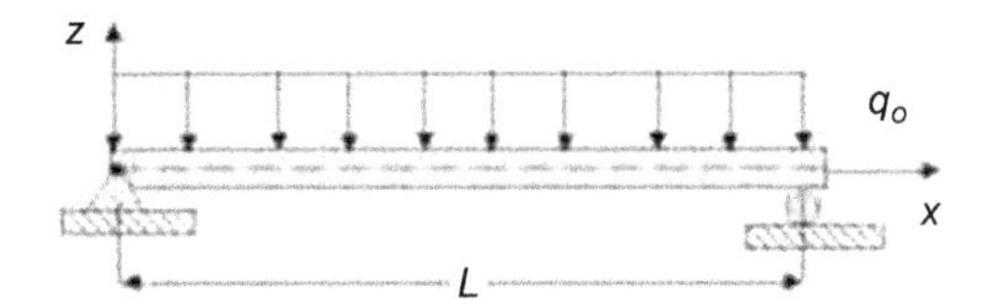

Figure 7.11 Uniformly loaded simply supported beam.

uniformly loaded, simply supported beam of Figure 7.11 is considered. Under these loading conditions, the governing Equations (7.39) and (7.40), excluding thermal and hygral effects, become

$$U_{0,xxx} = -\frac{q_o}{\bar{B}_x} \quad W_{0,xxxx} = -\frac{q_o}{\bar{D}_x}$$

The expression for Φ can be obtained from equations generated in the solutions for $U_{o,xxx}$ and $W_{o,xxxx}$. Integrating each of the above expressions results in

$$W_o(x) = -\frac{q_o}{24\bar{D}_x}x^4 + \frac{C_1}{6}x^3 + \frac{C_2}{2}x^2 + C_3x + C_4$$

$$U_o(x) = -\frac{q_o}{6\bar{B}_x}x^3 + \frac{C_5}{2}x^2 + C_6x + C_7$$

Using the third equation of (7.37), it can easily be shown that

$$\Phi_{,x} + W_{o,xx} = -\frac{q_o}{KA_{55}}$$

Integrating this expression results in

$$\Phi = -\frac{q_o x}{KA_{55}} - W_{o,x}$$

Substituting from the expression for W_o above yields

$$\Phi = -\frac{q_o}{KA_{55}}x - \frac{q_o}{6\bar{D}_x}x^3 - \frac{C_1}{2}x^2 - C_2x - C_3$$

The constants of integration are determined from the boundary conditions previously defined, namely

$$U_o(0) = W_o(0) = W_o(L) = N_x(0) = N_x(L) = M_x(0) = M_x(L) = 0$$

The equations defining $N_x(x)$ and $M_x(x)$ are different than those previously presented since the curvature is given as $\kappa_x = \Phi_{,x}$ for this case, as opposed to $\kappa_x = -W_{o,xx}$ for beams without shear effects. The expressions for $N_x(x)$ and $M_x(x)$ are

$$N_x(x) = A_x U_{o,x} + B_x \Phi_{,x} = A_x \left\{ -\frac{q_o}{2\bar{B}_x} x^2 + C_5 x + C_6 \right\} + B_x \left\{ -\frac{q_o}{KA_{55}} + \frac{q_o}{2\bar{D}_x} x^2 - C_1 x - C_2 \right\}$$

$$M_x(x) = B_x U_{o,x} + B_x \Phi_{,x} = B_x \left\{ -\frac{q_o}{2\bar{B}_x} x^2 + C_5 x + C_6 \right\} + D_x \left\{ -\frac{q_o}{KA_{55}} + \frac{q_o}{2\bar{D}_x} x^2 - C_1 x - C_2 \right\}$$

Solving for the constants of integration and substituting into the appropriate equations results in the following expressions for $U_o(x)$, $W_o(x)$, and $\Phi(x)$

$$U_o(x) = \frac{q_o L^3}{12\bar{B}_x} \left[-2\left(\frac{x}{L}\right)^3 + 3\left(\frac{x}{L}\right)^2 \right]$$

$$W_o(x) = \frac{q_o L^4}{24\bar{D}_x} \left[-\left(\frac{x}{L}\right)^4 + 2\left(\frac{x}{L}\right)^3 - \left(\frac{x}{L}\right) \right] + \frac{q_o L^2}{2KA_{55}} \left[\left(\frac{x}{L}\right) - \left(\frac{x}{L}\right)^2 \right]$$

$$\Phi(x) = \frac{q_o L^3}{24\bar{D}_x} \left[1 - 4\left(\frac{x}{L}\right)^3 - 6\left(\frac{x}{L}\right)^2 \right] - \frac{q_o L}{2KA_{55}}$$

The expression for $U_o(x)$ presented above is identical to that derived in Section 7.4. With the exception of the additional $(q_o L^2/2KA_{55})\left[(x/L) - (x/L)^2\right]$ term in $W_o(x)$ above, this expression is identical to the corresponding equation of Section 7.4. The expression for $\Phi(x)$ is needed for defining $\kappa_x = \Phi_{,x}$ and is required for evaluating stresses in the beam.

The effect of shear strain on the displacement $W_o(x)$ is best seen by closer examination of the shear term. For simplicity, assume the beam cross section is rectangular. For a rectangular section, the shear correction factor is $K = 1.2$, given in Ref. [8]. The effect of shear deformation depends on the material used in the beam. In particular, the magnitude of A_{55} as defined by Equation (6.21) with the coefficient C appearing in that equation has already been accounted for by K. Assuming the same beam size as described in the example of Section 7.4, and $G_{23} = 0.10 \times 10^6$ psi, $G_{13} = 1.0 \times 10^6$ psi, the definition of Q_{55} in Chapter 3 coupled with Equation (6.20) results in $A_{55} = 0.33 \times 10^6$. Based on this approximation for A_{55} and K, the shear strain related term in $W_o(x)$ is

$$1.26 \times 10^{-6}\, q_o L^2 \left[\left(\frac{x}{L}\right) - \left(\frac{x}{L}\right)^2 \right]$$

Combining this term with the remainder of the expression for $W_o(x)$ and assuming the same $\bar{D}_x$ as in previous examples results in

$$W_o(x) = q_o L^2 \left\{ 1.747 \times 10^{-6} \left[-\left(\frac{x}{L}\right)^4 + 2\left(\frac{x}{L}\right)^3 - \left(\frac{x}{L}\right) \right] + 1.26 \times 10^{-6} \left[\left(\frac{x}{L}\right) - \left(\frac{x}{L}\right)^2 \right] \right\}$$

The largest displacement is at $x/L = 0.5$, and will result in

$$W_o(L/2) = q_o L^2 \{ -5.46 \times 10^{-7} L^2 + 3.125 \times 10^{-7} \}$$

The second term in the bracket is the result of including shear strain. It is a trivial matter to show that in order for the shear term to affect the beam deflection by 10%, the beam length would be

$$L=\sqrt{\frac{3.125\times 10^{-7}}{0.10(5.46\times 10^{-7})}}=2.4\,\text{in.}$$

We conclude that for the simply supported beam considered, with the specified values of G_{23} and G_{13}, and the dimensions given, the effect of including shear is not significant. This does not imply that one should always neglect shear deformation. Different material properties and beam geometries would result in other numerical values. Both geometry (beam length, thickness, and cross section) and material are therefore important considerations when deciding whether or not to include shear deformation as part of beam analysis. Beams constructed from materials with small shear moduli (G_{23} and G_{13}) are affected more by shear deformation than those with large shear moduli. In addition to material considerations, the beam geometry, which influences the shear correction factor as well as $\bar{D}_x$, is an important consideration. Similarly, the beam loading and support conditions, including thermal and hygral effects, will influence considerations for defining an appropriate solution procedure for beam problems.

The normal stress distribution through the beam is determined as in previous sections from the relationship

$$(\sigma_x)_k=(E_x)_k\varepsilon_x=(E_x)_k\left[U_{o,x}+z\Phi_{,x}\right]$$

For this particular problem, Φ is independent of x. Therefore, the normal stress distribution is identical to that in Section 7.4, as is the expression for τ_{xz}, as represented by Equations (7.35) and (7.36).

7.7 Buckling

Failures associated with beam buckling are generally a result of a geometric instability rather than an over stressing condition. Development of the governing equations for the buckling of composite columns or beam-columns parallels corresponding developments for isotropic materials. Although buckling is typically associated with long slender columns, it can also occur in composite plates, which is a topic treated in various texts [2,3], as well as a monograph by Leissa [9], and is considered in the next chapter. The buckling problem considered herein pertains only to columns and beam-columns subjected to axially applied compressive loads, as well as distributed loads and moments. The effects of shear deformation are included, while thermal and hygral effects are ignored. The assumed deformed shape of segment *ACFD* of a beam-column and the associated loads acting on it are also shown in Figure 7.12.

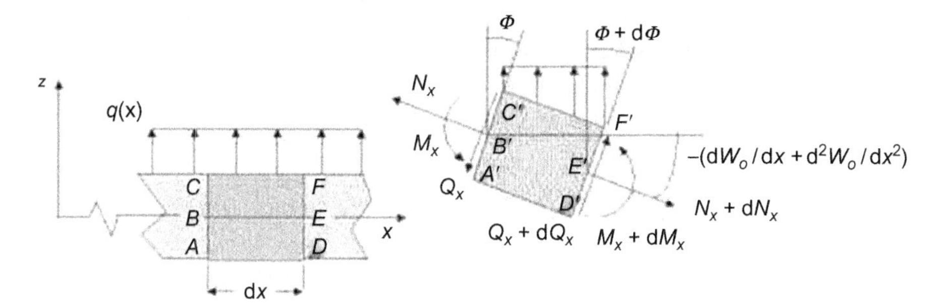

Figure 7.12 Beam-column buckling loads and deformed shape.

The distributed load $q(x)$ is assumed to be vertical, even in the deformed state. The governing equations of equilibrium are developed by first considering a summation of forces in the x-direction.

$$\sum F_x = 0$$
$$= (N_x + \mathrm{d}N_x)\cos\left(-\frac{\mathrm{d}W_o}{\mathrm{d}x} - \frac{\mathrm{d}^2W_o}{\mathrm{d}x^2}\right) - N_x\cos\left(-\frac{\mathrm{d}W_o}{\mathrm{d}x}\right) + (Q_x + \mathrm{d}Q_x)\sin(\Phi + \mathrm{d}\Phi) - Q_x\sin\Phi$$

Introducing the assumption of small angles results in

$$\cos\left(-\frac{\mathrm{d}W_o}{\mathrm{d}x} - \frac{\mathrm{d}^2W_o}{\mathrm{d}x^2}\right) \to 1 \quad \cos\left(-\frac{\mathrm{d}W_o}{\mathrm{d}x}\right) \to 1 \quad \sin(\Phi + \mathrm{d}\Phi) \to \Phi + \mathrm{d}\Phi \quad \sin(\Phi) \to \Phi$$

Substituting these into the $\sum F_x$ above yields

$$N_{x,x} + Q_x\Phi_{,x} + \Phi Q_{x,x} = 0 \tag{7.41}$$

Summation of forces in the z-direction produces

$$\sum F_z = 0 = q_o(x)\mathrm{d}x + N_x\sin\left(-\frac{\mathrm{d}W_o}{\mathrm{d}x}\right) - (N_x + \mathrm{d}N_x)\sin\left(-\frac{\mathrm{d}W_o}{\mathrm{d}x} - \frac{\mathrm{d}^2W_o}{\mathrm{d}x^2}\right)$$
$$+ (Q_x + \mathrm{d}Q_x)\cos(\Phi + \mathrm{d}\Phi) - Q_x\cos\Phi$$

Introducing the small angle assumption results in

$$Q_{x,x} + N_xW_{o,xx} + q(x) = 0 \tag{7.42}$$

Similarly, summation of moments about E' results in

$$M_{x,x} - Q_x = 0 \tag{7.43}$$

In general, we assume dx is small, therefore the $q(dx)^2/2$ term vanishes. The transverse shearing stresses and forces are generally small; therefore, we assume that the quadratic terms representing nonlinear interactions of transverse shear and rotations $(Q_x\Phi_{,x} + \Phi Q_{x,x})$ are negligible. This results in the governing equations of equilibrium being

$$\begin{aligned} &N_{x,x} = 0 \\ &Q_{x,x} + N_x W_{o,xx} + q(x) = 0 \\ &M_{x,x} - Q_x = 0 \end{aligned} \tag{7.44}$$

For most column or beam-column buckling problem, $N_x = -P$. Since $N_x = -P = A_x U_{o,x} + B_x \Phi_{,x}$ is assumed constant, $N_{x,x} = 0$. Therefore, the first two equations of (7.44) are

$$U_{o,xx} = -\frac{B_x}{A_x}\Phi_{,xx} \quad Q_{x,x} - PW_{o,xx} + q(x) = 0$$

Since $Q_x = KA_{55}(\Phi + W_{o,x})$, the second equation above can be written as

$$KA_{55}(\Phi_{,x} + W_{o,xx}) - PW_{o,xx} + q(x) = 0$$

This expression can be manipulated to result in

$$\Phi_{,x} = \left(\frac{P}{KA_{55}} - 1\right)W_{o,xx} - \frac{q(x)}{KA_{55}}$$

The moment equation is expressed in terms of $U_{o,x}$ and $\Phi_{,x}$ as $M_x = B_x U_{o,x} + D_x \Phi_{,x}$. Since $M_{x,x} - Q_x = 0$, differentiation of M_x results in $B_x U_{o,xx} - D_x \Phi_{,xx} = Q_x = KA_{55}(\Phi + W_{o,x}) = \bar{D}_x \Phi_{,xx}$. Using the expressions for $\Phi_{,x}$ and $U_{o,xx}$ previously generated, this equation can be expressed as

$$\bar{D}_x\left[\frac{P}{KA_{55}} - 1\right]W_{o,xxxx} - PW_{o,xx} = \frac{\bar{D}_x}{KA_{55}}q(x)_{,xx} - q(x) \tag{7.45}$$

An alternative form of this equation, which is more traditional in buckling analysis, is

$$W_{o,xxxx} + \left[\frac{P/\bar{D}_x}{1 - P/KA_{55}}\right]W_{o,xx} = \frac{q(x)_{,xx}}{P - KA_{55}} - \frac{q(x)/\bar{D}_x}{1 - P/KA_{55}}$$

Further reduction of this expression results by introducing the definition

$$\lambda^2 = \frac{P/\bar{D}_x}{1 - P/KA_{55}} \tag{7.46}$$

The governing equation relating W_o to the applied loads can therefore be written as

$$W_{o,xxxx} + \lambda^2 W_{o,xx} = \frac{q(x)_{,xx}}{P - KA_{55}} - \frac{q(x)/\bar{D}_x}{1 - P/KA_{55}} \tag{7.47}$$

The solution for this equation depends on the form of $q(x)$. For the general equation given in Equation (7.47), the solution consists of both particular and complementary solutions if a beam-column is considered $(q(x) \neq 0)$. For the special case of a column as shown in Figure 7.13, $q(x) = q(x)_{,xx} = 0$, and Equation (7.47) reduces to $W_{o,xxxx} + \lambda^2 W_{o,xx} = 0$.

The equation above is solved by assuming a Fourier series solution for W_o in the form $W_o(x) = C \sin n\pi x/L$. Taking the derivatives of $W_o(x)$ and substituting into the buckling equation yields

$$C\left[\left(\frac{n\pi}{L}\right)^4 + \lambda^2\left(\frac{n\pi}{L}\right)^2\right]\sin\frac{n\pi x}{L} = 0$$

Since $C \neq 0$, the solution of this equation must be of the form $\left[(n\pi/L)^4 + \lambda^2(n\pi/L)^2\right]\sin(n\pi x/L) = 0$. Introducing the definition of λ^2 given by Equation (7.46) results in $(n\pi/L)^2 + (P/\bar{D}_x)/(1 - P/KA_{55}) = 0$. Solving for the applied load P, which is termed the critical buckling load, P_{CR} results in

$$P_{CR} = \frac{\bar{D}_x n^2 \pi^2}{L^2 + \bar{D}_x n^2 \pi^2 / KA_{55}} \tag{7.48}$$

An alternate approach is to assume a Fourier series solution for $U_o(x)$, $\Phi(x)$, and $W_o(x)$ consistent with the imposed boundary conditions. For the column in Figure 7.13, the assumed solutions would be

$$U_o(x) = A \cos\frac{n\pi x}{L} \quad \Phi(x) = B \cos\frac{n\pi x}{L} \quad W_o(x) = C \cos\frac{n\pi x}{L}$$

where A, B, and C are determined from boundary conditions. These functions can be used in the previous derivations to obtain Equations (7.45) and (7.47) in terms of $n\pi$. Subsequently, λ can be determined in terms of P. The result is an explicit representation of $U_o(x)$, $W_o(x)$, $\Phi(x)$, and P_{CR}. In many applications, P_{CR} is more important in column buckling than the displacement fields, and the solution given by Equation (7.48) is typically all that is sought.

Figure 7.13 Simple column buckling.

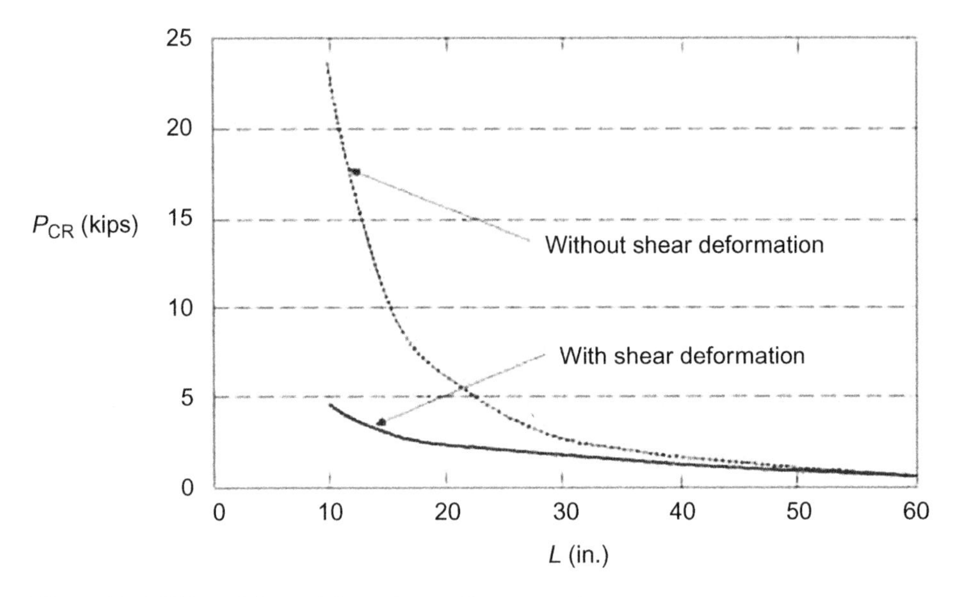

Figure 7.14 Effect of shear deformation on P_{CR} versus L.

The effect of shear deformation on P_{CR} is generally small, except for short columns as illustrated in Figure 7.14, where P_{CR} versus L is plotted from Equation (7.48) for the uniform width beam previously considered $(\bar{D}_x = 0.2386 \times 10^6)$ and $n = 1$. Two extreme cases are shown: one in which $KA_{55} = \infty$ (which corresponds to neglecting shear deformation), and the other for a very small A_{55} term (assumed to be $A_{55} = 0.5 \times 10^4$), with $K = 1.2$. The effects of including shear deformation for the material properties and geometry considered are easily assessed from this figure. It should be noted that the curve with shear deformation included is based on a small A_{55} and an assumed K. Different values for each of these would result in different magnitudes of P_{CR}. In addition, the effect of shear deformation is more pronounced for shorter columns.

In a similar manner, solutions for a beam-column $(q(x) \neq 0)$ can be obtained. The solution for the beam-column is more rigorous than for the simple column, since both particular and complementary solutions are required. The procedures for establishing stress and displacement fields in a composite beam-column again parallel those for an isotropic material. Solutions to composite beam-column problems are considered beyond the scope of this text.

7.8 Curved rings and beams

The analysis of curved laminated composite beams represents a special case of ring analysis. Therefore, the governing equations for ring analysis developed first, and subsequently reduced to the case of a curved beam. Figure 7.15 identifies the geometry of

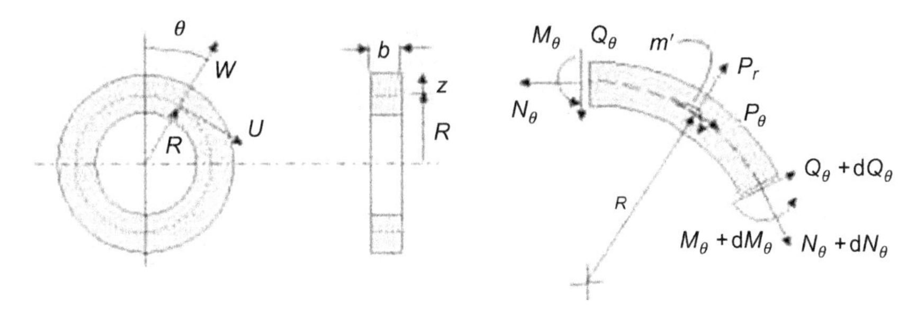

Figure 7.15 General ring geometry and element loads.

a general ring along with a representative volume element showing the possible loads. The displacements in the radial and circumferential directions are defined in terms of W and U, respectively. This notation is used for continuity between curved and straight beam analysis. The radius R defines the location of the neutral axis of the beam and not the mid-surface of the laminate. The coordinate z is used to define an arbitrary position relative to the neutral surface.

The normal forces, shear forces, and bending moments are designated with θ subscripts. Radial and circumferential loads P_r and P_θ as well as an applied moment m' are also defined. The P_r and P_θ terms can be modeled to include both concentrated and distributed loads (q) applied to the beam. Due to geometry, polar coordinates are used. The strain–displacement relations for polar coordinates can be found in most standard elasticity texts, such as Ref. [4]. Using the radial and circumferential displacements described above, the strains are:

$$\varepsilon_r = \frac{\partial W}{\partial r} \quad \varepsilon_\theta = \frac{1}{r}\left(\frac{\partial U}{\partial \theta} + W\right) \quad \gamma_{r\theta} = \frac{1}{r}\left(\frac{\partial W}{\partial \theta} - U\right) + \frac{\partial U}{\partial r}$$

The strain–displacements and corresponding stresses in polar coordinates are illustrated in Figure 7.16.

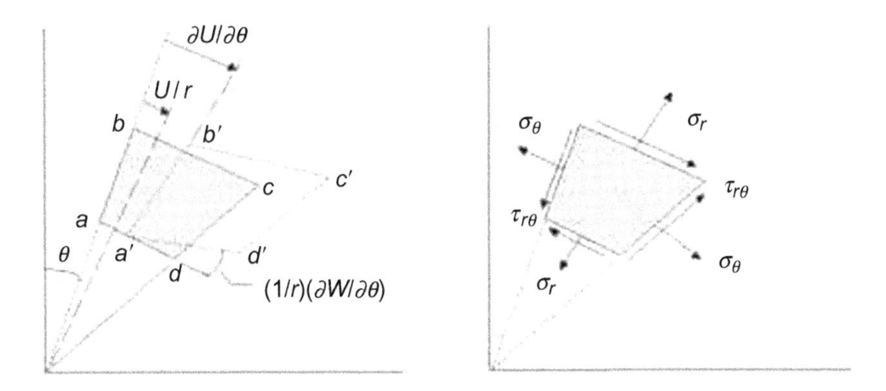

Figure 7.16 Normal and shear stresses in $r-\theta$ coordinate system.

Assumptions for admissible strains and displacements in the composite ring parallel to those for the composite beam. Only ε_θ and $\gamma_{r\theta}$ are allowed to exist. For the ring, the radial coordinate r is replaced by the coordinate $R+z$. Therefore, the relevant strains are expressed in terms of displacements as

$$\varepsilon_\theta = \frac{1}{R+z}\left(\frac{\partial U}{\partial \theta} + W\right) \quad \gamma_{r\theta} = \frac{1}{R+z}\left(\frac{\partial W}{\partial \theta} - U\right) + \frac{\partial U}{\partial r}$$

The displacement field for the composite ring is assumed to be

$$W = W_o \quad U = U_o + z\beta$$

where β is analogous to Φ for the straight beam and should not be confused with the coefficient of hygral expansion. If these displacement fields are substituted into the strain–displacement relations above, the following expressions are obtained

$$\begin{aligned} \varepsilon_\theta &= \frac{1}{R+z}\left(\frac{\partial U_o}{\partial \theta} + z\frac{\partial \beta}{\partial \theta} + W_o\right) \\ \gamma_{r\theta} &= \frac{1}{R+z}\left(\frac{\partial W_o}{\partial \theta} - U_o + R\beta\right) \end{aligned} \tag{7.49}$$

With the exception of beams with thin webs, the effects of shear stress and strain in curved beams are generally neglected. Therefore, assuming that $\gamma_{r\theta} = 0$

$$\beta = \frac{1}{R}\left(U_o - \frac{\partial W_o}{\partial \theta}\right)$$

Substitution of this expression for β into ε_θ generates

$$\varepsilon_\theta = \frac{1}{R}\frac{\partial U_o}{\partial \theta} + \frac{1}{R+z}\left(W_o - \frac{z}{R}\frac{\partial^2 W_o}{\partial \theta^2}\right)$$

This expression is identical to that given by Langhar [10], and is applicable to curved beams in which shear deformation is neglected.

For the derivations presented herein, shear deformation is included; therefore, it is assumed that $\gamma_{r\theta} \neq 0$. Equation (7.49) for ε_θ can be put into a form more compatible with classical lamination theory.

$$\varepsilon_\theta = \frac{1}{1+z/R}\left[\frac{1}{R}\left(\frac{\partial U_o}{\partial \theta} + W_o\right) + \frac{z}{R}\frac{\partial \beta}{\partial \theta}\right]$$

The mid-surface strain and curvature for the ring are defined as

$$\begin{aligned}\varepsilon_\theta^o &= \frac{1}{R}\left(\frac{\partial U_o}{\partial \theta} + W_o\right)\\ \kappa_\theta &= \frac{1}{R}\frac{\partial \beta}{\partial \theta}\end{aligned} \tag{7.50}$$

The expression for ε_θ comparable to conventional lamination theory is written as

$$\varepsilon_\theta = \frac{1}{1+z/R}\left[\varepsilon_\theta^o + \kappa_\theta\right] \tag{7.51}$$

The normal and shear stresses at any point in the laminated curved beam are expressed as $(\sigma_\theta)_k = (E_\theta)_k(\varepsilon_\theta)_k$ and $(\tau_{r\theta})_k = (G_{r\theta})_k(\gamma_{r\theta})_k$. Where $(E_\theta)_k$ is defined by Equation (7.17) with the subscript θ replacing the subscript x in that equation. The shear modulus $(G_{r\theta})_k$ is simply the shear modulus G_{13} for the kth lamina.

The resultant forces at a specified section in the beam are expressed as

$$N_\theta = \int \sigma_\theta \mathrm{d}A \quad M_\theta = \int z\sigma_\theta \mathrm{d}A \quad Q_\theta = \int \tau_{r\theta}\mathrm{d}A$$

For the laminated ring, the normal strain in the kth lamina, including thermal and hygral effects, is

$$\varepsilon_\theta = \frac{1}{1+z/R}\left(\varepsilon_\theta^o + \kappa_\theta\right) - (\alpha_\theta)_k\Delta T - (\beta_\theta)_k\bar{M}$$

where α_θ and β_θ for the ring are determined from Equations (3.22) and (3.31), respectively (with the x in these two equations replaced by θ). The β_θ term is the coefficient of hygral expansion and should not be confused with the β used in Equation (7.49). Using this expression for ε_θ, the normal force resultant is

$$\begin{aligned}N_\theta &= \int_{-h/2}^{h/2}(b_k\sigma_\theta)\mathrm{d}z = \sum_{k=1}^{N}\int_{h_{k-1}}^{h_k}(bE_\theta\varepsilon_\theta)_k\mathrm{d}z\\ &= \sum_{k=1}^{N}\int_{h_{k-1}}^{h_k}(bE_\theta)_k\left\{\frac{1}{1+z/R}\left(\varepsilon_\theta^o+\kappa_\theta\right) - (\alpha_\theta)_k\Delta T - (\beta_\theta)_k\bar{M}\right\}\mathrm{d}z\end{aligned}$$

The resultant moment is

$$\begin{aligned}M_\theta &= \int_{-h/2}^{h/2}(b_k\sigma_\theta)z\ \mathrm{d}z = \sum_{k=1}^{N}\int_{h_{k-1}}^{h_k}(bE_\theta\varepsilon_\theta)_k z\ \mathrm{d}z\\ &= \sum_{k=1}^{N}\int_{h_{k-1}}^{h_k}(bE_\theta)_k\left\{\frac{z}{1+z/R}\left(\varepsilon_\theta^o+\kappa_\theta\right) - z(\alpha_\theta)_k\Delta T - z(\beta_\theta)_k\bar{M}\right\}\mathrm{d}z\end{aligned}$$

Shear strain is not affected by thermal and hygral effects. It is a function of U_o, W_o, and β, which remain undetermined at this point. Therefore, the resultant shear force is

$$Q_\theta = \int_{-h/2}^{h/2} (b_k \tau_{r\theta}) \mathrm{d}z = \sum_{k=1}^{N} \int_{h_{k-1}}^{h_k} (bG_{r\theta}\gamma_{r\theta})_k \mathrm{d}z$$

These three equations can be expressed in a simplified manner as

$$\begin{aligned} N_\theta &= A_\theta \varepsilon_\theta^o + B_\theta \kappa_\theta - N_\theta^{\mathrm{T}} - N_\theta^{\mathrm{H}} \\ M_\theta &= B_\theta \varepsilon_\theta^o + D_\theta \kappa_\theta - M_\theta^{\mathrm{T}} - M_\theta^{\mathrm{H}} \\ Q_\theta &= K A_{r\theta} \gamma_{r\theta}^o \end{aligned} \tag{7.52}$$

where K represents the shear correction factor and $\gamma_{r\theta}^o$ is

$$\gamma_{r\theta}^o = \frac{W_{o,\theta} - U_o + R\beta}{R}$$

The coefficients A_θ, B_θ, D_θ, and $A_{r\theta}$ are defined through integration of the load–stress relationships over the laminate thickness, similar to defining the A, B, and D matrices of classical lamination theory. The resulting expressions can be simplified by noting the relationships between t_k, z_k, h_k, and h_{k-1} as illustrated in Chapter 6 for classical lamination theory and in Figure 7.17 for the laminated composite ring.

The resulting definitions of A_θ, B_θ, D_θ, and $A_{r\theta}$ are:

$$\begin{aligned} A_\theta &= \sum_{k=1}^{N} \int_{-h/2}^{h/2} \left(\frac{(bE_\theta)_k}{1 + z/R} \right) \mathrm{d}z = R \sum_{k=1}^{N} (bE_\theta)_k \ln \frac{R + h_k}{R + h_{k-1}} \\ B_\theta &= \sum_{k=1}^{N} \int_{-h/2}^{h/2} \left(\frac{(bE_\theta)_k}{1 + z/R} \right) z \, \mathrm{d}z = R \sum_{k=1}^{N} (bE_\theta)_k \left[t_k - R \ln \frac{R + h_k}{R + h_{k-1}} \right] \\ D_\theta &= \sum_{k=1}^{N} \int_{-h/2}^{h/2} \left(\frac{(bE_\theta)_k}{1 + z/R} \right)^2 \mathrm{d}z = R \sum_{k=1}^{N} (bE_\theta)_k \left[\bar{z}_k t_k - R t_k + R^2 \ln \frac{R + h_k}{R + h_{k-1}} \right] \\ A_{r\theta} &= R \sum_{k=1}^{N} (bG_{r\theta})_k \left\{ \ln \frac{R + h_k}{R + h_{k-1}} - \frac{4}{h^2} \left[\bar{z}_k t_k - R t_k + R^2 \ln \frac{R + h_k}{R + h_{k-1}} \right] \right\} \end{aligned} \tag{7.53}$$

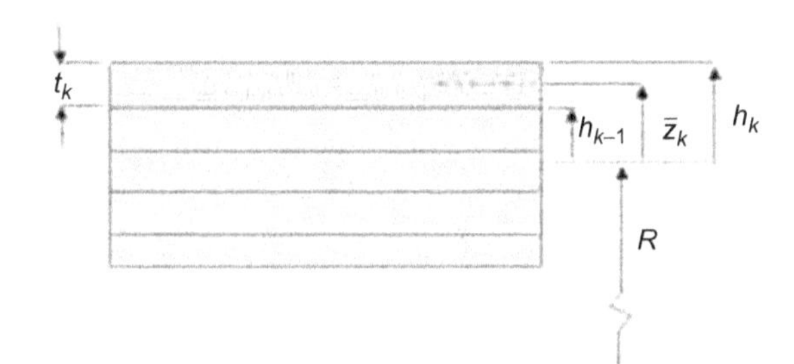

Figure 7.17 Definitions of t_k and $\bar{z}_k$ for a composite ring.

The thermal and hygral loads and moments can also be expressed in a similar manner as

$$N_\theta^{\mathrm{T}} = \sum_{k-1}^{N} (bE_\theta \alpha_\theta t)_k \Delta T \quad N_\theta^{\mathrm{H}} = \sum_{k-1}^{N} (bE_\theta \beta_\theta t)_k \bar{M}$$
$$M_\theta^{\mathrm{T}} = \sum_{k-1}^{N} (bE_\theta \alpha_\theta t\bar{z})_k \Delta T \quad M_\theta^{\mathrm{H}} = \sum_{k-1}^{N} (bE_\theta \beta_\theta t\bar{z})_k \bar{M} \tag{7.54}$$

The equations of motion for the general case, including dynamic effects, are obtained as they were in previous cases. They are similar to those presented for the straight beam and are

$$N_{\theta,\theta} + Q_\theta + RP_\theta = \overline{\rho}_1 \ddot{U}_o + \overline{\rho}_2 \ddot{\beta}$$
$$Q_{\theta,\theta} - N_\theta + RP_r = \overline{\rho}_1 \ddot{W}_o \tag{7.55}$$
$$M_{\theta,\theta} - RQ_\theta + Rm' = \overline{\rho}_1 \ddot{U}_o + \overline{\rho}_3 \ddot{\beta}$$

The $\overline{\rho}_1$, $\overline{\rho}_2$, and $\overline{\rho}_3$ terms are defined by Equations (7.27) through (7.29). For the static case, which is of primary importance to this text, the right-hand side of the Equation (7.55) is zero.

The resultant loads and moments for the composite ring in terms of displacements are:

$$N_\theta = \frac{A_\theta}{R}\left(U_{o,\theta} + W_o\right) + \frac{B_\theta}{R}\beta_{,\theta}$$
$$M_\theta = \frac{B_\theta}{R}\left(U_{o,\theta} + W_o\right) + \frac{D_\theta}{R}\beta_{,\theta} \tag{7.56}$$
$$Q_\theta = \frac{KA_{r\theta}}{R}\left(W_{o,\theta} - U_o + R\beta\right)$$

Taking the appropriate derivatives of these terms and substituting into Equation (7.55) with the right-hand side set equal to zero results in the following equations:

$$\frac{A_\theta}{R}\left(U_{o,\theta\theta} + W_{o,\theta}\right) + \frac{B_\theta}{R}\beta_{,\theta\theta} + \frac{KA_{r\theta}}{R}\left(W_{o,\theta} - U_o + R\beta\right) + RP_\theta = 0$$
$$\frac{B_\theta}{R}\left(U_{o,\theta\theta} + W_{o,\theta}\right) + \frac{D_\theta}{R}\beta_{,\theta\theta} + KA_{r\theta}\left(W_{o,\theta} - U_o + R\beta\right) + Rm' = 0 \tag{7.57}$$
$$\frac{KA_{r\theta}}{R}\left(W_{o,\theta\theta} - U_{o,\theta} + R\beta_{,\theta}\right) - \frac{A_\theta}{R}\left(U_{o,\theta} + W_o\right) - \frac{B_\theta}{R}\beta_{,\theta} + RP_r = 0$$

These equations can be used to determine U_o, W_o, and β. The coupling between terms makes this set of equations harder to solve than those previously presented. Techniques such as energy methods or numerical integration are more appropriate for solving these equations than the simple algebraic manipulation applicable to straight beams. Since energy methods and numerical techniques for laminated beam analysis are beyond the current scope of this text, the displacement fields are not developed. Instead, stress analysis similar to that presented for classical lamination theory is discussed.

7.8.1 Curved beams

The equations of motion defined by Equation (7.57) pertain to the laminated composite ring. For the special case of a curved beam, these equations require modification. In particular, the R in the ring equations is replaced by R_N for the curved beam. The R_N term denotes the radial location of the neutral axis. The location of the neutral axis is defined by an uncoupling of extension and bending deformations. Therefore, the expression for B_θ from Equation (7.53) is set equal to zero and the following equation results

$$\sum_{k=1}^{N}(bE_\theta t)_k - R_N\sum_{k=1}^{N}(bE_\theta)_k \ln\frac{R_c+h_k}{R_c+h_{k-1}} = 0$$

where R_c is the radial location of the centroid of the beam section and R_N is the radial location of the neutral surface of the beam

Solving this equation for R_N results in

$$R_N = \frac{\sum_{k=1}^{N}(bE_\theta t)_k}{\sum_{k=1}^{N}(bE_\theta)_k \ln\frac{R_c+h_k}{R_c+h_{k-1}}} \tag{7.58}$$

Solving Equation (7.57) with $R=R_N$ for a particular beam results in the governing equations of motion relating U_o, W_o, and β to applied loads. Solution of these equations results in explicit representations of ε_θ^o and κ_θ in terms of the displacement fields. As in the case of classical lamination theory, this procedure is not always required. This is particularly true when only stress analysis and not deflection is required. For cases in which only stresses are required, one can revert to the expression for load–strain, defined by Equation (7.52), and written as

$$\begin{Bmatrix} N_\theta \\ M_\theta \end{Bmatrix} = \begin{bmatrix} A_\theta & B_\theta \\ B_\theta & D_\theta \end{bmatrix} \begin{Bmatrix} \varepsilon_\theta^0 \\ \kappa_\theta \end{Bmatrix} - \begin{Bmatrix} N_\theta^{\mathrm{T}} \\ M_\theta^{\mathrm{T}} \end{Bmatrix} - \begin{Bmatrix} N_\theta^{\mathrm{H}} \\ M_\theta^{\mathrm{H}} \end{Bmatrix} \tag{7.59}$$

$$Q_\theta = KA_{r\theta}\gamma_{r\theta}^o$$

Introducing the definitions $\hat{N}_\theta = N_\theta + N_\theta^{\mathrm{T}} + N_\theta^{\mathrm{H}}$ and $\hat{M}_\theta = M_\theta + M_\theta^{\mathrm{T}} + M_\theta^{\mathrm{H}}$, Equation (7.59) can be solved for ε_θ^o and κ_θ as discussed in Chapter 6 for classical lamination theory. This results in

$$\begin{Bmatrix} \varepsilon_\theta^o \\ \kappa_\theta \end{Bmatrix} = \frac{1}{A_\theta D_\theta - B_\theta^2} \begin{bmatrix} D_\theta & -B_\theta \\ -B_\theta & A_\theta \end{bmatrix} \begin{Bmatrix} \hat{N}_\theta \\ \hat{M}_\theta \end{Bmatrix}$$
$$\gamma_{r\theta}^o = \frac{Q_\theta}{KA_{r\theta}} \tag{7.60}$$

The normal and shear stress in the kth lamina can be defined in terms of ε_θ^o, κ_θ, and $\gamma_{r\theta}^o$ as

$$(\sigma_\theta)_k = (E_\theta)_k \left[\varepsilon_\theta^0 + z\kappa_\theta - (\alpha_\theta)_k \Delta T - (\beta_\theta)_k \bar{M}\right]$$
$$(\tau_{r\theta})_k = (G_{r\theta})_k \gamma_{r\theta}^o \tag{7.61}$$

The laminated composite curved beam relations were developed based upon an assumption that extension and bending were uncoupled and $B_\theta = 0$. This assumption results in a simplified set of equations relating the mid-surface strains and curvatures to the applied loads and moments. These relationships are

$$\varepsilon_\theta^o = \frac{\hat{N}_\theta}{A_\theta} \quad \kappa_\theta = \frac{\hat{M}_\theta}{D_\theta} \tag{7.62}$$

The relationship of $\gamma_{r\theta}^o$ remains unchanged.

Example 7.6 As an example of curved beam stress analysis, consider the cantilevered curved beam shown in Figure E7.6-1. An end load F is applied as shown and $P_r = P_\theta = m' = 0$. The free-body diagram for a section of beam as shown can be used to define the resultant loads and moments at an arbitrary position along the beam. The equations of static equilibrium can be developed in either the $r - \theta$ or x–z components, and result in definitions of N_θ, M_θ, and Q_θ as functions of F, R, and θ. These relationships are

$$N_\theta = -F \sin\theta \quad Q_\theta = F \cos\theta \quad M_\theta = -FR \sin\theta$$

The stress distribution through the laminate at any radial and circumferential location is determined from Equation (7.61) once the beam geometry and material are defined. For simplicity, we assume that thermal and hygral effects are neglected. Furthermore, it is assumed that the beam has a uniform width and the relevant material properties (E_1, E_2, G_{12}, and ν_{12}) are the same as those presented in Section 7.4. For computational

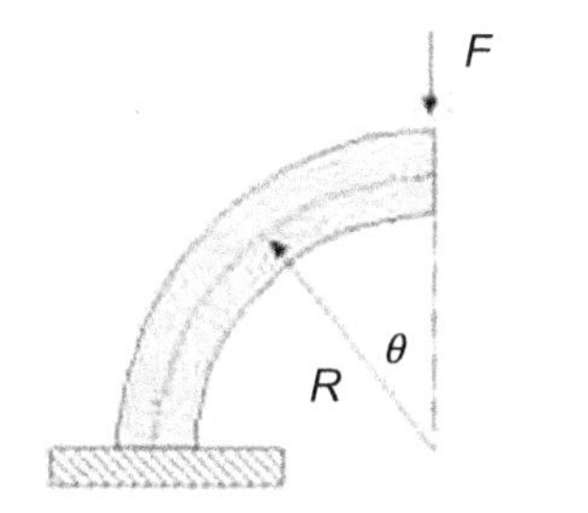

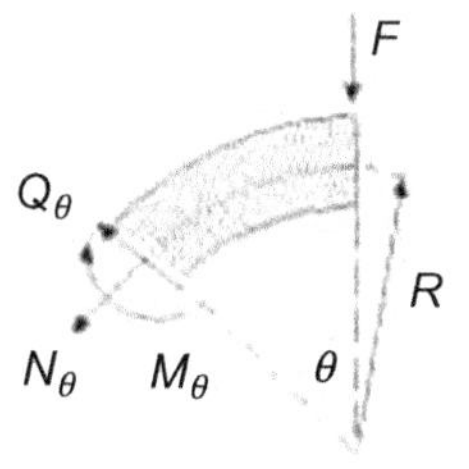

Figure E7.6-1 Cantilevered curved beam.

purposes, it is assumed that $G_{13}=1.0\times10^5$ psi and is constant through the beam thickness. The stacking sequence is $[0/\pm45/90]_s$ and each ply has a thickness of 0.10 in. The width of the beam is a constant $b=1.0$ in. and the centroid of the beam section is assumed to be located at $R_c=10.0$ in. from the center of curvature.

The elastic modulus (E_θ) of each lamina is, as before, $(E_\theta)_0=7.57\times10^6$ psi, $(E_\theta)_{\pm45}=2.92\times10^6$ psi, and $(E_\theta)_{90}=2.03\times10^6$ psi. The location of the neutral axis (R_N) is computed from Equation (8.58). For the simple rectangular section being considered, R_N will coincide with R_c. This, however, is a special case, and for many types of cross sections as well as other stacking arrangements and materials, $R_N\neq R_c$. The procedures for defining R_N are numerically demonstrated for the purpose of illustration. The numerator of Equation (7.58) is

$$\sum_{k=1}^{N}(bE_\theta t)_k=2bt_k\left[(E_\theta)_0+(E_\theta)_{90}+2(E_\theta)_{\pm45}\right]$$
$$=2(1.0)(0.10)\left[7.57\times10^6+2.03\times10^6+2\left(2.92\times10^6\right)\right]=3.09\times10^6$$

Numerically, this corresponds to the A_x term determined in Section 7.4.

The expression for the denominator of Equation (7.58) is not as simple as the numerator. It involves computations which must account for interfacial locations between adjacent lamina through h_k and h_{k-1}. These terms can be either positive or negative, depending upon the lamina being considered. For the lamina below R_c, they are negative, and for those above R_c, they are positive. For a laminate with uniform ply thickness, they can be expressed in terms of multiples of t_k. For example, for the uniform width, constant ply thickness beam under consideration, the R_c+h_k and R_c+h_{k-1} terms for the first ply (the 0° inside ply) are expressed as R_c-4t_k and R_c-3t_k, respectively. Continuing with this notation, the denominator of Equation (7.58) is evaluated from

$$\sum_{k=1}^{N}(bE_\theta)\ \mathrm{ln}\frac{R_c+h_k}{R_c+h_{k-1}}=b\left\{(E_\theta)_0\ \mathrm{ln}\frac{R_c-3t_k}{R_c-4t_k}+(E_\theta)_{45}\ \mathrm{ln}\frac{R_c-2t_k}{R_c-3t_k}+(E_\theta)_{-45}\ \mathrm{ln}\frac{R_c-t_k}{R_c-2t_k}\right.$$
$$\left.+(E_\theta)_{90}\ \mathrm{ln}\frac{R_c}{R_c-t_k}+(E_\theta)_{90}\ \mathrm{ln}\frac{R_c+t_k}{R_c}+(E_\theta)_{-45}\mathrm{ln}\frac{R_c+2t_k}{R_c+t_k}+\ldots\ldots\right\}$$
$$=1.0\left\{7.57\times10^6\ \mathrm{ln}\frac{9.7}{9.6}+2.92\times10^6\ \mathrm{ln}\frac{9.8}{9.7}+\ldots.+2.92\right.$$
$$\left.\times10^6\ \mathrm{ln}\frac{10.3}{10.2}+2.57\times10^6\ \mathrm{ln}\frac{10.4}{10.3}\right\}$$
$$=3.09\times10^5$$

Combining these two numerical values in the appropriate manner results in $R_N=10.0$ in. The numerical values of A_θ, D_θ, and $A_{r\theta}$ are also required. These are computed from Equation (7.53) with R_N (or R_c for this special case) replacing the R in Equation (7.53). The A_θ term can be determined directly from the computation of the denominator of Equation (7.58) by multiplying it by R_N, as can be easily seen by examination of both Equations (7.53) and (7.58). This results in $A_\theta = 3.09 \times 10^6$. The D_θ and $A_{r\theta}$ terms are more cumbersome to evaluate. Following directly from Equation (7.53), D_θ is

$$D_\theta = R\sum_{k=1}^{N}\left(bE_\theta\right)_k\left[\bar{z}_k t_k - Rt_k + R^2 \ln\frac{R+h_k}{R+h_{k-1}}\right] = 10.0(1.0)\left[\left(E_\theta\right)_0\left\{\left(z_0 - 10.0\right)(0.10) + (10)^2 \ln\frac{R_N - 3t_k}{R_N - 4t_k}\right\}\right.$$
$$+\left(E_\theta\right)_{45}\left\{\left(z_{45} - 10.0\right)(0.10) + (10)^2 \ln\frac{R_N - 2t_k}{R_N - 3t_k}\right\} + \left(E_\theta\right)_{-45}\left\{\left(z_{-45} - 10.0\right)(0.10) + (10)^2 \ln\frac{R_N - t_k}{R_N - 2t_k}\right\}$$
$$+\left(E_\theta\right)_{90}\left\{\left(z_{90} - 10.0\right)(0.10) + (10)^2 \ln\frac{R_N}{R_N - t_k}\right\} + \left(E_\theta\right)_{90}\left\{\left(z_{90} - 10.0\right)(0.10) + (10)^2 \ln\frac{R_N + t_k}{R_N}\right\} + \ldots\ldots$$
$$= 2.537 \times 10^5$$

Based on the assumption that $G_{r\theta} = G_{13} = 1.0 \times 10^5$ psi for each ply, and a total laminate thickness of $h = 0.80$ in., $A_{r\theta}$ is

$$A_{r\theta} = R\sum_{k=1}^{N}\left(bG_{r\theta}\right)_k\left\{\ln\frac{R+h_k}{R+h_{k-1}} - \frac{4}{h^2}\left[\bar{z}_k t_k - Rt_k + R^2 \ln\frac{R+h_k}{R+h_{k-1}}\right]\right\}$$
$$= 10.0(1.0)\left(1.0 \times 10^5\right)\left\{\ln\frac{R_N - 3t_k}{R_N - 4t_k} - \frac{4}{(0.8)^2}\left[(z_0 - 10.0)(0.10) + (10)^2 \ln\frac{R_N - 3t_k}{R_N - 4t_k}\right]\right.$$
$$\left.+\ln\frac{R_N - 2t_k}{R_N - 3t_k} - \frac{4}{(0.8)^2}\left[(z_{45} - 10.0)(0.10) + (10)^2 \ln\frac{R_N - 2t_k}{R_N - 3t_k}\right] + \ldots\ldots\right\}$$
$$= 5.0 \times 10^4$$

Using these numerical values of A_θ, D_θ, $A_{r\theta}$ and a shear correction factor of $K = 1.2$, the mid-surface strains and curvature for the beam under consideration are

$$\varepsilon_\theta^o = \frac{-F\sin\theta}{3.09 \times 10^6} = -3.23 \times 10^{-7} F\sin\theta \qquad \kappa_\theta = \frac{-FR_N \sin\theta}{2.537 \times 10^5} = -3.942 \times 10^{-5} F\sin\theta$$
$$\gamma_{r\theta}^o = \frac{F\cos\theta}{1.2\left(5.0 \times 10^4\right)} = 1.667 \times 10^{-5} F\cos\theta$$

The stress distribution through the laminate at an arbitrary θ position can be determined from Equation (7.61).

For beam sections which are not geometrically symmetric, the computations of A_θ, D_θ, and $A_{r\theta}$ become more cumbersome. For example, the beam section shown in Figure E7.6-2 is assumed to have an $R_c = 10.0$ in., but its location is not at the laminate mid-plane. The location of each interface between adjacent lamina $(R_N + h_k)$ and $(R_N + h_{k-1})$ for the $[0/\pm 45/90]_s$ laminate previously considered are indicated in Figure E7.6-2 as are the locations of $\bar{z}_k$. The $\bar{z}_k$ locations are accurate to three decimals, while the interface locations in this figure are rounded off at two decimal points.

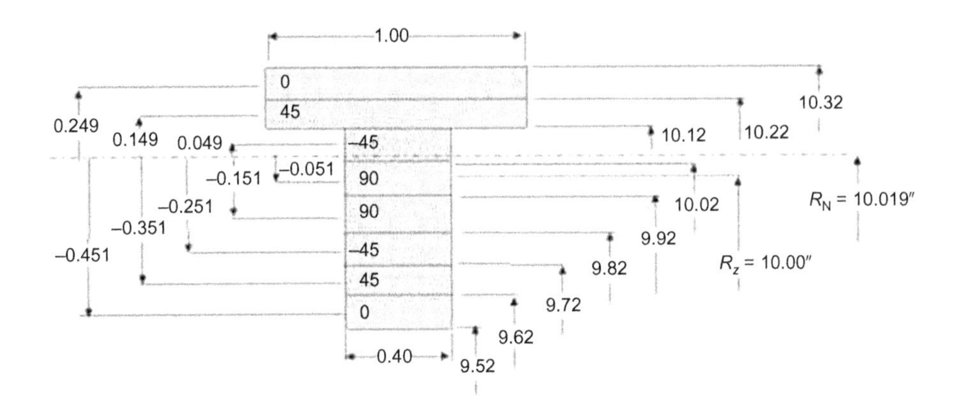

Figure E7.6-2 Curved T-section beam.

The actual values for interface locations are reflected in the following numerical computations. The thickness of each lamina is assumed to be constant ($t_k = 0.10$ in.), but the ply widths are variable.

The neutral axis location can be determined as before, from Equation (7.58). The numerator of this expression is

$$\sum_{k=1}^{N}(bE_\theta t)_k = (0.10)\{7.57\times10^6(0.4+1.0)+2.03\times10^6(0.4+0.4)$$
$$+2.92\times10^6(0.4+0.4+0.4+1.0)\} = 1.865\times10^6$$

The denominator of the equation is

$$\sum_{k=1}^{N}(bE_\theta)\ln\frac{R_c+h_k}{R_c+h_{k-1}} = 7.57\times10^6\left[(0.4)\ln\frac{9.618}{9.518}+(1.0)\ln\frac{10.318}{10.218}\right]$$
$$+2.92\times10^6\left[(0.40)\left\{\ln\frac{9.718}{9.618}+\ln\frac{9.818}{9.718}+\ln\frac{10.118}{10.018}\right\}+(1.0)\ln\frac{10.128}{10.118}\right]$$
$$+2.03\times10^6\left[(0.4)\left\{\ln\frac{9.918}{9.818}+\ln\frac{10.018}{9.918}\right\}\right] = 1.8598\times10^5$$

Combining these two values results in $R_N = 1.863\times10^6/1.8598\times10^5 = 10.019$. The neutral axis location is slightly above the centroidal axis location. Different lamina stacking arrangements, as well as different materials affects the location. Using Equation (7.53), with R_N replacing R results in $A_\theta = 1.863\times10^6$. In computing D_θ, z_k is measured from the centroidal axis, not the neutral axis. This is consistent with conventional curved beam analysis procedures [8]. Therefore, D_θ is given as

$$D_\theta = R\sum_{k=1}^{N}(bE_\theta)_k\left[\bar{z}_k t_k - Rt_k + R^2\ln\frac{R+h_k}{R+h_{k-1}}\right] = 1.3104\times10^5$$

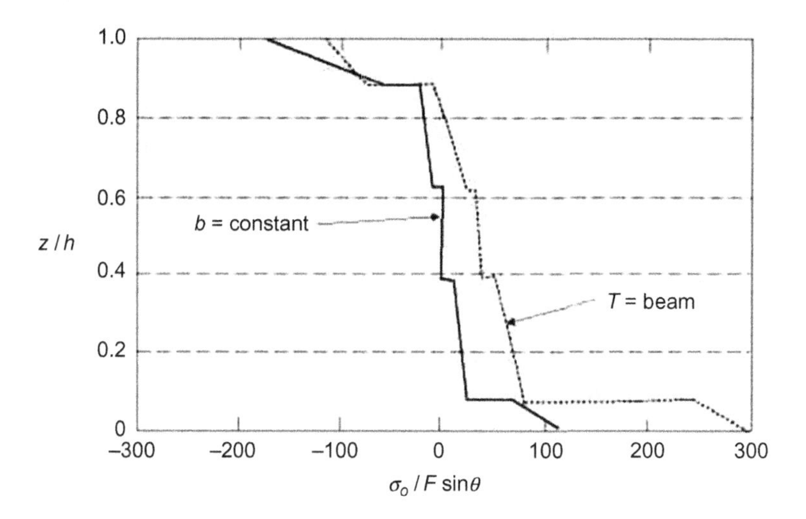

Figure E7.6-3 Normal stress distribution through curved cantilevered beam.

The $A_{r\theta}$ term is computed in a similar manner, and is $A_{r\theta} = 1.2186 \times 10^3$. These values of A_θ, D_θ, and $A_{r\theta}$ result in mid-surface strains and curvatures of

$$\varepsilon_\theta^o = -5.367 \times 10^{-7} F\sin\theta \quad \kappa_\theta = -7.645 \times 10^{-5} F\sin\theta$$

$$\gamma_{r\theta}^o = 6.838 \times 10^{-4} F\cos\theta$$

The normal stress, $(\sigma_\theta)_k = (E_\theta)_k\left[\varepsilon_\theta^0 + z\kappa_\theta\right]$, is determined from Equation (7.61). The stress ($\sigma_\theta/F \sin\theta$) for the constant width (b=constant) and variable width (T-beam) beam sections as a function of position (z/h) measured from the bottom of each beam section is shown in Figure E7.6-3. The shear stress distribution through a selected section of beam can be determined for each beam in the same manner as previously described. Changes in cross-section geometry, fiber orientation, materials, and loading conditions have different effects on both deformations and stress distributions through the curved composite beam. The analysis procedures remain the same.

7.9 Beam vibrations

The process of analyzing the vibration of composite beams is analogous to that of isotropic beams. Solutions to beam vibration problems generally consist of complimentary and particular solutions. The increased use of composites in the aircraft industry generated a need for the assessment of beam vibrations (rotary wing aircraft in particular). A number of significant contributions to the analysis of composite beam vibrations have been made over the years. These would include the anticipated free-vibration analysis as well as finite element analysis with both simple and higher order approximations, such as in Refs. [11,12]. The intent of this section is to illustrate the approach to composite beam analysis, not to explore all possible scenarios.

When considering beam vibrations, we begin with the equations of motion developed in Section 7.4 (Equations 7.24–7.26). Recall that the general forms of the equations of motion for a beam are

$$N_{x,x} + p_x(x,t) = \overline{\rho}_1 \ddot{U}_o + \overline{\rho}_2 \ddot{\Phi}$$

$$Q_{x,x} + q(x,t) = \overline{\rho}_1 \ddot{W}_o$$

$$M_{x,x} - Q_x + m_x(x,t) = \overline{\rho}_2 \ddot{U}_o + \overline{\rho}_3 \ddot{\Phi}$$

where $\overline{\rho}_1, \overline{\rho}_2, \overline{\rho}_3 = \sum_{k=1}^{N} \int_{h_{k-1}}^{h_k} b\left(1, z, z^2\right) \rho^{(k)} \mathrm{d}z$

In addition to the equations of motion, the load–deformation relationships are also needed.

$$N_x = A_x \varepsilon_x^o + B_x \kappa_x, \quad M_x = B_x \varepsilon_x^o + D_x \kappa_x, \quad Q_x = K A_{55} \gamma_{xz}$$

where $\varepsilon_x^o = U_{o,x}$, $\kappa_x = \Phi_{,x}$, and $\gamma_{xz} = \phi_x + W_{o,x}$

Thermal and hygral effects could be included if needed. They are neglected for this presentation. The next step in the procedure is to develop a relationship between displacements (U_o, W_o) and the natural frequency of the system.

7.9.1 Neglecting transverse shear

Neglecting the effects of transverse shear, we assume that $\gamma_{xz} = 0 \Rightarrow \Phi = -W_{o,x}$. Therefore, $\kappa_x = \Phi_{,x} = -W_{o,xx}$. The equations of motion are therefore expressed as

$$A_x U_{o,xx} - B_x W_{o,xxx} + p_x(x,t) = \overline{\rho}_1 \ddot{U}_o - \overline{\rho}_2 \ddot{W}_{o,x} \quad \text{(a)}$$

$$B_x U_{o,xxx} - D_x W_{o,xxxx} + q(x,t) + m_{x,x} = \overline{\rho}_1 \ddot{W} + \overline{\rho}_2 \ddot{U}_{o,x} - \overline{\rho}_3 \ddot{W}_{o,xx} \quad \text{(b)}$$

These two equations are (7.30) and (7.31) with the thermal and hygral effects eliminated. In order to solve these equations, we begin by taking the partial derivative of Equation (a) and rewriting it in the form

$$U_{o,xxx} = \frac{B_x}{A_x} W_{o,xxxx} - \frac{p_{x,x}}{A_x} + \frac{\overline{\rho}_1}{A_x} \ddot{U}_{o,x} - \frac{\overline{\rho}_2}{A_x} \ddot{W}_{o,xx} \quad \text{(c)}$$

Substituting this into Equation (b) and rearranging terms

$$\left(B_x^2 - A_xD_x\right)W_{o,xxxx} = B_xp_{x,x} - A_xq(x) - A_xm_{x,x} + A_x\overline{\rho}_1\ddot{W}_o + \left(A_x\overline{\rho}_2 - B_x\overline{\rho}_1\right)\ddot{U}_{o,x} + \left(B_x\overline{\rho}_2 - A_x\overline{\rho}_3\right)\ddot{W}_{o,xx} \tag{d}$$

Solving Equation (c) for $\ddot{U}_{o,x}$ and substituting it into Equation (d)

$$\left(\overline{\rho}_1B_x - \overline{\rho}_2A_x\right)U_{o,xxx} = \left(\overline{\rho}_1D_x - \overline{\rho}_2B_x\right)W_{o,xxxx} - \overline{\rho}_1q(x,t) + \overline{\rho}_2p_{x,x} - \overline{\rho}_1m_{x,x} + \overline{\rho}_1^2\ddot{W}_o + \left(\overline{\rho}_2^2 - \overline{\rho}_1\overline{\rho}_3\right)\ddot{W}_{o,xx} \tag{e}$$

Next, we take $\dfrac{\partial^2}{\partial t^2}$(e) and solve for $\ddot{U}_{o,xxx}$ and substituting this into $\dfrac{\partial^2}{\partial x^2}$(d) gives

$$\begin{aligned}&\left(B_x^2 - A_xD_x\right)W_{o,xxxxxx} - A_x\overline{\rho}_1\ddot{W}_{o,xx} - \left(2B_x\overline{\rho}_2 - A_x\overline{\rho}_3 - D_x\overline{\rho}_1\right)\ddot{W}_{o,xxxx} + \overline{\rho}_1^2\ddddot{W}_o \\ &- \left(\overline{\rho}_2^2 - \overline{\rho}_1\overline{\rho}_3\right)\ddddot{W}_{o,xx} - B_xp_{x,xxx} + A_xq_{,xx} + A_xm_{x,xx} + \overline{\rho}_1\ddot{q} - \overline{\rho}_2\ddot{p}_{x,x} + \overline{\rho}_1\ddot{m}_{x,x} = 0\end{aligned} \tag{7.63}$$

Similarly, we can eliminate W_o from the two expressions (a) and (b). The result is

$$\begin{aligned}&\left(B_x^2 - A_xD_x\right)U_{o,xxxxxx} + \left(D_x\overline{\rho}_1 + A_x\overline{\rho}_3 - 2B_x\overline{\rho}_2\right)\ddot{U}_{o,xxxx} - \left(\overline{\rho}_2^2 - \overline{\rho}_1\overline{\rho}_3\right)\ddddot{U}_{o,xx} - \overline{\rho}_1A_x\ddot{U}_{o,xx} \\ &+ \overline{\rho}_1^2\ddddot{U}_o - D_xp_{x,xxx} + B_x\left(q_{,xxx} + m_{x,xxxx}\right) + \overline{\rho}_3\ddot{p}_{x,xx} - \overline{\rho}_1\ddot{p} - \overline{\rho}_2\left(\ddot{q}_{,x} + \ddot{m}_{x,xx}\right) = 0\end{aligned} \tag{7.64}$$

The solution of these equations is typically addressed assuming an exponential solution for W_o and U_o. Prior to that, we will define two parameters, which depend on the loading condition.

$$A_1 \equiv A_x\left(q_{,xx} + m_{x,xx}\right) - B_xp_{x,xxx} + \overline{\rho}_1\left(\ddot{q} + \ddot{m}_{x,x}\right) - \overline{\rho}_2\ddot{p}_{x,x}$$

$$A_2 \equiv B_x\left(q_{,xxx} + m_{x,xxx}\right) - D_xp_{x,xxxx} - \overline{\rho}_1\ddot{p}_x - \overline{\rho}_2\left(\ddot{q}_{,x} + \ddot{m}_{x,xx}\right) + \overline{\rho}_3\ddot{p}_{x,xx}$$

For the case of free vibrations, the terms A_1 and A_2 are zero $(A_1 = A_2 = 0)$. We can assume a solution in the form $U_o(x,t) = g(x)e^{i\omega t}$ and $W_o(x,t) = f(x)e^{i\omega t}$. Substituting these into the equations of motion above results in

$$\begin{aligned}&\left\{\left(B_x^2 - A_xD_x\right)f_{,xxxxxx} + \omega^2\left(D_x\overline{\rho}_1 - 2B_x\overline{\rho}_2 + A_x\overline{\rho}_3\right)f_{,xxxx} + \omega^4\left(\overline{\rho}_1\overline{\rho}_3 - \overline{\rho}_2^2\right)f_{,xx}\right. \\ &\left. + \omega^4\overline{\rho}_1^2f - \omega^2A_x\overline{\rho}_1f_{,xx}\right\}e^{i\omega t} = 0\end{aligned} \tag{7.65}$$

$$\begin{aligned}&\left\{\left(B_x^2 - A_xD_x\right)g_{,xxxxxx} + \omega^2\left(D_x\overline{\rho}_1 - 2B_x\overline{\rho}_2 + A_x\overline{\rho}_3\right)g_{,xxxx}\right. \\ &\left. + \omega^4\left(\overline{\rho}_1\overline{\rho}_3 - \overline{\rho}_2^2\right)g_{,xx} + \omega^4\overline{\rho}_1^2g - \omega^2A_x\overline{\rho}_1g_{,xx}\right\}e^{i\omega t} = 0\end{aligned} \tag{7.66}$$

The solution of these equations depends on the boundary conditions, in general, the solution for free vibrations.

7.9.2 Including shear deformation

If shear deformation is being considered, we note that $Q_x = KA_{55}(\Phi + W_{o,x})$ and start with the general relationships

$$A_x U_{o,xx} + B_x \Phi_{,xx} + p_x(x) = \overline{\rho}_1 \ddot{U}_o + \overline{\rho}_2 \ddot{\Phi}$$

$$B_x U_{o,xx} + D_x \Phi_{,xx} + m(x,t) - Q_x = \overline{\rho}_2 \ddot{U}_o + \overline{\rho}_3 \ddot{\Phi}$$

$$KA_{55}(\Phi_{,x} + W_{o,xx}) + q(x,t) = \overline{\rho}_1 \ddot{W}_o$$

If $p_x = m_x = 0$ and the beam is symmetric $(B_x = \overline{\rho}_2 = 0)$, these equations reduce to

$$A_x U_{o,xx} = \overline{\rho}_1 \ddot{U}_o \quad \text{(a)}$$

$$D_x \Phi_{,xx} - KA_{55}(\Phi + W_{o,x}) = \overline{\rho}_3 \ddot{\Phi} \quad \text{(b)}$$

$$KA_{55}(\Phi_{,x} + W_{o,xx}) + q(x,t) = \overline{\rho}_1 \ddot{W}_o \quad \text{(c)}$$

Solving (c) for $\Phi_{,x}$

$$\Phi_{,x} = \frac{1}{KA_{55}}(\overline{\rho}_1 \ddot{W}_o - q) - W_{o,xx} \quad \text{(d)}$$

taking $\frac{\partial}{\partial x}[(\text{b})] = D_x \Phi_{,xxx} - KA_{55}(\Phi_{,x} + W_{o,xx}) = \overline{\rho}_3 \ddot{\Phi}_{,x}$. Using (d) and manipulating the equations finally results in

$$D_x W_{o,xxxx} + (\overline{\rho}_1 \ddot{W}_o - q) - \overline{\rho}_3 \ddot{W}_{o,xx} - \frac{D_x}{KA_{55}}(\overline{\rho}_1 \ddot{W}_o - q)_{,xx} + \frac{\overline{\rho}_3}{KA_{55}}(\overline{\rho}_1 \ddot{W}_o - q)_{,tt} = 0 \quad (7.67)$$

This equation can be thought of as a combination of contributing terms. These can be identified as

Bernoulli–Euler theory: $D_x W_{o,xxxx} + (\overline{\rho}_1 \ddot{W}_o - q)$

Principal rotary inertia: $\overline{\rho}_3 \ddot{W}_{o,xx}$

Principal SDT: $\frac{D_x}{KA_{55}}(\overline{\rho}_1 \ddot{W}_o - q)_{,xx}$

Combined rotary inertia and shear deformation: $\frac{\overline{\rho}_3}{KA_{55}}(\overline{\rho}_1 \ddot{W}_o - q)_{,tt}$

Example 7.7 Consider the free vibration of the simply supported beam shown in Figure E7.7-1.

Assume $L = 18''$, $h = 0.5''$, $b = 1.0''$, $\overline{\rho}_1 = 0.05$, and $KA_{55} = 0.50 \times 10^6$. Using solutions with and without shear deformation, determine the variation in first-mode natural frequencies for a beam for which D_x represents a range in laminate stacking arrangements from a 90° laminate $(D_x = 0.12 \times 10^6)$ to a 0° laminate $(D_x = 2.1 \times 10^6)$.

Figure E7.7-1 Simply supported beam.

For the simply supported beam shown, the boundary conditions are $W_o(0)=W_o(L)=0$, $M_x(0)=M_x(L)=0$, and $N_x(0)=N_x(L)=0$.

Without shear deformation: Assuming solutions for $W_o(x,t)$ and $U_o(x,t)$ that satisfy the boundary conditions, we can assume

$$W_o(x,t)=C_{1m}e^{i\omega_m t}\sin\left(\frac{m\pi x}{L}\right) \quad U_o(x,t)=C_{2m}e^{i\omega_m t}\cos\left(\frac{m\pi x}{L}\right)$$

where C_{1m} and C_{2m} are the maximum amplitude of vibration and m is the index of the wave number associated with the mth mode of vibration. Differentiating these displacements, the appropriate number of times and substituting into Equation (7.65) or (7.66), the frequency equation is developed. It is the same for each equation.

$$\left(B_x^2-A_xD_x\right)\left(\frac{m\pi}{L}\right)^6+\omega_m^2(D_x\overline{\rho}_1-2B_x\overline{\rho}_2+A_x\overline{\rho}_3)\left(\frac{m\pi}{L}\right)^4$$
$$+\omega_m^4\left(\overline{\rho}_1\overline{\rho}_3-\overline{\rho}_2^2\right)\left(\frac{m\pi}{L}\right)^2-\omega_m^4\overline{\rho}_1^2-\omega_m^2A_x\overline{\rho}_1\left(\frac{m\pi}{L}\right)^2=0$$

Letting $a_m=\left(\overline{\rho}_1\overline{\rho}_3-\overline{\rho}_2^2\right)\left(\frac{m\pi}{L}\right)^2-\overline{\rho}_1^2$, $b_m=(D_x\overline{\rho}_1-2B_x\overline{\rho}_2+A_x\overline{\rho}_3)\left(\frac{m\pi}{L}\right)^4-A_x\overline{\rho}_1\left(\frac{m\pi}{L}\right)^2$, and $c_m=\left(B_x^2-A_xD_x\right)\left(\frac{m\pi}{L}\right)^6$ we rewrite this as

$$a_m\omega_m^4+b_m\omega_m^2+c_m=0$$

The roots of this equation are $\omega_m^2=\frac{1}{2a_m}\left[-b_m\pm\sqrt{b_m^2-4a_mc_m}\right]$. Two sets of roots are obtained for each mode of vibration, the *extensional* and *flexural* (generally smaller than the extensional mode). This solution becomes simplified if extension-bending coupling does not exist ($B_x=0$), which results in $\overline{\rho}_2=0$. For vibrations which are considered large compared to either the characteristic wave length or beam thickness, the rotary inertia term is neglected ($\overline{\rho}_3=0$). Under these conditions ($B_x=0$, $\overline{\rho}_2=0$, $\overline{\rho}_3=0$), the equations of motion become

$$A_xU_{o,xx}+p_x=\overline{\rho}_1\ddot{U}_o \quad -D_xW_{o,xxxx}+q+m_{x,x}=\overline{\rho}_1\ddot{W}_o$$

These uncoupled equations are of the same form as a uniform bar or beam made from isotropic materials.

With shear deformation: For this case, we have $q=0$ and boundary conditions $W_o(0,t)=W_o(L,t)=0$ and $\Phi_{,x}(0,t)=\Phi_{,x}(L,t)=0$. This results in

$$\Phi_{,x}=\frac{\overline{\rho}_1}{KA_{55}}\ddot{W}_o-W_{o,xx}$$

Equation (7.67) becomes

$$D_x W_{o,xxxx} + \overline{\rho}_1 \ddot{W}_o - \left(\overline{\rho}_3 + \frac{\overline{\rho}_1 D_x}{KA_{55}}\right)\ddot{W}_{o,xx} + \frac{\overline{\rho}_1 \overline{\rho}_3}{KA_{55}} \ddddot{W}_o = 0$$

Assuming simple harmonic motion with $W_o(x, t) = W(x)\cos\omega t$, we get

$$\Phi_{,x} = -\left[W_{,xx} + \frac{\overline{\rho}_1 \omega^2}{KA_{55}} W\right]\cos\omega t$$

$$W_{o,xxxx} - \frac{\overline{\rho}_1 \omega^2}{D_x} W + \frac{1}{D_x}\left(\overline{\rho}_3 + \frac{\overline{\rho}_1 D_x}{KA_{55}}\right)\omega^2 W_{,xx} + \frac{\overline{\rho}_1 \overline{\rho}_3\, \omega^4}{KA_{55} D_x} W = 0$$

The boundary condition $\Phi_{,x} = 0$ implies $W_{,xx} = 0$ at $x = 0$ & L. Therefore, we assume a solution of the form $W_n(x) = C\sin\left(\frac{n\pi x}{L}\right)$, which results in

$$\left(\frac{n\pi}{L}\right)^4 - \frac{\overline{\rho}_1 \omega_n^2}{D_x} + \frac{1}{D_x}\left(\overline{\rho}_3 + \frac{\overline{\rho}_1 D_x}{KA_{55}}\right)\left(\frac{n\pi}{L}\right)^2 \omega_n^2 + \frac{\overline{\rho}_1 \overline{\rho}_3}{KA_{55} D_x}\omega_n^4 = 0$$

This can be manipulated to arrive at the relationship

$$\omega_n^4 - \left[\frac{KA_{55}}{\overline{\rho}_3} + \left(\frac{n\pi}{L}\right)^2\left(\frac{KA_{55}}{\overline{\rho}_1} + \frac{D_x}{\overline{\rho}_3}\right)\right]\omega_n^2 + \frac{KA_{55} D_x}{\overline{\rho}_1 \overline{\rho}_3}\left(\frac{n\pi}{L}\right)^4 = 0$$

The resulting solution to this equation is

$$\omega_n^2 = \frac{1}{2}\left(\frac{KA_{55}}{\overline{\rho}_3}\right)\left\{1 + \left(\frac{n\pi}{L}\right)^2\left(\frac{\overline{\rho}_3}{\overline{\rho}_1} + \frac{D_x}{KA_{55}}\right) \pm \sqrt{1 + 2\left(\frac{n\pi}{L}\right)^2\left(\frac{\overline{\rho}_3}{\overline{\rho}_1} + \frac{D_x}{KA_{55}}\right) + \left(\frac{n\pi}{L}\right)^4\left(\frac{\overline{\rho}_3}{\overline{\rho}_1} + \frac{D_x}{KA_{55}}\right)}\right\}$$

If the rotary inertia term is neglected ($\overline{\rho}_3 = 0$), we get

$$\left[KA_{55} + \left(\frac{n\pi}{L}\right)^2 D_x\right]\omega_n^2 + \frac{KA_{55} D_x}{\overline{\rho}_1}\left(\frac{n\pi}{L}\right)^4 = 0$$

The result is, the natural frequency can be written as

$$\omega_n^2 = \frac{\dfrac{D_x}{\overline{\rho}_1}\left(\dfrac{n\pi}{L}\right)^4}{1 + \left(\dfrac{n\pi}{L}\right)^2\left(\dfrac{D_x}{KA_{55}}\right)}$$

The difference between the first-mode natural frequencies for the desired range of D_x is shown in Figure E7.7-2. We note that the difference in predicted frequencies

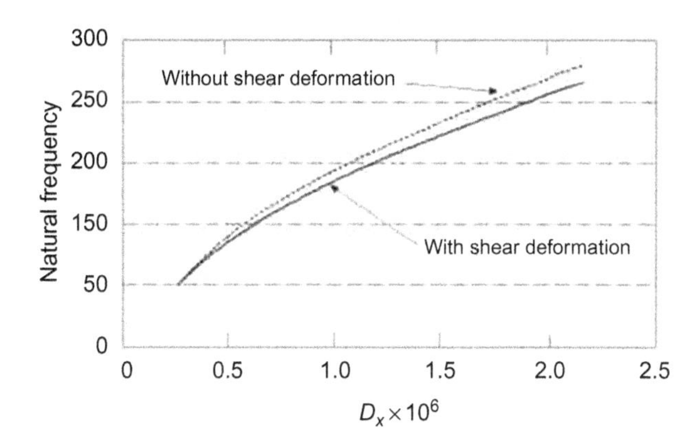

Figure E7.7-2 Difference in first-mode frequencies resulting from shear deformation.

between beams with shear deformation and those without increases with increasing D_x. For a 0° laminate $\left(D_x = 2.1 \times 10^6\right)$, the difference is approximately 6%.

7.10 Problems

7.1 Assume simple beam theory is applicable for the simply supported beam shown. Furthermore, assume the beam is made from a continuous fiber composite material with the properties $E_1 = 22 \times 10^6$ psi, $E_2 = 1.6 \times 10^6$ psi, $G_{12} = 0.8 \times 10^6$ psi, $\upsilon_{12} = 0.30$, and $t_k = 0.020$ in. The beam dimensions are $b = 1.0$ in. and $L = 12.0$ in.

(A) Derive the expression for beam deflection, $W(x)$.

(B) Assume the beam is constructed with lamina stacking sequences of $[0/\pm 45/90]_s$ and $[90/\pm 45/0]_s$. On the same graph, plot $W(x)/q_0$ versus x for both stacking sequences.

(C) Plot the stress distribution through the beam at its spanwise location where the bending moment is maximum.

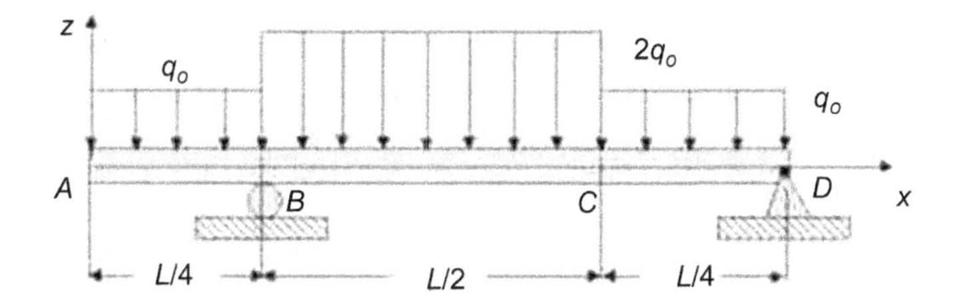

7.2 Assume simple beam theory is applicable for the propped cantilever beam shown. Furthermore, assume the beam is made from a continuous fiber composite material. Derive the expression for beam deflection, $W(x)$.

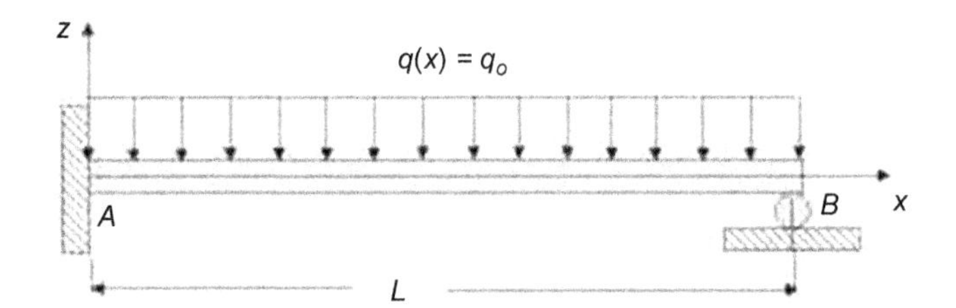

7.3 For the propped cantilever composite beam shown, determine expressions for $W(x)$ and $U(x)$. Assume simple beam theory.

(A) Integrate the expression $\mathrm{d}^4W(x)/\mathrm{d}x^4 = q(x)$ to determine $W(x)$. The beam width is b.

(B) Assume $L = 12''$, $b = 1''$, and D_{11} can have various values. Plot $D_{11}W(x)/q_0$ versus x.

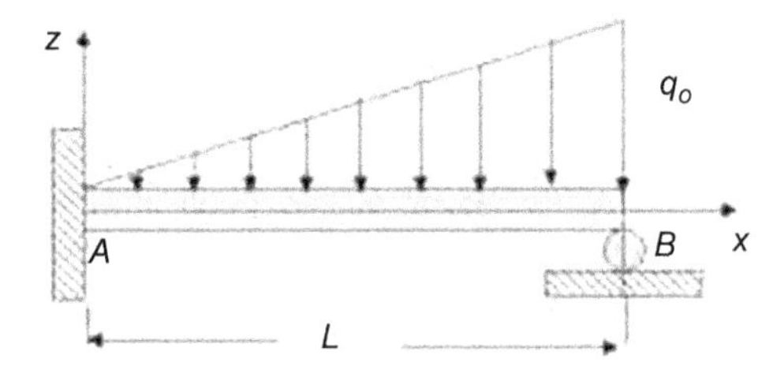

7.4 The simply supported beam shown is made from carbon/epoxy with material properties $E_1 = 21 \times 10^6$ psi, $E_2 = 1.76 \times 10^6$ psi, $G_{12} = 0.65 \times 10^6$ psi, and $\upsilon_{12} = 0.21$. The beam is made from 30 lamina oriented at 0° to the x-axis, and each lamina has a thickness of $t_k = 0.006$ in. The beam dimensions are $b = 1.0$ in. and $L = 12.0$ in. The applied load is $q_0 = 10\,\text{lb/in}$. Assume simple beam theory.

(A) Determine a general expression for the deflection of the beam in terms of the applied load q_0 and the length L.

(B) For the dimensions and load given, determine the maximum deflection of the beam.

(C) Determine the maximum normal stress in the beam.

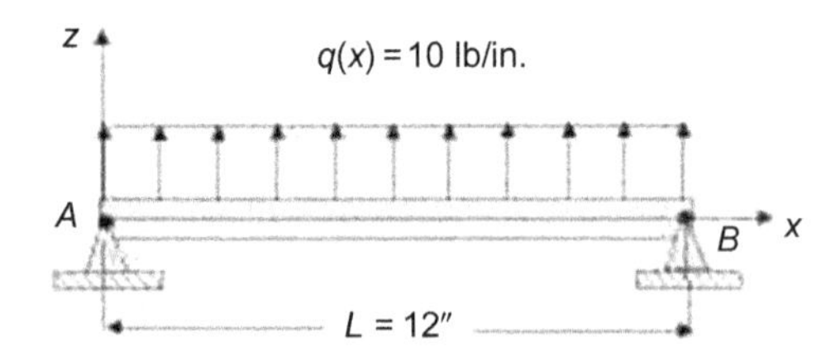

7.5 Assume simple beam theory is applicable for the beam shown and it is made from the same material as the beam described in problem 4. The beam length and width are $L = 20.0$ in. and $b = 1.0$ in., and $q_0 = 20\,\text{lb/in}$. Assuming that the maximum normal stress the material can sustain is $\sigma_{max} = 100$ ksi, determine the required beam height and the number of lamina required, assuming all lamina are oriented at 0° with respect to the x-axis.

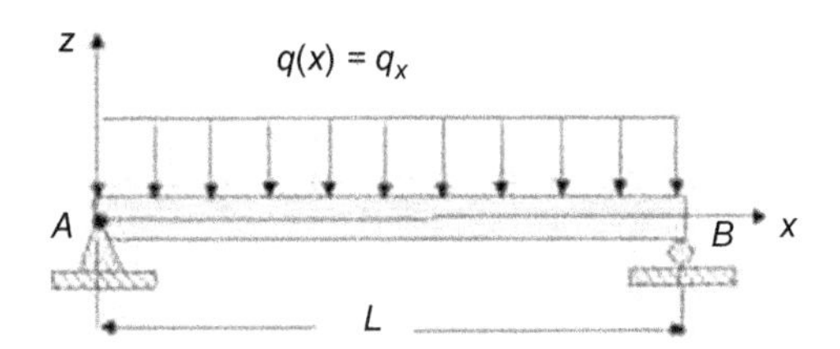

7.6 Consider the general laminated composite beam loaded and supported as shown. The beam has a uniform width of $b = 1.0$ in. Neglect shear deformation.

(A) Use advanced beam theory to derive expressions for $U_o(x)$, $W_o(x)$, $N_x(x)$, $M_x(x)$, $\sigma_x(x)$, and $\tau_{xz}(x)$.

(B) Assume that $P = 2qL$, $L = 18$ in., $a = 10$ in., $b = 8$ in., and the laminate stacking arrangement is $[0/30/45/60/90]_s$. Plot $U_o(x)/q$ and $W_o(x)/q$ along the span of the beam. In addition, plot through the thickness variation of $\sigma_x(x)/q$ and $\tau_{xz}(x)/q$ at $x = a = 10$ in. For this stacking arrangement, we know that $A_x = 20.415 \times 10^6$ and $D_x = 8.42 \times 10^4$.

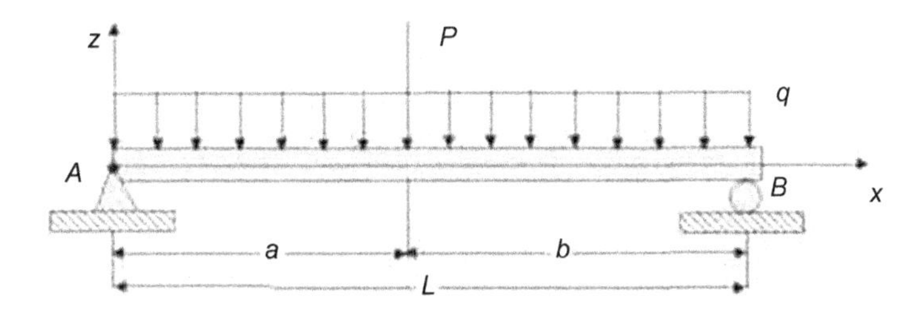

7.7 The beam loaded and supported as shown has a uniform width of $b = 1.0$ in. and a stacking arrangement of $[0/\pm 45/90]_s$. It is subjected to a uniform temperature change of $\Delta T = -200\,°F$ through its thickness and along its span. Use advanced beam theory, neglecting shear deformation, and plot $W_o(x)$ knowing that $A_x = 14.920 \times 10^6$ and $D_x = 4.890 \times 10^4$.

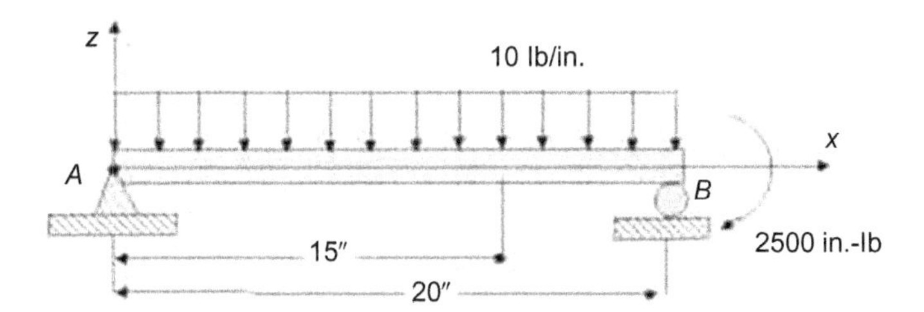

7.8. The simply supported composite beam shown is loaded with an eccentric axial force N at distance e above the mid-plane of the beam. Assume a general ply configuration for the laminate ($[B] \neq 0$). Neglecting shear deformation, use advance beam theory to determine expressions for $W_0(x)$ and $U_0(x)$.

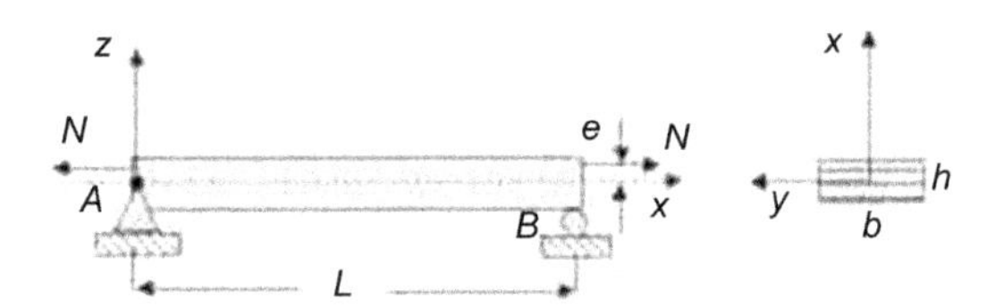

7.9 For the beam shown, the mid-surface displacements are given by

$$U_0(x) = \left(\frac{e}{\bar{B}_x} - \frac{1}{\bar{A}_x}\right) Nx \quad W_0(x) = \left(\frac{e}{2\bar{D}_x} - \frac{1}{2\bar{B}_x}\right)\left[\left(\frac{x}{L}\right)^2 - \left(\frac{x}{L}\right)\right] L^2 N$$

Assume two $[0/90_2/\pm45]_s$ laminate beams with cross sections as shown have been constructed from a material with the properties $E_1 = 7.8 \times 10^6$ psi, $E_2 = 2.6 \times 10^6$ psi, $G_{12} = 1.25 \times 10^6$ psi, $\nu_{12} = 0.25$, $t_k = 0.10$ in. The beam dimensions are such that $h = 1.0$ in., $e = h/4$, and $L = 24.0$ in. We can show that for the constant width beam $A_x = 3.88 \times 10^6$, $B_x = 0$, $D_x = 0.431 \times 10^6$, and for the variable width beam $A_x = 2.53 \times 10^6$, $B_x = 0$, $D_x = 0.394 \times 10^6$. Neglect shear deformation and develop an expression for the normal stress distribution through the beam in terms of $(\sigma_x)_k/N$. Plot this normal stress distribution through the laminate at $x = L/2$.

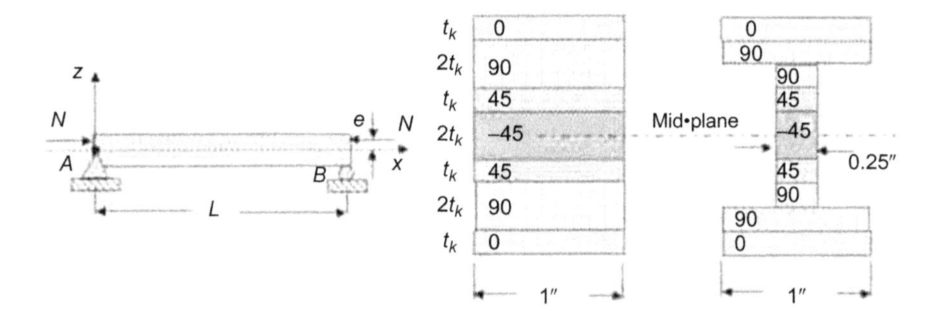

7.10 A $[0/\pm45_2/90_2]_s$ laminated composite beam is loaded and supported as shown. The beam is $b = 1.0''$ wide and each lamina has a thickness of $t_k = 0.05''$. The material properties for the composite material are $E_1 = 7.8 \times 10^6$ psi, $E_2 = 2.6 \times 10^6$ psi, $G_{12} = 1.25 \times 10^6$ psi, and $\nu_{12} = 0.25$. Plot, on the same graph, $W_o(x)/q_o$ for the beam assuming shear deformation is neglected, and shear deformation is considered, with $KA_{55} = 4 \times 10^5$.

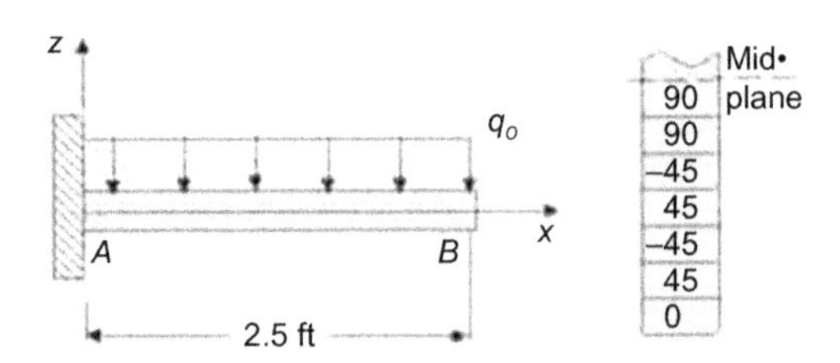

7.11 A simply supported, constant width (b), unidirectional laminate is subjected to the temperature profile shown along its entire span. Neglect shear deformation. The temperature distribution through the beam thickness is given as

$$\Delta T = \frac{T_o}{4}\left(1 + \frac{2z}{h}\right)^2$$

(A) Develop expressions for N_x^{T} and M_x^{T} in terms of E_x, b, α_x, T_o, and h.
(B) Develop expressions for $U_o(x)$, $W_o(x)$, $\sigma_x(x)$, and $\tau_{xz}(x)$ in terms of E_x, b, α_x, T_o, and h.

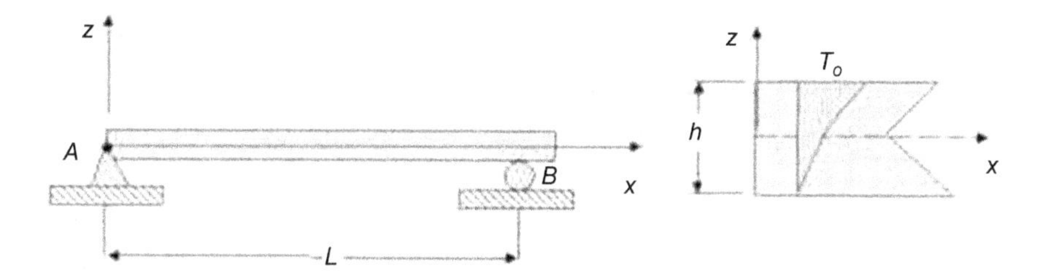

7.12 For the general laminated composite beam loaded and supported as shown, determine general expressions for $U_o(x)$ and $W_o(x)$.

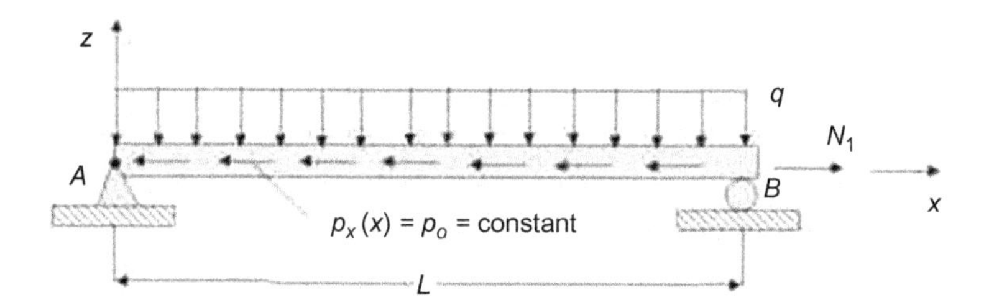

7.13 Determine the expression for the critical buckling load of a symmetric ($[B]=0$) fixed-simply supported laminated composite column. Neglect shear deformation.

7.14 Determine the relationships for ω_n and $U_n(x)$ for the free axial vibration a fixed–fixed beam as shown. Begin with $U_{o,xx} - \frac{\overline{p}_1}{A_x}\ddot{U}_o = 0$ and assume $U_o(x,t) = U(x)\cos(\omega t - \alpha)$.

7.15 Determine the expression for the natural frequency of free flexural vibration for a simply supported beam of length L. Begin with $W_{o,xxxx} + \frac{\overline{\rho}_1}{D_x}\ddot{W}_o = 0$ and assume harmonic motion with $W_o(x,t) = W(x)\cos(\omega t - \alpha)$.

7.16 Determine ω_n for the flexural and extensional modes of a hinged–hinged $[0/45/90]_s$ composite beam of uniform width $(b = 1.0'')$ knowing that $E_1 = 8.8 \times 10^6$ psi, $E_2 = 3.6 \times 10^6$ psi, $G_{12} = 1.75 \times 10^6$ psi, $\nu_{12} = 0.23$, $t_k = 0.10''$, and $\rho_k = 0.05\text{lb/in.}^3$.

References

[1] Ugural AC. Stresses in plates and shells. USA: McGraw-Hill; 1981.
[2] Vinson JR, Sierakowski RL. The behavior of structures composed of composite materials. The Netherlands: Martinus Nijhoff; 1987.
[3] Jones RM. Mechanics of composite materials. USA: Hemisphere; 1975.
[4] Timoshenko SP, Goodier JN. Theory of elasticity. USA: McGraw-Hill; 1970.
[5] Gere JM, Timoshenko SP. Mechanics of materials. USA: PWS-KENT; 1990.
[6] Popov EP. Engineering mechanics of solids. USA: Prentice Hall; 1990.
[7] Oden JT, Ripperger EA. Mechanics of elastic structures. USA: McGraw-Hill; 1981.
[8] Boresi AP, Sidebottom OM, Seely FB, Smith JO. Advanced mechanics of materials. USA: John Wiley and Sons; 1978.
[9] Leissa AW. Buckling of laminated composite plate and shell panels, AFWAL-TR-85-3069; 1985.
[10] Langhar HL. Energy methods in applied mechanics. USA: John Wiley and Sons; 1962.
[11] Hodges DH, Atilgan AR, Fulton MV, Rehfield LW. Free-vibration analysis of composite beams. J Am Helicopter Soc 1991;7:36–47.
[12] Shi G, Lam KY. Finite element vibration analysis of composite beams based on higher-order beam theory. J Sound Vib 1999;219:707–21.

Laminated composite plate analysis

8

8.1 Introduction

Plates are common structural members whose analysis is more complex than that of beams, rods, and bars. The complete analysis of structural members requires both stress and deformation analysis. Stress analysis for plates was discussed in Chapter 6 and is only briefly discussed herein. This chapter focuses on the three most common plate topics: bending, vibration, and buckling (stability). Each of these topics requires knowledge of the deflection of a plate to specific loading and boundary conditions. Relations between the load and deflection of plates are developed following procedures similar to those for beams. The analysis of plates as presented in Chapter 6 gave little consideration to the physical dimensions of the plate and focused on stress analysis. This chapter focuses on plate displacements and solving the equations of motion involves specific plate dimensions. Numerous references exist regarding general plate analysis, including Refs. [1–4], and an introduction to isotropic plate behavior should precede discussions of their behavior for laminated composites. The material herein is not intended to provide a complete set of laminated plate-related problems and solutions, but rather to act as a starting point for those interested in laminate plate analysis. The existing information regarding laminated plate analysis is extensive and includes numerous scholarly articles as well as segmented discussions presented in texts and dedicated texts, including Refs. [5–8]. The comprehensive text by Reddy [8] presents classical lamination and first-order shear deformation plate theories as well as nonlinear theories. It also includes finite element analysis and refined theories. The equations of motion generated for plate (and shell) analysis typically require numerical techniques to obtain complete solutions. For this reason, the discussions presented below are primarily confined to simplified laminates (specially orthotropic, angle ply, etc.).

8.2 Plate geometry and governing assumptions

In developing the equations of motion for finite dimensional plates, the dimensions and an arbitrary load shown in Figure 8.1 are used.

The Kirchoff–Love hypothesis cited in Chapter 6 is expanded to explicitly include reference to plate dimensions and itemize specific assumptions. These assumptions are:

1. The plate is made from an arbitrary number of orthotropic lamina bonded together. Each lamina may have an arbitrary fiber orientation with respect to the x axis of the plate.
2. Thin plate assumptions hold ($h \ll a$ and $h \ll b$).

Laminar Composites. http://dx.doi.org/10.1016/B978-0-12-802400-3.00008-8

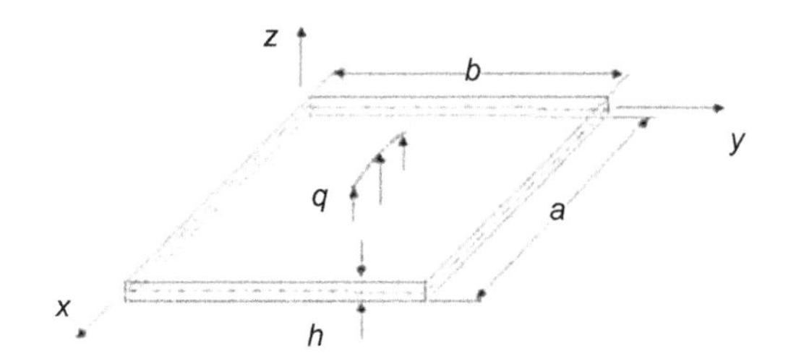

Figure 8.1 Plate dimensions and arbitrary loading.

3. The plate displacements in the x, y, and z directions (U, V, and W, respectively) are small compared to the plate thickness h.
4. In-plane strains ε_x, ε_y, and γ_{xy} are small compared to unity.
5. In order to include in-plane force effects, nonlinear terms involving products of stress and plate slopes are retained in the equations of motion. All other nonlinear terms are neglected.
6. Transverse shear strains γ_{xz} and γ_{yz} are neglected.
7. The displacements U and V are linear functions of z.
8. The out-of-plane strain ε_z is neglected.
9. Each lamina obeys Hooke's law.
10. The plate thickness is constant.
11. Rotary inertia terms are neglected.
12. Body forces are neglected.
13. The transverse shear stresses τ_{xz} and τ_{yz} vanish at $z=\pm h/2$.

The strain–displacement relationships developed as in Chapter 6 are applicable to this presentation and are reintroduced here. The assumed displacement fields in the x, y, and z directions are given by $U=U_o(x,y,t)+z\Phi(x,y,t)$, $V=V_o(x,y,t)+z\Psi(x,y,t)$, and $W=W_o(x,y,t)$. As before, $U_o(x,y,t)$, $V_o(x,y,t)$, and $W_o(x,y,t)$ are the displacement of the laminate mid-plane. Higher order theories for plate (and shell) analysis are available if required [8]. Although valuable, it is considered to be beyond the scope of this text. For convenience, the previously introduced notation for expressing partial derivatives $\left(\partial U/\partial x=U_{,x} \text{ and } \partial^2 U/\partial x^2=U_{,xx}\right)$ is used. The strain–displacement relations previously developed are repeated here for convenience: $\varepsilon_x=U_{o,x}+z\Phi_{,x}$, $\varepsilon_y=V_{o,y}+z\Psi_{,y}$, $\varepsilon_z=W_{o,z}$, $\gamma_{xy}=U_{o,y}+V_{o,x}+z\left\{\Phi_{,y}+\Psi_{,x}\right\}$, $\gamma_{xz}=W_{o,x}+\Phi$, and $\gamma_{yz}=W_{o,y}+\Psi$. The mid-surface strains are defined in Chapter 6 as $\varepsilon_x^o=U_{o,x}$, $\varepsilon_y^o=V_{o,y}$, and $\gamma_{xy}^o=U_{o,y}+V_{o,x}$. The assumption that $\gamma_{xz}=\gamma_{yz}=0$ leads directly to the definition of the curvatures being $\kappa_x=-W_{o,xx}$, $\kappa_y=-W_{o,yy}$, and $\kappa_{xy}=-2W_{o,xy}$. Without consideration of thermal or hygral effects, the strains are defined as $\{\varepsilon\}=\{\varepsilon^o\}+z\{\kappa\}$.

8.3 Equations of motion

The forces and moments applied to the plate are the same as those presented in Chapter 6. They are shown in Figure 8.2 and are defined in terms of stresses through the relationships.

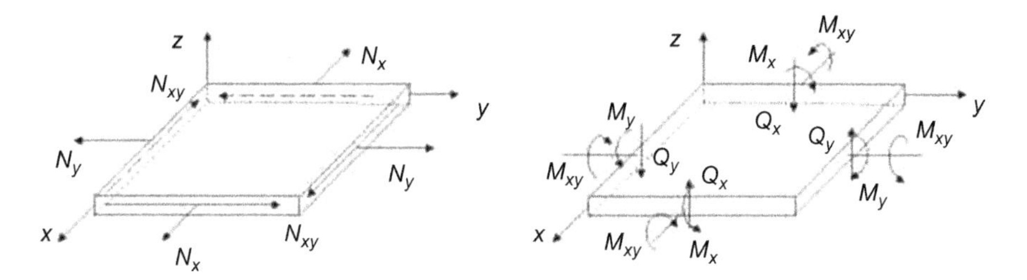

Figure 8.2 External forces and moments applied to plates.

$$\left(N_x, N_y, N_{xy}\right) = \int_{-h/2}^{h/2} \left(\sigma_x, \sigma_y, \tau_{xy}\right) \mathrm{d}z$$

$$\left(Q_x, Q_y\right) = \int_{-h/2}^{h/2} \left(\tau_{xz}, \tau_{yz}\right) \mathrm{d}z$$

$$\left(M_x, M_y, M_{xy}\right) = \int_{-h/2}^{h/2} \left(\sigma_x, \sigma_y, \tau_{xy}\right) z \, \mathrm{d}z$$

The general form of the equations of motion comes from the application of Newton's second law. These equations are initially presented in terms of stress [1]. They will then be cast into a form where the deflections are incorporated into the equations. In the x, y, and z directions, we have

$$x\text{-direction}: \quad \sigma_{x,x} + \tau_{xy,y} + \tau_{xz,z} = \rho_o \ddot{U} \tag{8.1}$$

$$y\text{-direction}: \quad \tau_{xy,x} + \sigma_{y,y} + \tau_{yz,z} = \rho_o \ddot{V} \tag{8.2}$$

$$z\text{-direction}: \quad q + \tau_{xz,x} + \tau_{yz,yt} + \sigma_{z,z} = \rho_o \ddot{W}$$

The third equation in this set must be augmented to include the fact that additional terms must be present to account for the coupling between normal and shear stresses in the x–z and y–z planes. This coupling is a result of assumption (5) above and is illustrated in Figure 8.3.

As a result of this coupling, we develop new expressions for τ_{xz}, τ_{yz}, and σ_z for use in the third equation above. These terms are

$$\tau_{xz} = \left(\tau_{xz}\right)_{\text{applied}} - \sigma_x \Phi - \tau_{xy} \Phi$$

$$\tau_{yz} = \left(\tau_{yz}\right)_{\text{applied}} - \tau_{xy} \Psi - \sigma_y \Psi$$

$$\sigma_z = \left(\sigma_z\right)_{\text{applied}} - \tau_{xz} \Phi - \tau_{yz} \Psi$$

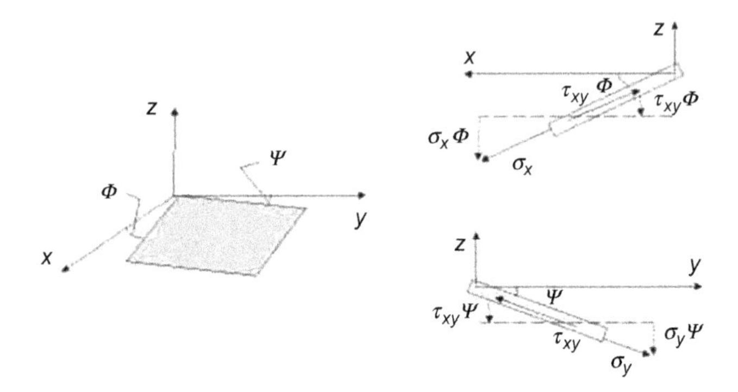

Figure 8.3 Coupling of normal and shear stress terms in the x–z and y–z planes.

Neglecting the effects of shear deformation, we assume $\gamma_{xz}=\gamma_{yz}=0$, which results in $\Phi=-W_{o,x}$ and $\Psi=-W_{o,y}$. Coupled with the terms above, the third equation of motion becomes

$$q+\frac{\partial}{\partial x}\left(\tau_{xz}+\sigma_x W_{o,x}+\tau_{xy}W_{o,y}\right)+\frac{\partial}{\partial y}\left(\tau_{yz}+\tau_{xy}W_{o,x}+\sigma_y W_{o,y}\right)$$
$$+\frac{\partial}{\partial z}\left(\sigma_z+\tau_{xz}W_{o,x}+\tau_{yz}W_{o,y}\right)=\rho_o\ddot{W} \tag{8.3}$$

Integrating Equation (8.1) between the limits $\pm h/2$, we get

$$\int_{-h/2}^{h/2}\sigma_{x,x}\mathrm{d}z+\int_{-h/2}^{h/2}\tau_{xy,y}\mathrm{d}z+\int_{-h/2}^{h/2}\tau_{xz,z}\mathrm{d}z=\int_{-h/2}^{h/2}\rho_o\ddot{U}\,\mathrm{d}z$$

Defining $\rho=\int_{-h/2}^{h/2}\rho_o\mathrm{d}z$ and noting that $\int_{-h/2}^{h/2}\tau_{xz,z}\mathrm{d}z=0$, we get

$$N_{x,x}+N_{xy,y}=\rho\ddot{U}$$

where $\ddot{U}=\ddot{U}_o+z\ddot{\Phi}=\ddot{U}_o-zW_{o,xtt}$. The term $W_{o,xtt}$ is the rotary inertia and via assumption (11) is neglected. Therefore

$$N_{x,x}+N_{xy,y}=\rho\ddot{U}_o \tag{8.4}$$

Applying similar techniques to Equation (8.2) results in

$$N_{xy,x}+N_{y,y}=\rho\ddot{V}_o \tag{8.5}$$

Similarly, Equation (8.3) can be integrated to obtain

$$N_xW_{o,xx}+2N_{xy}W_{o,xy}+N_yW_{o,yy}+Q_{x,x}+Q_{y,y}+W_{o,x}\left(N_{x,x}+N_{xy,y}\right)$$
$$+W_{o,y}\left(N_{xy,x}+N_{y,y}\right)+q=\rho\ddot{W}_o \tag{8.6}$$

where $q = \sigma_z(h/2) - \sigma_z(-h/2)$. The terms $N_{x,x} + N_{xy,y}$ and $N_{xy,x} + N_{y,y}$ vanish for the static case and are second-order terms in accordance with assumption (5) for the dynamic case. Therefore Equation (8.6) reduces to

$$N_x W_{o,xx} + 2N_{xy} W_{o,xy} + N_y W_{o,yy} + Q_{x,x} + Q_{y,y} + q = \rho \ddot{W}_o \tag{8.7}$$

Additional relationships can also be developed for the moments. Multiplying Equation (8.1) by z and integrating through the thickness results in

$$\int_{-h/2}^{h/2} z\sigma_{x,x} \mathrm{d}z + \int_{-h/2}^{h/2} z\tau_{xy,y} \mathrm{d}z + \int_{-h/2}^{h/2} z\tau_{xz,z} \mathrm{d}z = \int_{-h/2}^{h/2} \rho_o \ddot{U} z \, \mathrm{d}z$$

$$M_{x,x} + M_{xy,y} + \int_{-h/2}^{h/2} z\tau_{xz,z} \mathrm{d}z = \int_{-h/2}^{h/2} \rho_o z \left(\ddot{U}_o + z\ddot{\Phi}\right) \mathrm{d}z$$

Noting that

$$z\tau_{xz,z} = \frac{\partial}{\partial z}(z\tau_{xz}) - \tau_{xz} \rightarrow \int_{-h/2}^{h/2} z\tau_{xz,z} \mathrm{d}z = \int_{-h/2}^{h/2} \frac{\partial}{\partial z}(z\tau_{xz}) \mathrm{d}z - \int_{-h/2}^{h/2} \tau_{xz} \, \mathrm{d}z$$

Since $\int_{-h/2}^{h/2} \frac{\partial}{\partial z}(z\tau_{xz}) \mathrm{d}z = 0$ and $\int_{-h/2}^{h/2} \tau_{xz} \mathrm{d}z = Q_x$, the equation above reduces. In addition, $z\left(\ddot{U}_o + z\ddot{\Phi}\right)$ represents a rotary inertia term, which is neglected. Therefore, $M_{x,x} + M_{xy,y} - Q_x = 0$. Similarly, we have $M_{xy,x} + M_{y,y} - Q_y = 0$. These equations can be written as $Q_{x,x} = M_{x,xx} + M_{xy,xy}$ and $Q_{y,y} = M_{xy,xy} + M_{y,yy}$. Thus we can write Equation (8.7) as

$$N_x W_{o,xx} + 2N_{xy} W_{o,xy} + N_y W_{o,yy} + M_{x,xx} + 2M_{xy,xy} + M_{y,yy} + q = \rho \ddot{W}_o \tag{8.8}$$

The equations of motion for a plate are therefore represented by Equations (8.4), (8.5), and (8.8), as summarized below.

$$N_{x,x} + N_{xy,y} = \rho \ddot{U}_o$$

$$N_{xy,x} + N_{y,y} = \rho \ddot{V}_o$$

$$N_x W_{o,xx} + 2N_{xy} W_{o,xy} + N_y W_{o,yy} + M_{x,xx} + 2M_{xy,xy} + M_{y,yy} + q = \rho \ddot{W}_o$$

From classical lamination theory, we recall that

$$\begin{Bmatrix} N \\ M \end{Bmatrix} = \begin{bmatrix} A & B \\ B & D \end{bmatrix} \begin{Bmatrix} \varepsilon^o \\ \kappa \end{Bmatrix} - \begin{Bmatrix} N \\ M \end{Bmatrix}^{\mathrm{T}} - \begin{Bmatrix} N \\ M \end{Bmatrix}^{\mathrm{H}}$$

where, as before

$$(A_{ij}, B_{ij}, D_{ij}) = \int_{-h/2}^{h/2} [\bar{Q}_{ij}]_k (1, z, z^2) \mathrm{d}z$$

$$(N^{\mathrm{T}}, M^{\mathrm{T}}) = \sum_{k=1}^{N} \int_{h_{k-1}}^{h_k} [\bar{Q}]_k \{\alpha\}_k (1, z) \Delta T \, \mathrm{d}z$$

$$(N^{\mathrm{H}}, M^{\mathrm{H}}) = \sum_{k=1}^{N} \int_{h_{k-1}}^{h_k} [\bar{Q}]_k \{\beta\}_k (1, z) \bar{M} \, \mathrm{d}z$$

Using the definitions $\varepsilon_x^o = U_{o,x}, \varepsilon_y^o = V_{o,x}, \gamma_{xy}^o = U_{o,y} + V_{o,x}, \kappa_x = -W_{o,xx}, \kappa_y = -W_{o,yy}$, and $\kappa_{xy} = -2W_{o,xy}$, we can write the forces and moments in matrix form as

$$\begin{Bmatrix} N_x \\ N_y \\ N_{xy} \\ M_x \\ M_y \\ M_{xy} \end{Bmatrix} = \begin{bmatrix} A_{11} & A_{12} & A_{16} & B_{11} & B_{12} & B_{16} \\ A_{12} & A_{22} & A_{26} & B_{12} & B_{22} & B_{26} \\ A_{16} & A_{26} & A_{66} & B_{16} & B_{26} & B_{66} \\ B_{11} & B_{12} & B_{16} & D_{11} & D_{12} & D_{16} \\ B_{12} & B_{22} & B_{26} & D_{12} & D_{22} & D_{26} \\ B_{16} & B_{26} & B_{66} & D_{16} & D_{26} & D_{66} \end{bmatrix} \begin{Bmatrix} U_{o,x} \\ V_{o,y} \\ U_{o,y} + V_{o,x} \\ -W_{o,xx} \\ -W_{o,yy} \\ -2W_{o,xy} \end{Bmatrix} - \begin{Bmatrix} N_x^{\mathrm{T}} \\ N_y^{\mathrm{T}} \\ N_{xy}^{\mathrm{T}} \\ M_x^{\mathrm{T}} \\ M_y^{\mathrm{T}} \\ M_{xy}^{\mathrm{T}} \end{Bmatrix} - \begin{Bmatrix} N_x^{\mathrm{H}} \\ N_y^{\mathrm{H}} \\ N_{xy}^{\mathrm{H}} \\ M_x^{\mathrm{H}} \\ M_y^{\mathrm{H}} \\ M_{xy}^{\mathrm{H}} \end{Bmatrix}$$

The shear forces Q_x and Q_y are more conveniently expressed in equation form as

$$\begin{aligned} Q_x = {} & B_{11}U_{o,xx} + 2B_{16}U_{o,xy} + B_{66}U_{o,yy} + B_{16}V_{o,xx} + (B_{12} + B_{66})V_{o,xy} + B_{26}V_{o,yy} \\ & - D_{11}W_{o,xxx} - 3D_{16}W_{o,xxy} - (D_{12} + 2D_{66})W_{o,xyy} - D_{26}W_{o,yyy} - M_{x,x}^{\mathrm{T}} \\ & - M_{x,x}^{\mathrm{H}} - M_{xy,y}^{\mathrm{T}} - M_{xy,y}^{\mathrm{H}} \end{aligned}$$

$$\begin{aligned} Q_y = {} & B_{16}U_{o,xx} + (B_{12} + B_{66})U_{o,xy} + B_{26}U_{o,yy} + B_{66}V_{o,xx} + 2B_{26}V_{o,xy} + B_{22}V_{o,yy} \\ & - D_{16}W_{o,xxx} - (D_{12} + 2D_{66})W_{o,xxy} - 3D_{26}W_{o,xyy} - D_{22}W_{o,yyy} - M_{xy,x}^{\mathrm{T}} \\ & - M_{xy,x}^{\mathrm{H}} - M_{y,y}^{\mathrm{T}} - M_{y,y}^{\mathrm{H}} \end{aligned}$$

Substituting these into Equations (8.4), (8.5), and (8.8) results in the three equations of motion below

$$\begin{aligned} & A_{11}U_{o,xx} + 2A_{16}U_{o,xy} + A_{66}U_{o,yy} + A_{16}V_{o,xx} + (A_{12} + A_{66})V_{o,xy} + A_{26}V_{o,yy} \\ & - B_{11}W_{o,xxx} - 3B_{16}W_{o,xxy} - (B_{12} + 2B_{66})W_{o,xyy} - B_{26}W_{o,yyy} - N_{x,x}^{\mathrm{T}} \\ & - N_{xy,y}^{\mathrm{T}} - N_{x,x}^{\mathrm{H}} - N_{xy,y}^{\mathrm{H}} = \rho \ddot{U}_o \end{aligned} \tag{8.9}$$

$$\begin{aligned} & A_{16}U_{o,xx} + (A_{12} + A_{66})U_{o,xy} + A_{26}U_{o,yy} + A_{66}V_{o,xx} + 2A_{26}V_{o,xy} + A_{22}V_{o,yy} \\ & - B_{16}W_{o,xxx} - (B_{12} + 2B_{66})W_{o,xxy} - 3B_{26}W_{o,xyy} - B_{22}W_{o,yyy} - N_{xy,x}^{\mathrm{T}} \\ & - N_{y,y}^{\mathrm{T}} - N_{xy,x}^{\mathrm{H}} - N_{y,y}^{\mathrm{H}} = \rho \ddot{V}_o \end{aligned} \tag{8.10}$$

$$\begin{aligned}&D_{11}W_{o,xxxx}+4D_{16}W_{o,xxxy}+2(D_{12}+2D_{66})W_{o,xxyy}+4D_{26}W_{o,xyyy}\\&+D_{22}W_{o,yyyy}-B_{11}U_{o,xxx}-3B_{16}U_{o,xxy}-(B_{12}+2B_{66})U_{o,xyy}-B_{26}U_{o,yyy}\\&-B_{16}V_{o,xxx}-(B_{12}+2B_{66})V_{o,xxy}-3B_{26}V_{o,xyy}-B_{22}V_{o,yyy}+M^{\mathrm{T}}_{x,xx}\\&+M^{\mathrm{H}}_{x,xx}+2M^{\mathrm{T}}_{xy,xy}+2M^{\mathrm{H}}_{xy,xy}+M^{\mathrm{T}}_{y,yy}+M^{\mathrm{H}}_{y,yy}+\rho\ddot{W}_o=q\end{aligned}\tag{8.11}$$

The exact form of each of these equations will depend on the laminate under consideration. For symmetric laminates $[B]=0$, while for other types of laminates (e.g., specially orthotropic), some of the terms presented above may vanish.

The stress in each lamina of a laminate can also be expressed in terms of the displacements. The stress in the kth lamina of a laminate can be written as $\{\sigma\}_k=[\bar{Q}]_k(\varepsilon^o+z\kappa)$. Using the definitions of ε^o and κ given above, we write

$$\begin{aligned}(\sigma_x)_k=&(\bar{Q}_{11})_kU_{o,x}+(\bar{Q}_{16})_k(U_{o,y}+V_{o,x})+(\bar{Q}_{12})_kV_{o,y}\\&-z\left[(\bar{Q}_{11})_kW_{o,xx}+2(\bar{Q}_{16})_kW_{o,xy}+(\bar{Q}_{12})_kW_{o,yy}\right]\end{aligned}$$

$$\begin{aligned}(\sigma_y)_k=&(\bar{Q}_{12})_kU_{o,x}+(\bar{Q}_{26})_k(U_{o,y}+V_{o,x})+(\bar{Q}_{22})_kV_{o,y}\\&-z\left[(\bar{Q}_{11})_kW_{o,xx}+2(\bar{Q}_{26})_kW_{o,xy}+(\bar{Q}_{22})_kW_{o,yy}\right]\end{aligned}$$

$$\begin{aligned}(\tau_{xy})_k=&(\bar{Q}_{16})_kU_{o,x}+(\bar{Q}_{66})_k(U_{o,y}+V_{o,x})+(\bar{Q}_{26})_kV_{o,y}\\&-z\left[(\bar{Q}_{16})_kW_{o,xx}+2(\bar{Q}_{66})_kW_{o,xy}+(\bar{Q}_{26})_kW_{o,yy}\right]\end{aligned}$$

The interlaminar shear stresses can also be expressed in terms of the displacements. The interlaminar shear stresses are determined from $(\tau_{xz})_k=-\int(\sigma_{x,x}+\tau_{xy,y})_k\mathrm{d}z$ and $(\tau_{yz})_k=-\int(\tau_{xy,x}+\sigma_{y,y})_k\mathrm{d}z$. Using the expressions above, and noting that after integration we will have constants of integration, we get

$$\begin{aligned}(\tau_{xz})_k=&\frac{z^2}{2}\left[(\bar{Q}_{11})_kW_{o,xxx}+3(\bar{Q}_{16})_kW_{o,xxy}+\left((\bar{Q}_{12})_k+2(\bar{Q}_{66})_k\right)W_{o,xyy}+(\bar{Q}_{26})_kW_{o,yyy}\right]\\&-z\left[(\bar{Q}_{11})_kU_{o,xx}+2(\bar{Q}_{16})_kU_{o,xy}+(\bar{Q}_{66})_kU_{o,yy}+(\bar{Q}_{16})_kV_{o,xx}\right.\\&\left.+\left((\bar{Q}_{12})_k+(\bar{Q}_{66})_k\right)V_{o,xy}+(\bar{Q}_{26})_kV_{o,yy}\right]+f_k(x,y)\end{aligned}$$

$$\begin{aligned}(\tau_{yz})_k=&\frac{z^2}{2}\left[(\bar{Q}_{16})_kW_{o,xxx}+\left((\bar{Q}_{12})_k+2(\bar{Q}_{66})_k\right)W_{o,xxy}+3(\bar{Q}_{26})_kW_{o,xyy}+(\bar{Q}_{22})_kW_{o,yyy}\right]\\&-z\left[(\bar{Q}_{16})_kU_{o,xx}+\left((\bar{Q}_{12})_k+2(\bar{Q}_{66})_k\right)U_{o,xy}+(\bar{Q}_{26})_kU_{o,yy}\right.\\&\left.+(\bar{Q}_{66})_kV_{o,xx}+2(\bar{Q}_{26})_kV_{o,xy}+(\bar{Q}_{22})_kV_{o,yy}\right]+g_k(x,y)\end{aligned}$$

where $f_k(x, y)$ and $g_k(x, y)$ are constants of integration for each lamina. Solving the equations of motion requires knowing the boundary conditions generally encountered in plate analysis.

8.4 Boundary conditions

In order to solve the equations of motion, the boundary conditions must be understood. The nomenclature used to identify boundary conditions for plate analysis is somewhat different than that used for beam analysis and follow that of Ref. [9]. In Figure 8.4, a plate with dimensions a and b and thickness h is shown. In this figure, "n" refers to the normal direction and "t" refers to the tangential direction in the x–y plane. In addition, they refer to the particular direction which is supported. The boundary conditions in terms of "n" and "t" as well as the traditional x and y coordinates for various conditions are presented below.

Simple supports: For the simple support shown in Figure 8.5, it is generally assumed that the support extends the entire width (b) of the plate. The boundary conditions in terms of n and t are $N_n = N_{nt} = W_o = M_n = 0$. The deflection and moment at the support are obviously both zero. The support is actually more like a roller than a pinned connection; therefore, the normal forces in the n and t directions are both zero too. Expressing these in terms of x and y coordinates we have, at the support $N_x = N_y = W_o = M_x = 0$. In a similar manner, the support could be along the entire length (a) of the plate. This does not alter the fact that normal forces, plate deflection, and bending moment (M_y in this case) are zero.

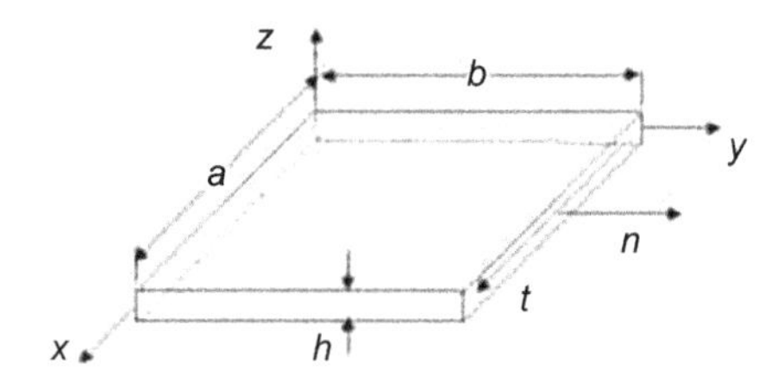

Figure 8.4 Plate geometry and nomenclature for identifying boundary conditions.

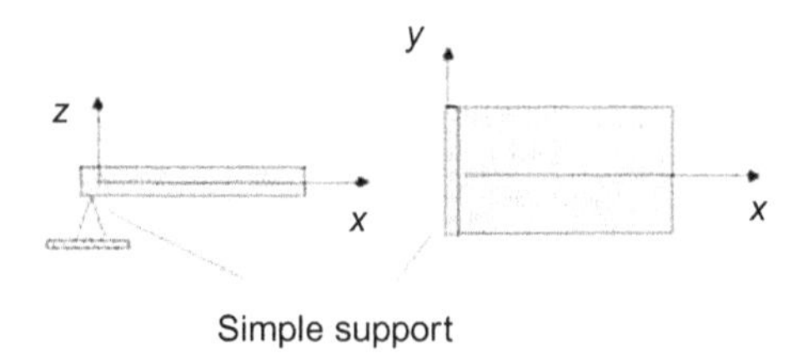

Figure 8.5 Schematic of a simply supported plate.

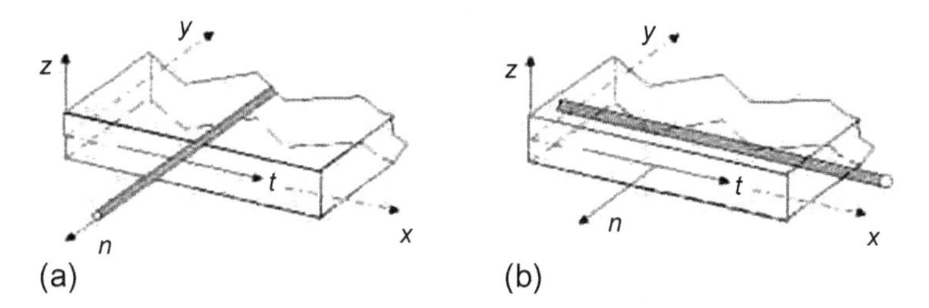

Figure 8.6 Schematic of hinged-free plate support in the (a) normal and (b) tangential directions.

Hinged-free supports: There are two possible directions for a hinged-free support, normal and tangential. These are shown in Figure 8.6 and the associated boundary conditions for each are:

Normal direction: For this type of support, the boundary conditions are $N_n = U_{ot} = W_o = M_t = 0$. In terms of x and y coordinates, these are $N_y = U_o = W_o = M_{xy} = 0$.
Tangential direction: For this type of support, the boundary conditions are $N_n = U_{on} = W_o = M_n = 0$. In terms of x and y coordinates, these are $N_x = V_o = W_o = M_y = 0$.

Clamped support: A clamped support is as shown in Figure 8.7. The displacements and rotations are zero. This means $U_{on} = V_{ot} = W_o = \partial W_o / \partial x = 0$. It is a simple matter to see that these can be expressed as $U_o = V_o = W_o = \partial W_o / \partial x = 0$.

Elastic edge supports: Elastic edge supports are somewhat more difficult to define, since spring rates are required as illustrated in Figure 8.8. Two spring rates

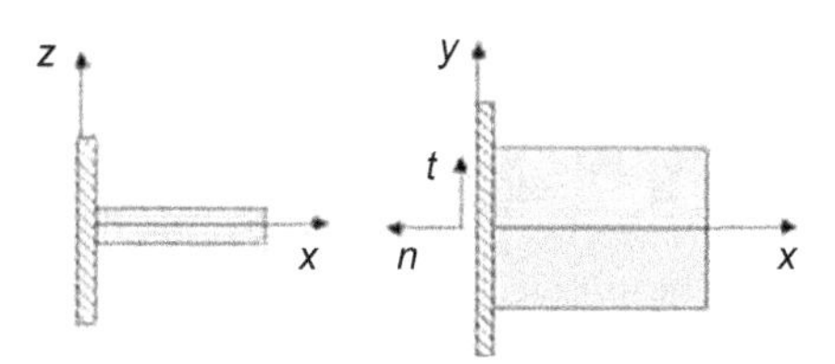

Figure 8.7 Schematic of clamped plate support.

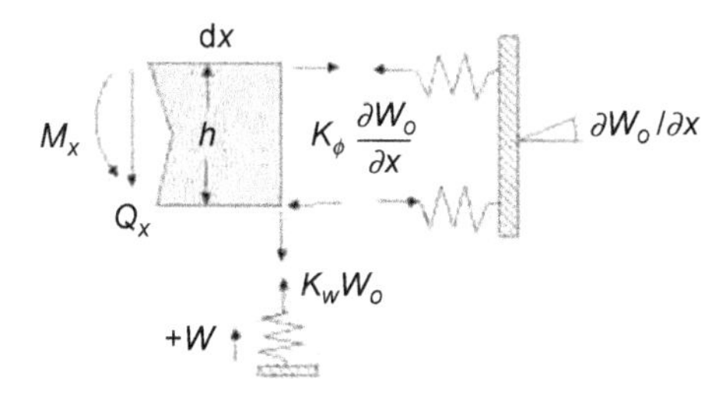

Figure 8.8 Schematic of elastic edge supports of a plate.

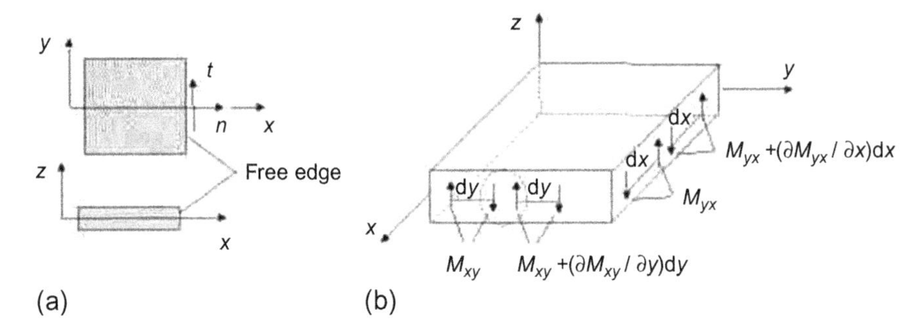

Figure 8.9 Schematic of the free edge of a plate illustrating the (a) basic model and (b) detailed model.

(K_w and K_ϕ) are generally defined. The boundary conditions are expressed as $Q_x = -K_w W_o$ and $M_x = K_\phi(\partial W_o/\partial x)$.

Free edge: The free edge of a plate requires more attention to detail than one might initially expect. Figure 8.9a illustrates the basic model, with $N_n = M_n = 0$ (or $N_x = M_x = 0$), and (b) illustrates a detailed model. For the detailed model, also define an effective shear resultant which is also zero. It is related to two additional terms Q_x and $\partial M_{xy}/\partial y$.

Considering the free edge of a laminate in the x direction, we have $M_x = M_{xy} = Q_x = 0$. A total of six boundary conditions must be satisfied on the x = constant lines. The plate equation is a fourth-order equation and only two boundary conditions can be satisfied on each edge. Note that including shear deformation results in a sixth-order equation and this problem is eliminated. Knowing that M_{xy} consists of two forces a distance dy apart and that M_{xy} is not constant, we can, by considering the two forces inside the dashed line on the x face shown, write

$$-M_{xy} + M_{xy} + \frac{\partial M_{xy}}{\partial y}\mathrm{d}y + Q_x\mathrm{d}y = 0 \Rightarrow \frac{\partial M_{xy}}{\partial y} + Q_x = 0$$

Defining an "effective shear resultant" as V_n (sometimes denoted as Q_x'), we generate the boundary condition

$$V_n = Q_x' = \frac{\partial M_{xy}}{\partial y} + Q_x = 0$$

Similarly, in the y direction, we get

$$V_t = Q_y' = \frac{\partial M_{yx}}{\partial y} + Q_y = 0$$

8.5 Plate bending

In general, plate bending is emphasized more than extension or compression. As a result, Equation (8.11) is the only applicable equation of motion. For plate bending, we assume thermal and hygral effects are neglected and only static analysis $\left(\ddot{W}_o = 0\right)$

is considered. In addition, we will consider specially orthotropic ($D_{16} = D_{26} = 0$) symmetric laminates ($[B] = 0$). Under these conditions, Equation (8.11) reduces to

$$D_{11}W_{o,xxxx} + 2(D_{12} + 2D_{66})W_{o,xxyy} + D_{22}W_{o,yyyy} = q(x, y) \tag{8.12}$$

This equation is sometimes seen in a more familiar form as

$$D_{11}\frac{\partial^4 W_o}{\partial x^4} + 2(D_{12} + 2D_{66})\frac{\partial^4 W_o}{\partial x^2 \partial y^2} + D_{22}\frac{\partial^4 W_o}{\partial y^4} = q(x, y)$$

Consider a plate simply supported on all sides and subjected to a uniformly distributed load $q(x,y)$ as illustrated in Figure 8.1. For this plate, the boundary conditions are

$$\text{At } x = 0, a: \quad W_o = 0, \ M_x = 0 = -D_{11}\frac{\partial^2 W_o}{\partial x^2} - D_{12}\frac{\partial^2 W_o}{\partial y^2}$$

$$\text{At } y = 0, b: \quad W_o = 0, \ M_y = 0 = -D_{12}\frac{\partial^2 W_o}{\partial x^2} - D_{22}\frac{\partial^2 W_o}{\partial y^2}$$

In order to solve the problem, we must define the load distribution $q(x,y)$. The standard procedure for this is to use Naviers solution, for which we assume that the load can be expressed as a double trigonometric (Fourier) series [8] such as

$$q(x, y) = \sum_{m=1}^{\infty}\sum_{n=1}^{\infty} q_{mn}\sin\frac{m\pi x}{a}\sin\frac{n\pi y}{b}$$

Multiplying each side by $\sin\frac{m'\pi x}{a}\sin\frac{n'\pi y}{b}\mathrm{d}x\mathrm{d}y$ and integrating

$$\int_0^b\int_0^a q(x, y)\sin\frac{m'\pi x}{a}\sin\frac{n'\pi y}{b}\mathrm{d}x\mathrm{d}y$$
$$= \sum_{m=1}^{\infty}\sum_{n=1}^{\infty} q_{mn}\int_0^b\int_0^a \sin\frac{m\pi x}{a}\sin\frac{n\pi y}{b}\sin\frac{m'\pi x}{a}\sin\frac{n'\pi y}{b}\mathrm{d}x\mathrm{d}y$$

By direct integration, it can be shown that

$$\int_0^a \sin\frac{m\pi x}{a}\sin\frac{m'\pi x}{a}\mathrm{d}x = \begin{cases} 0 & m \neq m' \\ a/2 & m = m' \end{cases}$$

$$\int_0^b \sin\frac{n\pi y}{b}\sin\frac{n'\pi y}{b}\mathrm{d}y = \begin{cases} 0 & n \neq n' \\ b/2 & n = n' \end{cases}$$

As a result, we can write

$$q_{mn} = \frac{4}{ab}\int_0^b\int_0^a q(x, y)\sin\frac{m\pi x}{a}\sin\frac{n\pi y}{b}\mathrm{d}x\mathrm{d}y \tag{8.13}$$

For a plate simply supported on all sides, we can satisfy the displacement conditions by assuming

$$W_o = \sum_{m=1}^{\infty}\sum_{n=1}^{\infty} A_{mn} \sin\frac{m\pi x}{a}\sin\frac{n\pi y}{b}$$

The coefficients A_{mn} must be determined. This assumed displacement field satisfies the condition that $W_o = 0$ along all edges, but has some finite value at all interior points. Substituting this expression into the equation of motion yields

$$D_{11}\sum\sum A_{mn}\left(\frac{m\pi}{a}\right)^4 \sin\frac{m\pi x}{a}\sin\frac{n\pi y}{b} + D_{22}\sum\sum A_{mn}\left(\frac{n\pi}{b}\right)^4 \sin\frac{m\pi x}{a}\sin\frac{n\pi y}{b}$$
$$+2(D_{12}+2D_{66})\sum\sum A_{mn}\left(\frac{m\pi}{a}\right)^2\left(\frac{n\pi}{b}\right)^2 \sin\frac{m\pi x}{a}\sin\frac{n\pi y}{b} = \sum\sum q_{mn}\sin\frac{m\pi x}{a}\sin\frac{n\pi y}{b}$$

This reduces to

$$\sum_{m=1}^{\infty}\sum_{n=1}^{\infty}\left\{A_{mn}\left[D_{11}\left(\frac{m\pi}{a}\right)^4 + 2(D_{12}+2D_{66})\left(\frac{m\pi}{a}\right)^2\left(\frac{n\pi}{b}\right)^2 + D_{22}\left(\frac{n\pi}{b}\right)^4\right] - q_{mn}\right\}\sin\frac{m\pi x}{a}\sin\frac{n\pi y}{b} = 0$$

Since this expression must be satisfied for all x and y coordinates

$$A_{mn}\left[D_{11}\left(\frac{m}{a}\right)^4 + 2(D_{12}+2D_{66})\left(\frac{mn}{ab}\right)^2 + D_{22}\left(\frac{n}{b}\right)^4\right]\pi^4 - q_{mn} = 0$$

Defining

$$D_{mn} = D_{11}m^4 + 2(D_{12}+2D_{66})(mnR)^2 + D_{22}(nR)^4 \tag{8.14}$$

where $R = a/b$, results in

$$A_{mn} = \frac{a^4\, q_{mn}}{\pi^4 D_{mn}} \tag{8.15}$$

With this expression for A_{mn}, the infinite series solution for displacement is

$$W_o = \left(\frac{a}{\pi}\right)^4 \sum_{m=1}^{\infty}\sum_{n=1}^{\infty}\frac{q_{mn}}{D_{mn}}\sin\frac{m\pi x}{a}\sin\frac{n\pi y}{b} \tag{8.16}$$

Using this expression for the displacement, the resultant moments and stresses in the plate can be expressed as an infinite series. The moments are written as

$$M_x = \left(\frac{a}{\pi}\right)^2 \sum_{m=1}^{\infty}\sum_{n=1}^{\infty} \frac{q_{mn}}{D_{mn}} \left(m^2 D_{11} + n^2 R^2 D_{12}\right) \sin\frac{m\pi x}{a} \sin\frac{n\pi y}{b}$$

$$M_y = \left(\frac{a}{\pi}\right)^2 \sum_{m=1}^{\infty}\sum_{n=1}^{\infty} \frac{q_{mn}}{D_{mn}} \left(m^2 D_{12} + n^2 R^2 D_{22}\right) \sin\frac{m\pi x}{a} \sin\frac{n\pi y}{b} \tag{8.17}$$

$$M_{xy} = -2RD_{66}\left(\frac{a}{\pi}\right)^2 \sum_{m=1}^{\infty}\sum_{n=1}^{\infty} \frac{q_{mn}}{D_{mn}} mn\cos\frac{m\pi x}{a} \cos\frac{n\pi y}{b}$$

The in-plane stresses in each lamina are defined by

$$(\sigma_x)_k = \left(\frac{a}{\pi}\right)^2 z \sum_{m=1}^{\infty}\sum_{n=1}^{\infty} \frac{q_{mn}}{D_{mn}} \left(m^2 (\bar{Q}_{11})_k + n^2 R^2 (\bar{Q}_{12})_k\right) \sin\frac{m\pi x}{a} \sin\frac{n\pi y}{b}$$

$$(\sigma_y)_k = \left(\frac{a}{\pi}\right)^2 z \sum_{m=1}^{\infty}\sum_{n=1}^{\infty} \frac{q_{mn}}{D_{mn}} \left(m^2 (\bar{Q}_{12})_k + n^2 R^2 (\bar{Q}_{22})_k\right) \sin\frac{m\pi x}{a} \sin\frac{n\pi y}{b} \tag{8.18}$$

$$(\tau_{xy})_k = -2R(\bar{Q}_{66})_k \left(\frac{a}{\pi}\right)^2 z \sum_{m=1}^{\infty}\sum_{n=1}^{\infty} \frac{q_{mn}}{D_{mn}} mn\cos\frac{m\pi x}{a} \cos\frac{n\pi y}{b}$$

In addition, the interlaminar shear stresses are written as

$$(\tau_{xz})_k = -\frac{a}{\pi} \sum_{m=1}^{\infty}\sum_{n=1}^{\infty} \int_{-h/2}^{z_k} \frac{mq_{mn}}{D_{mn}} \left(m^2 (\bar{Q}_{11})_k + n^2 R^2 \left[(\bar{Q}_{12})_k + 2(\bar{Q}_{66})_k\right]\right) \cos\frac{m\pi x}{a} \sin\frac{n\pi y}{b} z\mathrm{d}z$$

$$(\tau_{yz})_k = \frac{a}{\pi} \sum_{m=1}^{\infty}\sum_{n=1}^{\infty} \int_{-h/2}^{z_k} \frac{nq_{mn}}{D_{mn}} \left(m^2 \left[(\bar{Q}_{12})_k + 2(\bar{Q}_{66})_k\right] + n^2 R^2 (\bar{Q}_{22})_k\right) \sin\frac{m\pi x}{a} \cos\frac{n\pi y}{b} z\mathrm{d}z \tag{8.19}$$

8.5.1 Specific loading conditions

In order to complete the solutions described by Equations (8.16)–(8.19), specific loading conditions to describe q_{mn} must be specified. Two specific loading conditions are presented below and a more complete set can be found in Ref. [8].

Uniform load: For a uniform load ($q(x,y) = q_o$), we write Equation (8.13) as

$$q_{mn} = \frac{4}{ab}\int_0^b \int_0^a q_o \sin\frac{m\pi x}{a} \sin\frac{n\pi y}{b} \mathrm{d}x\mathrm{d}y = \frac{4q_o}{ab}\left(\frac{a}{m\pi}\right)\left(\frac{b}{n\pi}\right)(1-\cos m\pi)(1-\cos n\pi)$$

$$q_{mn} = \begin{cases} \dfrac{16q_o}{\pi^2 mn} & m,n \text{ odd} \\ 0 & m,n \text{ even} \end{cases}$$

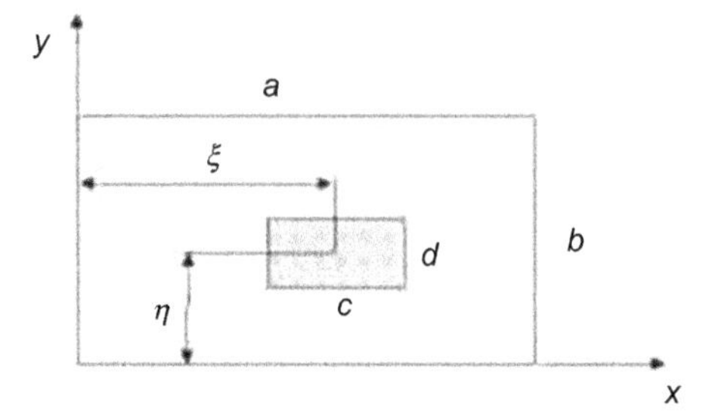

Figure 8.10 Plate subjected to a distributed load $q(x,y)$ over a finite region.

Load distributed over a finite region: Assume the applied load is distributed over a finite region with dimensions $c \times d$ as illustrated in Figure 8.10. The center of the region is located at (ξ,η) from the x and y axes, as indicated.

Assuming that the load is uniform, with magnitude P, we can express the load distribution as $q_o = P/cd$. Therefore, we can write Equation (8.13) as

$$q_{mn} = \frac{4}{ab}\left(\frac{P}{cd}\right)\int_{\xi-c/2}^{\xi+c/2}\int_{\eta-d/2}^{\eta+d/2} \sin\frac{m\pi x}{a}\sin\frac{n\pi y}{b} dxdy$$
$$= \frac{16P}{\pi^2 mncd}\sin\frac{m\pi\xi}{a}\sin\frac{n\pi\eta}{b}\sin\frac{m\pi c}{2a}\sin\frac{n\pi d}{2b}$$

Taking the limit as $c \to 0$ and $d \to 0$, this expression represents a concentrated load P positioned at the plate coordinates $x=\xi$, $y=\eta$ and is written as

$$q_{mn} = \frac{4P}{ab}\sin\frac{m\pi\xi}{a}\sin\frac{n\pi\eta}{b}$$

Example 8.1 Consider a simply supported, uniformly loaded, specially orthotropic ($D_{16}=D_{26}=0$) laminated plate with a stacking arrangement of $[0/90]_s$. The lamina is made from a material with the characteristics of $\nu_{12}=0.25$, $E_2=1.5G_{12}$, and $E_1/E_2=$ variable. We wish to develop relations which allow us to explore the maximum displacement of the plate at $x=a/2$ and $y=b/2$ in terms of material properties and dimensions. Since the plate load is uniform, we know that q_{mn} is

$$q_{mn} = \begin{cases} \dfrac{16q_o}{\pi^2 mn} & m,n \text{ odd} \\ 0 & m,n \text{ even} \end{cases}$$

Therefore, the maximum displacement is given as

$$W_{\max} = \left(\frac{a}{\pi}\right)^4 \sum_{m=1,3,..}^{\infty}\sum_{n=1,3,..}^{\infty} \frac{16q_o}{D_{mn}\pi^2 mn}\sin\frac{m\pi}{2}\sin\frac{n\pi}{2}$$

Knowing that $D_{mn}=f(D_{11},D_{12},D_{22},D_{66},m,n,a,b)$ and that D_{11},D_{12}, etc. are determined using Equation (6.22), the maximum displacement can be written in terms of E_2,G_{12},h^3, and the plate dimension a. Developing such an expression can be a lengthy process, depending on the complexity of the ply orientations within the laminate. For the [0/90]$_s$ laminate being considered, we know that D_{ij} depends on $\bar{Q}_{ij}$, and for this four ply laminate, we know that

$$(\bar{Q}_{11})_0=Q_{11}=\frac{E_1}{1-\nu_{12}\nu_{21}},\ (\bar{Q}_{11})_{90}=Q_{22}=\frac{E_2}{1-\nu_{12}\nu_{21}},\ (\bar{Q}_{66})_0=(\bar{Q}_{66})_{90}=Q_{66}=G_{12},$$

$$(\bar{Q}_{12})_0=(\bar{Q}_{12})_{90}=Q_{12}=\frac{\nu_{12}E_2}{1-\nu_{12}\nu_{21}}=\frac{\nu_{21}E_1}{1-\nu_{12}\nu_{21}},\ \text{etc.}$$

Assume all laminas have the same thickness ($t_k=h/4$). We know that $|\bar{z}|_1=|\bar{z}|_4=3t_k/2$ and $|\bar{z}|_2=|\bar{z}|_3=t_k/2$. From Equation (6.22), the D_{11} term for the [0/90]$_s$ laminate becomes

$$D_{11}=2(\bar{Q}_{11})_0\left[t_k\left(\frac{3t_k}{2}\right)^2+\frac{t_k^3}{12}\right]+2(\bar{Q}_{11})_{90}\left[t_k\left(\frac{t_k}{2}\right)^2+\frac{t_k^3}{12}\right]$$

$$=\frac{t_k^3}{6}\left\{28(\bar{Q}_{11})_0+4(\bar{Q}_{11})_{90}\right\}=\frac{(h/4)^3}{6}\left\{28\frac{E_1}{1-\nu_{12}\nu_{21}}+4\frac{E_2}{1-\nu_{12}\nu_{21}}\right\}$$

Knowing that $\nu_{21}=(E_2/E_1)\nu_{12}=E_2/4E_1$ and that $1-\nu_{21}\nu_{12}=1-(E_2/4E_1)(0.25)=(16E_1-E_2)/16E_1$, we can write the above expression as

$$D_{11}=\frac{h^3}{384}\left\{28\frac{16E_1^2}{16E_1-E_2}+4\frac{16E_1E_2}{16E_1-E_2}\right\}=\frac{h^3E_1}{24(16E_1-E_2)}\{28E_1+4E_2\}$$

Since $E_1/E_2=$ variable, we can define the ratio of the elastic moduli as some parameter (r), where $E_1=rE_2$, so that the above expression becomes

$$D_{11}=\frac{h^3rE_2\{28rE_2+4E_2\}}{24(16rE_2-E_2)}=\frac{h^3(28r+4)}{24(16r-1)}rE_2=\frac{h^3(7r+1)}{6(16r-1)}rE_2$$

Similar expressions can be developed for the other D_{ij} terms. This approximation allows us to define each remaining D_{ij} in terms of E_2, r, and h as

$$D_{22}=\frac{h^3(7+r)}{6(16r-1)}E_2,\ D_{12}=\frac{h^3}{3(16r-1)}rE_2,\ D_{66}=\frac{G_{12}h^3}{12}=\frac{E_2h^3}{18}$$

An expression for D_{mn} in terms of E_2 and h are now developed. The maximum deflection is expressed as

$$W_{\max}=\frac{16q_oa^4}{\pi^6}\sum_{m=1,3,..}^{\infty}\sum_{n=1,3,..}^{\infty}\frac{(-1)^{\left(\frac{m+n+2}{2}\right)}}{D_{mn}nm}$$

The complexity of the solution will depend on the number of terms taken, recalling that both m and n are odd. For example, assume we take two terms for m and two terms for $n (m,n=1,1\ \ 1,3\ \ 3,1\ \ 3,3)$, we get

$$W_{\max}=\frac{16q_o a^4}{\pi^6}\left(\frac{1}{D_{11}}-\frac{1}{3D_{13}}-\frac{1}{3D_{31}}+\frac{1}{9D_{33}}\right)$$

Each D_{mn} term is determined form Equation (8.14). This introduces the plate aspect ratio $(R=a/b)$ as well as the ratio of elastic moduli $(r=E_1/E_2)$ into the maximum deflection equation. For each combination of m and n, we get

$$m,n=1,1:\ D_{11}=\frac{h^3E_2}{(16r-1)}\left[\frac{(7r+1)r}{6}+\frac{65r-3}{27}R^2+\frac{(7+r)}{6}R^4\right]$$

$$m,n=1,3:\ D_{13}=\frac{h^3E_2}{(16r-1)}\left[\frac{r(7r+1)}{2}+(38r-2)R^2+\frac{81(7+r)}{6}R^4\right]$$

$$m,n=3,1:\ D_{31}=\frac{E_2h^3}{(16r-1)}\left[\frac{81r(7r+1)}{6}+(38r-2)R^2+\frac{(7+r)}{6}R^4\right]$$

$$m,n=3,3:\ D_{33}=\frac{E_2h^3}{(16r-1)}\left[\frac{81r(7r+1)}{6}+(198r-9)R^2+\frac{81(7+r)}{6}R^4\right]$$

This type of presentation allows one to vary specific parameters such as plate dimensions and elastic moduli ratio in order to obtain a desired deflection. A more complete definition of the material being used as well as plate dimensions and loading is required for a thorough analysis. This is better demonstrated in the next example.

Example 8.2 Consider a simply supported laminated composite plate subjected to the same uniform loading as the previous example. In this case, however, we consider two specific laminates, a $[0/90_2]_s$ and a $[90_2/0]_s$ laminate made form a material with $E_1=30\times10^6$, $E_2=1.5\times10^6$, $G_{12}=0.75\times10^6$, and $\nu_{12}=0.28$. For these laminates, we determine

$$[0/90_2]_s:\ [D]=\begin{bmatrix}39.0 & 0.760 & 0\\ 0.760 & 18.0 & 0\\ 0 & 0 & 1.35\end{bmatrix}\times10^4\quad [90_2/0]_s:$$

$$[D]=\begin{bmatrix}4.62 & 0.759 & 0\\ 0.759 & 52.3 & 0\\ 0 & 0 & 1.35\end{bmatrix}\times10^4$$

Recalling that $D_{mn}=D_{11}m^4+2(D_{12}+2D_{66})(mnR)^2+D_{22}(nR)^4$, it is obvious that the solution depends on the number of terms taken. This series converges after approximately 13 terms. The previous examples illustrated the computation of D_{mn} in terms of variables. This example illustrates how D_{mn} is determined using actual material properties. For the $[0/90_2]_s$ laminate, we consider three terms:

$$\begin{aligned}\text{1 term}: \ m=1,n=1: \ D_{mn} &= \left[39(1)^4+2(0.76+2(1.35))((1)(1)(a/b))^2 \right.\\ &\quad \left. +18((1)(a/b))^4\right]\times 10^4\\ &= \left[39+6.92(a/b)^2+18(a/b)^4\right]\times 10^4\end{aligned}$$

$$\begin{aligned}\text{2 terms}: \ m=1,n=3: \ D_{mn} &= \left[39(1)^4+2(0.76+2(1.35))((1)(3)(a/b))^2 \right.\\ &\quad \left. +18((3)(a/b))^4\right]\times 10^4\\ &= \left[39+20.76(a/b)^2+1458(a/b)^4\right]\times 10^4\end{aligned}$$

$$\begin{aligned}\text{3 terms}: \ m=3,n=3: \ D_{mn} &= \left[39(3)^4+2(0.76+2(1.35))((3)(3)(a/b))^2 \right.\\ &\quad \left. +18((3)(a/b))^4\right]\times 10^4\\ &= \left[3159+62.28(a/b)^2+1458(a/b)^4\right]\times 10^4\end{aligned}$$

Assuming, for simplicity that one term is sufficient, the maximum displacement of each laminate is given by

$$[0/90_2]_s\text{laminate}: \ D_{mn} = \left[39+6.92(a/b)^2+18(a/b)^4\right]\times 10^4$$

$$W_{\max} = \frac{16a^4q_o}{\pi^6}\left\{\frac{10^{-4}}{39+6.92(a/b)^2+18(a/b)^4}\right\}$$

$$[90_2/0]_s\text{laminate}: \ D_{mn} = \left[4.62+6.92(a/b)^2+52.3(a/b)^4\right]\times 10^4$$

$$W_{\max} = \frac{16a^4q_o}{\pi^6}\left\{\frac{10^{-4}}{4.62+6.92(a/b)^2+52.3(a/b)^4}\right\}$$

The maximum displacement is dependent upon the ratio a/b. Defining a normalized displacement as $W_n = \left(\pi^6 W_{\max}/16a^4q_o\right)10^6$, a plot of W_n versus a/b for each laminate is shown in Figure E8.2.

As one would anticipate, there is more deflection associated with the $[90_2/0]_s$ laminate than the $[0/90_2]_s$ laminate, which has its stiffer (0°) fibers farther from the mid-surface, thus increasing the flexural rigidity of the laminate.

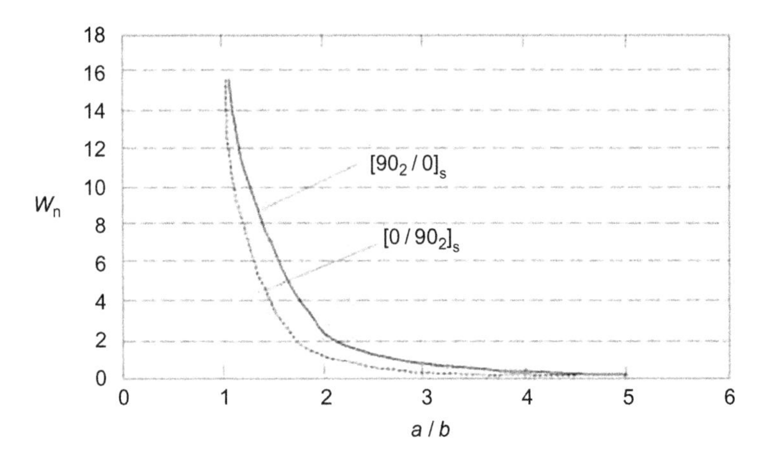

Figure E8.2 Normalized displacement for $[0/90_2]_s$ and $[90_2/0]_s$ laminated plates.

8.5.2 Solutions for a simply supported general laminate

For the general case of a simply supported symmetric laminated plate which is not specially orthotropic, we have $D_{16} \neq D_{26} \neq 0$, which complicates the solution. We begin with the same approximation for displacement

$$W_o = \sum_{m=1}^{\infty}\sum_{n=1}^{\infty} A_{mn} \sin\frac{m\pi x}{a} \sin\frac{n\pi y}{b}$$

Due to the laminate being symmetric, and assuming we neglect thermal and hygral effects, Equation (8.11) reduces to

$$D_{11}W_{o,xxxx} + 4D_{16}W_{o,xxxy} + 2(D_{12} + 2D_{66})W_{o,xxyy} + 4D_{26}W_{o,xyyy} + D_{22}W_{o,yyyy} - q = 0$$

Taking the appropriate derivatives of our assumed displacement and substituting into this equation of motion results in the final equation of motion for the plate being

$$\sum_{m=1}^{\infty}\sum_{n=1}^{\infty} A_{mn}\left\{\left[D_{11}\left(\frac{m\pi}{a}\right)^4 + 2(D_{12} + 2D_{66})\left(\frac{m\pi}{a}\right)^2\left(\frac{n\pi}{b}\right)^2 + D_{22}\left(\frac{n\pi}{b}\right)^4 - q_{mn}\right]\right.$$
$$\left.\sin\frac{m\pi x}{a}\sin\frac{n\pi y}{b} - 4\left[D_{16}\left(\frac{m\pi}{a}\right)^3\left(\frac{n\pi}{b}\right) + D_{26}\left(\frac{m\pi}{a}\right)\left(\frac{n\pi}{b}\right)^3\right]\cos\frac{m\pi x}{a}\cos\frac{n\pi y}{b}\right\} = 0$$

The solution of this equation is obviously more complicated than for the specially orthotropic case. Beyond the simple specially orthotropic case, a numerical solution is generally required, as presented in Ref. [8].

8.5.2.1 Simply supported rectangular plates (Levy's solution)

Levy's solution was proposed in 1899 and has been applied to numerous plate problems by many researches, including Refs. [2,4,8]. This procedure is applicable to problems in which particular boundary conditions exist on two opposite edges and arbitrary conditions on the remaining edges. This solution assumes that the total displacement can be written as $W_o = W_h + W_p$, where W_h is the homogeneous solution and W_p is the particular solution. For the previously applicable assumptions of symmetric specially orthotropic laminates ($[B]=0$ and $D_{16}=D_{26}=0$), the equation of motion remains unchanged

$$D_{11}\frac{\partial^4 W_o}{\partial x^4} + 2(D_{12}+2D_{66})\frac{\partial^4 W_o}{\partial x^2 \partial y^2} + D_{22}\frac{\partial^4 W_o}{\partial y^4} = q(x,y)$$

For the homogeneous solution, $q(x,y)=0$ and W_h can be expressed as a single series. The form of W_h depends on the problem being solved. For example, the simply supported plate is shown in Figure 8.11, in which the assumed form of W_h depends on the edges which are simply supported.

Functions $\Phi_m(y)$ and $\Phi_n(x)$ must be obtained so that conditions at $y=0,b$ or $x=0,a$ are satisfied. The sin and cos terms result from support conditions. The procedure for using Levy's solution is illustrated using an example of a plate that is simply supported along edges $y=0$ and $y=b$. For the homogeneous solution, we assume a solution that satisfies the simple support boundary conditions. The solution selected is

$$W_h = \sum_{n=1}^{\infty} \Phi_n(x) \sin\frac{n\pi y}{b}$$

Substituting this into the equation of motion results in

$$\sum_{n=1}^{\infty}\left[D_{11}\frac{d^4\Phi_n}{dx^4} - 2(D_{12}+2D_{66})\left(\frac{n\pi}{b}\right)^2\frac{d^2\Phi_n}{dx^2} + D_{22}\left(\frac{n\pi}{b}\right)^4\Phi_n\right]\sin\frac{n\pi y}{b} = 0$$

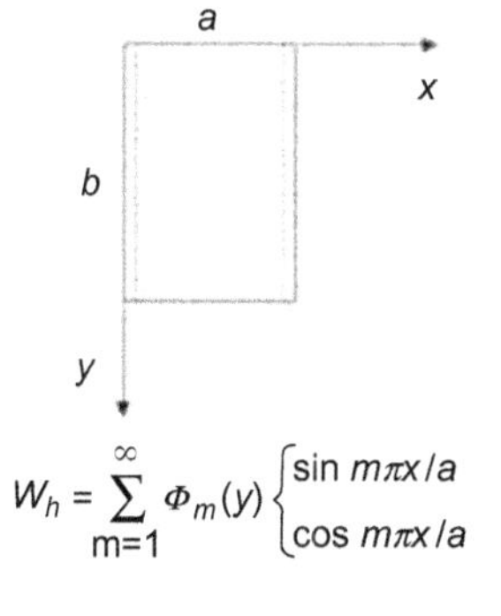

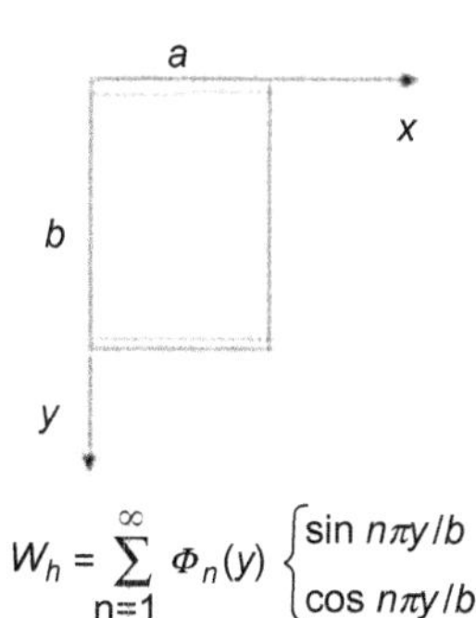

Figure 8.11 Possible forms of the homogeneous solution for a simply supported plate.

Therefore, $\Phi_n(x)$ must satisfy

$$D_{11}\frac{d^4\Phi_n}{dx^4} - 2(D_{12}+2D_{66})\left(\frac{n\pi}{b}\right)^2\frac{d^2\Phi_n}{dx^2} + D_{22}\left(\frac{n\pi}{b}\right)^4\Phi_n = 0$$

Solutions to this differential equation are of the form $\Phi_n(x) = e^{\frac{n\pi\lambda}{b}x}$. Substituting this into the equation above results in

$$\left[D_{11}\left(\frac{n\pi}{b}\right)^4\lambda^4 - 2(D_{12}+2D_{66})\left(\frac{n\pi}{b}\right)^4\lambda^2 + D_{22}\left(\frac{n\pi}{b}\right)^4\right]e^{\frac{n\pi\pi}{b}x} = 0$$

$$D_{11}\lambda^4 - 2(D_{12}+2D_{66})\lambda^2 + D_{22} = 0$$

The roots to this equation are

$$\lambda^2 = \frac{1}{D_{11}}\left[D_{12}+2D_{66} \pm \sqrt{(D_{12}+2D_{66})^2 - D_{11}D_{22}}\right] \tag{8.20}$$

The solution to Equation (8.20) can be written in terms of four arbitrary constants A_n, B_n, C_n, D_n. The actual form of the solution depends on the roots determined from Equation (8.20).

Case I: Roots of Equation (8.20) are real and unequal: The roots are denoted as $\pm\lambda_1$ and $\pm\lambda_2$ $(\lambda_1, \lambda_2 > 0)$. The solution for $\Phi_n(x)$ can be written as

$$\Phi_n(x) = A_n\cosh\frac{n\pi\lambda_1 x}{b} + B_n\sinh\frac{n\pi\lambda_1 x}{b} + C_n\cosh\frac{n\pi\lambda_2 x}{b} + D_n\sinh\frac{n\pi\lambda_2 x}{b}$$

The homogeneous solution is therefore

$$W_h = \sum_{n=1}^{\infty}\left\{A_n\cosh\frac{n\pi\lambda_1 x}{b} + B_n\sinh\frac{n\pi\lambda_1 x}{b} + C_n\cosh\frac{n\pi\lambda_2 x}{b} + D_n\sinh\frac{n\pi\lambda_2 x}{b}\right\}\sin\frac{n\pi y}{b}$$

Case II: Roots of Equation (8.20) are real and equal: The roots are denoted as $\pm\lambda$ $(\lambda > 0)$. The solution for $\Phi_n(x)$ can be written as

$$\Phi_n(x) = (A_n + B_n x)\cosh\frac{n\pi\lambda x}{b} + (C_n + D_n x)\sinh\frac{n\pi\lambda x}{b}$$

The homogeneous solution is therefore

$$W_h = \sum_{n=1}^{\infty}\left\{(A_n + B_n x)\cosh\frac{n\pi\lambda x}{b} + (C_n + D_n x)\sinh\frac{n\pi\lambda x}{b}\right\}\sin\frac{n\pi y}{b}$$

Case III: Roots of Equation (8.20) are complex: The roots are denoted as $\lambda_1 \pm i\lambda_2$ and $-\lambda_1 \pm i\lambda_2$ $(\lambda_1, \lambda_2 > 0)$. The solution for $\Phi_n(x)$ can be written as

$$\Phi_n(x) = \left(A_n \cos\frac{n\pi\lambda_2 x}{b} + B_n \sin\frac{n\pi\lambda_2 x}{b}\right)\cosh\frac{n\pi\lambda_1 x}{b} + \left(C_n \cos\frac{n\pi\lambda_2 x}{b} + D_n \sin\frac{n\pi\lambda_2 x}{b}\right)\sinh\frac{n\pi\lambda_1 x}{b}$$

The homogeneous solution is therefore

$$W_h = \sum_{n=1}^{\infty}\left[\left(A_n \cos\frac{n\pi\lambda_2 x}{b} + B_n \sin\frac{n\pi\lambda_2 x}{b}\right)\cosh\frac{n\pi\lambda_1 x}{b} + \left(C_n \cos\frac{n\pi\lambda_2 x}{b} + D_n \sin\frac{n\pi\lambda_2 x}{b}\right)\sinh\frac{n\pi\lambda_1 x}{b}\right]\sin\frac{n\pi y}{b}$$

The particular solution must also be considered. Along the edges $y = 0, b$, the boundary conditions can be satisfied by

$$W_p = \sum_{n=1}^{\infty} k_n(x)\sin\frac{n\pi y}{b} \tag{a}$$

where k_n is a function of x only. Expanding $q(x, y)$ in a single Fourier series

$$q(x, y) = \sum_{n=1}^{\infty} q_n(x)\sin\frac{n\pi y}{b} \tag{b}$$

where

$$q_n(x) = \frac{2}{b}\int_0^b q(x, y)\sin\frac{n\pi y}{b}\mathrm{d}y \tag{c}$$

Substituting (a), (b), and (c) into the equation of motion results in

$$\sum_{n=1}^{\infty}\left[D_{11}\frac{\mathrm{d}^4 k_n}{\mathrm{d}x^4} - 2(D_{12} + 2D_{66})\left(\frac{n\pi}{b}\right)^2\frac{d^2 k_n}{dx^2} + D_{22}\left(\frac{n\pi}{b}\right)^4 k_n\right]\sin\frac{n\pi y}{b} = \sum_{n=1}^{\infty} q_n(x)\sin\frac{n\pi y}{b}$$

In this expression, $q_n(x)$ is defined by Equation (c) above. The solution of this equation for k_n gives the explicit form of the particular solution, which is combined with the homogeneous solution to give the complete solution.

8.5.3 Bending of clamped rectangular plates

Bending of completely clamped plates as shown in Figure 8.12 is traditionally addressed by using the Ritz method, which is an approximate method known to yield acceptable results.

We assume the applied load is uniformly distributed so that $q(x, y) = q_o$. The Ritz method is based on the minimization of potential energy, and is the simplest method

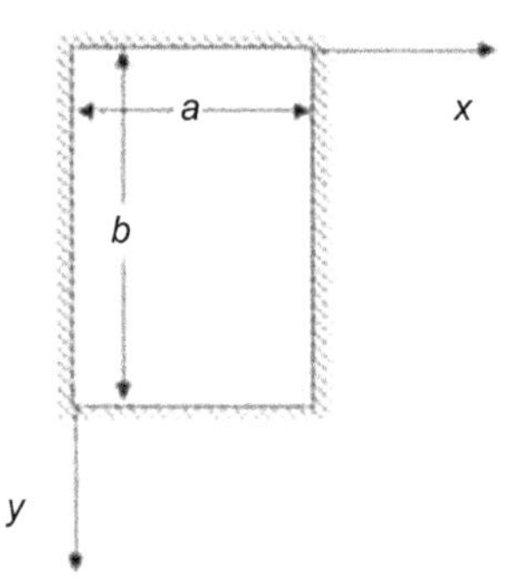

Figure 8.12 Completely clamped rectangular plate.

for solving plate and shell deflections using hand calculations. Mixed edge conditions are easy to handle using the Ritz method. The potential energy is defined as $\Pi = U + W$, where U = elastic strain energy and W = work done by external forces. In general

$$U = \frac{1}{2}\iiint\limits_V (\sigma_x\varepsilon_x + \sigma_y\varepsilon_y + \sigma_z\varepsilon_z + \ldots\ldots + \tau_{xy}\gamma_{xy})\mathrm{d}x\mathrm{d}y\mathrm{d}z$$

Based on thin plate assumptions $(\sigma_z = \tau_{xz} = \tau_{yz} = 0)$, and treating strains in terms of displacements, the strain energy for a specially orthotropic ($[B] = 0$) laminated plate is

$$U = \frac{1}{2}\int_0^a\int_0^b \left[D_{11}\left(\frac{\partial^2 W_o}{\partial x^2}\right)^2 + 2D_{12}\frac{\partial^2 W_o}{\partial x^2}\frac{\partial^2 W_o}{\partial y^2} + D_{22}\left(\frac{\partial^2 W_o}{\partial y^2}\right)^2 + 4D_{66}\left(\frac{\partial^2 W_o}{\partial x \partial y}\right)^2\right]\mathrm{d}x\mathrm{d}y$$

For a uniformly loaded plate with $q(x, y) = q_o$, the work is

$$W = -\int_0^a\int_0^b q_o W_o \mathrm{d}x\mathrm{d}y$$

In order to find an approximate solution for the plate deflection, we assume a series solution

$$W_o(x, y) = \sum_{m=1}^{M}\sum_{n=1}^{N} A_{mn}X_m(x)Y_n(y)$$

The functions $X_m(x)$ and $Y_n(y)$ must be chosen such that they satisfy the boundary conditions. For the completely clamped plate, we have

$$x = 0, a: \; W_o = \frac{\partial W_o}{\partial x} = 0 \;\text{ and }\; y = 0, b: \; W_o = \frac{\partial W_o}{\partial y} = 0$$

In order to satisfy the conditions of minimum potential energy,

$$\frac{\partial U}{\partial W_o} = -\frac{\partial W}{\partial A_{mn}} \quad \begin{cases} m = 1,2,3,\ldots M \\ n = 1,2,3,\ldots N \end{cases}$$

Using the assumed displacements and the minimum potential energy expression with the knowledge that $m = 1,2,3,\ldots M$ and $n = 1,2,3,\ldots N$ results in

$$\sum_{i=1}^{M}\sum_{j=1}^{N} A_{ij}\left\{D_{11}\left[\int_0^a \frac{d^2X_i}{dx^2}\frac{d^2X_m}{dx^2}dx\int_0^b Y_jY_n dy\right] + 2D_{12}\left[\int_0^a X_m\frac{d^2X_i}{dx^2}dx\int_0^b Y_j\frac{d^2Y_n}{dy^2}dy\right.\right.$$
$$\left. + \int_0^a X_i\frac{d^2X_m}{dx^2}dx\int_0^b Y_n\frac{d^2Y_j}{dy^2}dy\right] + D_{22}\left[\int_0^a X_iX_m dx\int_0^b \frac{d^2Y_j}{dy^2}\frac{d^2Y_n}{dy^2}dy\right]$$
$$\left. + 4D_{66}\left[\int_0^a \frac{dX_i}{dx}\frac{dX_m}{dx}dx\int_0^b \frac{dY_j}{dy}\frac{dY_n}{dy}dy\right]\right\} = q_o\int_0^a X_m dx\int_0^b Y_n dy$$

There are many possible choices for X_m and Y_n. Two are considered for illustrative purposes.

Case I: Assume a polynomial for X_m and Y_n so that

$$X_m(x) = \left(x^2 - ax\right)^2 X^{M-1} \quad \text{and} \quad Y_n(y) = \left(y^2 - by\right)^2 Y^{N-1}$$

As a first approximation, assume $M = N = 1$, which results in

$$\begin{array}{ll} X_1 = x^4 - 2ax^3 + a^2x^2 & Y_1 = y^4 - 2by^3 + b^2y^2 \\ X_{1,x} = 4x^3 - 6ax^2 + 2a^2x & Y_{1,y} = 4y^3 - 6by^2 + 2a^2y \\ X_{1,xx} = 12x^2 - 12ax + 2a^2 & Y_{1,yy} = 12y^2 - 12by + 2b^2 \end{array}$$

Substituting into the equation above results in

$$A_{11}\left\{D_{11}\left[\int_0^a \left(12x^2 - 12ax + 2a^2\right)^2 dx\int_0^b \left(y^4 - 2by^3 + b^2y^2\right)^2 dy\right]\right.$$
$$+ 2D_{12}\left[\int_0^a \left(x^4 - 2ax^3 + a^2x^2\right)\left(12x^2 - 12ax + 2a^2\right)dx\right.$$
$$\int_0^b \left(y^4 - 2by^3 + b^2y^2\right)\left(12y^2 - 12by + 2b^2\right)dy$$
$$+ \int_0^a \left(x^4 - 2ax^3 + a^2x^2\right)\left(12x^2 - 12ax + 2a^2\right)dx$$
$$\left.\int_0^b \left(y^4 - 2by^3 + b^2y^2\right)\left(12y^2 - 12by + 2b^2\right)dy\right]$$
$$+ D_{22}\left[\int_0^a \left(x^4 - 2ax^3 + a^2x^2\right)^2 dx\int_0^b \left(12y^2 - 12by + 2b^2\right)^2 dy\right]$$
$$\left. + 4D_{66}\left[\int_0^a \left(4x^3 - 6ax^2 + 2a^2x\right)^2 dx\int_0^b \left(4y^3 - 6by^2 + 2b^2y\right)^2 dy\right]\right\}$$
$$= q_o\int_0^a \left(x^4 - 2ax^3 + a^2x^2\right)dx\int_0^b \left(y^4 - 2by^3 + b^2y^2\right)dy$$

After reducing this expression, we arrive at

$$A_{11} = \frac{6.125 q_o}{7D_{11}b^4 + 4(D_{12} + 2D_{66})a^2b^2 + 7D_{22}a^4}$$

Therefore, the deflection is given as

$$W_o(x,y) = \frac{6.125 q_o (x^2 - ax)^2 (y^2 - by)^2}{7D_{11}b^4 + 4(D_{12} + 2D_{66})a^2b^2 + 7D_{22}a^4}$$

The maximum deflection occurs at the center of the plate $(x = a/2, y = b/2)$ and is

$$W_{\max} = W_o(a/2, b/2) = \frac{0.00342 q_o a^4}{D_{11} + 0.571(D_{12} + 2D_{66})R^2 + D_{22}R^4}$$

where $R = a/b$

Case II: Assume the characteristic shape function for a solution to the natural vibration problem for a beam with clamped end. This gives

$$X_m(x) = \gamma_m \cos\frac{\lambda_m x}{a} - \gamma_m \cosh\frac{\lambda_m x}{a} + \gamma_m \sin\frac{\lambda_m x}{a} - \gamma_m \sinh\frac{\lambda_m x}{a}$$

$$Y_n(y) = \gamma_n \cos\frac{\lambda_n y}{b} - \gamma_n \cosh\frac{\lambda_n y}{b} + \gamma_n \sin\frac{\lambda_n y}{b} - \gamma_n \sinh\frac{\lambda_n y}{b}$$

where λ_m and λ_n are roots of the frequency equation $\cos\lambda_i \cosh\lambda_i = 1$ and $\gamma_i = \dfrac{\cos\lambda_i - \cosh\lambda_i}{\sin\lambda_i + \sinh\lambda_i}$. Solving the free vibration problem for a beam, we get $\lambda_1 = 4.73$, $\lambda_2 = 7.85$, $\lambda_3 = 10.99$, $\lambda_4 = 14.14$, and $\lambda_5 = 17.28$. An approximation exists for $i > 5$: $\cosh\lambda_i \approx e^{\lambda_i}/2$ and $\cos\lambda_i \approx 2e^{-\lambda_i}$. Therefore,

$$\lambda_i \approx (2i+1)\pi/2$$

The functions X_m and Y_n satisfy the boundary conditions. For $M = N = 1$, we get $\lambda_1 = 4.73$ and $\gamma_1 = 0.9825$ so that

$$X_m(x) = 0.9825\left[\cos\frac{4.73x}{a} - \cosh\frac{4.73x}{a}\right] + \sin\frac{4.73x}{a} - \sinh\frac{4.73x}{a}$$

$$Y_n(x) = 0.9825\left[\cos\frac{4.73y}{b} - \cosh\frac{4.73y}{b}\right] + \sin\frac{4.73y}{b} - \sinh\frac{4.73y}{b}$$

Substituting these into Equation (a) above, we get

$$A_{11}\left[500.56\left(D_{11} + D_{22}R^4\right) + 302.71R^2\left(D_{12} + 2D_{66}\right)\right] = 0.6903 q_o a^4$$

$$A_{11} = \frac{0.6903 q_o a^4}{500.56[D_{11} + D_{22}R^4 + 0.6047R^2(D_{12} + 2D_{66})]}$$

The maximum deflection can be determined by evaluating $X_1(x)$ and $Y_1(y)$ at the center of the plate $(x = a/2, y = b/2)$. The maximum deflection will be

$$W_{\max} = \frac{0.00348 q_o a^4}{D_{11} + D_{22}R^4 + 0.6047R^2(D_{12} + 2D_{66})}$$

Comparing this solution to the previous case (the polynomial solution), we see that they are almost identical. If we assume the plate is isotropic $(D_{11} = D_{22} = D_{12} + 2D_{66} = D)$ and the plate is square $(R = 1)$, we have $W_{\max} = 0.00133 q_o a^4/D$ for the polynomial and $W_{\max} = 0.00134 q_o a^4/D$ for the case above. Using a larger number of terms in our approximate solution using the beam mode shape results in $W_{\max} = 0.00126 q_o a^4/D$.

The maximum moment occurs at the middle of each edge and can be calculated by differentiating W_o in conjunction with

$$M_x = -D\left(\frac{\partial^2 W_o}{\partial x^2} + \frac{\partial^2 W_o}{\partial y^2}\right)$$

Using case I (polynomial solution), $M_{\max} = -0.0425 q_o a^2$.
Using case II (beam functions), $M_{\max} = -0.0384 q_o a^2$.
Using a larger number of terms, $M_{\max} = -0.0513 q_o a^2$.

Although a one-term approximation gives adequate correlation between solutions for the plate deflection, it does not work well for the bending moments.

8.6 Plate vibrations

Several cases of plate vibration and solution techniques are presented in this section and they generally refer to variations of Equation (8.11). Only relatively simple cases are considered, since complete vibration analysis requires extensive computational approaches. The discussions herein are limited to specially orthotropic plates for which $[B] = 0$ and the only nonzero terms of $[A]$ and $[D]$ are A_{11}, A_{12}, A_{22}, A_{66}, D_{11}, D_{12}, D_{22}, and D_{66}. This presentation is intended to illustrate approaches and procedures involved in exploring plate vibration problems.

8.6.1 Free vibrations of simply supported plates

Consider a specially orthotropic simply supported rectangular plate with $q = N_x = N_y = N_{xy} = 0$ and no thermal or hygral effects. The equation of motion is determine from Equation (8.11) to be

$$D_{11}\frac{d^4W_o}{dx^4}+2(D_{12}+2D_{66})\frac{d^4W_o}{dx^2dy^2}+D_{22}\frac{d^4W_o}{dy^4}+\rho\frac{d^2W_o}{dt^2}=0$$

In order to satisfy boundary and initial conditions, we assume a solution of the form $W_o=W(x,y)e^{i\omega t}$, where ω is the natural frequency of vibration of the plate. The equation of motion becomes

$$D_{11}W_{,xxxx}+2(D_{12}+2D_{66})W_{,xxyy}+D_{22}W_{,yyyy}-\rho\omega^2W=0$$

The boundary conditions at $x=0,a$ are $W(x,y)=M_x=-D_{11}W_{,xx}-D_{12}W_{,yy}=0$, and at $y=0,b$: $W(x,y)=M_x=-D_{12}W_{,xx}-D_{22}W_{,yy}=0$.

Assuming a displacement field of $W(x,y)=A_{mn}\sin\frac{m\pi x}{a}\sin\frac{n\pi y}{b}$, where m and n are integers, we get

$$D_{11}\left(\frac{m\pi}{a}\right)^4+2(D_{12}+2D_{66})\left(\frac{m\pi}{a}\right)^2\left(\frac{n\pi}{b}\right)^2+D_{22}\left(\frac{n\pi}{b}\right)^4-\rho\omega_{mn}^2=0$$

Defining the ratio $R=a/b$, we can determine that

$$\omega_{mn}=\frac{\pi^2}{R^2b^2}\frac{1}{\sqrt{\rho}}\sqrt{D_{11}m^4+2(D_{12}+2D_{66})m^2n^2R^2+D_{22}n^4R^4}$$

For the case of $m=n=1$, we get

$$\omega_{11}=\frac{\pi^2}{R^2b^2}\frac{1}{\sqrt{\rho}}\sqrt{D_{11}+2(D_{12}+2D_{66})R^2+D_{22}R^4}$$

The solution for an isotropic plate is $\omega_{11}=\frac{\pi^2}{b^2}\sqrt{\frac{D}{\rho}}\left[\left(\frac{b}{a}\right)^2+1\right]$.

Consider a square orthotropic plate $(R=1)$ with $D_{11}/D_{22}=10$ and $(D_{12}+2D_{66})/D_{22}=1$, a square isotropic plate as initially introduced by Jones [6]. The normalized vibration frequencies and mode shapes of the first four modes for both cases are presented in Table 8.1.

8.6.2 Rectangular angle-ply plates

Solutions for a rectangular angle ply plate can be difficult to obtain without numerical techniques involving computer algorithms. For a symmetric angle-ply laminate, $D_{16}\neq0$ and $D_{26}\neq0$. Therefore, Equation (8.11) reduces to

$$D_{11}W_{o,xxxx}+4D_{16}W_{o,xxxy}+2(D_{12}+2D_{66})W_{o,xxyy}+D_{26}W_{o,xyyy}+D_{22}W_{o,yyyy}+\rho W_{o,tt}=0$$

In order to satisfy boundary and initial conditions, we assume a solution of the form $W_o=W(x,y)e^{i\omega t}$, where ω is the natural frequency of vibration of the plate. The equation of motion becomes

Table 8.1 **Normalized vibration frequencies and mode shapes for simply supported square specially orthotropic and isotropic plates (after Ref. [6])**

	Orthotropic: $\omega = \frac{k\pi^2}{b^2}\sqrt{\frac{D_{22}}{\rho}}$				Isotropic: $\omega = \frac{k\pi^2}{b^2}\sqrt{\frac{D}{\rho}}$			
Mode	m	n	k	**Mode shape**	m	n	k	**Mode shape**
1	1	1	3.61		1	1	2.0	
2	1	2	5.83		1	2	5.0	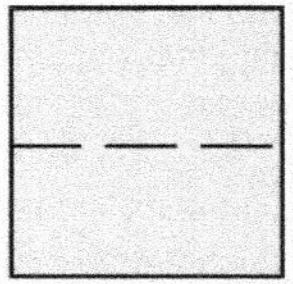
3	1	3	10.44	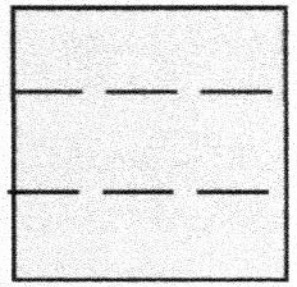	2	1	5.0	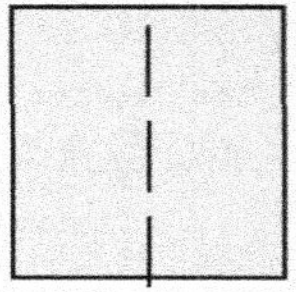
4	2	1	13.0	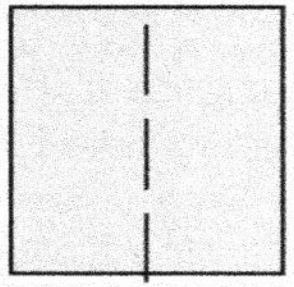	2	2	8.0	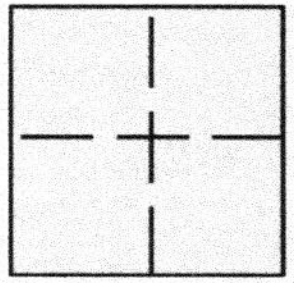

$$D_{11}W_{,xxxx}+4D_{16}W_{,xxxy}+2(D_{12}+2D_{66})W_{,xxyy}+D_{26}W_{,xyyy}+D_{22}W_{,yyyy}-\rho\omega^2W=0$$

The boundary conditions at $x=0,a$ are $W(x,y)=M_x=-D_{11}W_{,xx}-D_{12}W_{,yy}-2D_{16}W_{xy}=0$, and at $y=0,b$: $W(x,y)=M_x=-D_{12}W_{,xx}-D_{22}W_{,yy}-2D_{26}W_{xy}=0$.

Assuming $W(x,y)=A_{mn}\sin\frac{m\pi x}{a}\sin\frac{n\pi y}{b}$, where m and n are integers, we get

$$D_{11}\left(\frac{m\pi}{a}\right)^4+4D_{16}\left(\frac{m\pi}{a}\right)^3\left(\frac{n\pi}{b}\right)2(D_{12}+2D_{66})\left(\frac{m\pi}{a}\right)^2\left(\frac{n\pi}{b}\right)^2$$
$$+4D_{26}\left(\frac{m\pi}{a}\right)\left(\frac{n\pi}{b}\right)^3+D_{22}\left(\frac{n\pi}{b}\right)^4-\rho\omega_{mn}^2=0$$

Defining $R=a/b$, we determine

$$\omega_{mn}=\frac{\pi^2}{R^2b^2}\frac{1}{\sqrt{\rho}}\sqrt{D_{11}m^4+4D_{16}m^3nR+2(D_{12}+2D_{66})m^2n^2R^2+4D_{26}mn^3R^3+D_{22}n^4R^4}$$

Once again we considering a square orthotropic plate ($R=1$) with the appropriate terms of $[D]$ approximated to be $D_{11}/D_{22}=10$, $(D_{12}+2D_{66})/D_{22}=1$, $4D_{16}/D_{22}=-0.3$, and $4D_{26}/D_{22}=-0.03$. The frequency equation above therefore becomes

$$\omega_{mn}=\frac{\pi^2}{b^2}\frac{1}{\sqrt{\rho}}\sqrt{D_{22}(10m^4-0.3m^3n+2m^2n^2-0.03mn^3+n^4)}$$

For the case of $m=n=1$, we get

$$\omega_{11}=\frac{\pi^2}{b^2}\frac{1}{\sqrt{\rho}}\sqrt{11.67D_{22}}\approx\frac{3.56\pi^2}{b^2}\sqrt{\frac{D_{22}}{\rho}}$$

For the specially orthotropic laminate, we had $\omega_{11}=(3.62\pi^2/b^2)\sqrt{D_{22}/\rho}$. Although the difference between these two solutions does not appear significant, it does indicate the influence of material behavior (the $[D]$ matrix) on the solution.

8.7 Effects of shear deformation

The effects of shear deformation on laminated composite plates have been explored by many investigators, including Refs. [10–13]. In order to include the effects of shear deformation, we relax the assumption that $\gamma_{xz}=\gamma_{yz}=0$. The result is that the interlaminar shear strains are $\gamma_{xz}=W_{o,x}+\Phi$ and $\gamma_{yz}=W_{o,y}+\Psi$. All other strain–displacement relations are unchanged. The stress and strains in any k^{th} lamina are again related through $\{\sigma\}_k=[\bar{Q}]_k\{\varepsilon\}_k$, and the applied forces and moments are defined as before by

$$(N_x, N_y, N_{xy}) = \int_{-h/2}^{h/2} \left[(\sigma_x)_k, (\sigma_y)_k, (\tau_{xy})_k\right] \mathrm{d}z$$

$$(M_x, M_y, M_{xy}) = \int_{-h/2}^{h/2} \left[(\sigma_x)_k, (\sigma_y)_k, (\tau_{xy})_k\right] z\mathrm{d}z$$

$$(Q_x, Q_y) = \int_{-h/2}^{h/2} \left[(\tau_{xz})_k (\tau_{yz})_k\right] \mathrm{d}z$$

In addition, we continue to employ the relations

$$\begin{Bmatrix} N \\ M \end{Bmatrix} = \begin{bmatrix} A & B \\ B & D \end{bmatrix} \begin{Bmatrix} \varepsilon^o \\ \kappa \end{Bmatrix} \quad \begin{Bmatrix} Q_x \\ Q_y \end{Bmatrix} = K \begin{bmatrix} A_{55} & A_{45} \\ A_{45} & A_{44} \end{bmatrix} \begin{Bmatrix} \gamma_{xz} \\ \gamma_{yz} \end{Bmatrix}$$

$$(A_{ij}, B_{ij}, D_{ij}) = \int_{-h/2}^{h/2} \left[\bar{Q}_{ij}\right]_k (1, z, z^2) \mathrm{d}z$$

$$A_{ij} = \sum_{k=1}^{N} \left[\bar{Q}_{ij}\right]_k \left[(h_k - h_{k-1}) - \frac{4}{3}(h_k^3 - h_{k-1}^3)/h^2\right] \quad (i, j = 4, 5)$$

Relaxing the restrictions on the interlaminar shear terms results in governing equations being

$$N_{x,x} + N_{xy,y} = \bar{\rho}_1 \ddot{U}_o + \bar{\rho}_2 \ddot{\Phi}$$

$$N_{xy,x} + N_{y,y} = \bar{\rho}_1 \ddot{V}_o + \bar{\rho}_2 \ddot{\Psi}$$

$$M_{x,x} + M_{xy,y} - Q_x = \bar{\rho}_2 \ddot{U}_o + \bar{\rho}_3 \ddot{\Phi}$$

$$M_{xy,x} + M_{y,y} - Q_y = \bar{\rho}_2 \ddot{V}_o + \bar{\rho}_3 \ddot{\Psi}$$

$$Q_{x,x} + Q_{y,y} + N_x W_{o,xx} + 2N_{xy} W_{o,xy} + N_y W_{o,yy} + q = \bar{\rho}_1 \ddot{W}_o$$

where $(\bar{\rho}_1, \bar{\rho}_2, \bar{\rho}_3) \int_{-h/2}^{h/2} \rho_k (1, z, z^2) \mathrm{d}z$. Using these relations, we can develop the plate equations of motion including shear deformation, but neglecting thermal and hygral behavior. Inclusion of thermal and hygral effects simply adds terms to the equations of motion. In most cases, the effects of these are negligible. The equations of motion are

$$\begin{aligned} &A_{11} U_{o,xx} + 2A_{16} U_{o,xy} + A_{66} U_{o,yy} + A_{16} V_{o,xx} + (A_{12} + A_{66}) V_{o,xy} + A_{26} V_{o,yy} \\ &+ B_{11} \Phi_{,xx} + 2B_{16} \Phi_{,xy} + B_{66} \Phi_{,yy} + B_{16} \Psi_{,xx} + (B_{12} + B_{66}) \Psi_{,xy} + B_{26} \Psi_{,yy} = \bar{\rho}_1 \ddot{U}_o + \bar{\rho}_2 \ddot{\Phi} \end{aligned} \tag{8.21}$$

$$\begin{aligned} &A_{16} U_{o,xx} + (A_{12} + A_{66}) U_{o,xy} + A_{26} U_{o,yy} + A_{66} V_{o,xx} + 2A_{26} V_{o,xy} + A_{22} V_{o,yy} \\ &+ B_{16} \Phi_{,xx} + (B_{12} + B_{66}) \Phi_{,xy} + B_{26} \Phi_{,xy} + B_{66} \Psi_{,xx} + 2B_{26} \Psi_{,xy} + B_{22} \Psi_{,yy} = \bar{\rho}_1 \ddot{V}_o + \bar{\rho}_2 \ddot{\Psi} \end{aligned} \tag{8.22}$$

$$\begin{aligned}&B_{11}U_{o,xx}+2B_{16}U_{o,xy}+B_{66}U_{o,yy}+B_{16}V_{o,xx}+(B_{12}+B_{66})V_{o,xy}+B_{26}V_{o,yy}\\&+D_{11}\Phi_{,xx}+2D_{16}\Phi_{,xy}+D_{66}\Phi_{,yy}+D_{16}\Psi_{,xx}+(D_{12}+D_{66})\Psi_{,xy}+D_{26}\Psi_{,yy}\\&-K\left[A_{55}\left(\Phi+W_{o,x}\right)+A_{45}\left(\Psi+W_{o,y}\right)\right]=\overline{\rho}_2\ddot{U}_o+\overline{\rho}_3\ddot{\Phi}\end{aligned}\tag{8.23}$$

$$\begin{aligned}&B_{16}U_{o,xx}+(B_{12}+B_{66})U_{o,xy}+B_{26}U_{o,yy}+B_{66}V_{o,xx}+2B_{26}V_{o,xy}+B_{22}V_{o,yy}\\&+D_{16}\Phi_{,xx}+(D_{12}+D_{66})\Phi_{,xy}+D_{26}\Phi_{,yy}+D_{66}\Psi_{,xx}+2D_{26}\Psi_{,xy}+D_{22}\Psi_{,yy}\\&-K\left[A_{45}\left(\Phi+W_{o,x}\right)+A_{55}\left(\Psi+W_{o,y}\right)\right]=\overline{\rho}_2\ddot{V}_o+\overline{\rho}_3\ddot{\Psi}\end{aligned}\tag{8.24}$$

$$\begin{aligned}&K\left[A_{55}\left(\Phi_{,x}+W_{o,xx}\right)+A_{45}\left(\Phi_{,x}+\Psi_{,y}+2W_{o,xx}\right)+A_{44}\left(\Psi_{,y}+W_{o,yy}\right)\right]+q\\&+N_xW_{o,xx}+2N_{xy}W_{o,xy}+N_yW_{o,yy}=\overline{\rho}_1\ddot{W}_o\end{aligned}\tag{8.25}$$

Special cases of these equations result for specific laminates, such as angle-ply laminates $\left([\pm\theta]_n\right)$, for which $A_{16}=A_{26}=B_{11}=B_{22}=B_{12}=B_{66}=D_{16}=D_{26}=0$. The resulting equations of motion are reduced to

$$\begin{aligned}&A_{11}U_{o,xx}+A_{66}U_{o,yy}+(A_{12}+A_{66})V_{o,xy}+A_{26}V_{o,yy}+2B_{16}\Phi_{,xy}+B_{16}\Psi_{,xx}+B_{26}\Psi_{,yy}\\&=\overline{\rho}_1\ddot{U}_o+\overline{\rho}_2\ddot{\Phi}\\&(A_{12}+A_{66})U_{o,xy}+A_{66}V_{o,xx}+A_{22}V_{o,yy}+B_{16}\Phi_{,xx}+B_{26}\Phi_{,xy}+2B_{26}\Psi_{,xy}\\&=\overline{\rho}_1\ddot{V}_o+\overline{\rho}_2\ddot{\Psi}\\&2B_{16}U_{o,xy}+B_{16}V_{o,xx}+B_{26}V_{o,yy}+D_{11}\Phi_{,xx}+D_{66}\Phi_{,yy}+(D_{12}+D_{66})\Psi_{,xy}\\&-KA_{55}\left(\Phi+W_{o,x}\right)=\overline{\rho}_2\ddot{U}_o+\overline{\rho}_3\ddot{\Phi}\end{aligned}$$

$$\begin{aligned}&B_{16}U_{o,xx}+B_{26}U_{o,yy}+2B_{26}V_{o,xy}+(D_{12}+D_{66})\Phi_{,xy}+D_{66}\Psi_{,xx}+D_{22}\Psi_{,yy}\\&-KA_{55}\left(\Psi+W_{o,y}\right)=\overline{\rho}_2\ddot{V}_o+\overline{\rho}_3\ddot{\Psi}\\&K\left[A_{55}\left(\Phi_{,x}+W_{o,xx}\right)+A_{44}\left(\Psi_{,y}+W_{o,yy}\right)\right]+q+N_xW_{o,xx}+2N_{xy}W_{o,xy}\\&+N_yW_{o,yy}=\overline{\rho}_1\ddot{W}_o\end{aligned}$$

In a similar manner, Equations (8.21)–(8.25) are simplified if the laminated plate is symmetric, with $[B]=0$ and $\overline{\rho}_2=0$. In this case, the equations of motion reduce to

$$A_{11}U_{o,xx}+2A_{16}U_{o,xy}+A_{66}U_{o,yy}+A_{16}V_{o,xx}+(A_{12}+A_{66})V_{o,xy}+A_{26}V_{o,yy}=\overline{\rho}_1\ddot{U}_o$$

$$A_{16}U_{o,xx}+(A_{12}+A_{66})U_{o,xy}+A_{26}U_{o,yy}+A_{66}V_{o,xx}+2A_{26}V_{o,xy}+A_{22}V_{o,yy}=\overline{\rho}_1\ddot{V}_o$$

$$\begin{aligned}&+D_{11}\Phi_{,xx}+2D_{16}\Phi_{,xy}+D_{66}\Phi_{,yy}+D_{16}\Psi_{,xx}+(D_{12}+D_{66})\Psi_{,xy}+D_{26}\Psi_{,yy}\\&-K\left[A_{55}\left(\Phi+W_{o,x}\right)+A_{45}\left(\Psi+W_{o,y}\right)\right]=\overline{\rho}_3\ddot{\Phi}\end{aligned}$$

$$\begin{aligned}&+D_{16}\Phi_{,xx}+(D_{12}+D_{66})\Phi_{,xy}+D_{26}\Phi_{,yy}+D_{66}\Psi_{,xx}+2D_{26}\Psi_{,xy}+D_{22}\Psi_{,yy}\\&-K\left[A_{45}\left(\Phi+W_{o,x}\right)+A_{55}\left(\Psi+W_{o,y}\right)\right]=\overline{\rho}_3\ddot{\Psi}\end{aligned}$$

$$K\left[A_{55}\left(\Phi_{,x}+W_{o,xx}\right)+A_{45}\left(\Phi_{,x}+\Psi_{,y}+2W_{o,xx}\right)+A_{44}\left(\Psi_{,y}+W_{o,yy}\right)\right]+q$$
$$+N_x W_{o,xx}+2N_{xy}W_{o,xy}+N_y W_{o,yy}=\overline{\rho}_1\ddot{W}_o$$

The stresses within each ply of the laminated plate can be determined using

$$\begin{Bmatrix}\sigma_x\\ \sigma_y\\ \tau_{xy}\end{Bmatrix}_k=\begin{bmatrix}\bar{Q}_{11} & \bar{Q}_{12} & \bar{Q}_{16}\\ \bar{Q}_{12} & \bar{Q}_{22} & \bar{Q}_{26}\\ \bar{Q}_{16} & \bar{Q}_{26} & \bar{Q}_{66}\end{bmatrix}_k\begin{Bmatrix}U_{o,x}\\ V_{o,y}\\ U_{o,y}+V_{o,x}\end{Bmatrix}+z\begin{bmatrix}\bar{Q}_{11} & \bar{Q}_{12} & \bar{Q}_{16}\\ \bar{Q}_{12} & \bar{Q}_{22} & \bar{Q}_{26}\\ \bar{Q}_{16} & \bar{Q}_{26} & \bar{Q}_{66}\end{bmatrix}_k\begin{Bmatrix}\Phi_{,x}\\ \Psi_{,y}\\ \Phi_{,y}+\Psi_{,x}\end{Bmatrix}$$

$$\begin{Bmatrix}\tau_{yz}\\ \tau_{xz}\end{Bmatrix}_k=\begin{bmatrix}\bar{Q}_{44} & \bar{Q}_{45}\\ \bar{Q}_{45} & \bar{Q}_{55}\end{bmatrix}_k\begin{Bmatrix}\Psi+W_{o,y}\\ \Phi+W_{o,x}\end{Bmatrix}$$

In order to solve these problems, the boundary conditions must be known. They are generally specified in the directions normal (n) and tangential (t) to the plate as previously presented.

8.7.1 Bending and free vibration of angle-ply plates

For an angle-ply-laminated plate $\left([\pm\theta]_n\right)$, $A_{16}=A_{26}=B_{11}=B_{22}=B_{12}=B_{66}=D_{16}=D_{26}=0$. The only nonvanishing extension-bending coupling terms are B_{16} and B_{26}. If we assume that the in-plane forces N_x, N_y, and N_{xy} are zero, the last equation of motion is reduced. For this case, the equations of motion are

$$A_{11}U_{o,xx}+A_{66}U_{o,yy}+(A_{12}+A_{66})V_{o,xy}+A_{26}V_{o,yy}+2B_{16}\Phi_{,xy}+B_{16}\Psi_{,xx}+B_{26}\Psi_{,yy}$$
$$=\overline{\rho}_1\ddot{U}_o+\overline{\rho}_2\ddot{\Phi}$$

$$(A_{12}+A_{66})U_{o,xy}+A_{66}V_{o,xx}+A_{22}V_{o,yy}+B_{16}\Phi_{,xx}+B_{26}\Phi_{,xy}+2B_{26}\Psi_{,xy}$$
$$=\overline{\rho}_1\ddot{V}_o+\overline{\rho}_2\ddot{\Psi}$$

$$2B_{16}U_{o,xy}+B_{16}V_{o,xx}+B_{26}V_{o,yy}+D_{11}\Phi_{,xx}+D_{66}\Phi_{,yy}+(D_{12}+D_{66})\Psi_{,xy}$$
$$-KA_{55}\left(\Phi+W_{o,x}\right)=\overline{\rho}_2\ddot{U}_o+\overline{\rho}_3\ddot{\Phi}$$

$$B_{16}U_{o,xx}+B_{26}U_{o,yy}+2B_{26}V_{o,xy}+(D_{12}+D_{66})\Phi_{,xy}+D_{66}\Psi_{,xx}+D_{22}\Psi_{,yy}$$
$$-KA_{55}\left(\Psi+W_{o,y}\right)=\overline{\rho}_2\ddot{V}_o+\overline{\rho}_3\ddot{\Psi}$$

$$K\left[A_{55}\left(\Phi_{,x}+W_{o,xx}\right)+A_{44}\left(\Psi_{,y}+W_{o,yy}\right)\right]+q=\overline{\rho}_1\ddot{W}_o$$

Assume a plate with hinged edges that are free in the tangential direction. The following boundary conditions apply:

$$x=0,a:\quad U_o=N_{xy}=A_{66}\left(U_{o,y}+V_{o,x}\right)+B_{16}\Phi_{,x}+B_{26}\Psi_{,y}=0$$
$$W_o=\Psi=M_x=B_{16}\left(U_{o,y}+V_{o,x}\right)+D_{11}\Phi_{,x}+D_{12}\Psi_{,y}=0$$

$$y=0,b:\quad V_o=N_{xy}=A_{66}\left(U_{o,y}+V_{o,x}\right)+B_{16}\Phi_{,x}+B_{26}\Psi_{,y}=0$$
$$W_o=\Phi=M_y=B_{26}\left(U_{o,y}+V_{o,x}\right)+D_{12}\Phi_{,x}+D_{22}\Psi_{,y}=0$$

Assuming the plate is loaded with a distributed load $q(x,y) = q_o \sin(m\pi x/a)\sin(n\pi y/b)$, we assume a solution that can be expressed as

$$U_o = \sum_{m=1}^{\infty}\sum_{n=1}^{\infty} U_{mn} \sin\frac{m\pi x}{a} \cos\frac{n\pi y}{b} e^{i\omega_{mn}t}, V_o = \sum_{m=1}^{\infty}\sum_{n=1}^{\infty} V_{mn} \cos\frac{m\pi x}{a} \sin\frac{n\pi y}{b} e^{i\omega_{mn}t}$$

$$\Phi = \sum_{m=1}^{\infty}\sum_{n=1}^{\infty} X_{mn} \cos\frac{m\pi x}{a} \sin\frac{n\pi y}{b} e^{i\omega_{mn}t}, \Psi = \sum_{m=1}^{\infty}\sum_{n=1}^{\infty} Y_{mn} \sin\frac{m\pi x}{a} \cos\frac{n\pi y}{b} e^{i\omega_{mn}t}$$

$$W_o = \sum_{m=1}^{\infty}\sum_{n=1}^{\infty} W_{mn} \sin\frac{m\pi x}{a} \sin\frac{n\pi y}{b} e^{i\omega_{mn}t}$$

Substituting these into the equations of motion, and assuming that each ply is made from the same material ($\overline{\rho}_2 = 0$), we can express the solution in matrix form as

$$\begin{bmatrix} (K_{11}-\lambda) & K_{12} & K_{13} & K_{14} & 0 \\ K_{12} & (K_{22}-\lambda) & K_{14} & K_{24} & 0 \\ K_{13} & K_{14} & (K_{33}-h^2\lambda/12) & K_{34} & K_{35} \\ K_{14} & K_{24} & K_{34} & (K_{44}-h^2\lambda/12) & K_{45} \\ 0 & 0 & K_{35} & K_{45} & (K_{55}-\lambda) \end{bmatrix} \begin{Bmatrix} U_{mn} \\ V_{mn} \\ X_{mn} \\ Y_{mn} \\ W_{mn} \end{Bmatrix} = \begin{Bmatrix} 0 \\ 0 \\ 0 \\ 0 \\ \bar{q}_o \end{Bmatrix}$$

where:

$$K_{11} = A_{11}m^2 + A_{66}n^2R^2$$

$$K_{12} = (A_{12} + A_{66})mnR$$

$$K_{13} = 2B_{16}mnR$$

$$K_{14} = B_{16}m^2 + B_{26}n^2R^2$$

$$K_{22} = A_{66}m^2 + A_{22}n^2R^2$$

$$K_{24} = 2B_{26}mnR$$

$$K_{33} = D_{11}m^2 + D_{66}n^2R^2 + KA_{55}a^2/\pi^2$$

$$K_{34} = (D_{12} + D_{66})mnR$$

$$K_{35} = KA_{55}ma/\pi$$

$$K_{44} = D_{66}m^2 + D_{22}n^2R^2 + KA_{55}a^2/\pi^2$$

$$K_{45}=KA_{55}nRa/\pi$$

$$K_{55}=K\left(A_{55}m^2+A_{44}n^2\right)$$

$$\lambda=\overline{\rho}_1\omega^2/\pi^2$$

$$\bar{q}_o=q_oa^2/\pi^2$$

$$R=a/b$$

$$K=\text{shear correction factor}$$

For the static case $(\lambda=0)$, U_{mn}, V_{mn}, X_{mn}, Y_{mn}, and W_{mn} can be solved for. Assuming $m=n=1$, the applied load is $q(x,y)=q_o\sin(\pi x/a)\sin(\pi y/b)$, and the maximum deflection occurs at $x=a/2$, $y=b/2$. For a $[\pm 45]$ laminated plate made from a material with assumed elastic modulus ratios $E_1/E_2=40$, $G_{12}/E_2=0.6$, $G_{13}/E_2=0.5$, $G_{13}=G_{23}$, and $\nu_{12}=0.25$, we can plot the deflection as a function of the plate size a/h, as done in Ref. [10]. Comparing the displacement from classical plate theory (CPT) to that for shear deformation theory (SDT) and defining $W_{\text{comp}}=W_{\max}Eh^3\times 10^3/q_oa^4$ as the comparison displacement, Figure 8.13 illustrates the results assuming the shear correction factor is $K=5/6$.

In addition, we can examine the case of free vibration $(q=0)$. For a nontrivial solution, λ is chosen so that the determinant of the coefficient matrix vanishes. For correlation between CPT and SDT, we define
$\omega_{\text{cor}}=\omega a^2\sqrt{\rho/E_2h^3}$ and plot this versus the ratio a/h for a $[\pm 45]$ laminate as done in Ref. [10] in Figure 8.14.

As noted in our discussions of beams, the shear deformation becomes significant in special cases. For the $[\pm 45]$ laminated plates considered, the CPT and SDT theories are relatively close for ratios of plate width to thickness of $a/h>20$. Obviously, a

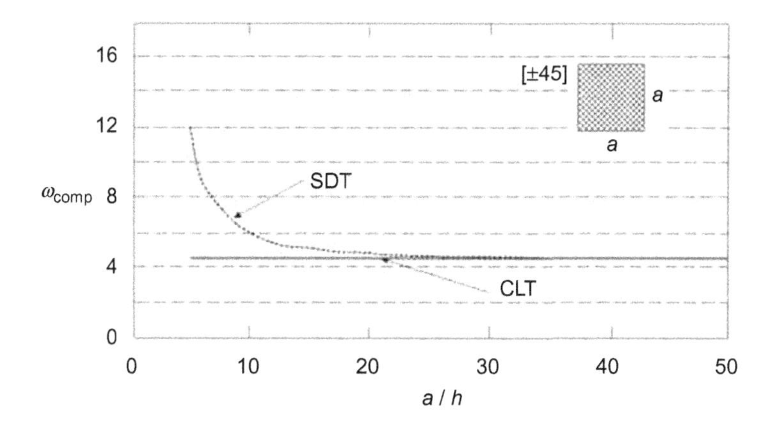

Figure 8.13 Comparison of CPT and SDT for a static case (after Ref. [10]).

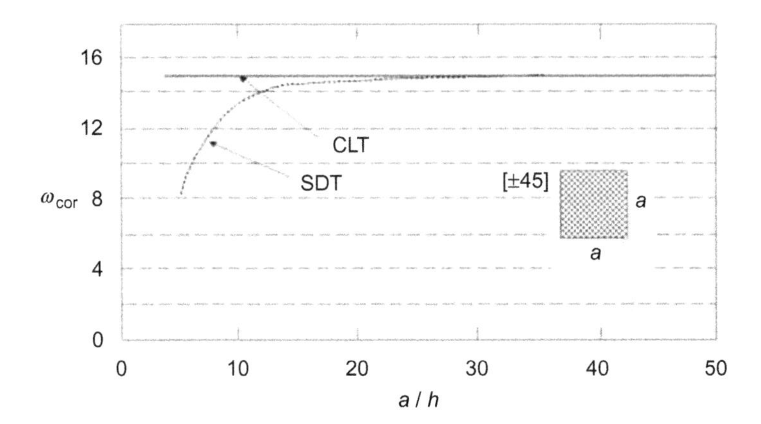

Figure 8.14 Free vibration frequency of a CLT and SDT solution (after Ref. [10]).

different stacking arrangement, plate dimensions a and b, thickness h, as well as material would yield different results. With beams, we noted that shear deformation becomes influential for beams with small length to height ratios. A similar observation has now been made for plates.

8.8 Stability

The effect of compressive loads on a plate can cause an instability, which in turn can cause the plate to buckle. The effects of plate aspect ratio (a/b), overall thickness, material property ratios, etc. have been illustrated in many references, including Refs. [14,15]. In setting up a general approach to solving a stability (buckling) problem, consider a rectangular plate (simply supported) compressed by uniform loads N_x and N_y as illustrated in Figure 8.15.

For the static analysis of a symmetric laminate with $q(x, y) = 0$, the governing equation is

$$D_{11}\frac{\partial^4 W_o}{\partial x^4} + 2(D_{12} + 2D_{66})\frac{\partial^4 W_o}{\partial x^2 \partial y^2} + D_{22}\frac{\partial^4 W_o}{\partial y^4} = N_x\frac{\partial^2 W_o}{\partial x^2} + N_y\frac{\partial^2 W_o}{\partial y^2} \tag{8.26}$$

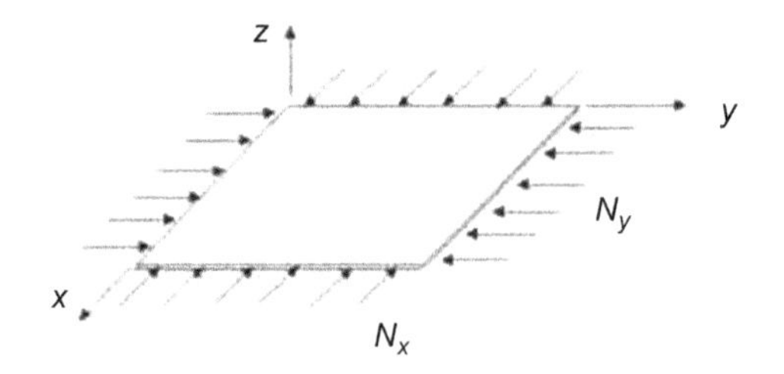

Figure 8.15 Rectangular plate subjected to compressive loads in two directions.

Since the ends are simply supported ends, the boundary equations required to solve the equation above are

$$x=0,a: \quad W_o=M_x=-D_{11}\frac{\partial^2 W_o}{\partial x^2}-D_{12}\frac{\partial^2 W_o}{\partial y^2}=0$$

$$y=0,b: \quad W_o=M_y=-D_{12}\frac{\partial^2 W_o}{\partial x^2}-D_{12}\frac{\partial^2 W_o}{\partial x^2}=0$$

The boundary conditions are satisfied by a deflection of the form

$$W_o=\sum_{m=1}^{\infty}\sum_{n=1}^{\infty}A_{mn}\sin\frac{m\pi x}{a}\sin\frac{n\pi y}{b}$$

where m and n are integers. Defining the ratio of plate dimensions as $R=a/b$ and substituting into Equation (8.26) results in

$$\pi^2 A_{mn}\left[D_{11}m^4+2(D_{12}+2D_{66})m^2n^2R^2+D_{22}n^4R^4\right]=-A_{mn}a^2\left\{N_x m^2+N_y n^2R^2\right\}$$

The trivial solution of $A_{mn}=0$ is not of interest, so we focus on

$$\pi^2\left[D_{11}m^4+2(D_{12}+2D_{66})m^2n^2R^2+D_{22}n^4R^4\right]=-a^2\left\{N_x m^2+N_y n^2R^2\right\}$$

Assuming the applied forces N_x and N_y are not the same, we can denote their ratio by a parameter $k=N_y/N_x$. Therefore, we can write $N_x=-N_o$ and $N_y=-kN_o$, where $N_o>0$. Substituting this into the equation above and solving for N_o results in

$$N_o=\frac{\pi^2[D_{11}m^4+2(D_{12}+2D_{66})m^2n^2R^2+D_{22}n^4R^4]}{a^2(m^2+kn^2R^2)} \tag{8.27}$$

The relationship above provides the values of N_o (for a given k) for any configuration of m and n. Although an infinite number of values for N_o can be determined, unique values exist for each combination of m and n. Several particular cases are considered below.

Case I: $k=0$: In this case, $N_y=0$ and Equation (8.27) becomes

$$N_o=\frac{\pi^2}{a^2m^2}\left[D_{11}m^4+2(D_{12}+2D_{66})m^2n^2R^2+D_{22}n^4R^4\right]$$

The smallest value of N_o occurs when $n=1$. The buckling load is given by the smallest value of the relationship for N_o as m varies and $n=1$. This relationship is

$$N_o=\frac{\pi^2}{a^2m^2}\left[D_{11}m^4+2(D_{12}+2D_{66})m^2R^2+D_{22}R^4\right]$$

For given values of D_{11}, D_{12}, D_{66}, D_{22}, and R, we determine the critical buckling load from the value of m which yields the lowest value of N_o. For example, consider an orthotropic plate for which the following are valid $D_{11}/D_{22}=10$ and $(D_{12}+2D_{66})/D_{22}=1$. The equation above becomes

$$N_o=\frac{\pi^2 D_{22}}{a^2 m^2}\left[10m^4+2m^2R^2+R^4\right]$$

The minimum value of N_o occurs when $m=1$. At certain aspect rations, two possible buckled mode shapes are possible. These occur when

$$R=[m(m+1)]^{1/2}\left(\frac{D_{11}}{D_{22}}\right)^{1/4}=1.78[m(m+1)]^{1/2}$$

The two buckled shapes are $W_o=A_{m1}\sin\frac{m\pi x}{a}\sin\frac{n\pi y}{b}$ and $W_o=A_{(m+1)1}\sin\frac{m\pi x}{a}\sin\frac{n\pi y}{b}$. These lead to identical values of the critical buckling load N_{CR}. In addition, the absolute minimum buckling load occurs for $m=i$ such that $R=i^4(D_{11}/D_{22})^{1/4}=1.78i$. In this case,

$$N_{CR}=8.32\frac{\pi^2}{a^2}D_{22}$$

Case II: $k=1, a=b$: This corresponds to a square plate under biaxial compression. For this case,

$$N_o=\frac{\pi^2\left[D_{11}m^4+2(D_{12}+2D_{66})m^2n^2+D_{22}n^4\right]}{a^2(m^2+n^2)}$$

The critical load occurs when $m=1$, provided $D_{11}\geq D_{12}$. In this case,

$$N_{CR}=\frac{\pi^2}{a^2}D_{22}\frac{\left[\frac{D_{11}}{D_{22}}+2\frac{(D_{12}+2D_{66})}{D_{22}}n^2+n^4\right]}{(1+n^2)}$$

For the numerical values considered in Case I, this becomes, for $n=1$

$$N_{CR}=6.5\frac{\pi^2}{a^2}D_{22}$$

For $D_{11}/D_{22}=15$ and $(D_{12}+2D_{66})/D_{22}=1$, for $n=2$ we get $N_{CR}=7.8\frac{\pi^2}{a^2}D_{22}$

Case III: $k<0$: This corresponds to the case of N_y being tensile. For a square plate with $k=-0.5$

$$N_o=\frac{2\pi^2\left[D_{11}m^4+2(D_{12}+2D_{66})m^2n^2+D_{22}n^4\right]}{a^2(2m^2-n^2)}$$

Since $k < 0$, this equation indicates that the critical buckling load will be larger than for the case of $N_y = 0$. Using the numerical values considered in Case I, with $m = n = 1$, the critical buckling load is

$$N_{\mathrm{CR}} = 26\frac{\pi^2}{a^2}D_{22}$$

Therefore, for a tensile N_y that is half the compressive value of N_x, the critical buckling load is twice that obtained for $N_y = 0$. The effect of the tensile force is to increase the effective bending stiffness of the plate.

8.9 Problems

8.1 Given the solution for the deflection of a simply supported plate using Navier's method (as defined by Equation (8.16)), determine the stress distribution at $(a/4, b/4)$ and $(a/2, b/2)$ using a one-term series $(m = n = 1)$ and a four-term series $(m, n = 1,1 \;\; 1,3 \;\; 3,1 \;\; 3,3)$. Plot σ_x/q_o versus z and τ_{xz}/q_o versus z for both the one- and four-term solutions. The plate is a $[(0/90)_2]_s$ laminate with $E_1 = 11.0 \times 10^6, E_2 = 0.798 \times 10^6, G_{12} = 0.334 \times 10^6, \nu_{12} = 0.34$, $\rho = 0.09\ \mathrm{lb/in^3}$, $t_{\mathrm{ply}} = 0.01$ in, $a = b = 12''$.

8.2 Obtain the Navier solution for a simply supported specially orthotropic rectangular plate subjected to a temperature distribution defined by $T(x, y, z) = T_o(x, y) + zT_1(x, y)$. Both T_o and T_1 are assumed to be known functions of x and y only, which can be expanded in double Fourier series in the same way as the mechanical loading $q(x, y)$.

8.3 Use the one-parameter Ritz approximation $W(x, y) \approx C_1 x(a - x)y(b - y)$ to determine the deflection of a simply supported specially orthotropic rectangular plate with a uniform load q_o.

8.4 For a specially orthotropic plate, the minimum frequency is given as

$$\omega = \left(\frac{\pi}{Rb}\right)^2 \frac{1}{\sqrt{\rho}} \sqrt{\left[D_{11}m^4 + 2(D_{12} + 2D_{66})m^2n^2R^2 + D_{22}n^4R^4\right]}$$

For a homogeneous isotropic plate, the frequency is given as

$$\omega_{11} = \left(\frac{\pi}{b}\right)^2 \sqrt{\frac{D}{\rho}}\left(1 + \frac{1}{R^2}\right) \quad \text{where } D = \frac{Eh^3}{12(1 - \nu^2)}$$

Plot the ratio of ω/ω_{11} as a function of R for the laminate stacking arrangement of problem 1 with $m = n = 1$, assuming that the isotropic plate has the properties $E = 10 \times 10^6$, $\nu = 0.25$, $\rho = 0.10\ \mathrm{lb/in^3}$.

8.5 For a simply supported uniformly loaded plate with $(q(x, y) = -q_o)$, determine expressions for the particular and homogeneous forms of W. Assuming the roots for Case I of Section 8.5.2.1 when determining the answer.

8.6 The plate shown is a square, symmetric, specially orthotropic, and is loaded with a constant twisting moment along its edges as shown. If W_o is the deflection in the z-direction, find an expression for the deflection $W(x,y)$. List all boundary conditions used and note that $M_{xy} = M_1$, $M_{yx} = -M_1$, and $M_{xy} = -M_{yx}$.

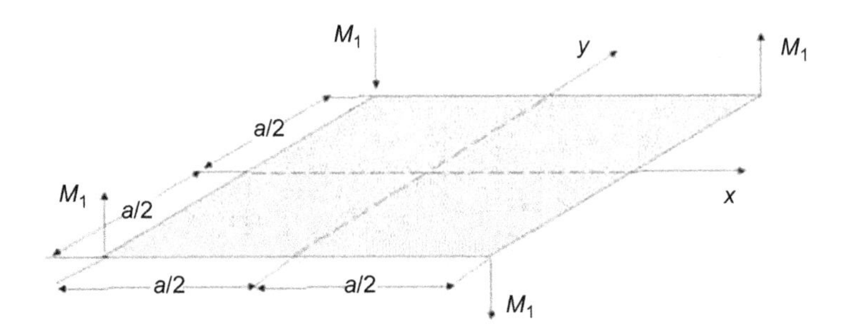

8.7 A plate with $b \gg a$ is treated as a long narrow strip. If the plate is uniformly loaded with a load q_o, the variation of W_o in the y-direction vanishes, with $W_{o,y} = W_{o,xy} = 0$. This type of plate is termed a cylindrical beam. For a cylindrical beam, the displacement can be expressed as

$$W_o = \frac{A_{11}}{24(A_{11}D_{11} - B_{11}^2)} q_o (x^3 - 2ax^2 + a^3)$$

where q_o is applied in the z-direction.

Assume the same material properties and stacking arrangement as in problem 12. Assume $a = 6''$ and plot W_o for the equation above and W_o for the simply supported beam shown (neglect shear deformation).

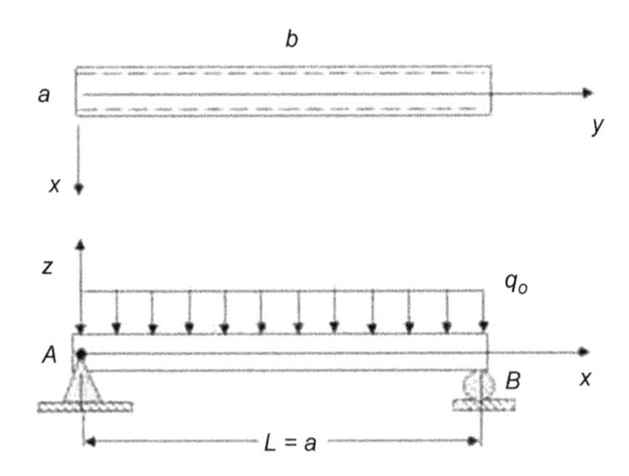

References

[1] Timoshenko SP, Woinowsky-Krieger S. Theory of plates and shells. New York: McGraw-Hill; 1959.
[2] Ugural AC. Stresses in plates and shells. New York: McGraw-Hill; 1981.
[3] Panc V. Theories of elastic plates. The Netherlands: Noordhoff; 1975.
[4] Reddy JN. Theory and analysis of elastic plates and shells. New York: CRC Press; 2006.
[5] Ashton JE, Whitney JM. Theory of laminated plates. Westport, CT: Technomic; 1970.
[6] Jones RM. Mechanics of composite material. New York: Hemisphere; 1975.

[7] Ambartsumyan SA. Theory of anisotropic plates [Translated from Russian by T. Cheron]. Stamford, CT: Technomic; 1969.
[8] Reddy JN. Mechanics of laminated composite plates theory and analysis. New York: CRC Press; 1997.
[9] Reissner E. On the derivation of boundary conditions for plate theory. Proc R Soc Lond Ser A Math Phys Sci 1963;276:178–86.
[10] Whitney JM. The effect of transverse shear deformation on the bending of laminated plates. J Compos Mater 1969;36:534–47.
[11] Whitney JM, Pagano NJ. Shear deformation in heterogeneous anisotropic plates. J Appl Mech 1970;37:1031–6.
[12] Reissner E. A consistent treatment of transverse shear deformation in laminated anisotropic plates. AIAA J 1972;10:716–8.
[13] Reissner E. Note on the effect of transverse shear deformation in laminated anisotropic plates. Comput Methods Appl Mech Eng 1979;20:203–9.
[14] Jones RM. Buckling and vibration of rectangular unsymmetrically laminated cross-ply plates. AIAA J 1973;11:1626–32.
[15] Jones RM, Morgan HS, Whitney JM. Buckling and vibration of antisymmetric laminated angle-ply rectangular plates. J Appl Mech 1973;40:1143–4.

Analysis of laminated composite shells

9

9.1 Introduction

Shells are common structural elements used in many engineering applications such as ship hulls, aircraft, pressure vessels, etc. They are typically considered to be curved plates. In order to account for the curvature, a coordinate system similar to that used for rings and curved beams in Chapter 7 is used. The theory behind laminated shells is based on classical shell, flat plate, and curved beam theory. Many of the original theories for laminated shell analysis are based on the Kirchoff–Love kinematic hypotheses and were developed in the 1950s, 1960s, and 1970s. A small sampling of the pertinent references for developing laminated shell theory is presented in Refs. [1–7]. As with Chapter 8, the complexity of shell problems typically requires numerical techniques for all but the simplest of problems. A short list of articles related to the numerical solution of composite shell problems is given in Refs. [8–10]. Reddy's initial text for plates [11] has been extended to include shells [12]. This text includes higher order and numerical solutions and represents similar to his plates text and is an excellent reference. The primary focus of this chapter is to present the relevant equations of motion required for the analysis of laminated shells. As with the previous chapter, these are expressed in terms of displacement, which in turn is based on assumed displacement fields.

9.2 Strain–displacement relations for cylindrical shells

Considering a segment of a cylinder, we define the coordinate system used, the size of a section of the shell, and the forces and moment that can act on the shell. The shells are assumed to have a thickness h, and the mid-surface of the shell is located at a distance R from the center of curvature of the shell, as shown in Figure 9.1.

In classical shell theory, the assumed displacement field is similar to that assumed for plates. The nomenclature is slightly different, and the assumed displacement field is expressed as

$$\begin{aligned} U(x,\theta,t) &= U_o(x,\theta,t) + z\beta_x(x,\theta,t) \\ V(x,\theta,t) &= V_o(x,\theta,t) + z\beta_\theta(x,\theta,t) \\ W(x,\theta,t) &= W_o(x,\theta,t) \end{aligned} \tag{9.1}$$

These are similar to the displacement fields in classical lamination theory in that U_o, V_o, and W_o represent the displacement of the mid-plane of the shell in the x, θ, and z

Laminar Composites. http://dx.doi.org/10.1016/B978-0-12-802400-3.00009-X

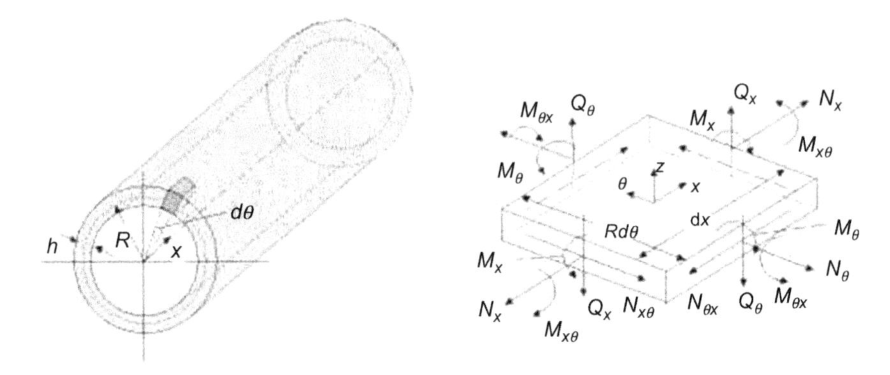

Figure 9.1 Coordinate system and loads acting on a shell.

directions, respectively. In developing the equations for shells, we replace the y coordinate from classical lamination theory with θ and the terms Φ and Ψ are replaced with β_x and β_θ. The strain–displacement relationships in cylindrical coordinates are

$$\begin{aligned} &\varepsilon_x = U_{,x} && \varepsilon_\theta = \frac{1}{R+z}\left(W + V_{,\theta}\right) \\ &\gamma_{x\theta} = \frac{U_{,\theta}}{R+z} + V_{,x} && \gamma_{xz} = U_{,z} + W_{,x} \qquad \gamma_{\theta z} = V_{,z} + \frac{1}{R+z}\left(W_{,\theta} - V\right) \end{aligned} \tag{9.2}$$

Substituting the displacement field from Equation (9.1) into the strain–displacement relations of Equation (9.2) and manipulating results produces a convenient form of the strain–displacement relations

$$\begin{aligned} &\varepsilon_x = U_{o,x} + z\beta_{x,x} \quad \varepsilon_\theta = \frac{1}{1+z/R}\left[\frac{1}{R}\left(W_o + V_{o,\theta}\right) + \frac{z}{R}\beta_{\theta,\theta}\right] \\ &\gamma_{x\theta} = \frac{1}{1+z/R}\left(\frac{1}{R}\right)U_{o,\theta} + V_{o,x} + z\left(\frac{1}{R+z}\left(\frac{1}{R}\right)\beta_{x,\theta} + \beta_{\theta,x}\right) \\ &\gamma_{xz} = \beta_x + W_{o,x} \quad \gamma_{\theta z} = \frac{1}{1+z/R}\left\{\beta_\theta + \frac{1}{R}\left(W_{o,\theta} - V_o\right)\right\} \end{aligned} \tag{9.3}$$

These can be expressed in an alternate form similar to that of classical lamination theory

$$\begin{aligned} &\varepsilon_x = \varepsilon_x^o + z\kappa_x \qquad \varepsilon_\theta = \varepsilon_\theta^o + z\kappa_\theta \\ &\gamma_{x\theta} = \gamma_{x\theta}^o + z\kappa_{x\theta} \quad \gamma_{xz} = \beta_x + W_{o,x} \quad \gamma_{\theta z} = \frac{1}{1+z/R}\left\{\beta_\theta + \frac{1}{R}\left(W_{o,\theta} - V_o\right)\right\} \end{aligned}$$

where

$$\varepsilon_x^o = U_{o,x} \quad \varepsilon_\theta^o = \frac{1}{1+z/R}\left(\frac{1}{R}\right)(W_o + V_{o,\theta}) \quad \gamma_{x\theta}^o = \frac{1}{1+z/R}\left(\frac{1}{R}\right)U_{o,\theta} + V_{o,x}$$

$$\kappa_x = \beta_{x,x} \quad \kappa_\theta = \frac{1}{1+z/R}\left(\frac{\beta_{\theta,\theta}}{R}\right) \quad \kappa_{x\theta} = \frac{1}{R+z}\left(\frac{1}{R}\right)\beta_{x,\theta} + \beta_{\theta,x}$$

In many applications involving composite shells, the shells are thin and the ratio of the shell thickness to its radius is very small ($z/R \ll 1$). Therefore some of the equations above can be expressed in a slightly different form

$$\begin{aligned} &\varepsilon_x = \varepsilon_x^o + z\kappa_x \quad &&\varepsilon_\theta = \varepsilon_\theta^o + z\kappa_\theta \quad &&\gamma_{x\theta} = \gamma_{x\theta}^o + z\kappa_{x\theta} \\ &\gamma_{xz} = \beta_x + W_{o,x} \quad &&\gamma_{\theta z} = \beta_\theta + \frac{1}{R}(W_{o,\theta} - V_o) \end{aligned} \tag{9.4}$$

where

$$\begin{aligned} &\varepsilon_x^o = U_{o,x} \quad &&\varepsilon_\theta^o = \left(\frac{1}{R}\right)(W_o + V_{o,\theta}) \quad &&\gamma_{x\theta}^o = \left(\frac{1}{R}\right)U_{o,\theta} + V_{o,x} \\ &\kappa_x = \beta_{x,x} \quad &&\kappa_\theta = \frac{\beta_{\theta,\theta}}{R} \quad &&\kappa_{x\theta} = \frac{\beta_{x,\theta}}{R} + \beta_{\theta,x} \end{aligned} \tag{9.5}$$

9.3 Equations of motion

The equations of motion are derived based upon the strain–displacement relations previously discussed as well as the external forces and internal forces acting on the shell. The external surface forces and internal shear force (Q_x, Q_θ, $N_{x\theta}$, and $N_{\theta x}$), normal forces (N_x and N_θ), torques ($M_{x\theta}$ and $M_{\theta x}$) , and bending moments (M_x and M_θ) are defined as indicated in Figure 9.2.

In developing the equations of motion, we must assume that the internal forces across a representative volume element of shell are not constant. As a result of this,

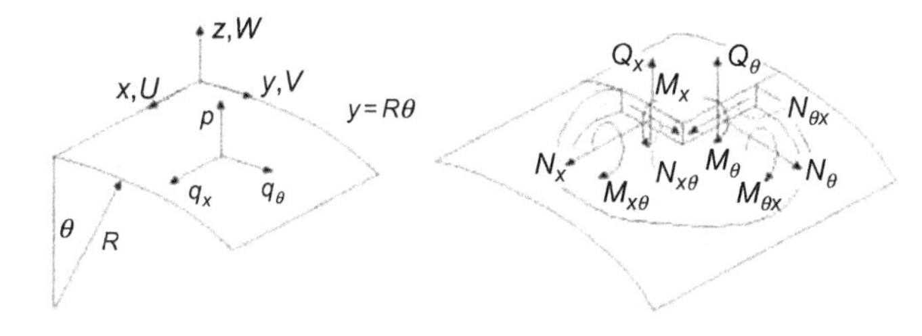

Figure 9.2 External and internal forces acting on a shell.

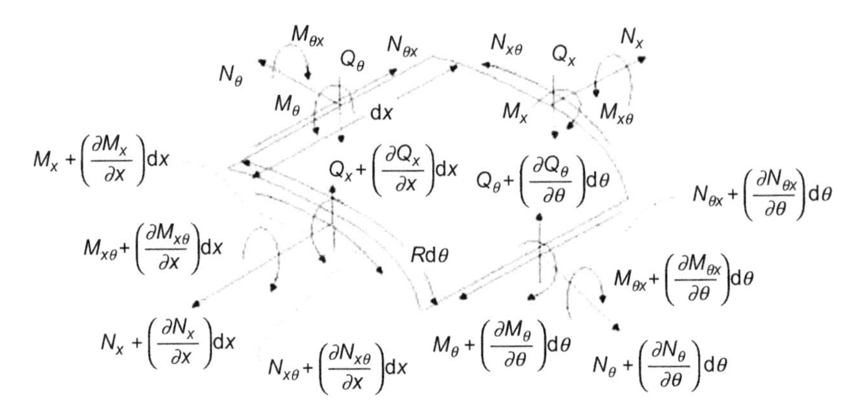

Figure 9.3 Forces and moments acting on an element of shell.

the in-plane normal forces, out-of-plane shear forces, bending moments, and torques are modeled as shown in Figure 9.3.

The forces (normal and shear) and moments are expressed as pounds (or Newton's) per unit length of the section. Satisfying equilibrium in the x direction results in

$$N_{x,x}(\mathrm{d}xR\mathrm{d}\theta) + N_{\theta x,\theta}(\mathrm{d}x\mathrm{d}\theta) + q_x(R\mathrm{d}\theta\mathrm{d}x) = \overline{\rho}_1 U_{o,tt} + \overline{\rho}_2 \beta_{x,tt}$$

This reduces to

$$N_{x,x} + \frac{N_{\theta x,\theta}}{R} + q_x = \overline{\rho}_1 U_{o,tt} + \overline{\rho}_2 \beta_{x,tt} \tag{9.6}$$

In the θ direction, we end up with

$$\frac{N_{\theta,\theta}}{R} + N_{x\theta,x} + \frac{Q_\theta}{R} + q_\theta = \overline{\rho}_1 V_{o,tt} + \overline{\rho}_2 \beta_{\theta,tt} \tag{9.7}$$

Similarly, in the z direction, we obtain

$$Q_{x,x} + \frac{Q_{\theta,\theta}}{R} - \frac{N_\theta}{R} + p = \overline{\rho}_1 W_{o,tt} \tag{9.8}$$

Summing moments about the x axis yields

$$M_{x\theta,x} + \frac{M_{\theta,\theta}}{R} - (Q_\theta - m_\theta) = \overline{\rho}_2 U_{o,tt} + \overline{\rho}_3 \beta_{x,tt} \tag{9.9}$$

The final two equations result from taking moments about the y and z axes, respectively. These are

$$\frac{M_{\theta x,\theta}}{R} + M_{x,x} - (Q_x - m_x) = \overline{\rho}_2 V_{o,tt} + \overline{\rho}_3 \beta_{\theta,tt} \tag{9.10}$$

$$N_{x\theta} - N_{\theta x} - \frac{M_{\theta x}}{R} = 0 \tag{9.11}$$

In these equations, we have introduced several terms, which are defined below as

$$\begin{aligned}
q_x &= \tau_{zx}\left(\frac{h}{2}\right) - \tau_{zx}\left(\frac{-h}{2}\right) = \tau_{1x} - \tau_{2x} \\
q_\theta &= \tau_{z\theta}\left(\frac{h}{2}\right) - \tau_{z\theta}\left(\frac{-h}{2}\right) = \tau_{1\theta} - \tau_{2\theta} \\
m_x &= \frac{h}{2}\left[\tau_{zx}\left(\frac{h}{2}\right) - \tau_{zx}\left(\frac{-h}{2}\right)\right] = \frac{h}{2}(\tau_{1x} + \tau_{2x}) \\
m_\theta &= \frac{h}{2}\left[\tau_{z\theta}\left(\frac{h}{2}\right) - \tau_{z\theta}\left(\frac{-h}{2}\right)\right] = \frac{h}{2}(\tau_{1\theta} + \tau_{2\theta})
\end{aligned} \tag{9.12}$$

In these equations, the terms τ_{1x}, τ_{2x}, $\tau_{1\theta}$, and $\tau_{2\theta}$ refer to the shear stresses in the x and θ directions on the upper (subscript 1) and lower (subscript 2) surfaces of the shell, as initially introduced in Chapter 7. The equations for a general cylindrical shell are given in (9.6)–(9.11). For a composite shell, it is generally assumed that $z/R \ll 1$, and that $N_{x\theta} = N_{\theta x}$ and $M_{x\theta} = M_{\theta x}$. As a result of this, Equation (9.11) vanishes and the remaining equations are written as

$$N_{x,x} + \frac{N_{\theta x,\theta}}{R} = -q_x + \overline{\rho}_1 U_{o,tt} + \overline{\rho}_2 \beta_{x,tt} \tag{9.13}$$

$$\frac{N_{\theta,\theta}}{R} + N_{x\theta,x} + \frac{Q_\theta}{R} = -q_\theta + \overline{\rho}_1 V_{o,tt} + \overline{\rho}_2 \beta_{\theta,tt} \tag{9.14}$$

$$Q_{x,x} + \frac{Q_{\theta,\theta}}{R} - \frac{N_\theta}{R} = -p + \overline{\rho}_1 W_{o,tt} \tag{9.15}$$

$$M_{x\theta,x} + \frac{M_{\theta,\theta}}{R} - Q_\theta = -m_\theta + \overline{\rho}_2 U_{o,tt} + \overline{\rho}_3 \beta_{x,tt} \tag{9.16}$$

$$\frac{M_{\theta x,\theta}}{R} + M_{x,x} - Q_x = -m_x + \overline{\rho}_2 V_{o,tt} + \overline{\rho}_3 \beta_{\theta,tt} \tag{9.17}$$

The forces and moment are related to the stresses through

$$\begin{Bmatrix} N_x \\ N_\theta \\ N_{x\theta} \end{Bmatrix} = \sum_{k=1}^{N} \int_{h_{k-1}}^{h_k} \begin{Bmatrix} \sigma_x \\ \sigma_\theta \\ \tau_{x\theta} \end{Bmatrix} \mathrm{d}z \quad \begin{Bmatrix} M_x \\ M_\theta \\ M_{x\theta} \end{Bmatrix} = \sum_{k=1}^{N} \int_{h_{k-1}}^{h_k} \begin{Bmatrix} \sigma_x \\ \sigma_\theta \\ \tau_{x\theta} \end{Bmatrix} z\mathrm{d}z \quad \begin{Bmatrix} Q_x \\ Q_\theta \end{Bmatrix} = \sum_{k=1}^{N} \int_{h_{k-1}}^{h_k} \begin{Bmatrix} \tau_{xz} \\ \tau_{x\theta} \end{Bmatrix}$$

As previously done, we define $(\overline{\rho}_1, \overline{\rho}_2, \overline{\rho}_3) = \sum_{k=1}^{N} \int_{h_{k-1}}^{h_k} \rho_k (1, z, z^2) \mathrm{d}z$.

9.4 Unidirectional laminate, axisymmetric loading: Static analysis

A unidirectional laminate is symmetric and the extension-bending coupling does not exist ($[B]=0$), and there are no other coupling terms. This means that $\varepsilon_x=\varepsilon_x^o$, $\varepsilon_\theta=\varepsilon_\theta^o$, etc. In addition, assume an axisymmetric loading in which all terms differentiated with respect to θ (N_θ, $N_{x\theta}$, $N_{\theta x}$, M_θ, $M_{x\theta}$, $M_{\theta x}$, Q_θ, and q_θ) drop out. For static analysis, the right-hand side of Equations (9.13–9.17) vanishes. The result is three equations of equilibrium.

$$
\begin{aligned}
&N_{x,x}+q_x=0\\
&Q_{x,x}-N_\theta/R+p=0\\
&M_{x,x}-Q_x=0
\end{aligned}
\tag{9.18}
$$

The strain displacements relations become

$$\varepsilon_x^o=U_{o,x},\quad \varepsilon_\theta^o=\frac{W_o}{R},\quad \gamma_{x\theta}^o=V_{o,x},\quad \kappa_x=\beta_{x,x},\quad \kappa_\theta=0,\quad \kappa_{x\theta}=\beta_{\theta,x}$$

In addition, $\gamma_{xz}=0\rightarrow\beta_x=-W_{o,x}$ and $\gamma_{\theta z}=0\rightarrow\beta_\theta=\frac{V_o}{R}$. Therefore, we can write

$$\varepsilon_x^o=U_{o,x},\quad \varepsilon_\theta^o=\frac{W_o}{R},\quad \gamma_{x\theta}^o=V_{o,x},\quad \kappa_x=-W_{o,xx},\quad \kappa_\theta=0,\quad \kappa_{x\theta}=\frac{V_{o,x}}{R}$$

In defining a set of stress–strain relationships, it is convenient to define elastic constants in terms of x and θ as illustrated in Figure 9.4. In general, the elastic constants E_1, E_2, G_{12}, and ν_{12} are known. We can define E_x, E_θ, and $\nu_{x\theta}$ in terms of these elastic constants using a variation Equation (3.19). Recalling that we have defined $m=\cos\theta$ and $n=\sin\theta$, the expressions required are

$$\frac{1}{E_x}=\frac{m^4}{E_1}+\left(\frac{1}{G_{12}}-\frac{2\nu_{12}}{E_1}\right)m^2n^2+\frac{n^4}{E_2}$$

$$\frac{1}{E_\theta}=\frac{n^4}{E_1}+\left(\frac{1}{G_{12}}-\frac{2\nu_{12}}{E_1}\right)m^2n^2+\frac{m^4}{E_2}$$

$$\nu_{x\theta}=E_x\left[\nu_{12}\frac{m^4+n^4}{E_1}-\left(\frac{1}{E_1}+\frac{1}{E_2}-\frac{1}{G_{12}}\right)m^2n^2\right]$$

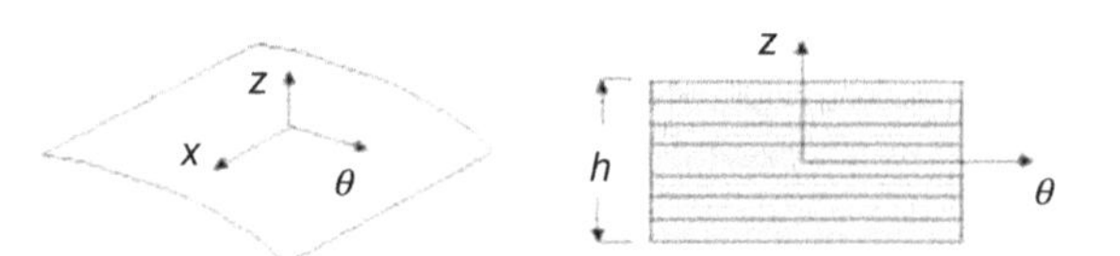

Figure 9.4 Primary direction of interest for defining elastic constants.

Neglecting hygral effects (which would simply add $\beta_x \bar{M}$ and $\beta_\theta \bar{M}$ terms), the stress–strain relations, including thermal effects, can be written as

$$\varepsilon_x = \frac{1}{E_x}(\sigma_x - \nu_{x\theta}\sigma_\theta) + \alpha_x \Delta T \quad \varepsilon_\theta = \frac{1}{E_\theta}(\sigma_\theta - \nu_{\theta x}\sigma_x) + \alpha_\theta \Delta T$$

or

$$\sigma_x = \frac{E_x}{1 - \nu_{x\theta}\nu_{\theta x}}[(\varepsilon_x + \nu_{\theta x}\varepsilon_\theta) - (\alpha_x + \nu_{\theta x}\alpha_\theta)\Delta T]$$

$$\sigma_\theta = \frac{E_\theta}{1 - \nu_{x\theta}\nu_{\theta x}}[(\varepsilon_\theta + \nu_{x\theta}\varepsilon_x) - (\alpha_\theta + \nu_{x\theta}\alpha_x)\Delta T]$$

where $\nu_{x\theta}/E_x = \nu_{\theta x}/E_\theta$. In addition, we have $N_x = \int_{-h/2}^{h/2} \sigma_x \mathrm{d}z$, $N_\theta = \int_{-h/2}^{h/2} \sigma_\theta \mathrm{d}z$, $M_x = \int_{-h/2}^{h/2} \sigma_x z \mathrm{d}z$, and $M_\theta = \int_{-h/2}^{h/2} \sigma_\theta z \mathrm{d}z$. Incorporating the stress relationships above, including thermal and hygral effects, and recalling that all lamina have the same fiber orientation, we express these as

$$N_x = \frac{E_x h}{1 - \nu_{x\theta}\nu_{\theta x}}\left(\varepsilon_x^o + \nu_{\theta x}\varepsilon_\theta^o\right) - N_x^{\mathrm{T}} - N_x^{\mathrm{H}} \quad N_\theta = \frac{E_\theta h}{1 - \nu_{x\theta}\nu_{\theta x}}\left(\varepsilon_\theta^o + \nu_{x\theta}\varepsilon_x^o\right) - N_\theta^{\mathrm{T}} - N_\theta^{\mathrm{H}}$$

$$M_x = \frac{E_x h^3}{12(1 - \nu_{x\theta}\nu_{\theta x})}\kappa_x - M_x^{\mathrm{T}} - M_x^{\mathrm{H}} \qquad M_\theta = \frac{E_\theta h^3}{12(1 - \nu_{x\theta}\nu_{\theta x})}\kappa_\theta - M_\theta^{\mathrm{T}} - M_\theta^{\mathrm{H}}$$

From the third equilibrium equation in (9.18), we have $Q_x = \partial M_x / \partial x$. In addition, we define thermal and hygral loads and moments as

$$N_x^{\mathrm{T}} = \frac{E_x}{1 - \nu_{x\theta}\nu_{\theta x}}\int_{-h/2}^{h/2} (\alpha_x + \nu_{\theta x}\alpha_\theta)\Delta T \mathrm{d}z \quad N_\theta^{\mathrm{T}} = \frac{E_\theta}{1 - \nu_{x\theta}\nu_{\theta x}}\int_{-h/2}^{h/2} (\alpha_\theta + \nu_{\theta x}\alpha_x)\Delta T \mathrm{d}z$$

$$N_x^{\mathrm{H}} = \frac{E_x}{1 - \nu_{x\theta}\nu_{\theta x}}\int_{-h/2}^{h/2} (\beta_x + \nu_{\theta x}\beta_\theta)\bar{M} \mathrm{d}z \quad N_\theta^{\mathrm{H}} = \frac{E_\theta}{1 - \nu_{x\theta}\nu_{\theta x}}\int_{-h/2}^{h/2} (\beta_\theta + \nu_{\theta x}\beta_x)\bar{M} \mathrm{d}z$$

$$M_x^{\mathrm{T}} = \frac{E_x}{1 - \nu_{x\theta}\nu_{\theta x}}\int_{-h/2}^{h/2} (\alpha_x + \nu_{\theta x}\alpha_\theta)\Delta T z \mathrm{d}z \quad M_\theta^{\mathrm{T}} = \frac{E_\theta}{1 - \nu_{x\theta}\nu_{\theta x}}\int_{-h/2}^{h/2} (\alpha_\theta + \nu_{\theta x}\alpha_x)\Delta T z \mathrm{d}z$$

$$M_x^{\mathrm{H}} = \frac{E_x}{1 - \nu_{x\theta}\nu_{\theta x}}\int_{-h/2}^{h/2} (\beta_x + \nu_{\theta x}\beta_\theta)\bar{M} z \mathrm{d}z \quad M_\theta^{\mathrm{H}} = \frac{E_\theta}{1 - \nu_{x\theta}\nu_{\theta x}}\int_{-h/2}^{h/2} (\beta_\theta + \nu_{\theta x}\beta_x)\bar{M} z \mathrm{d}z$$

For convenience, we define four constants: $K_x \equiv \frac{E_x h}{1 - \nu_{x\theta}\nu_{\theta x}}$, $K_\theta \equiv \frac{E_\theta h}{1 - \nu_{x\theta}\nu_{\theta x}}$, $D_x \equiv \frac{E_x h^3}{12(1 - \nu_{x\theta}\nu_{\theta x})}$, and $D_\theta \equiv \frac{E_\theta h^3}{12(1 - \nu_{x\theta}\nu_{\theta x})}$. Using these, the normal force, bending moment, and shear equations are expressed as

$$N_x = K_x\left(U_{o,x} + \nu_{\theta x}\frac{W_o}{R}\right) - N_x^{\mathrm{T}} - N_x^{\mathrm{H}} \qquad N_\theta = K_\theta\left(\frac{W_o}{R} + \nu_{x\theta}U_{o,x}\right) - N_\theta^{\mathrm{T}} - N_\theta^{\mathrm{H}} \tag{9.19}$$

$$M_x = -D_x W_{o,xx} - M_x^{\mathrm{T}} - M_x^{\mathrm{H}} \quad M_\theta = -\nu_{\theta x} D_x W_{o,xx} - M_\theta^{\mathrm{T}} - M_\theta^{\mathrm{H}} \tag{9.20}$$

$$Q_x = M_{x,x} = -D_x W_{o,xxx} - M_{x,x}^{\mathrm{T}} - M_{x,x}^{\mathrm{H}} \tag{9.21}$$

Substituting these into the equilibrium equations (9.18) yields the two governing equation below

$$K_x\left(U_{o,xx} + \frac{\nu_{\theta x}}{R}W_{o,x}\right) = -q_x + N_{x,x}^{\mathrm{T}} + N_{x,x}^{\mathrm{H}} \tag{9.22}$$

$$-D_x W_{o,xxxx} - \frac{K_\theta}{R}\left(\nu_{x\theta}U_{o,x} + \frac{W_o}{R}\right) = -p - \frac{N_\theta^{\mathrm{T}}}{R} - \frac{N_\theta^{\mathrm{H}}}{R} + M_{x,xx}^{\mathrm{T}} + M_{x,xx}^{\mathrm{H}} \tag{9.23}$$

These can be reduced to a single equation. By integrating Equation (9.22) to get

$$K_x\left(U_{o,x} + \frac{\nu_{\theta x}}{R}W_o\right) = -\int q_x \mathrm{d}x + N_x^{\mathrm{T}} + N_x^{\mathrm{H}} + C$$

Comparing this to the expression for N_x in (9.19) using limits on the integral we find that

$$N_x = -\int_0^x q_x \mathrm{d}x + C$$

At $x = 0$, $N_x(0) = C$. Thus we can write $N_x(x) = -\int_0^x q_x \mathrm{d}x + N_x(0)$. Solving the equation above for $U_{o,x}$

$$U_{o,x} = \frac{1}{K_x}\left[-\int_0^x q_x \mathrm{d}x + N_x^{\mathrm{T}} + N_x^{\mathrm{H}} + N_x(0)\right] - \frac{\nu_{\theta x}}{R}W_o$$

Substituting into the remaining equilibrium equation

$$-D_x W_{o,xxxx} - \frac{K_\theta}{R}\left(\nu_{x\theta}\left\{\frac{1}{K_x}\left[-\int_0^x q_x \mathrm{d}x + N_x^{\mathrm{T}} + N_x^{\mathrm{H}} + N_x(0)\right] - \frac{\nu_{\theta x}}{R}W_o\right\} + \frac{W_o}{R}\right)$$
$$= -p - \frac{N_\theta^{\mathrm{T}}}{R} + M_{x,xx}^{\mathrm{T}} + M_{x,xx}^{\mathrm{H}}$$

or

$$-D_x W_{o,xxxx} + \frac{W_o}{R^2}K_\theta(1 - \nu_{x\theta}\nu_{\theta x}) = -p - \frac{N_\theta^{\mathrm{T}}}{R} + M_{x,xx}^{\mathrm{T}} + M_{x,xx}^{\mathrm{H}}$$
$$+ \frac{\nu_{x\theta}K_\theta}{RK_x}\left[\int_0^x q_x \mathrm{d}x - N_x^{\mathrm{T}} - N_x^{\mathrm{H}} - N_x(0)\right]$$

This can be written as

$$W_{o,xxxx}+4K^4W_o=\frac{1}{D_x}\left\{p+\frac{N_\theta^{\mathrm{T}}}{R}-M_{x,xx}^{\mathrm{T}}-M_{x,xx}^{\mathrm{H}}-\frac{\nu_{x\theta}K_\theta}{RK_x}\left[\int_0^x q_x\mathrm{d}x-N_x^{\mathrm{T}}-N_x^{\mathrm{H}}-N_x(0)\right]\right\} \tag{9.24}$$

where

$$4K^4=\frac{K_\theta(1-\nu_{x\theta}\nu_{\theta x})}{R^2D_x}=\frac{\left(\frac{E_\theta h}{1-\nu_{x\theta}\nu_{\theta x}}\right)(1-\nu_{x\theta}\nu_{\theta x})}{R^2\left(\frac{E_xh^3}{12(1-\nu_{x\theta}\nu_{\theta x})}\right)}=\frac{12E_\theta(1-\nu_{x\theta}\nu_{\theta x})}{R^2E_xh^2}$$

or

$$K^4=\frac{3E_\theta(1-\nu_{x\theta}\nu_{\theta x})}{R^2E_xh^2} \tag{9.25}$$

The solution of Equation (9.24) obviously depends on the magnitude of K. The dominant terms in Equation (9.25) are E_θ and E_x. Knowing the material properties radius and thickness of the shell, one can get a feeling for the variation of K with θ and thus plan an appropriate ply orientation. We note that Equation (9.24) is the same general equation as that representing the bending of a prismatic beam on an elastic foundation. Assuming $q_x=N_x^{\mathrm{T}}=N_\theta^{\mathrm{T}}=M_x^{\mathrm{T}}=M_\theta^{\mathrm{T}}=N_x^{\mathrm{H}}=N_\theta^{\mathrm{H}}=M_x^{\mathrm{H}}=M_\theta^{\mathrm{H}}=0$ and $N_x(x)=$ constant, this expression can be reduced. If one of the boundary conditions is unconstrained against axial deformation, say for example, $N_x(0)=0$, then $N_x=0$ throughout the entire shell. With this simplification,

$$N_x=K_x\left(U_{o,x}+\nu_{\theta x}\frac{W_o}{R}\right)=0\rightarrow U_{o,x}=-\nu_{\theta x}\frac{W_o}{R}$$

$$\begin{aligned}N_\theta&=K_\theta\left(\frac{W_o}{R}+\nu_{x\theta}U_{o,x}\right)=K_\theta\left(\frac{W_o}{R}+\nu_{x\theta}\left(-\nu_{\theta x}\frac{W_o}{R}\right)\right)\\&=\frac{E_\theta h}{1-\nu_{x\theta}\nu_{\theta x}}\left(\frac{W_o}{R}\right)(1-\nu_{x\theta}\nu_{\theta x})=\frac{E_\theta hW_o}{R}\end{aligned}$$

This is a simple relationship between N_θ and the mid-plane displacement W_o.

In order to establish the homogeneous solution to Equation (9.24), we set the right-hand side of the equation equal to zero to get $W_{o,xxxx}+4K^4W_o=0$. For this type of equation, it is customary to assume a solution of the form $W_o=A\mathrm{e}^{mx}$. Substituting this assumed solution into this equation results in $m^4+4K^4=0$. The resulting roots to this equation are $m=\pm K(1+i)$, $\pm K(1-i)$, which gives

$$W_o(x)=\mathrm{e}^{-Kx}(C_1\cos Kx+C_2\sin Kx)+\mathrm{e}^{Kx}(C_3\cos Kx+C_4\sin Kx) \tag{9.26}$$

where $C_1 \rightarrow C_4$ are constants. The particular solution depends on the loading condition, and the constants are determined from applying the appropriate boundary conditions.

9.4.1 Semi-infinite cylindrical shells

The complete solution to Equation (9.24) involves four integration constants, which constitutes a two-point boundary value problem. Within the class of axisymmetrically loaded cylindrical shells, there are many cases for which the forces at one boundary do not significantly affect those at the other boundary. Such shells are termed semi-infinite. The term e^{Kx} will grow very large with increasing x unless $C_3 = C_4 = 0$, and e^{-Kx} will decrease with increasing x. Therefore, for a semi-infinite cylindrical shell, we set $C_3 = C_4 = 0$, and the solution becomes

$$W_o(x) = e^{-Kx}(C_1 \cos Kx + C_2 \sin Kx)$$

where C_1 and C_2 are determined from boundary conditions at $x = 0$. For example, consider a cylindrical shell as depicted in Figure 9.5. There is a shear force Q_o and a bending moment M_o applied uniformly around the circumference at $x = 0$. Neglecting thermal and hygral effects, we have

$$M_o = M_x(0) = -D_x W_{o,xx}(0)$$

$$Q_o = Q_x(0) = -D_x W_{o,xxx}(0)$$

Taking the appropriate derivatives of the expression for $W_o(x)$ above, we find that

$$W_{o,xx}(0) = -\frac{M_o}{D_x} = -2C_2K^2$$

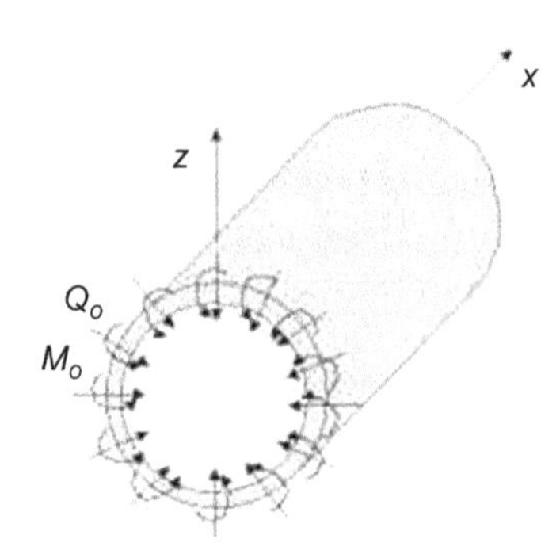

Figure 9.5 Cylindrical shell with uniform shear and bending at $x = 0$.

$$W_{o,xxx}(0) = -\frac{Q_o}{D_x} = 2K^3(C_1 + C_2)$$

Solving for the two constants, we get $C_1 = -\frac{1}{2D_xK^3}(Q_o + KM_o)$ and $C_2 = \frac{M_o}{2D_xK^2}$.

A convenient way to express the displacement and its associated derivatives is by first defining the functions $F_1 \rightarrow F_4$ below

$$\begin{aligned} F_1(Kx) &= e^{-Kx}(\cos Kx + \sin Kx) \\ F_2(Kx) &= e^{-Kx}(\cos Kx - \sin Kx) \\ F_3(Kx) &= e^{-Kx}\cos Kx = (F_1 + F_2)/2 \\ F_4(Kx) &= e^{-Kx}\sin Kx = (F_1 - F_2)/2 \end{aligned} \tag{9.27}$$

Using these definitions, we can express $W_o(x)$, and its various derivatives as

$$\begin{Bmatrix} W_o \\ W_{o,x} \\ W_{o,xx} \\ W_{o,xxx} \end{Bmatrix} = \frac{1}{D_x}\begin{bmatrix} 0 & -M_o/2K^2 & -Q_o/2K^3 & 0 \\ Q_o/2K^2 & 0 & M_o/K & 0 \\ -M_o & 0 & 0 & -Q_o/K \\ 0 & -Q_o & 0 & 2KM_o \end{bmatrix}\begin{Bmatrix} F_1(Kx) \\ F_2(Kx) \\ F_3(Kx) \\ F_4(Kx) \end{Bmatrix} \tag{9.28}$$

The variation of $F_1 \rightarrow F_4$ as a function of Kx is shown in Figure 9.6. As seen in this figure, the edge loads will produce insignificant effects for $Kx > 3$.

In the literature, it is common to find that $L \geq \pi/K$ (designated as the length) for which a given shell is considered to be a long shell, and treatable by the semi-infinite approach. The length L is called the half-wave length of bending and

$$L = \frac{\pi}{K} = \frac{\pi}{\sqrt[4]{3E_\theta(1 - \nu_{\theta x}\nu_{x\theta})/E_x}}\sqrt{Rh}$$

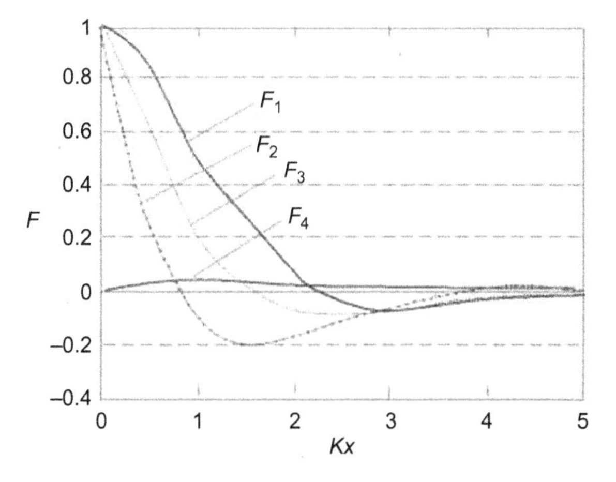

Figure 9.6 Variation of $F_1 \rightarrow F_4$ with Kx.

The length L is often used as a measure of the penetration of the effect of various singularities, such as concentrated loads, holes, discontinuities, etc., into the interior of a shell.

Recalling from Equation (9.25) that $K^4 = 3E_\theta(1-\nu_{x\theta}\nu_{\theta x})/R^2E_xh^2$, we can define $f(\theta) = E_\theta(1-\nu_{x\theta}\nu_{\theta x})/E_x$, which results in

$$K^4 = \frac{3}{R^2h^2}f(\theta) \rightarrow K = \frac{1.316}{\sqrt{Rh}}\sqrt[4]{f(\theta)}$$

Knowing that the effects of end loads are insignificant after $Kx > 3$, we can determine the distance x from

$$x = \frac{3\sqrt{Rh}}{1.316\sqrt[4]{f(\theta)}} = \frac{2.2796\sqrt{Rh}}{\sqrt[4]{f(\theta)}}$$

The distance x will vary according to $f(\theta)$ as well as the ratio of shell thickness (h) to radius (R). Values of R/h are variable, and application dependent. Introducing $R = nh$ into the equation above we get

$$x = \frac{2.2796\sqrt{nh^2}}{\sqrt[4]{f(\theta)}} = \frac{2.2796h\sqrt{n}}{\sqrt[4]{f(\theta)}} \rightarrow \frac{x}{h} = \frac{2.2796\sqrt{n}}{\sqrt[4]{f(\theta)}}$$

We can define explicit expressions for the stress resultants in terms of the functions $F_1 \rightarrow F_4$. These are

$$M_x(x) = -D_xW_{o,xx} = M_oF_1(Kx) + \frac{Q_o}{K}F_4(Kx)$$

$$M_\theta(x) = \nu_{\theta x}M_x(x) = \nu_{\theta x}\left[M_oF_1(Kx) + \frac{Q_o}{K}F_4(Kx)\right]$$

$$Q_x(x) = -D_xW_{o,xxx} = Q_oF_2(Kx) - 2KM_oF_4(Kx)$$

$$N_\theta(x) = \frac{E_\theta h}{R}W_o(x) = -\frac{E_\theta h}{2RK^2D_x}\left[M_oF_2(Kx) + \frac{Q_o}{K}F_3(Kx)\right]$$

Example 9.1 Consider a long unidirectional laminated cylinder subjected to internal pressure P and fixed to a wall. We want to establish the displacement as a function of distance from the free end. We want to express this in terms of the pressure by plotting $W_o(x)/P$ vs. x for laminates with ply orientations of 0, 45, and 90 degrees. Assume $E_1 = 20 \times 10^6$, $E_2 = 1.8 \times 10^6$, $G_{12} = 0.7 \times 10^6$, $\nu_{12} = 0.21$, $h = 0.125''$, and $R = 20h$. Because of the fixed wall, the boundary conditions at the wall are a circumferential displacement as well as shear and bending moment, as illustrated in Figure E9.1-1.

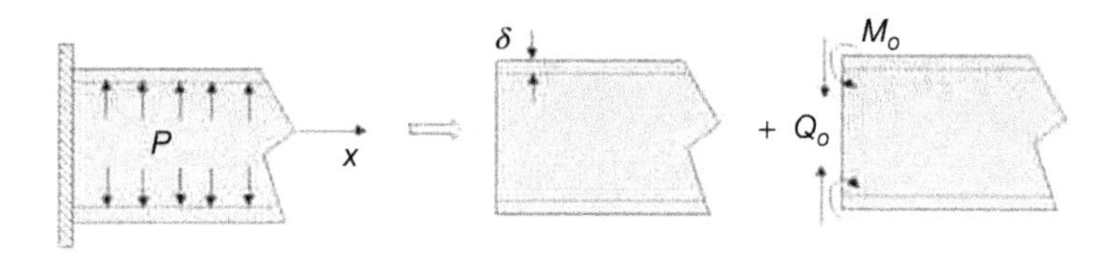

Figure E9.1-1 Boundary conditions at the wall of a cylinder fixed to a wall at one end.

We assume that $N_x = 0$ resulting in $\sigma_x = 0$, and the internal pressure causes stress

$$\sigma_\theta = \frac{PR}{h}$$

In addition to the σ_θ stress, there is an end moment M_o and shear Q_o, both of which are applied uniformly around the circumference at $x = 0$. The displacement δ is related to the internal pressure through

$$\delta = \varepsilon_\theta R = \frac{R}{E_\theta}\sigma_\theta = \frac{PR^2}{E_\theta h}$$

At $x = 0$, $W_o(0) = \dfrac{PR^2}{E_\theta h}$

From the previous discussion pertaining to $F_1 \rightarrow F_4$, we know that $F_1(0) = F_2(0) = F_3(0) = F_4(0) = 1$. Therefore,

$$W_o(0) = -\frac{1}{D_x}\left[\frac{M_o}{2K^2} + \frac{Q_o}{2K^3}\right] = \frac{PR^2}{E_\theta h}$$

From the matrix presented in Equation (9.28), $W_o(0)_{,x} = 0 \Rightarrow \dfrac{M_o}{K} + \dfrac{Q_o}{2K^3} = 0 \Rightarrow Q_o = -2M_o K$

$$-\frac{1}{D_x}\left[\frac{M_o}{2K^2} + \frac{-2M_o K}{2K^3}\right] = \frac{PR^2}{E_\theta h} \Rightarrow \frac{PR^2}{E_\theta h} = \frac{M_o}{2K^2 D_x}$$

$$M_o = \frac{2K^2 D_x R^2}{E_\theta h}P \quad Q_o = -\frac{4K^3 D_x R^2}{E_\theta h}P$$

The component of the displacement due to end loads M_o and Q_o is

$$(W_o(x))_{\text{end loads}} = e^{-Kx}(C_1 \cos Kx + C_2 \sin Kx) = C_1 F_2(Kx) + C_2 F_3(Kx)$$

where $F_2(Kx) = e^{-Kx}(\cos Kx - \sin Kx)$ and $F_3(Kx) = e^{-Kx}\cos Kx$. Therefore,

$$C_1 = -\frac{1}{2K^3 D_x}(Q_o + KM_o) = \frac{PR^2}{E_\theta h} \text{ and } C_2 = \frac{M_o}{2K^2 D_x} = \frac{PR^2}{E_\theta h}$$

$$(W_o(x))_{\text{end loads}} = \frac{PR^2}{E_\theta h} e^{-Kx}[(\cos Kx - \sin Kx) + \cos Kx]$$
$$= \frac{PR^2}{E_\theta h} e^{-Kx}(2\cos Kx - \sin Kx)$$

The total displacement must account for the internal pressure. The displacement due to internal pressure was given previously as

$$(W_o(x))_{\text{pressure}} = \frac{PR^2}{E_\theta h}$$

The total deflection is a combination of these. The displacement due to pressure tends to expand the cylinder, while the end loads tend to compress it. Therefore, the total displacement is given as

$$W_o(x) = (W_o(x))_{\text{end loads}} + (W_o(x))_{\text{pressure}} = \frac{PR^2}{E_\theta h}\left[1 - e^{-Kx}(2\cos Kx - \sin Kx)\right]$$

From our previously presented discussions of K, we have

$$K^4 = \frac{3}{R^2 h^2} f(\theta) \rightarrow K = \frac{1.316}{\sqrt{Rh}} \sqrt[4]{f(\theta)}$$

where $f(\theta) = E_\theta(1 - \nu_{x\theta}\nu_{\theta x})/E_x$. Using the material properties above and assuming $R = 20h$, we can express K and $W_o(x)$ as

$$K = \frac{1.316}{\sqrt{20h^2}} \sqrt[4]{f(\theta)} = \frac{0.294}{\mathrm{h}} \sqrt[4]{\mathrm{f}(\theta)}$$

$$W_o(x) = \frac{P(20h)^2}{E_\theta h}\left[1 - e^{-Kx}(2\cos Kx - \sin Kx)\right]$$
$$= \frac{400hP}{E_\theta}\left[1 - e^{-Kx}(2\cos Kx - \sin Kx)\right]$$

Assume the thickness of the shell $h = 0.125''$, in which case $K = 2.352\sqrt[4]{f(\theta)}$ and

$$W_o(x) = \frac{50P}{E_\theta}\left[1 - e^{-Kx}(2\cos Kx - \sin Kx)\right]$$

Plotting $W_o(x)/P$ vs. x for ply orientations of 0, 45, and 90 degrees, we see from Figure E9.1-2 that the end effects have different effects on the displacement as a function of the ply orientation angle. The 90° laminate (fibers wrapped around the circumference of the cylinder) is least affected, while the 0° laminate is most affected. In all cases, the magnitudes of the displacement are quite small, with $W_o(x)/P = 35 \times 10^{-6}$

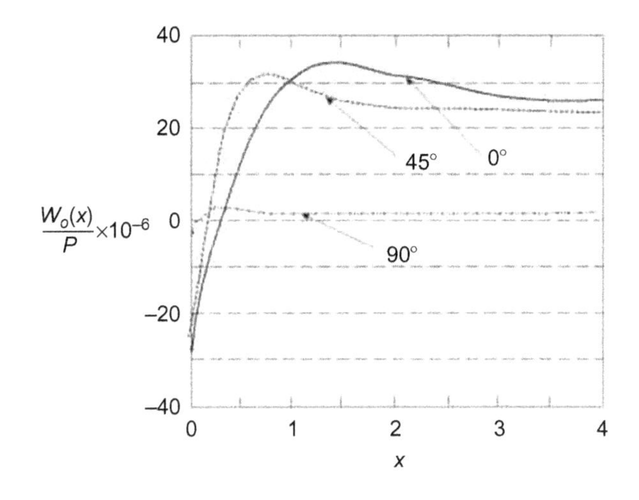

Figure E9.1-2 $W_o(x)/P$ vs. x for three fiber orientations.

in the worst case. The effects of the end loads are seen to become negligible after approximately 3.5″ in all three cases. Different materials would yield different magnitudes, but the trend would remain similar.

Example 9.2 Assume a circumferential line load is applied to a unidirectional laminated cylindrical shell at a location $x=0$. We want to determine expressions for $W_o(0)$, $M_x(0)$, $M_\theta(0)$, $Q_x(0)$, and $N_\theta(0)$. We assume that the shell extends a distance of at least π/K in each direction from the point of application of the load P. The circumferential line load P has units of force/length. We consider the section of the cylindrical shell adjacent to the point of load application and define this as the origin ($x=0$) of the coordinate system as modeled in Figure E9.2 ($W_o(0), M_x(0), M_\theta(0), Q_x(0), N_\theta(0)$).

From this free-body diagram indicated that at $x=0$, we have $Q_o = Q_x(0) = -P/2$ and $W_{o,x}(0)=0$. From the matrix relationship in Equation (9.28), we can write

$$W_{o,x}(x) = \frac{Q_o}{2D_xK^2}F_1(Kx) + \frac{M_o}{D_xK}F_3(Kx) = 0 \rightarrow M_o = -\frac{Q_o}{2K} = \frac{P}{4K}$$

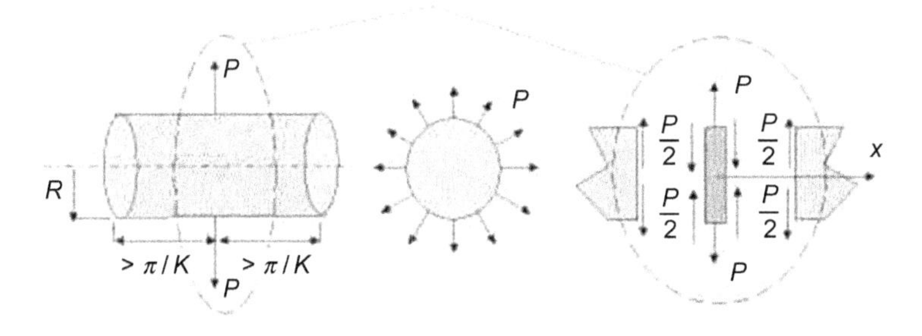

Figure E9.2 Circumferential line load on a cylindrical shell with free body diagram.

Similarly, from Equation (9.28), we can write

$$W_o(x) = -\frac{M_o}{2D_xK^2}F_2(Kx) - \frac{Q_o}{2D_xK^3}F_3(Kx) = -\frac{P}{4D_xK^3}\left[\frac{1}{2}F_2(Kx) - F_3(Kx)\right]$$

Since $F_3(Kx) = (F_1 + F_2)/2$,

$$W_o(x) = \frac{P}{8D_xK^3}F_1(Kx)$$

Similarly,

$$M_x(x) = -D_xW_{o,xx} = M_oF_1(Kx) + \frac{Q_o}{K}F_4(Kx) = \frac{P}{4K}F_1(Kx) - \frac{P}{2K}F_4(Kx)$$

Since $F_4(Kx) = (F_1 - F_2)/2$,

$$M_x(x) = \frac{P}{4K}F_2(Kx)$$

In addition, we can now determine

$$M_\theta(x) = \nu_{\theta x}M_x(x) = \frac{P}{4K}\nu_{\theta x}F_2(Kx)$$

Similarly,

$$Q_x(x) = -D_xW_{o,xxx} = -Q_oF_2(Kx) + 2M_oKF_4(Kx) = -\frac{P}{2}F_2(Kx) - \frac{P}{4K}(2K)F_4(Kx)$$
$$= -\frac{P}{2}[F_2(Kx) + F_4(Kx)]$$

Since $F_2(Kx) + F_4(Kx) = F_3(Kx)$, we have $Q_x(x) = -\frac{P}{2}F_3(Kx)$. Finally, we can determine

$$N_\theta(x) = \frac{E_\theta h}{R}W_o(x) = \frac{PE_\theta h}{8D_xRK^3}F_1(Kx)$$

In summary, we have the general relationships

$$W_o(x) = \frac{P}{8D_xK^3}F_1(kx),\;\; M_x(x) = \frac{P}{4K}F_2(Kx),\;\; M_\theta(x) = \frac{P}{4K}\nu_{\theta x}F_2(Kx)$$
$$Q_x(x) = -\frac{P}{2}F_3(Kx),\qquad N_\theta(x) = \frac{PE_\theta h}{8D_xRK^3}F_1(Kx)$$

At the origin, these become

$$W_o(0) = \frac{P}{8D_xK^3},\quad M_x(0) = \frac{P}{4K},\quad M_\theta(0) = \frac{P}{4K}\nu_{\theta x},\quad Q_x(0) = -\frac{P}{2},$$
$$N_\theta(0) = \frac{PE_\theta h}{8D_xRK^3} = \frac{PKR}{2}$$

Example 9.3 Consider a built-in unidirectional laminated shell subjected to internal pressure and a temperature gradient as illustrated in Figure E9.3-1. The temperature gradient is linear through the thickness of the shell, with an outside temperature T_1 and an internal temperature $T_2 (T_1 > T_2)$. Both the temperature gradient and the pressure are constant along the length of the shell. The ends of the shell are fixed and cannot translate or rotate. It is assumed that the shell length is sufficiently large that semi-infinite analysis is valid. Assume $P = 100$psi, $R = 10$ in., $h = 0.1$ in., $\alpha_1 = 0.1 \times 10^{-6}$ in./in./°F, and $\alpha_2 = 15 \times 10^{-6}$ in./in./°F. We wish to consider fiber orientations of 0°, 45°, and 90°. For the material used, these fiber orientations result in $(E_\theta)_0 = 1.5 \times 10^6$, $(E_\theta)_{45} = 2.37 \times 10^6$, and $(E_\theta)_{90} = 20 \times 10^6$. We want to plot $W_o(0)$ vs. Kx for two average temperatures: $T_{\text{average}} = T_s = (T_1 + T_2)/2 = 0°, 200°$.

The linear temperature gradient can be defined in terms of its symmetric (T_s) and antisymmetric (T_a) components, with $T(z) = T_s + zT_a$. The temperature distribution through the wall of the shell is modeled as shown in Figure E9.3-2.

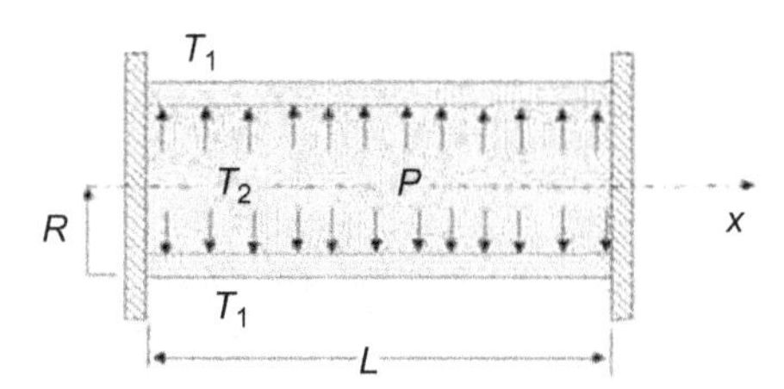

Figure E9.3-1 Fixed-fixed shell with internal pressure and a temperature gradient.

$T_s = (T_1 + T_2)/2$

$T_a = (T_1 - T_2)/2$

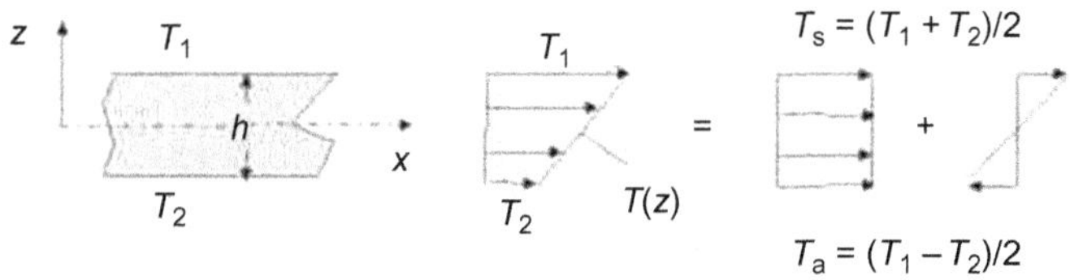

Figure E9.3-2 Model of temperature gradient through fixed-fixed shell.

We can write the temperature distribution as

$$T(z) = T_s + zT_a,$$

where $T_s = (T_1 + T_2)/2$ and $T_a = (T_1 - T_2)/2$

The homogeneous solution to the shell displacement is

$$(W_o(x))_h = e^{-Kx}(C_1 \cos Kx + C_2 \sin Kx) = C_1 F_3(Kx) + C_2 F_4(Kx) \tag{a}$$

The particular solution consists of three components. The first component reflects the internal pressure and is selected to be

$$(W_o(x))_{p1} = \frac{p}{4K^4 D_x} = \frac{pR^2}{E_\theta h} \tag{b}$$

In order to obtain the contribution of the temperature gradient, we need to determine N_T and M_T in terms of $T(z) = T_s + zT_a$. Therefore, we write

$$N_x^T = \frac{E_x(\alpha_x + \nu_{\theta x}\alpha_\theta)}{1 - \nu_{\theta x}\nu_{x\theta}} \int_{-h/2}^{h/2} (T_s + zT_a)dz = \frac{E_x h(\alpha_x + \nu_{\theta x}\alpha_\theta)}{1 - \nu_{\theta x}\nu_{x\theta}} T_s$$

$$N_\theta^T = \frac{E_\theta(\alpha_\theta + \nu_{x\theta}\alpha_x)}{1 - \nu_{\theta x}\nu_{x\theta}} \int_{-h/2}^{h/2} (T_s + zT_a)dz = \frac{E_\theta h(\alpha_\theta + \nu_{x\theta}\alpha_x)}{1 - \nu_{\theta x}\nu_{x\theta}} T_s$$

$$M_x^T = \frac{E_x(\alpha_x + \nu_{\theta x}\alpha_\theta)}{1 - \nu_{\theta x}\nu_{x\theta}} \int_{-h/2}^{h/2} (T_s + zT_a)z dz = \frac{E_x h^3(\alpha_x + \nu_{\theta x}\alpha_\theta)}{12(1 - \nu_{\theta x}\nu_{x\theta})} T_a$$

$$M_\theta^T = \frac{E_\theta(\alpha_\theta + \nu_{x\theta}\alpha_x)}{1 - \nu_{\theta x}\nu_{x\theta}} \int_{-h/2}^{h/2} (T_s + zT_a)z dz = \frac{E_\theta h^3(\alpha_\theta + \nu_{x\theta}\alpha_x)}{12(1 - \nu_{\theta x}\nu_{x\theta})} T_a$$

Next, from our general equation, we have

$$\begin{aligned} 4K^2(W_o(x))_{p2} &= \frac{1}{RD_x}\left(N_\theta^T - \nu_{\theta x} N_x^T\right) = \frac{1}{RD_x}\frac{hT_s}{1 - \nu_{\theta x}\nu_{x\theta}}(E_\theta\alpha_\theta - \nu_{\theta x}\nu_{x\theta}E_\theta\alpha_\theta) \\ &= \frac{hT_s E_\theta \alpha_\theta}{RD_x} \end{aligned}$$

Recalling our definition $K^4 = \frac{E_\theta h}{4D_x R^2}$, this reduces to

$$(W_o(x))_{p2} = R\alpha_\theta T_s \tag{c}$$

Since $M_{T,xx} = q_x = 0$ in this case, no particular solution is required for these terms. Finally, we have the axial stress resultant $N_x(0)$ which gives

$$4K^2(W_o(x))_{p3} = -\frac{\nu_{\theta x}}{RD_x}N_x(0)$$

$$(W_o(x))_{p3} = -\frac{\nu_{\theta x}R}{E_\theta h}N_x(0) \tag{d}$$

Combining (b), (c), and (d), we determine the particular solution to be

$$(W_o(x))_p = (W_o(x))_{p1} + (W_o(x))_{p2} + (W_o(x))_{p3} = \frac{pR^2}{E_\theta h} + \alpha_\theta RT_s - \frac{\nu_{\theta x}R}{E_\theta h}N_x(0)$$

The general solution is the sum of the homogeneous and particular solutions, expressed as

$$\begin{aligned} W_o(x) &= (W_o(x))_h + (W_o(x))_p \\ &= C_1F_3(Kx) + C_2F_4(Kx) + \frac{pR^2}{E_\theta h} + \alpha_\theta RT_s - \frac{\nu_{\theta x}R}{E_\theta h}N_x(0) \end{aligned}$$

where $F_3(Kx) = e^{-Kx}\cos Kx = (F_1 + F_2)/2$ and $F_4(Kx) = e^{-Kx}\sin Kx = (F_1 - F_2)/2$. From the boundary condition, $W_o(0) = 0$ we get

$$C_1 + \frac{pR^2}{E_\theta h} + \alpha_\theta RT_s - \frac{\nu_{\theta x}R}{E_\theta h}N_x(0) = 0$$

Similarly, using the boundary condition $W_o(0)_{,x} = 0$, we get $K(-C_1 + C_2) = 0$, or $C_1 = C_2$, which results in

$$C_1 = -\left[\frac{pR^2}{E_\theta h} + \alpha_\theta RT_s - \frac{\nu_{\theta x}R}{E_\theta h}N_x(0)\right]$$

In order to evaluate $N_x(0)$ explicitly, we integrate the previously defined expression for $U_{o,x}$ to get

$$U_o(x) = \int_0^x U_{o,x}\mathrm{d}x + U_o(0)$$

with

$$U_{o,x}(0) = \frac{1 - \nu_{\theta x}\nu_{x\theta}}{E_x h}\left[N_x^{\mathrm{T}} + N_x(0)\right]$$

The restraints at the end of the shell ($U_o(0) = U_{o,x}(0) = 0$) combined with

$$N_x(0) = -N_x^{\mathrm{T}} = -\frac{E_x h(\alpha_x + \nu_{\theta x}\alpha_\theta)}{1 - \nu_{\theta x}\nu_{x\theta}} T_s$$

yields, upon substitution into the expression for C_1

$$C_1 = -\left[\frac{pR^2}{E_\theta h} + \alpha_\theta R T_s + \left(\frac{\nu_{\theta x} R}{E_\theta h}\right)\frac{E_x h(\alpha_x + \nu_{\theta x}\alpha_\theta)}{1 - \nu_{\theta x}\nu_{x\theta}} T_s\right]$$
$$= -\left[\frac{pR^2}{E_\theta h} + \frac{1 + \nu_{x\theta}}{1 - \nu_{\theta x}\nu_{x\theta}}\alpha_\theta R T_s\right]$$

Knowing the explicit form of $N_x(0)$, the particular solution can now be consolidated into

$$(W_o(x))_p = \frac{pR^2}{E_\theta h} + \left(\frac{1 + \nu_{x\theta}}{1 - \nu_{\theta x}\nu_{x\theta}}\right)\alpha_\theta R T_s$$

Furthermore, we can now express $W_o(x)$ as

$$W_o(x) = \left(\frac{pR^2}{E_\theta h} + \left(\frac{1 + \nu_{x\theta}}{1 - \nu_{\theta x}\nu_{x\theta}}\right)\alpha_\theta R T_s\right)[1 - F_3(Kx) - F_4(Kx)]$$

Recalling that $\alpha_\theta = n^2\alpha_1 + m^2\alpha_2$, we can determine that $(\alpha_\theta)_0 = 15 \times 10^{-6}$ in./in./°F, $(\alpha_\theta)_{45} = 7.55 \times 10^{-6}$ in./in./°F, and $(\alpha_\theta)_{90} = 0.1 \times 10^{-6}$ in./in./°F. For each lamina, $F_3(Kx) = \mathrm{e}^{-Kx}\cos Kx$ and $F_4(Kx) = \mathrm{e}^{-Kx}\sin Kx$; therefore, we can establish the relationships

$$(W_o(x))_0 = \left(6.67 \times 10^{-2} + 1.9314 \times 10^{-4} T_s\right)\left[1 - \mathrm{e}^{-Kx}\cos(Kx) - \mathrm{e}^{-Kx}\sin(Kx)\right]$$

$$(W_o(x))_{45} = \left(4.219 \times 10^{-2} + 9.2525 \times 10^{-5} T_s\right)\left[1 - \mathrm{e}^{-Kx}\cos(Kx) - \mathrm{e}^{-Kx}\sin(Kx)\right]$$

$$(W_o(x))_{90} = \left(0.5 \times 10^{-2} + 1.023 \times 10^{-6} T_s\right)\left[1 - \mathrm{e}^{-Kx}\cos(Kx) - \mathrm{e}^{-Kx}\sin(Kx)\right]$$

The displacement $W_o(x)$ vs. Kx for each fiber orientation is shown in Figure E9.3-3 for $T_s = 0°$ and in Figure E9.3-4 for $T_s = 200°$.

From these figures, we can conclude that the $T_s = 200°$ has the most impact upon the displacements, and the 0° laminate will experience the most deformation due to temperature because all fibers are aligned with the longitudinal axis of the shell and do nothing to reinforce the circumferential direction.

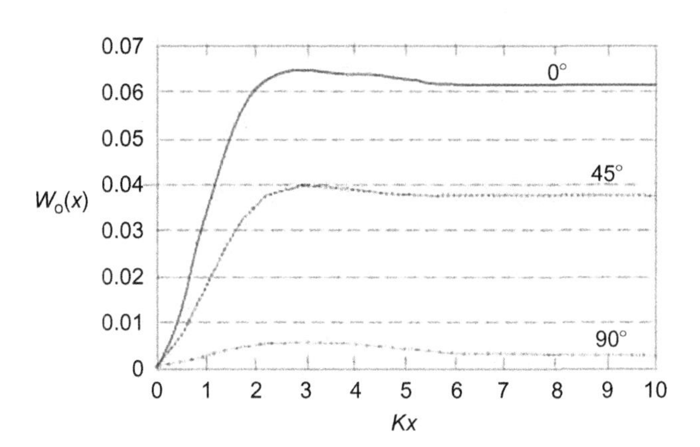

Figure E9.3-3 $W_o(x)$ vs. Kx for $T_s = 0$°F

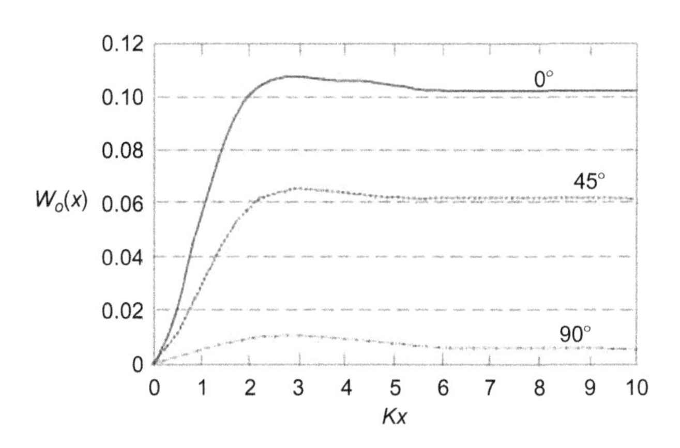

Figure E9.3-4 $W_o(x)$ vs. Kx for $T_s = 200$°F

9.4.2 Short cylindrical shells

A short cylindrical shell as illustrated in Figure 9.7 is characterized by having its distance between boundary points, L, being such that $L < \pi/K$. The semi-infinite assumption is no longer valid, and the general solution is used.

The homogeneous solution is written as

$$(W_o(x))_h = C_1F_5(Kx) + C_2F_6(Kx) + C_3F_7(Kx) + C_4F_8(Kx)$$

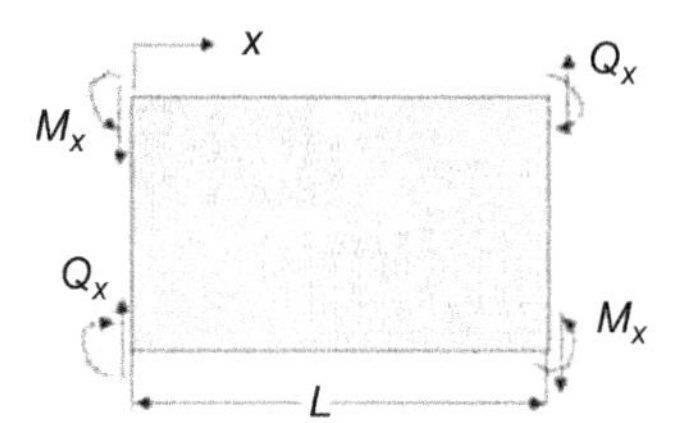

Figure 9.7 Short cylindrical shell with end moments and shear.

where

$$F_5(Kx) = \mathrm{e}^{-Kx}\cos(Kx) \quad F_6(Kx) = \mathrm{e}^{-Kx}\sin(Kx)$$

$$F_7(Kx) = \mathrm{e}^{Kx}\cos(Kx) \quad F_8(Kx) = \mathrm{e}^{Kx}\sin(Kx)$$

The variation of $F_5 \rightarrow F_8$ as a function of Kx is shown in Figure 9.8 for $Kx < 2$. The terms F_7 and F_8 grow exponentially and become very large as Kx increases. Since we are considering short cylinders, the value of Kx will generally be small.

The boundary conditions are

$$M_x(0) = M_\mathrm{A} = -D_x W_{o,xx}(0) \quad M_x(L) = M_\mathrm{B} = -D_x W_{o,xx}(L)$$
$$Q_x(0) = Q_\mathrm{A} = -D_x W_{o,xxx}(0) \quad Q_x(L) = Q_\mathrm{B} = -D_x W_{o,xxx}(L)$$

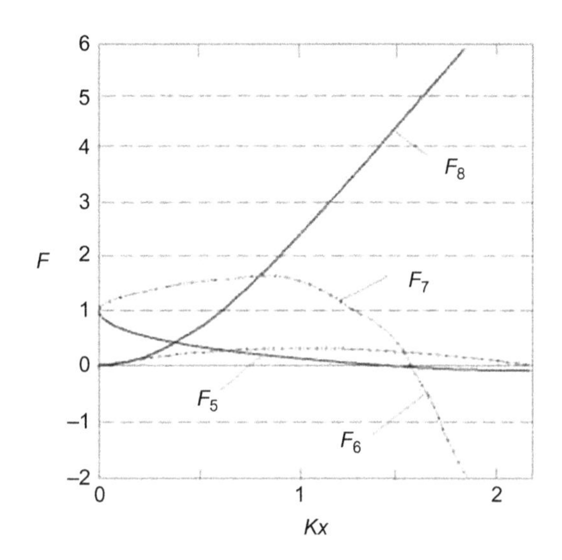

Figure 9.8 Variation of $F_5 \rightarrow F_8$ vs. Kx for a short cylindrical shell.

This can be written in matrix form as

$$\begin{bmatrix} 0 & -1 & 0 & 1 \\ K & K & -K & K \\ F_6(KL) & -F_5(KL) & -F_8(KL) & F_7(KL) \\ K\begin{bmatrix} F_5(KL) \\ -F_6(KL) \end{bmatrix} & K\begin{bmatrix} F_5(KL) \\ +F_6(KL) \end{bmatrix} & -K\begin{bmatrix} F_7(KL) \\ +F_8(KL) \end{bmatrix} & K\begin{bmatrix} F_7(KL) \\ -F_8(KL) \end{bmatrix} \end{bmatrix} \begin{Bmatrix} C_1 \\ C_2 \\ C_3 \\ C_4 \end{Bmatrix}$$

$$= -\frac{1}{2K^2 D_x} \begin{Bmatrix} M_{\mathrm{A}} \\ Q_{\mathrm{A}} \\ M_{\mathrm{B}} \\ Q_{\mathrm{B}} \end{Bmatrix} \tag{9.29}$$

In addition, we can express W_o, $W_{o,x}$, $W_{o,xx}$, and $W_{o,xxx}$ in terms of the constants $C_1 \rightarrow C_4$ in matrix form as

$$\begin{Bmatrix} W_o \\ W_{o,x} \\ W_{o,xx} \\ W_{o,xxx} \end{Bmatrix} = \begin{bmatrix} C_1 & C_2 & C_3 & C_4 \\ K(C_2 - C_1) & -K(C_1 + C_2) & K(C_3 + C_4) & K(C_4 - C_3) \\ -2K^2 C_2 & 2K^2 C_1 & 2K^2 C_4 & -2K^2 C_3 \\ 2K^3(C_1 + C_2) & 2K^3(C_2 - C_1) & 2K^3(C_4 - C_3) & 2K^3(C_3 + C_4) \end{bmatrix}$$

$$\begin{Bmatrix} F_5(KL) \\ F_6(KL) \\ F_7(KL) \\ F_8(KL) \end{Bmatrix} \tag{9.30}$$

Under general loading conditions, we have

$$Q_\theta = M_{x\theta,x} + \frac{1}{R} M_{\theta,\theta} \quad Q_x = M_{x,x} + \frac{1}{R} M_{x\theta,\theta}$$

Therefore, our original set of five equations given by (9.13)–(9.17) reduces to three equations, given by

$$\begin{aligned} & N_{x,x} + \frac{1}{R} N_{x\theta,\theta} + q_x = 0 \\ & N_{x\theta,x} + \frac{1}{R} N_{\theta,\theta} + \frac{1}{R} M_{x\theta,x} + \frac{1}{R^2} M_{\theta,\theta} + q_\theta = 0 \\ & M_{x,xx} + \frac{1}{R^2} M_{\theta,\theta\theta} + \frac{z}{R} M_{x\theta,x\theta} - \frac{1}{R} N_\theta + p = 0 \end{aligned} \tag{9.31}$$

The stress resultants are expressed in terms of displacements by

$$
\begin{aligned}
N_x &= \frac{E_x h}{1-\nu_{x\theta}\nu_{\theta x}}\left[U_{o,x}+\frac{\nu_{\theta x}}{R}\left(V_{o,\theta}+W_o\right)\right]-N_x^{\mathrm{T}}-N_x^{\mathrm{H}} \\
N_\theta &= \frac{E_\theta h}{1-\nu_{x\theta}\nu_{\theta x}}\left[\nu_{x\theta}U_{o,x}+\frac{1}{R}\left(V_{o,\theta}+W_o\right)\right]-N_\theta^{\mathrm{T}}-N_\theta^{\mathrm{H}} \\
N_{x\theta} &= G_{x\theta}h\left[V_{o,x}+\frac{1}{R}U_{o,\theta}\right] \\
M_x &= -\frac{E_x h^3}{12(1-\nu_{x\theta}\nu_{\theta x})}\left[W_{o,xx}+\frac{\nu_{\theta x}}{R^2}\left(W_{o,\theta\theta}-V_{o,\theta}\right)\right]-M_x^{\mathrm{T}}-M_x^{\mathrm{H}} \\
M_\theta &= -\frac{E_\theta h^3}{12(1-\nu_{x\theta}\nu_{\theta x})}\left[\nu_{x\theta}W_{o,xx}+\frac{1}{R^2}\left(W_{o,\theta\theta}-V_{o,\theta}\right)\right]-M_x^{\mathrm{T}}-M_x^{\mathrm{H}} \\
M_{x\theta} &= \frac{G_{x\theta}h^3}{12}\left[\frac{V_{o,x}}{R}-\frac{W_{o,x\theta}}{R}\right]
\end{aligned}
\tag{9.32}
$$

Substituting these into the equations of motion and manipulating them to reduce each to a convenient form results in

$$
\begin{aligned}
&U_{o,xx}+\frac{G_{x\theta}(1-\nu_{x\theta}\nu_{\theta x})}{E_x R^2}U_{o,\theta\theta}+\frac{\nu_{x\theta}E_\theta+G_{x\theta}(1-\nu_{x\theta}\nu_{\theta x})}{E_x R}V_{o,x\theta}+\frac{\nu_{x\theta}E_\theta}{E_x R}W_{o,x} \\
&\quad=\frac{1-\nu_{x\theta}\nu_{\theta x}}{E_x h}\left(-q_x+N_{x,x}^{\mathrm{T}}+N_{x,x}^{\mathrm{H}}\right)+\frac{\rho(1-\nu_{x\theta}\nu_{\theta x})}{E_x}U_{o,tt}
\end{aligned}
\tag{9.33}
$$

$$
\begin{aligned}
&\frac{G_{x\theta}(1-\nu_{x\theta}\nu_{\theta x})}{E_\theta}V_{o,xx}+\frac{1}{R^2}V_{o,\theta\theta}+\frac{\nu_{x\theta}E_\theta+G_{x\theta}(1-\nu_{x\theta}\nu_{\theta x})}{E_\theta R}U_{o,x\theta}+\frac{1}{R^2}W_{o,\theta} \\
&+\frac{h^2}{12R^2}\left\{\frac{G_{x\theta}(1-\nu_{x\theta}\nu_{\theta x})}{E_\theta}V_{o,xx}+\frac{1}{R^2}V_{o,\theta\theta}-\frac{\nu_{x\theta}E_\theta+G_{x\theta}(1-\nu_{x\theta}\nu_{\theta x})}{E_\theta}W_{o,xx\theta}-\frac{1}{R^2}W_{o,\theta\theta\theta}\right\} \\
&=\left(\frac{1-\nu_{x\theta}\nu_{\theta x}}{E_\theta h}\right)\left(-q_\theta+\frac{N_{\theta,\theta}^{\mathrm{T}}+N_{\theta,\theta}^{\mathrm{H}}}{R}+\frac{M_{\theta,\theta}^{\mathrm{T}}+M_{\theta,\theta}^{\mathrm{H}}}{R^2}\right)+\frac{\rho(1-\nu_{x\theta}\nu_{\theta x})}{E_\theta}V_{o,tt}
\end{aligned}
\tag{9.34}
$$

$$
\begin{aligned}
&W_{o,xxxx}+2\left[\frac{\nu_{\theta x}E_x+G_{x\theta}(1-\nu_{x\theta}\nu_{\theta x})}{E_x R^2}\right]W_{o,xx\theta\theta}+\frac{E_\theta}{R^4 E_x}W_{o,\theta\theta\theta\theta}-\frac{\nu_{\theta x}E_x+2G_{x\theta}(1-\nu_{x\theta}\nu_{\theta x})}{E_x R^2}V_{o,xx\theta} \\
&\quad-\frac{E_x}{R^4 E_\theta}V_{o,\theta\theta\theta}+\frac{12E_\theta}{E_x h^2}\left\{\frac{\nu_{x\theta}}{R}U_{o,x}+\frac{1}{R^2}\left(V_{o,\theta}+W_o\right)\right\} \\
&\quad=\frac{12(1-\nu_{x\theta}\nu_{\theta x})}{E_x h^3}\left(p+\frac{N_\theta^{\mathrm{T}}+N_\theta^{\mathrm{H}}}{R}-\frac{M_{\theta,\theta\theta}^{\mathrm{T}}+M_{\theta,\theta\theta}^{\mathrm{H}}}{R^2}-\left(M_{x,xx}^{\mathrm{T}}+M_{x,xx}^{\mathrm{H}}\right)\right) \\
&\quad+\frac{12\rho(1-\nu_{x\theta}\nu_{\theta x})}{E_x h^2}W_{o,tt}
\end{aligned}
\tag{9.35}
$$

Example 9.4 Consider the short cylindrical shell in Figure 9.8. Assume that its length is $L = 3\pi/4K$, and the end loads are identical ($M_A = M_B = M_o$ and $Q_A = Q_B = Q_o$) We wish to determine the constants $C_1 \rightarrow C_4$ in terms of the end moments and shear forces, M_o and Q_o.

At $x = L$, $F_5 \rightarrow F_8$ have the values $F_5(KL) = -0.067$, $F_6(KL) = 0.067$, $F_7(KL) = -7.459$, and $F_8(KL) = 7.459$. Therefore, Equation (9.29) becomes

$$\begin{bmatrix} 0 & -1 & 0 & 1 \\ K & K & -K & K \\ 0.067 & 0.067 & -7.459 & -7.459 \\ -0.134K & 0 & 0 & -14.918K \end{bmatrix} \begin{Bmatrix} C_1 \\ C_2 \\ C_3 \\ C_4 \end{Bmatrix} = -\frac{1}{2K^2 D_x} \begin{Bmatrix} M_A \\ Q_A \\ M_B \\ Q_B \end{Bmatrix} = -\frac{1}{2K^2 D_x} \begin{Bmatrix} M_o \\ Q_o \\ M_o \\ Q_o \end{Bmatrix}$$

Solving for the constants, we find

$$C_1 = \frac{1}{K^3 D_x}(-1.2451 Q_o) + \frac{1}{K^2 D_x}(-0.8888 M_o)$$

$$C_2 = \frac{1}{K^3 D_x}(0.07821 Q_o) + \frac{1}{K^2 D_x}(1.00798 M_o)$$

$$C_3 = \frac{1}{K^3 D_x}(-0.08868 Q_o) + \frac{1}{K^2 D_x}(0.1272 M_o)$$

$$C_4 = \frac{1}{K^3 D_x}(0.07812 Q_o) + \frac{1}{K^2 D_x}(0.00798 M_o)$$

9.5 Anisotropic cylindrical shells

For an anisotropic cylindrical shell, the forces and moments are related to the mid-plane strains and curvatures through the standard relationships

$$\begin{Bmatrix} N_x \\ N_\theta \\ N_{x\theta} \\ M_x \\ M_\theta \\ M_{x\theta} \end{Bmatrix} = \begin{bmatrix} A_{11} & A_{12} & A_{16} & B_{11} & B_{12} & B_{16} \\ A_{12} & A_{22} & A_{26} & B_{12} & B_{22} & B_{26} \\ A_{16} & A_{26} & A_{66} & B_{16} & B_{26} & B_{66} \\ B_{11} & B_{12} & B_{16} & D_{11} & D_{12} & D_{16} \\ B_{12} & B_{22} & B_{26} & D_{12} & D_{22} & D_{26} \\ B_{16} & B_{26} & B_{66} & D_{16} & D_{26} & D_{66} \end{bmatrix} \begin{Bmatrix} \varepsilon_x^o \\ \varepsilon_\theta^o \\ \gamma_{x\theta}^o \\ \kappa_x \\ \kappa_\theta \\ \kappa_{x\theta} \end{Bmatrix} \quad \text{and}$$

$$\begin{Bmatrix} Q_x \\ Q_\theta \end{Bmatrix} = \begin{bmatrix} A_{55} & A_{45} \\ A_{45} & A_{44} \end{bmatrix} \begin{Bmatrix} \gamma_{xz} \\ \gamma_{\theta z} \end{Bmatrix}$$

where the A_{ij}, B_{ij}, and D_{ij} terms are as defined by CLT in Chapter 6, and the strains and curvatures are defined by Equation (9.5). Using the displacement field representations of the strains and curvatures, the stress resultants are

$$N_x = A_{11}U_{o,x} + \frac{A_{12}}{R}\left(W_o + V_{o,\theta}\right) + A_{16}\left(\frac{U_{o,\theta}}{R} + V_{o,x}\right) + B_{11}\beta_{x,x} + \frac{B_{12}}{R}\beta_{\theta,\theta} + B_{16}\left(\frac{\beta_{x,\theta}}{R} + \beta_{\theta,x}\right)$$

$$N_\theta = A_{12}U_{o,x} + \frac{A_{22}}{R}\left(W_o + V_{o,\theta}\right) + A_{26}\left(\frac{U_{o,\theta}}{R} + V_{o,x}\right) + B_{12}\beta_{x,x} + \frac{B_{22}}{R}\beta_{\theta,\theta} + B_{26}\left(\frac{\beta_{x,\theta}}{R} + \beta_{\theta,x}\right)$$

$$N_{x\theta} = A_{16}U_{o,x} + \frac{A_{26}}{R}\left(W_o + V_{o,\theta}\right) + A_{66}\left(\frac{U_{o,\theta}}{R} + V_{o,x}\right) + B_{16}\beta_{x,x} + \frac{B_{26}}{R}\beta_{\theta,\theta} + B_{66}\left(\frac{\beta_{x,\theta}}{R} + \beta_{\theta,x}\right)$$

$$M_x = B_{11}U_{o,x} + \frac{B_{12}}{R}\left(W_o + V_{o,\theta}\right) + B_{16}\left(\frac{U_{o,\theta}}{R} + V_{o,x}\right) + D_{11}\beta_{x,x} + \frac{D_{12}}{R}\beta_{\theta,\theta} + D_{16}\left(\frac{\beta_{x,\theta}}{R} + \beta_{\theta,x}\right)$$

$$M_\theta = B_{12}U_{o,x} + \frac{B_{22}}{R}\left(W_o + V_{o,\theta}\right) + B_{26}\left(\frac{U_{o,\theta}}{R} + V_{o,x}\right) + D_{12}\beta_{x,x} + \frac{D_{22}}{R}\beta_{\theta,\theta} + D_{26}\left(\frac{\beta_{x,\theta}}{R} + \beta_{\theta,x}\right)$$

$$M_{x\theta} = B_{16}U_{o,x} + \frac{B_{26}}{R}\left(W_o + V_{o,\theta}\right) + B_{66}\left(\frac{U_{o,\theta}}{R} + V_{o,x}\right) + D_{16}\beta_{x,x} + \frac{D_{26}}{R}\beta_{\theta,\theta} + D_{66}\left(\frac{\beta_{x,\theta}}{R} + \beta_{\theta,x}\right)$$

$$Q_x = A_{55}\left(\beta_x + W_{o,x}\right) + A_{45}\left(\beta_\theta + \frac{W_{o,\theta} - V_o}{R}\right)$$

$$Q_\theta = A_{45}\left(\beta_x + W_{o,x}\right) + A_{44}\left(\beta_\theta + \frac{W_{o,\theta} - V_o}{R}\right)$$

Substituting into the equilibrium equation, we get a large set of equations as noted in Refs. [3] and [6]. Representing these in matrix form allows us to write

$$\begin{bmatrix} K_{11} & K_{12} & K_{13} & K_{14} & K_{15} \\ K_{21} & K_{22} & K_{23} & K_{24} & K_{25} \\ K_{31} & K_{32} & K_{33} & K_{34} & K_{35} \\ K_{41} & K_{42} & K_{43} & K_{44} & K_{45} \\ K_{51} & K_{52} & K_{53} & K_{54} & K_{55} \end{bmatrix} \begin{Bmatrix} U_o \\ V_o \\ \beta_x \\ \beta_\theta \\ W_o \end{Bmatrix} + \begin{Bmatrix} q_x \\ q_\theta \\ m_x \\ m_\theta \\ p \end{Bmatrix} = \begin{bmatrix} \overline{\rho}_1 & 0 & \overline{\rho}_2 & 0 & 0 \\ 0 & \overline{\rho}_1 & 0 & \overline{\rho}_2 & 0 \\ \overline{\rho}_2 & 0 & \overline{\rho}_3 & 0 & 0 \\ 0 & \overline{\rho}_2 & 0 & \overline{\rho}_3 & 0 \\ 0 & 0 & 0 & 0 & \overline{\rho}_1 \end{bmatrix} \begin{Bmatrix} \ddot{U}_o \\ \ddot{V}_o \\ \ddot{\beta}_x \\ \ddot{\beta}_\theta \\ \ddot{W}_o \end{Bmatrix} \tag{9.36}$$

The K_{ij} terms above are explicitly written as

$$K_{11} = A_{11}\frac{\partial^2}{\partial x^2} + 2\frac{A_{16}}{R}\frac{\partial^2}{\partial\theta\partial x} + \frac{A_{66}}{R^2}\frac{\partial^2}{\partial\theta^2}$$

$$K_{12} = A_{16}\frac{\partial^2}{\partial x^2} + (A_{12} + A_{66})\frac{1}{R}\frac{\partial^2}{\partial\theta\partial x} + \frac{A_{26}}{R^2}\frac{\partial^2}{\partial\theta^2}$$

$$K_{13} = B_{11}\frac{\partial^2}{\partial x^2} + 2\frac{B_{16}}{R}\frac{\partial^2}{\partial\theta\partial x} + \frac{B_{66}}{R^2}\frac{\partial^2}{\partial\theta^2}$$

$$K_{14} = B_{16}\frac{\partial^2}{\partial x^2} + (B_{12} + B_{66})\frac{1}{R}\frac{\partial^2}{\partial\theta\partial x} + \frac{B_{26}}{R^2}\frac{\partial^2}{\partial\theta^2}$$

$$K_{15} = \frac{A_{12}}{R}\frac{\partial}{\partial x} + \frac{A_{26}}{R^2}\frac{\partial}{\partial\theta}$$

$$K_{21} = A_{16}\frac{\partial^2}{\partial x^2} + (A_{12} + A_{66})\frac{1}{R}\frac{\partial^2}{\partial\theta\partial x} + \frac{A_{26}}{R^2}\frac{\partial^2}{\partial\theta^2}$$

$$K_{22} = A_{66}\frac{\partial^2}{\partial x^2} + 2\frac{A_{26}}{R}\frac{\partial^2}{\partial\theta\partial x} + \frac{A_{22}}{R^2}\frac{\partial^2}{\partial\theta^2} - \frac{A_{44}}{R^2}$$

$$K_{23} = B_{16}\frac{\partial^2}{\partial x^2} + (B_{12} + B_{66})\frac{1}{R}\frac{\partial^2}{\partial\theta\partial x} + \frac{B_{26}}{R^2}\frac{\partial^2}{\partial\theta^2} + \frac{A_{45}}{R}$$

$$K_{24} = B_{66}\frac{\partial^2}{\partial x^2} + 2B_{26}\frac{1}{R}\frac{\partial^2}{\partial\theta\partial x} + \frac{B_{22}}{R^2}\frac{\partial^2}{\partial\theta^2} + \frac{A_{44}}{R}$$

$$K_{25} = \left(\frac{A_{12} + A_{55}}{R}\right)\frac{\partial}{\partial x} + \left(\frac{A_{26} + A_{45}}{R^2}\right)\frac{\partial}{\partial\theta}$$

$$K_{31} = B_{11}\frac{\partial^2}{\partial x^2} + 2\frac{B_{16}}{R}\frac{\partial^2}{\partial\theta\partial x} + \frac{B_{66}}{R^2}\frac{\partial^2}{\partial\theta^2}$$

$$K_{32} = B_{16}\frac{\partial^2}{\partial x^2} + (B_{12} + B_{66})\frac{1}{R}\frac{\partial^2}{\partial\theta\partial x} + \frac{B_{26}}{R^2}\frac{\partial^2}{\partial\theta^2}$$

$$K_{33} = D_{11}\frac{\partial^2}{\partial x^2} + 2\frac{D_{16}}{R}\frac{\partial^2}{\partial\theta\partial x} + \frac{D_{66}}{R^2}\frac{\partial^2}{\partial\theta^2} - A_{55}$$

$$K_{34} = D_{16}\frac{\partial^2}{\partial x^2} + (D_{12} + D_{66})\frac{1}{R}\frac{\partial^2}{\partial\theta\partial x} + \frac{D_{26}}{R^2}\frac{\partial^2}{\partial\theta^2} - A_{45}$$

$$K_{35} = \left(\frac{B_{12}}{R} - A_{55}\right)\frac{\partial}{\partial x} + \left(\frac{B_{26}}{R} - A_{45}\right)\frac{\partial}{\partial\theta}$$

$$K_{41} = B_{16}\frac{\partial^2}{\partial x^2} + (B_{12} + B_{66})\frac{1}{R}\frac{\partial^2}{\partial\theta\partial x} + \frac{B_{26}}{R^2}\frac{\partial^2}{\partial\theta^2}$$

$$K_{42} = B_{66}\frac{\partial^2}{\partial x^2} + 2B_{26}\frac{1}{R}\frac{\partial^2}{\partial\theta\partial x} + \frac{B_{22}}{R^2}\frac{\partial^2}{\partial\theta^2} + \frac{A_{44}}{R}$$

$$K_{43} = D_{16}\frac{\partial^2}{\partial x^2} + (D_{12} + D_{66})\frac{1}{R}\frac{\partial^2}{\partial\theta\partial x} + \frac{D_{26}}{R^2}\frac{\partial^2}{\partial\theta^2} - A_{45}$$

$$K_{44} = D_{66}\frac{\partial^2}{\partial x^2} + 2\frac{D_{26}}{R}\frac{\partial^2}{\partial\theta\partial x} + \frac{D_{22}}{R^2}\frac{\partial^2}{\partial\theta^2} - A_{44}$$

$$K_{45} = \left(\frac{B_{22}}{R} - A_{45}\right)\frac{\partial}{\partial x} + \left(\frac{B_{22}}{R} - A_{44}\right)\frac{1}{R}\frac{\partial}{\partial\theta}$$

$$K_{51} = -\frac{A_{12}}{R}\frac{\partial}{\partial x} - \frac{A_{26}}{R^2}\frac{\partial}{\partial\theta}$$

$$K_{52} = -\left(\frac{A_{26} + A_{45}}{R}\right)\frac{\partial}{\partial x}\left(\frac{A_{22} + A_{44}}{R^2}\right)\frac{\partial}{\partial\theta}$$

$$K_{53} = \left(A_{55} - \frac{B_{12}}{R}\right)\frac{\partial}{\partial x} + \left(A_{45} - \frac{B_{26}}{R}\right)\frac{1}{R}\frac{\partial}{\partial\theta}$$

$$K_{54} = \left(A_{45} - \frac{B_{26}}{R}\right)\frac{\partial}{\partial x} + \left(A_{44} - \frac{B_{22}}{R}\right)\frac{1}{R}\frac{\partial}{\partial\theta}$$

$$K_{55} = A_{55}\frac{\partial^2}{\partial x^2} + 2\frac{A_{45}}{R}\frac{\partial^2}{\partial\theta\partial x} - \frac{A_{22}}{R^2}\frac{\partial^2}{\partial\theta^2}$$

The matrix form of the equations of motion lends themselves to numerical solution with currently available software. Alternately, Equation (9.36) can be expressed in an expanded form as

$$\begin{aligned} &A_{11}U_{o,xx} + 2\frac{A_{16}}{R}U_{o,\theta x} + \frac{A_{66}}{R^2}U_{o,\theta\theta} + A_{16}V_{o,xx} + \frac{(A_{12} + A_{66})}{R}V_{o,\theta x} + \frac{A_{26}}{R^2}V_{o,\theta\theta} \\ &+ B_{11}\beta_{x,xx} + 2\frac{B_{16}}{R}\beta_{x,\theta x} + \frac{B_{66}}{R^2}\beta_{x,\theta\theta} + B_{16}\beta_{\theta,xx} + \frac{(B_{12} + B_{66})}{R}\beta_{\theta,\theta x} + \frac{B_{26}}{R^2}\beta_{\theta,\theta\theta} \\ &+ \frac{A_{12}}{R}W_{o,x} + \frac{A_{26}}{R^2}W_{o,\theta} + q_x = \overline{\rho}_1\ddot{U}_o + \overline{\rho}_2\ddot{\beta}_x \end{aligned}$$

$$A_{16}U_{o,xx}+\frac{(A_{12}+A_{66})}{R}U_{o,\theta x}+\frac{A_{26}}{R^2}U_{o,\theta\theta}+A_{66}V_{o,xx}+2\frac{A_{26}}{R}V_{o,\theta x}+\frac{A_{22}}{R^2}V_{o,\theta\theta}$$
$$+B_{16}\beta_{x,xx}+\frac{(B_{12}+B_{66})}{R}\beta_{x,\theta x}+\frac{B_{26}}{R^2}\beta_{x,\theta\theta}+\frac{A_{45}}{R}\beta_x+B_{66}\beta_{\theta,xx}+2\frac{B_{26}}{R}\beta_{\theta,\theta x}$$
$$+\frac{B_{22}}{R^2}\beta_{\theta,\theta\theta}+\frac{A_{44}}{R}\beta_\theta+\frac{(A_{12}+A_{55})}{R}W_{o,x}+\frac{(A_{26}+A_{45})}{R^2}W_{o,\theta}+q_\theta=\overline{\rho}_1\ddot{V}_o+\overline{\rho}_2\ddot{\beta}_\theta$$

$$B_{11}U_{o,xx}+2\frac{B_{16}}{R}U_{o,\theta x}+\frac{B_{66}}{R^2}U_{o,\theta\theta}+B_{16}V_{o,xx}+\frac{(B_{12}+B_{66})}{R}V_{o,\theta x}+\frac{B_{26}}{R^2}V_{o,\theta\theta}$$
$$+D_{11}\beta_{x,xx}+2\frac{D_{16}}{R}\beta_{x,\theta x}+\frac{D_{66}}{R^2}\beta_{x,\theta\theta}-A_{55}\beta_x+D_{16}\beta_{\theta,xx}+\frac{(D_{12}+D_{66})}{R}\beta_{\theta,\theta x}$$
$$+\frac{D_{26}}{R^2}\beta_{\theta,\theta\theta}-A_{45}\beta_\theta+\left(\frac{B_{12}}{R}-A_{55}\right)W_{o,x}+\left(\frac{B_{26}}{R}-A_{45}\right)W_{o,\theta}+m_x=\overline{\rho}_2\ddot{U}_o+\overline{\rho}_3\ddot{\beta}_x$$

$$B_{16}U_{o,xx}+\frac{(B_{12}+B_{66})}{R}U_{o,\theta x}+\frac{B_{26}}{R^2}U_{o,\theta\theta}+B_{66}V_{o,xx}+2\frac{B_{26}}{R}V_{o,\theta x}+\frac{B_{22}}{R^2}V_{o,\theta\theta}$$
$$+\frac{A_{44}}{R}V_o+D_{16}\beta_{x,xx}+\frac{(D_{12}+D_{66})}{R}\beta_{x,\theta x}+\frac{D_{26}}{R^2}\beta_{x,\theta\theta}-A_{45}\beta_x+D_{66}\beta_{\theta,xx}$$
$$+2\frac{D_{26}}{R}\beta_{\theta,\theta x}+\frac{D_{22}}{R^2}\beta_{\theta,\theta\theta}-A_{44}\beta_\theta+\left(\frac{B_{22}}{R}-A_{45}\right)W_{o,x}+\left(\frac{B_{22}}{R}-A_{44}\right)W_{o,\theta}$$
$$+m_\theta=\overline{\rho}_2\ddot{V}_o+\overline{\rho}_3\ddot{\beta}_\theta$$

$$-\frac{A_{12}}{R}U_{o,x}-\frac{A_{26}}{R^2}U_{o,\theta}-\left(\frac{A_{26}+A_{45}}{R}\right)V_{o,x}+\left(\frac{A_{22}+A_{44}}{R^2}\right)V_{o,\theta}+\left(A_{55}-\frac{B_{12}}{R}\right)\beta_{x,x}$$
$$+\left(A_{45}-\frac{B_{26}}{R}\right)\left(\frac{1}{R}\right)\beta_{x,\theta}+\left(A_{45}-\frac{B_{26}}{R}\right)\beta_{\theta,x}+\left(A_{44}-\frac{B_{22}}{R}\right)\left(\frac{1}{R}\right)\beta_{\theta,\theta}+A_{55}W_{o,xx}$$
$$+2\frac{A_{45}}{R}W_{o,\theta x}-\frac{A_{22}}{R^2}W_{o,\theta\theta}+p=\overline{\rho}_1\ddot{W}_o$$

Both the matrix and expanded form of the equations of motion represent the most general case. Simplifications result for special cases, such as a symmetric laminates in which $[B]=0$ and $\overline{\rho}_2=0$. For this case, we get several terms within (9.35) to be zero, and several terms that reduce in complexity.

$$\begin{bmatrix} K_{11} & K_{12} & 0 & 0 & K_{15} \\ K_{21} & K_{22} & K_{23} & K_{24} & K_{25} \\ 0 & 0 & K_{33} & K_{34} & K_{35} \\ 0 & K_{42} & K_{43} & K_{44} & K_{45} \\ K_{51} & K_{52} & K_{53} & K_{54} & K_{55} \end{bmatrix} \begin{Bmatrix} U_o \\ V_o \\ \beta_x \\ \beta_\theta \\ W_o \end{Bmatrix} + \begin{Bmatrix} q_x \\ q_\theta \\ m_x \\ m_\theta \\ p \end{Bmatrix} = \begin{bmatrix} \overline{\rho}_1 & 0 & 0 & 0 & 0 \\ 0 & \overline{\rho}_1 & 0 & 0 & 0 \\ 0 & 0 & \overline{\rho}_3 & 0 & 0 \\ 0 & 0 & 0 & \overline{\rho}_3 & 0 \\ 0 & 0 & 0 & 0 & \overline{\rho}_1 \end{bmatrix} \begin{Bmatrix} \ddot{U}_o \\ \ddot{V}_o \\ \ddot{\beta}_x \\ \ddot{\beta}_\theta \\ \ddot{W}_o \end{Bmatrix}$$

The terms that reduce are

$$K_{23}=\frac{A_{45}}{R},\ K_{24}=\frac{A_{44}}{R},\ K_{35}=-A_{55}\frac{\partial}{\partial x}-A_{45}\frac{\partial}{\partial\theta},\ K_{42}=\frac{A_{44}}{R}$$

$$K_{45}=-A_{45}\frac{\partial}{\partial x}-\frac{A_{44}}{R}\frac{\partial}{\partial\theta},\ K_{53}=A_{55}\frac{\partial}{\partial x}+\frac{A_{45}}{R}\frac{\partial}{\partial\theta},\ K_{54}=A_{45}\frac{\partial}{\partial x}+\frac{A_{44}}{R}\frac{\partial}{\partial\theta}$$

Similarly, the equations of motion reduce to

$$A_{11}U_{o,xx}+2\frac{A_{16}}{R}U_{o,\theta x}+\frac{A_{66}}{R^2}U_{o,\theta\theta}+A_{16}V_{o,xx}+\frac{(A_{12}+A_{66})}{R}V_{o,\theta x}$$
$$+\frac{A_{26}}{R^2}V_{o,\theta\theta}+\frac{A_{12}}{R}W_{o,x}+\frac{A_{26}}{R^2}W_{o,\theta}+q_x=\overline{\rho}_1\ddot{U}_o$$

$$A_{16}U_{o,xx}+\frac{(A_{12}+A_{66})}{R}U_{o,\theta x}+\frac{A_{26}}{R^2}U_{o,\theta\theta}+A_{66}V_{o,xx}+2\frac{A_{26}}{R}V_{o,\theta x}+\frac{A_{22}}{R^2}V_{o,\theta\theta}$$
$$+\frac{A_{44}}{R}\beta_\theta+\frac{(A_{12}+A_{55})}{R}W_{o,x}+\frac{(A_{26}+A_{45})}{R^2}W_{o,\theta}+q_\theta=\overline{\rho}_1\ddot{V}_o$$

$$D_{11}\beta_{x,xx}+2\frac{D_{16}}{R}\beta_{x,\theta x}+\frac{D_{66}}{R^2}\beta_{x,\theta\theta}-A_{55}\beta_x+D_{16}\beta_{\theta,xx}+\frac{(D_{12}+D_{66})}{R}\beta_{\theta,\theta x}+\frac{D_{26}}{R^2}\beta_{\theta,\theta\theta}$$
$$-A_{45}\beta_\theta-A_{55}W_{o,x}-A_{45}W_{o,\theta}+m_x=\overline{\rho}_3\ddot{\beta}_x$$

$$\frac{A_{44}}{R}V_o+D_{16}\beta_{x,xx}+\frac{(D_{12}+D_{66})}{R}\beta_{x,\theta x}+\frac{D_{26}}{R^2}\beta_{x,\theta\theta}-A_{45}\beta_x+D_{66}\beta_{\theta,xx}+2\frac{D_{26}}{R}\beta_{\theta,\theta x}$$
$$+\frac{D_{22}}{R^2}\beta_{\theta,\theta\theta}-A_{44}\beta_\theta-A_{45}W_{o,x}-A_{44}W_{o,\theta}+m_\theta=\overline{\rho}_3\ddot{\beta}_\theta$$

$$-\frac{A_{12}}{R}U_{o,x}-\frac{A_{26}}{R^2}U_{o,\theta}-\left(\frac{A_{26}+A_{45}}{R}\right)V_{o,x}+\left(\frac{A_{22}+A_{44}}{R^2}\right)V_{o,\theta}+A_{55}\beta_{x,x}+\left(\frac{A_{45}}{R}\right)\beta_{x,\theta}$$
$$+A_{45}\beta_{\theta,x}+\left(\frac{A_{44}}{R}\right)\beta_{\theta,\theta}+A_{55}W_{o,xx}+2\frac{A_{45}}{R}W_{o,\theta x}-\frac{A_{22}}{R^2}W_{o,\theta\theta}+p=\overline{\rho}_1\ddot{W}_o$$

Obviously, either the equations of motion defined in Equation (9.35) or those reduced by symmetry ($[B]=0$ and $\overline{\rho}_2=0$) require numerical techniques such as finite elements [8–10,12] to obtain reliable solutions.

9.6 Problems

9.1 Consider a 0.125″ thick single-layer cylindrical shell made from a material with $E_1=30\times10^6$, $E_2=2.1\times10^6$, $G_{12}=1.2\times10^6$, and $\nu_{12}=0.25$. Shell radius to thickness ratios (R/h) being considered are of 10, 15, and 20. Plot the length vs. fiber orientation for which the shell can be treated as semi-infinite for each of these potential diameters.

9.2 A long laminated cylinder is subjected to internal pressure of $P=100$ psi and is fixed to a wall. All lamina within the laminate have the same fiber orientation θ. The vessel, a thickness of $h=0.375''$ and a 20″ diameter, is made from a material with $E_1=24\times10^6$,

$E_2 = 1.7 \times 10^6$, $G_{12} = 0.9 \times 10^6$, and $\nu_{12} = 0.30$. Determine and plot the moment and magnitude of the shear force at the fixed end of the cylinder as a function of ply orientation.

9.3 For the cylinder described in problem 9.2, plot the end displacement as a function of ply orientation for $x = 0.05,\ 0.10,\ 0.15''$.

9.4 Extend the results of Example 9.2 to model a distributed load as illustrated. Assume the distributed load shown on the section of pipe can be modeled as if it were a concentrated load of magnitude $P_x = p\mathrm{d}x$ and on a section dx and apply the solution of Example 9.2 to determine an expressions for.

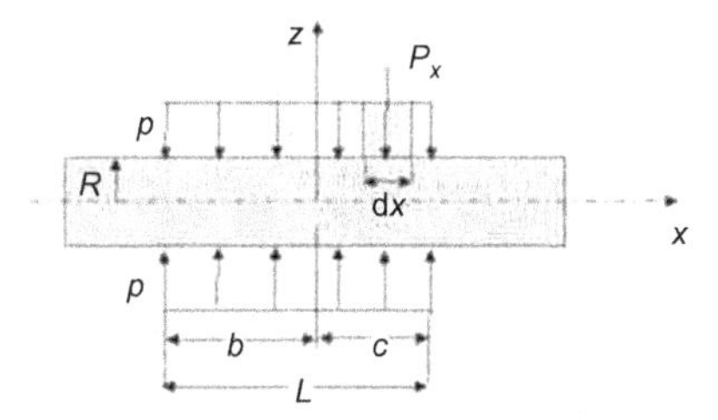

9.5 A unidirectional short cylindrical shell as shown in Figure 9.8 has a length of $L = 5\pi/8K$ is made from a material with $E_1 = 30 \times 10^6$, $E_2 = 2.1 \times 10^6$, $G_{12} = 1.2 \times 10^6$, and $\nu_{12} = 0.25$. The cylinder has a 24″ diameter, is 0.25″ thick, and is subjected to the end loads $M_A = M_o$, $M_B = 2M_o$, $Q_A = Q_o$, and $Q_B = Q_o$, where $M_o = 20$ and $Q_o = 10$. Plot the homogeneous deflection $(W_o(x))_h$ at a spamwise location of $x = 5\pi/16K$ for fiber orientations of $0 \leq \theta \leq 90°$.

9.6 For the cylindrical shell in problem 9.5, plot the homogeneous deflection $(W_o(x))_h$ at a spamwise location of $x = 5\pi/16K$ for fiber orientations of $0 \leq \theta \leq 90°$.

References

[1] Kuhn P. Stress in aircraft and shell structures. New York: McGraw-Hill; 1956.
[2] Novozhilov VV. The theory of thin shells. The Netherlands: Nordhoff; 1959.
[3] Ambartsumyan SA. Theory of anisotropic shells. NASA report TT F-118; 1964.
[4] Kraus H. Thin elastic shells. New York: John Wiley; 1967.
[5] Liberrescu L. Elastostatics and kinetics of anisotropic and heterogenous shell-type structures. The Netherlands: Nordhoff; 1975.
[6] Dong SB, Pister KS, Taylor RL. On the theory of laminated anisotropic shells and plates. J Aerospace Sci 1962;29:969–75.
[7] Dong SB, Tso KW. On a laminated orthotropic shell theory including transverse shear deformation. J Appl Mech 1976;39:1091–6.
[8] Owen DRJ, Figueiras JA. Anisotropic elasto-plastric finite element analysis of thick and thin plates and shells. Int J Numer Methods Eng 1983;19:541–66.
[9] Chao WC, Reddy JN. Analysis of laminated composite shells using a degenerated 3-D element. Int J Numer Methods Eng 1984;20:1991–2007.
[10] Carrera E. Theories and finite elements for multilayered, anisotropic, composite plates and shells. Arch Comput Methods Eng 2002;9:87–140.
[11] Reddy JN. Mechanics of laminated composite plates theory and analysis. New York: CRC Press; 1997.
[12] Reddy JN. Mechanics of laminated composite plates and shells. 2nd ed. New York: CRC Press; 2004.

Appendix A: Generalized transformations

The stress and strain transformations presented in Chapter 2 are valid only for rotations about the z-axis. Although this represents the most commonly used transformation from one coordinate system to another, it may not be general enough for all applications. Therefore, the stress and strain transformations applicable to arbitrary rotations are presented herein. It is assumed that x, y, and z represent the original coordinate system, and x', y', and z' the transformed coordinate system. Transformations from the unprimed to the primed system are assumed to be in accordance with the possible rotations shown in Figure A.1. Development of the general transformation equations is not presented, since they are well established and typically are available in many mathematics and engineering texts.

The direction cosines relating the primed and unprimed coordinate systems are presented next. As with conventional transformations, the designations l, m, and n are used to represent the direction cosines of the transformed axis with the original axis. The direction cosines defined here are similar to those used in defining the orientation of a vector in introductory statics courses. For example, if α, β, and γ are used to define the angles from the x-, y-, and z-axes, respectively, to a specified vector, the direction cosines are defined as:

$$l = \cos\alpha \quad m = \cos\beta \quad n = \cos\gamma$$

The difference between these direction cosines and those used for a general transformation is that three possibilities exist for each axis. The direction cosines relating each primed and unprimed axis are:

Direction cosines

	x	y	z
x'	l_1	m_1	n_1
y'	l_2	m_2	n_2
z'	l_3	m_3	n_3

Stress transformations from the unprimed to the primed system are defined in terms of a transformation matrix as:

$$\begin{Bmatrix} \sigma'_x \\ \sigma'_y \\ \sigma'_z \\ \tau'_{yz} \\ \tau'_{xz} \\ \tau'_{xy} \end{Bmatrix} = [T_\sigma] \begin{Bmatrix} \sigma_x \\ \sigma_y \\ \sigma_z \\ \tau_{yz} \\ \tau_{xz} \\ \tau_{xy} \end{Bmatrix}$$

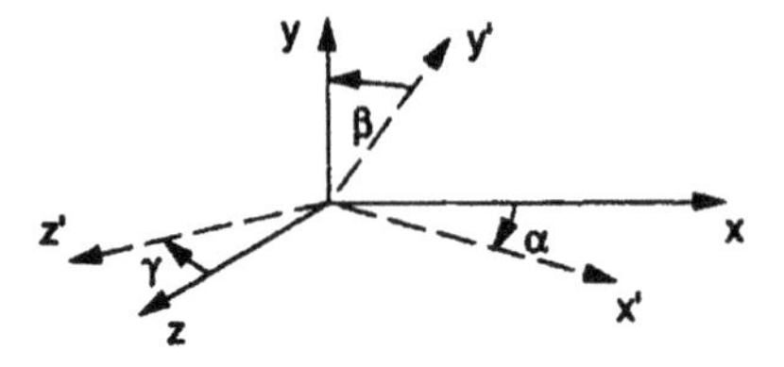

Figure A.1 General coordinate rotations.

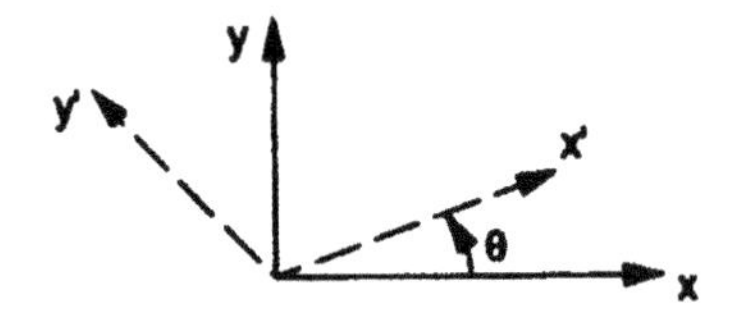

Figure A.2 Coordinate rotations about the z-axis.

where $[T_\sigma]$ is defined in terms of direction cosines as:

$$[T_\sigma]=\begin{bmatrix} l_1^2 & m_1^2 & n_1^2 & 2m_1n_1 & 2n_1l_1 & 2l_1m_1 \\ l_2^2 & m_2^2 & n_2^2 & 2m_2n_2 & 2n_2l_2 & 2l_2m_2 \\ l_3^2 & m_3^2 & n_3^2 & 2m_3n_3 & 2n_3l_3 & 2l_3m_3 \\ l_1l_3 & m_1m_3 & n_1n_3 & (m_1n_3+m_3n_1) & (l_1n_3+l_3n_1) & (l_1m_3+l_3m_1) \\ l_2l_3 & m_2m_3 & n_2n_3 & (m_2n_3+m_3n_2) & (l_2n_3+l_3n_2) & (l_2m_3+l_3m_2) \\ l_1l_2 & m_1m_2 & n_1n_2 & (m_1n_2+m_2n_1) & (l_1n_2+l_2n_1) & (l_1m_2+l_2m_1) \end{bmatrix}$$

For a case of plane stress where $\sigma_z=\tau_{yz}=\tau_{xz}$ and all rotations are about the z-axis (as shown in Figure A.2), the direction cosines are $l_1=m_2=\cos\theta$, $m_1=\sin\theta$, $l_2=-\sin\theta$, $n_3=1$, and $n_1=n_2=l_3=m_3=0$. Using these, the stress transformation matrix in Chapter 2 results:

$$\begin{Bmatrix} \sigma'_x \\ \sigma'_y \\ \tau'_{xy} \end{Bmatrix}=\begin{bmatrix} \cos^2\theta & \sin^2\theta & 2\sin\theta\cos\theta \\ \sin^2\theta & \cos^2\theta & -2\sin\theta\cos\theta \\ -\sin\theta\cos\theta & \sin\theta\cos\theta & \cos^2\theta-\sin^2\theta \end{bmatrix}\begin{Bmatrix} \sigma_x \\ \sigma_y \\ \tau_{xy} \end{Bmatrix}$$

The generalized form of strain transformation is similar to that of stress transformation, except that $[T_\sigma]$ is replaced by $[T_\varepsilon]$ and is:

$$\begin{Bmatrix} \varepsilon'_x \\ \varepsilon'_y \\ \varepsilon'_z \\ \gamma'_{yz} \\ \gamma'_{xz} \\ \gamma'_{xy} \end{Bmatrix}=[T_\varepsilon]\begin{Bmatrix} \varepsilon_x \\ \varepsilon_y \\ \varepsilon_z \\ \gamma_{yz} \\ \gamma_{xz} \\ \gamma_{xy} \end{Bmatrix}$$

Assuming that strains are allowed to be transformed via coordinate axes as defined by Figure A.1, $[T_\varepsilon]$ is expressed as:

$$[T_\varepsilon] = \begin{bmatrix} l_1^2 & m_1^2 & n_1^2 & 2m_1n_1 & 2n_1l_1 & 2l_1m_1 \\ l_2^2 & m_2^2 & n_2^2 & 2m_2n_2 & 2n_2l_2 & 2l_2m_2 \\ l_3^2 & m_3^2 & n_3^2 & 2m_3n_3 & 2n_3l_3 & 2l_3m_3 \\ 2l_1l_3 & 2m_1m_3 & 2n_1n_3 & 2(m_1n_3+m_3n_1) & 2(l_1n_3+l_3n_1) & 2(l_1m_3+l_3m_1) \\ 2l_2l_3 & 2m_2m_3 & 2n_2n_3 & 2(m_2n_3+m_3n_2) & 2(l_2n_3+l_3n_2) & 2(l_2m_3+l_3m_2) \\ 2l_1l_2 & 2m_1m_2 & 2n_1n_2 & 2(m_1n_2+m_2n_1) & 2(l_1n_2+l_2n_1) & 2(l_1m_2+l_2m_1) \end{bmatrix}$$

As demonstrated for the case of plane stress, a simplification of this matrix is also possible. The result, using direction cosines previously defined for the plane stress case, is a strain transformation matrix identical to that defined in Chapter 2.

Appendix B: Summary of useful equations

B.1 Lamina

Stress and strain transformations:

$$\begin{Bmatrix} \varepsilon_x' \\ \varepsilon_y' \\ \gamma_{xy}' \end{Bmatrix} = [T_\varepsilon] \begin{Bmatrix} \varepsilon_x \\ \varepsilon_y \\ \gamma_{xy} \end{Bmatrix} \quad \begin{Bmatrix} \sigma_x' \\ \sigma_y' \\ \tau_{xy}' \end{Bmatrix} = [T_\sigma] \begin{Bmatrix} \sigma_x \\ \sigma_y \\ \tau_{xy} \end{Bmatrix}$$

where

$$[T_\varepsilon] = \begin{bmatrix} m^2 & n^2 & mn \\ n^2 & m^2 & -mn \\ -2mn & 2mn & m^2 - n^2 \end{bmatrix} \tag{2.1}$$

$$[T_\sigma] = \begin{bmatrix} m^2 & n^2 & 2mn \\ n^2 & m^2 & -2mn \\ -mn & mn & m^2 - n^2 \end{bmatrix} \tag{2.3}$$

and $m = \cos\theta$ and $n = \sin\theta$.

Off-axis stress–strain relationships:

$$\begin{Bmatrix} \sigma_x \\ \sigma_y \\ \sigma_z \\ \tau_{yz} \\ \tau_{xz} \\ \tau_{xy} \end{Bmatrix} = \begin{bmatrix} \bar{Q}_{11} & \bar{Q}_{12} & \bar{Q}_{13} & 0 & 0 & \bar{Q}_{16} \\ \bar{Q}_{12} & \bar{Q}_{22} & \bar{Q}_{23} & 0 & 0 & \bar{Q}_{26} \\ \bar{Q}_{13} & \bar{Q}_{23} & \bar{Q}_{33} & 0 & 0 & \bar{Q}_{36} \\ 0 & 0 & 0 & \bar{Q}_{44} & \bar{Q}_{45} & 0 \\ 0 & 0 & 0 & \bar{Q}_{45} & \bar{Q}_{55} & 0 \\ \bar{Q}_{16} & \bar{Q}_{26} & \bar{Q}_{36} & 0 & 0 & \bar{Q}_{66} \end{bmatrix} \begin{Bmatrix} \varepsilon_x \\ \varepsilon_y \\ \varepsilon_z \\ \gamma_{yz} \\ \gamma_{xz} \\ \gamma_{xy} \end{Bmatrix} \tag{3.7}$$

where

$$\begin{aligned}
\bar{Q}_{11} &= Q_{11}m^4 + 2(Q_{12} + 2Q_{66})m^2n^2 + Q_{22}n^4 \\
\bar{Q}_{12} &= (Q_{11} + Q_{22} - 4Q_{66})m^2n^2 + Q_{12}(m^4 + n^4) \\
\bar{Q}_{13} &= Q_{13}m^2 + Q_{23}n^2 \\
\bar{Q}_{16} &= -Q_{22}mn^3 + Q_{11}m^3n - (Q_{12} + 2Q_{66})mn(m^2 - n^2) \\
\bar{Q}_{22} &= Q_{11}n^4 + 2(Q_{12} + 2Q_{66})m^2n^2 + Q_{22}m^4 \\
\bar{Q}_{23} &= Q_{13}n^2 + Q_{23}m^2 \\
\bar{Q}_{26} &= -Q_{22}m^3n + Q_{11}mn^3 + (Q_{12} + 2Q_{66})mn(m^2 - n^2) \\
\bar{Q}_{33} &= Q_{33} \\
\bar{Q}_{36} &= (Q_{13} - Q_{23})mn \\
\bar{Q}_{44} &= Q_{44}m^2 + Q_{55}n^2 \\
\bar{Q}_{45} &= (Q_{55} - Q_{44})mn \\
\bar{Q}_{55} &= Q_{55}m^2 + Q_{44}n^2 \\
\bar{Q}_{66} &= (Q_{11} + Q_{22} - 2Q_{12})m^2n^2 + Q_{66}(m^2 - n^2)^2
\end{aligned} \tag{3.8}$$

Plane stress stiffness matrix:

$$[Q] = \begin{bmatrix} Q_{11} & Q_{12} & 0 \\ Q_{12} & Q_{22} & 0 \\ 0 & 0 & Q_{66} \end{bmatrix} \tag{3.9}$$

$$\begin{aligned}
Q_{11} &= E_1/(1 - \nu_{12}\nu_{21}) \\
Q_{22} &= E_2/(1 - \nu_{12}\nu_{21}) \\
Q_{12} &= \nu_{12}E/(1 - \nu_{12}\nu_{21}) \\
&= \nu_{21}E_1/(1 - \nu_{12}\nu_{21}) \\
Q_{66} &= G_{12}
\end{aligned}$$

Off-axis relationship:

$$\begin{Bmatrix} \sigma_x \\ \sigma_y \\ \tau_{xy} \end{Bmatrix} = \begin{bmatrix} \bar{Q}_{11} & \bar{Q}_{12} & \bar{Q}_{16} \\ \bar{Q}_{12} & \bar{Q}_{22} & \bar{Q}_{26} \\ \bar{Q}_{16} & \bar{Q}_{26} & \bar{Q}_{66} \end{bmatrix} \begin{Bmatrix} \varepsilon_x \\ \varepsilon_y \\ \gamma_{xy} \end{Bmatrix} \tag{3.10}$$

Plane stress compliance matrix:

$$\begin{Bmatrix} \varepsilon_1 \\ \varepsilon_2 \\ \gamma_{12} \end{Bmatrix} = \begin{bmatrix} S_{11} & S_{12} & 0 \\ S_{12} & S_{22} & 0 \\ 0 & 0 & S_{66} \end{bmatrix} \begin{Bmatrix} \sigma_1 \\ \sigma_2 \\ \tau_{12} \end{Bmatrix} \tag{3.13}$$

$$\begin{aligned}
S_{11} &= 1/E_1 \\
S_{12} &= -\nu_{12}/E_1 = -\nu_{21}/E_2 \\
S_{22} &= 1/E_2 \\
S_{66} &= 1/G_{12}
\end{aligned} \tag{3.14}$$

Off-axis strain–stress relationship:

$$\begin{Bmatrix} \varepsilon_x \\ \varepsilon_y \\ \gamma_{xy} \end{Bmatrix} = \begin{bmatrix} \bar{S}_{11} & \bar{S}_{12} & \bar{S}_{16} \\ \bar{S}_{12} & \bar{S}_{22} & \bar{S}_{26} \\ \bar{S}_{16} & \bar{S}_{26} & \bar{S}_{66} \end{bmatrix} \begin{Bmatrix} \sigma_x \\ \sigma_y \\ \tau_{xy} \end{Bmatrix} \tag{3.15}$$

where

$$\begin{aligned} \bar{S}_{11} &= S_{11}m^4 + (2S_{12}+S_{66})m^2n^2 + S_{22}n^4 \\ \bar{S}_{12} &= (S_{11}+S_{22}-S_{66})m^2n^2 + S_{12}(m^4+n^4) \\ \bar{S}_{16} &= (2S_{11}-2S_{12}-S_{66})m^3n - (2S_{22}-2S_{12}-S_{66})mn^3 \\ \bar{S}_{22} &= S_{11}n^4 + (2S_{12}+S_{66})m^2n^2 + S_{22}m^4 \\ \bar{S}_{26} &= (2S_{11}-2S_{12}-S_{66})mn^3 - (2S_{22}-2S_{12}-S_{66})m^3n \\ \bar{S}_{66} &= 2(2S_{11}+2S_{22}-4S_{12}-S_{66})m^2n^2 + S_{66}(m^4+n^4) \end{aligned} \tag{3.16}$$

Relationships between on-axis and off-axis properties (3.19):

$$\frac{1}{E_x} = \frac{m^4}{E_1} + \left(\frac{1}{G_{12}} - \frac{2v_{12}}{E_1}\right)m^2n^2 + \frac{n^4}{E_2}$$

$$\frac{1}{E_y} = \frac{n^4}{E_1} + \left(\frac{1}{G_{12}} - \frac{2v_{12}}{E_1}\right)m^2n^2 + \frac{m^4}{E_2}$$

$$\frac{1}{G_{xy}} = 2m^2n^2\left(\frac{2}{E_1} + \frac{2}{E_2} + \frac{4v_{12}}{E_1} - \frac{1}{G_{12}}\right) + \frac{(n^4+m^4)}{G_{12}}$$

$$v_{xy} = E_x\left[\frac{v_{12}(n^4+m^4)}{E_1} - m^2n^2\left(\frac{1}{E_1} + \frac{1}{E_2} - \frac{1}{G_{12}}\right)\right]$$

$$\eta_{xy,x} = E_x\left[C_1m^3n - C_2mn^3\right] \quad \eta_{xy,x} = E_x\left[C_1mn^3 - C_2nm^3\right]$$

where

$$C_1 = \frac{2}{E_1} + \frac{2v_{12}}{E_1} - \frac{1}{G_{12}} \quad C_2 = \frac{2}{E_2} + \frac{2v_{12}}{E_1} - \frac{1}{G_{12}}$$

Thermal and hygral effects for plane stress applications:

$$\begin{aligned} \alpha_x &= m^2\alpha_1 + n^2\alpha_2 \\ \alpha_y &= n^2\alpha_1 + m^2\alpha_2 \\ \alpha_z &= \alpha_3 \\ \alpha_{xy} &= 2mn(\alpha_1 - \alpha_2) \end{aligned} \tag{3.22}$$

$$\begin{aligned} \beta_x &= m^2\beta_1 + n^2\beta_2 \\ \beta_y &= n^2\beta_1 + m^2\beta_2 \\ \beta_z &= \beta_3 \\ \beta_{xy} &= 2mn(\beta_1 - \beta_2) \end{aligned} \tag{3.31}$$

On-axis stress–strain relations including thermal and hygral effects:

$$\begin{Bmatrix}\sigma_1\\\sigma_2\\\sigma_6\end{Bmatrix}=\begin{bmatrix}Q_{11}&Q_{12}&0\\Q_{12}&Q_{22}&0\\0&0&Q_{66}\end{bmatrix}\left(\begin{Bmatrix}\varepsilon_1\\\varepsilon_2\\\lambda_{12}\end{Bmatrix}-\begin{Bmatrix}\alpha_1\\\alpha_2\\0\end{Bmatrix}\Delta T-\begin{Bmatrix}\beta_1\\\beta_2\\0\end{Bmatrix}\bar{M}\right) \tag{3.34}$$

Off-axis stress–strain relations including thermal and hygral effects:

$$\begin{Bmatrix}\sigma_x\\\sigma_y\\\tau_{xy}\end{Bmatrix}=\begin{bmatrix}\bar{Q}_{11}&\bar{Q}_{12}&\bar{Q}_{16}\\\bar{Q}_{12}&\bar{Q}_{22}&\bar{Q}_{26}\\\bar{Q}_{16}&\bar{Q}_{26}&\bar{Q}_{66}\end{bmatrix}\left(\begin{Bmatrix}\varepsilon_x\\\varepsilon_y\\\gamma_{xy}\end{Bmatrix}-\begin{Bmatrix}\alpha_x\\\alpha_y\\\alpha_{xy}\end{Bmatrix}\Delta T-\begin{Bmatrix}\beta_x\\\beta_y\\\beta_{xy}\end{Bmatrix}\bar{M}\right) \tag{3.35}$$

Rule-of-mixtures approximations:

$$E_1=E_f\upsilon_f+E_m\upsilon_m \tag{3.36}$$

$$E_2=\frac{E_fE_m}{E_m\upsilon_f+E_f\upsilon_m} \tag{3.37}$$

$$G_{12}=\frac{G_fG_m}{\upsilon_fG_m+\upsilon_mG_f} \tag{3.38}$$

$$\nu_{12}=\upsilon_f\nu_f+\upsilon_m\nu_m \tag{3.39}$$

B.2 Failure theories

$$\text{Tsai–Hill}: \sigma_1^2-\sigma_1\sigma_2+\left(\frac{X}{Y}\right)^2\sigma_2^2+\left(\frac{X}{S}\right)^2\tau_{12}^2=X^2 \tag{5.5}$$

$$\text{Tsai–Wu}: F_{11}\sigma_1^2+2F_{12}\sigma_1\sigma_2+F_{22}\sigma_2^2+F_{66}\tau_{12}^2+F_1\sigma_1+F_2\sigma_2=1 \tag{5.7}$$

B.3 Classical lamination theory

$$\begin{aligned}[A_{ij}]&=\sum_{k=1}^{N}[\bar{Q}_{ij}]_k t_k\\ [B_{ij}]&=\sum_{k=1}^{N}[\bar{Q}_{ij}]_k t_k\bar{z}_k\\ [D_{ij}]&=\sum_{k=1}^{N}[\bar{Q}_{ij}]_k\left(t_k\bar{z}_k^2+\frac{t_k^3}{12}\right)\end{aligned} \tag{6.22}$$

For transverse shear, the analogous expression is:

$$[A_{ij}] = c\sum_{k=1}^{N} [\bar{Q}_{ij}]_k t_k \left\{1 + \frac{4}{h^2}\left(\bar{z}_k^2 + \frac{t_k^3}{12}\right)\right\} \tag{6.23}$$

Thermal and hygral effects:

$$\{N^{\mathrm{T}}\} = \begin{Bmatrix} N_x^{\mathrm{T}} \\ N_y^{\mathrm{T}} \\ N_{xy}^{\mathrm{T}} \end{Bmatrix} = \sum_{k=1}^{N} [\bar{Q}]_k \begin{Bmatrix} \alpha_x \\ \alpha_y \\ \alpha_{xy} \end{Bmatrix}_k \Delta T t_k \tag{6.25}$$

$$\{M^{\mathrm{T}}\} = \begin{Bmatrix} M_x^{\mathrm{T}} \\ M_y^{\mathrm{T}} \\ M_{xy}^{\mathrm{T}} \end{Bmatrix} = \sum_{k=1}^{N} [\bar{Q}]_k \begin{Bmatrix} \alpha_x \\ \alpha_y \\ \alpha_{xy} \end{Bmatrix}_k \Delta T (t_k \bar{z}_k) \tag{6.26}$$

$$\{N^{\mathrm{H}}\} = \begin{Bmatrix} N_x^{\mathrm{H}} \\ N_y^{\mathrm{H}} \\ N_{xy}^{\mathrm{H}} \end{Bmatrix} = \sum_{k=1}^{N} [\bar{Q}]_k \begin{Bmatrix} \beta_x \\ \beta_y \\ \beta_{xy} \end{Bmatrix}_k \bar{M} t_k \tag{6.29}$$

$$\{M^{\mathrm{H}}\} = \begin{Bmatrix} M_x^{\mathrm{H}} \\ M_y^{\mathrm{H}} \\ M_{xy}^{\mathrm{H}} \end{Bmatrix} = \sum_{k=1}^{N} [\bar{Q}]_k \begin{Bmatrix} \beta_x \\ \beta_y \\ \beta_{xy} \end{Bmatrix}_k \bar{M} (t_k \bar{z}_k) \tag{6.30}$$

Load, strain–curvature, and $[A]$, $[B]$, $[D]$ relationships:

$$\begin{aligned} \{\hat{N}\} &= \{N\} + \{N^{\mathrm{T}}\} + \{N^{\mathrm{H}}\} \\ \{\hat{M}\} &= \{M\} + \{M^{\mathrm{T}}\} + \{M^{\mathrm{H}}\} \end{aligned} \tag{6.32}$$

$$\begin{Bmatrix} \hat{N} \\ --- \\ \hat{M} \end{Bmatrix} = \begin{bmatrix} A & | & B \\ -- & -- & -- \\ B & | & D \end{bmatrix} \begin{Bmatrix} \varepsilon^0 \\ --- \\ \kappa \end{Bmatrix} \tag{6.33}$$

$$\begin{Bmatrix} \varepsilon^0 \\ --- \\ \kappa \end{Bmatrix} = \begin{bmatrix} A' & | & B' \\ -- & -- & -- \\ C' & | & D' \end{bmatrix} \begin{Bmatrix} \hat{N} \\ --- \\ \hat{M} \end{Bmatrix} \tag{6.35}$$

Off-axis stresses in k-th lamina:

$$\begin{Bmatrix} \sigma_x \\ \sigma_y \\ \tau_{xy} \end{Bmatrix}_k = [\bar{Q}]_k \left(\begin{Bmatrix} \varepsilon_x^0 \\ \varepsilon_y^0 \\ \gamma_{xy}^0 \end{Bmatrix} + z \begin{Bmatrix} \kappa_x \\ \kappa_y \\ \kappa_{xy} \end{Bmatrix} - \begin{Bmatrix} \alpha_x \\ \alpha_y \\ \alpha_{xy} \end{Bmatrix}_k \Delta T - \begin{Bmatrix} \beta_x \\ \beta_y \\ \beta_{xy} \end{Bmatrix}_k \bar{M} \right) \tag{6.36}$$

On-axis strains and k-th-layer stresses:

$$\begin{Bmatrix} \varepsilon_1 \\ \varepsilon_2 \\ \gamma_{12} \end{Bmatrix}_k = [T_\varepsilon] \begin{Bmatrix} \varepsilon_x \\ \varepsilon_y \\ \gamma_{xy} \end{Bmatrix}_k = [T_\varepsilon] \left(\begin{Bmatrix} \varepsilon_x^0 \\ \varepsilon_y^0 \\ \gamma_{xy}^0 \end{Bmatrix} + z \begin{Bmatrix} \kappa_x \\ \kappa_y \\ \kappa_{xy} \end{Bmatrix} \right) \tag{6.37}$$

$$\begin{Bmatrix} \sigma_1 \\ \sigma_2 \\ \tau_{12} \end{Bmatrix}_k = [Q]_k \begin{Bmatrix} \varepsilon_1 - \alpha_1 \Delta T - \beta_1 \bar{M} \\ \varepsilon_2 - \alpha_2 \Delta T - \beta_2 \bar{M} \\ \gamma_{12} \end{Bmatrix}_k \tag{6.38}$$

B.4 Laminated composite beam analysis

Elementary beam theory:

$$\begin{aligned} &\frac{dN_x}{dx} + \tau_{1x} - \tau_{2x} = 0 \\ &\frac{dM_x}{dx} - Q_x + \frac{h}{2}(\tau_{1x} - \tau_{2x}) = 0 \\ &\frac{dQ_x}{dx} + p_1 - p_2 = 0 \end{aligned} \tag{7.9}$$

$$(\sigma_x)_k = (\bar{Q}_{11})_k \left(\frac{dU_o}{dx} - z \frac{d^2 W_o}{dx^2} \right) \tag{7.15}$$

Advanced beam theory:

$$(\sigma_x)_k = (E_x)_k \{\varepsilon_x\}_k \tag{7.16}$$

$$\frac{1}{(E_x)_k} = \frac{\cos^4\theta_k}{(E_1)_k} + \left(\frac{1}{G_{12}} - \frac{2\nu_{12}}{E_1} \right)_k \sin^2\theta_k \cos^2\theta_k + \frac{\sin^4\theta_k}{(E_2)_k} \tag{7.17}$$

$$(\varepsilon_x)_k = \varepsilon_x^o + z\kappa_x - (\alpha_x)_k \Delta T - (\beta_x)_k \bar{M} \tag{7.18}$$

$$N_x = A_x \varepsilon_x^o + B_x \kappa_x - N_x^{\mathrm{T}} - N_x^{\mathrm{H}} \tag{7.19}$$

$$M_x = B_x \varepsilon_x^o + D_x \kappa_x - M_x^{\mathrm{T}} - M_x^{\mathrm{H}} \tag{7.20}$$

$$Q_x = \sum_{k=1}^{N} \int_{h_{k-1}}^{h_k} (\tau_{xz})_k b_k dz \tag{7.21}$$

$$A_x = \sum_{k=1}^{N} (bE_x)_k t_k, \quad B_x = \sum_{k=1}^{N} (bE_x)_k t_k \bar{z}_k, \quad D_x = \sum_{k=1}^{N} (bE_x)_k \left(t_k \bar{z}_k^2 + \frac{t_k^3}{12} \right) \tag{7.22}$$

$$N_x^{\mathrm{T}} = \sum_{k=1}^{N} (bE_x\alpha_x)_k t_k \Delta T \quad M_x^{\mathrm{T}} = \sum_{k=1}^{N} (bE_x\alpha_x)_k t_k \bar{z}_k \Delta T$$
$$N_x^{\mathrm{H}} = \sum_{k=1}^{N} (bE_x\beta_x)_k t_k \bar{M} \quad M_x^{\mathrm{H}} = \sum_{k=1}^{N} (bE_x\beta_x)_k t_k \bar{z}_k \bar{M} \tag{7.23}$$

$$N_{x,x} + p_x(x) = \overline{\rho}_1 \ddot{U}_o + \overline{\rho}_2 \ddot{\Phi} \tag{7.24}$$

$$Q_{x,x} + q(x) = \overline{\rho}_1 \ddot{W}_o \tag{7.25}$$

$$M_{x,x} - Q_x + m_x(x) = \overline{\rho}_2 \ddot{U}_o + \overline{\rho}_3 \ddot{\Phi} \tag{7.26}$$

$$A_x U_{0,xx} + B_x W_{o,xxx} + p_x(x) - N_{x,x}^{\mathrm{T}} - N_{x,x}^{\mathrm{H}} = \overline{\rho}_1 \ddot{U}_o + \overline{\rho}_2 \ddot{\Phi} \tag{7.30}$$

$$\begin{aligned} B_x U_{o,xxx} - D_x W_{o,xxxx} + m_x(x)_{,x} + q(x) - M_{x,xx}^{\mathrm{T}} - M_{x,xx}^{\mathrm{H}} \\ = \overline{\rho}_1 \ddot{W}_o + \overline{\rho}_2 \ddot{U}_{o,x} + \overline{\rho}_3 \ddot{W}_{o,xx} \end{aligned} \tag{7.31}$$

$$\bar{A}_x \equiv \left(\frac{A_x D_x - B_x^2}{D_x} \right) \quad \bar{B}_x \equiv \left(\frac{A_x D_x - B_x^2}{B_x} \right) \quad \bar{D}_x \equiv \left(\frac{A_x D_x - B_x^2}{A_x} \right) \tag{7.32}$$

$$U_{o,xxx} = \frac{1}{\bar{B}_x} \left[q(x) + m_x(x)_{,x} - M_{x,xx}^{\mathrm{T}} - M_{x,xx}^{\mathrm{H}} \right] - \frac{1}{\bar{A}_x} \left[p_x(x)_{,x} - N_{x,xx}^{\mathrm{T}} - N_{x,xx}^{\mathrm{H}} \right] \tag{7.33}$$

$$W_{o,xxxx} = \frac{1}{\bar{D}_x} \left[q(x) + m_x(x)_{,x} - M_{x,xx}^{\mathrm{T}} - M_{x,xx}^{\mathrm{H}} \right] - \frac{1}{\bar{B}_x} \left[p_x(x)_{,x} - N_{x,xx}^{\mathrm{T}} - N_{x,xx}^{\mathrm{H}} \right] \tag{7.34}$$

$$(\sigma_x)_k = (E_x)_k \left[U_{0,x} - zW_{o,xx} - (\alpha_x)_k \Delta T - (\beta_x)_k \bar{M} \right] \tag{7.35}$$

$$(\tau_{xz})_k = \frac{1}{b_k} \int_{h_o}^{z} (\sigma_{x,x}) \mathrm{d}A \tag{7.36}$$

Beams with shear deformation:

$$\begin{aligned} &A_x U_{o,xx} + B_x \Phi_{,xx} - N_{x,x}^{\mathrm{T}} - N_{x,x}^{\mathrm{H}} + p_x(x) = 0 \\ &B_x U_{o,xx} + D_x \Phi_{,xx} - M_{x,x}^{\mathrm{T}} - M_{x,x}^{\mathrm{H}} - Q_x + m_x(x) = 0 \\ &KA_{55} \left(\Phi_{,x} + W_{o,xx} \right) + q(x) = 0 \end{aligned} \tag{7.37}$$

$$\Phi_{,xxx} = -\frac{q(x)}{\bar{D}_x} + \frac{p_x(x)_{,x}}{\bar{B}_x} - \frac{m(x)_{,x}}{\bar{D}_x} - \frac{N_{x,xx}^{\mathrm{T}} + N_{x,xx}^{\mathrm{H}}}{\bar{B}_x} + \frac{M_{x,xx}^{\mathrm{T}} + M_{x,xx}^{\mathrm{H}}}{\bar{D}_x} \tag{7.38}$$

$$W_{0,xxxx} = \frac{q(x)}{\bar{D}_x} - \frac{q(x)_{,xx}}{KA_{55}} - \frac{p_x(x)_{,x}}{\bar{B}_x} + \frac{m(x)_{,x}}{\bar{D}_x} + \frac{N^{\mathrm{T}}_{x,xx} + N^{\mathrm{H}}_{x,xx}}{\bar{B}_x} - \frac{M^{\mathrm{T}}_{x,xx} + M^{\mathrm{H}}_{x,xx}}{\bar{D}_x} \tag{7.39}$$

$$U_{0,xxx} = -\frac{p_x(x)_{,x}}{\bar{A}_x} + \frac{q(x)}{\bar{B}_x} + \frac{m_x(x)_{,x}}{\bar{B}_x} + \frac{N^{\mathrm{T}}_{x,xx} + N^{\mathrm{H}}_{x,xx}}{\bar{A}_x} - \frac{M^{\mathrm{T}}_{x,xx} + M^{\mathrm{H}}_{x,xx}}{\bar{B}_x} \tag{7.40}$$

Buckling:

$$\bar{D}_x \left[\frac{P}{KA_{55}} - 1 \right] W_{o,xxxx} - PW_{o,xx} = \frac{\bar{D}_x}{KA_{55}} q(x)_{,xx} - q(x) \tag{7.45}$$

$$\lambda^2 = \frac{P/\bar{D}_x}{1 - P/KA_{55}} \tag{7.46}$$

$$W_{o,xxxx} + \lambda^2 W_{o,xx} = \frac{q(x)_{,xx}}{P - KA_{55}} - \frac{q(x)/\bar{D}_x}{1 - P/KA_{55}} \tag{7.47}$$

$$P_{CR} = \frac{\bar{D}_x n^2 \pi^2}{L^2 + \bar{D}_x n^2 \pi^2 / KA_{55}} \tag{7.48}$$

Curved rings and beams:

$$\varepsilon_\theta = \frac{1}{R+z} \left(\frac{\partial U_o}{\partial \theta} + z \frac{\partial \beta}{\partial \theta} + W_o \right), \quad \gamma_{r\theta} = \frac{1}{R+z} \left(\frac{\partial W_o}{\partial \theta} - U_o + R\beta \right) \tag{7.49}$$

$$\varepsilon^o_\theta = \frac{1}{R} \left(\frac{\partial U_o}{\partial \theta} + W_o \right), \quad \kappa_\theta = \frac{1}{R} \frac{\partial \beta}{\partial \theta} \tag{7.50}$$

$$\varepsilon_\theta = \frac{1}{1 + z/R} \left[\varepsilon^o_\theta + \kappa_\theta \right] \tag{7.51}$$

$$\begin{aligned} N_\theta &= A_\theta \varepsilon^o_\theta + B_\theta \kappa_\theta - N^{\mathrm{T}}_\theta - N^{\mathrm{H}}_\theta \\ M_\theta &= B_\theta \varepsilon^o_\theta + D_\theta \kappa_\theta - M^{\mathrm{T}}_\theta - M^{\mathrm{H}}_\theta \\ Q_\theta &= KA_{r\theta} \gamma^o_{r\theta} \end{aligned} \tag{7.52}$$

$$\begin{aligned} A_\theta &= R \sum_{k=1}^{N} (bE_\theta)_k \ln \frac{R + h_k}{R + h_{k-1}}, \quad B_\theta = R \sum_{k=1}^{N} (bE_\theta)_k \left[t_k - R \ln \frac{R + h_k}{R + h_{k-1}} \right], \\ D_\theta &= R \sum_{k=1}^{N} (bE_\theta)_k \left[\bar{z}_k t_k - R t_k + R^2 \ln \frac{R + h_k}{R + h_{k-1}} \right] \\ A_{r\theta} &= R \sum_{k=1}^{N} (bG_{r\theta})_k \left\{ \ln \frac{R + h_k}{R + h_{k-1}} - \frac{4}{h^2} \left[\bar{z}_k t_k - R t_k + R^2 \ln \frac{R + h_k}{R + h_{k-1}} \right] \right\} \end{aligned} \tag{7.53}$$

$$
\begin{aligned}
N_\theta^{\mathrm{T}} &= \sum_{k-1}^{N} (bE_\theta \alpha_\theta t)_k \Delta T \quad N_\theta^H = \sum_{k-1}^{N} (bE_\theta \beta_\theta t)_k \bar{M} \\
M_\theta^{\mathrm{T}} &= \sum_{k-1}^{N} (bE_\theta \alpha_\theta t \bar{z})_k \Delta T \quad M_\theta^H = \sum_{k-1}^{N} (bE_\theta \beta_\theta t \bar{z})_k \bar{M}
\end{aligned}
\tag{7.54}
$$

$$
\begin{aligned}
& N_{\theta,\theta} + Q_\theta + RP_\theta = \bar{\rho}_1 \ddot{U}_o + \bar{\rho}_2 \ddot{\beta} \\
& Q_{\theta,\theta} - N_\theta + RP_r = \bar{\rho}_1 \ddot{W}_o \\
& M_{\theta,\theta} - RQ_\theta + Rm' = \bar{\rho}_1 \ddot{U}_o + \bar{\rho}_3 \ddot{\beta}
\end{aligned}
\tag{7.55}
$$

$$
\begin{aligned}
N_\theta &= \frac{A_\theta}{R}\left(U_{o,\theta} + W_o\right) + \frac{B_\theta}{R}\beta_{,\theta} \\
M_\theta &= \frac{B_\theta}{R}\left(U_{o,\theta} + W_o\right) + \frac{D_\theta}{R}\beta_{,\theta} \\
Q_\theta &= \frac{KA_{r\theta}}{R}\left(W_{o,\theta} - U_o + R\beta\right)
\end{aligned}
\tag{7.56}
$$

$$
\begin{aligned}
& \frac{A_\theta}{R}\left(U_{o,\theta\theta} + W_{o,\theta}\right) + \frac{B_\theta}{R}\beta_{,\theta\theta} + \frac{KA_{r\theta}}{R}\left(W_{o,\theta} - U_o + R\beta\right) + RP_\theta = 0 \\
& \frac{B_\theta}{R}\left(U_{o,\theta\theta} + W_{o,\theta}\right) + \frac{D_\theta}{R}\beta_{,\theta\theta} + KA_{r\theta}\left(W_{o,\theta} - U_o + R\beta\right) + Rm' = 0 \\
& \frac{KA_{r\theta}}{R}\left(W_{o,\theta\theta} - U_{o,\theta} + R\beta_{,\theta}\right) - \frac{A_\theta}{R}\left(U_{o,\theta} + W_o\right) - \frac{B_\theta}{R}\beta_{,\theta} + RP_r = 0
\end{aligned}
\tag{7.57}
$$

$$
R_N = \frac{\sum_{k=1}^{N} (bE_\theta t)_k}{\sum_{k=1}^{N} (bE_\theta)_k \ln \dfrac{R_c + h_k}{R_c + h_{k-1}}}
\tag{7.58}
$$

$$
\begin{Bmatrix} N_\theta \\ M_\theta \end{Bmatrix} = \begin{bmatrix} A_\theta & B_\theta \\ B_\theta & D_\theta \end{bmatrix} \begin{Bmatrix} \varepsilon_\theta^0 \\ \kappa_\theta \end{Bmatrix} - \begin{Bmatrix} N_\theta^{\mathrm{T}} \\ M_\theta^{\mathrm{T}} \end{Bmatrix} - \begin{Bmatrix} N_\theta^{\mathrm{H}} \\ M_\theta^{\mathrm{H}} \end{Bmatrix}, \quad Q_\theta = KA_{r\theta}\gamma_{r\theta}^o
\tag{7.59}
$$

$$
\begin{Bmatrix} \varepsilon_\theta^o \\ \kappa_\theta \end{Bmatrix} = \frac{1}{A_\theta D_\theta - B_\theta^2} \begin{bmatrix} D_\theta & -B_\theta \\ -B_\theta & A_\theta \end{bmatrix} \begin{Bmatrix} \hat{N}_\theta \\ \hat{M}_\theta \end{Bmatrix}, \quad \gamma_{r\theta}^o = \frac{Q_\theta}{KA_{r\theta}}
\tag{7.60}
$$

$$
(\sigma_\theta)_k = (E_\theta)_k \left[\varepsilon_\theta^0 + z\kappa_\theta - (\alpha_\theta)_k \Delta T - (\beta_\theta)_k \bar{M}\right], \quad (\tau_{r\theta})_k = (G_{r\theta})_k \gamma_{r\theta}^o
\tag{7.61}
$$

Beam vibrations—without transverse shear:

$$
\begin{aligned}
& \left(B_x^2 - A_x D_x\right) W_{o,xxxxxx} - A_x \bar{\rho}_1 \ddot{W}_{o,xx} - \left(2B_x \bar{\rho}_2 - A_x \bar{\rho}_3 - D_x \bar{\rho}_1\right) \ddot{W}_{o,xxxx} + \bar{\rho}_1^2 \ddddot{W}_o \\
& - \left(\bar{\rho}_2^2 - \bar{\rho}_1 \bar{\rho}_3\right) \ddddot{W}_{o,xx} - B_x p_{x,xxx} + A_x q_{,xx} + A_x m_{x,xx} + \bar{\rho}_1 \ddot{q} - \bar{\rho}_2 \ddot{p}_{x,x} + \bar{\rho}_1 \ddot{m}_{x,x} = 0
\end{aligned}
\tag{7.63}
$$

$$\left(B_x^2 - A_xD_x\right)U_{o,xxxxxx} + \left(D_x\overline{\rho}_1 + A_x\overline{\rho}_3 - 2B_x\overline{\rho}_2\right)\ddot{U}_{o,xxxx} - \left(\overline{\rho}_2^2 - \overline{\rho}_1\overline{\rho}_3\right)\dddot{U}_{o,xx} - \overline{\rho}_1A_x\ddot{U}_{o,xx} + \overline{\rho}_1^2\dddot{U}_o - D_xp_{x,xxx} + B_x\left(q_{,xxx} + m_{x,xxxx}\right) + \overline{\rho}_3\ddot{p}_{x,xx} - \overline{\rho}_1\ddot{p} - \overline{\rho}_2\left(\ddot{q}_{,x} + \ddot{m}_{x,xx}\right) = 0 \tag{7.64}$$

$$\left\{\left(B_x^2 - A_xD_x\right)f_{,xxxxxx} + \omega^2\left(D_x\overline{\rho}_1 - 2B_x\overline{\rho}_2 + A_x\overline{\rho}_3\right)f_{,xxxx} + \omega^4\left(\overline{\rho}_1\overline{\rho}_3 - \overline{\rho}_2^2\right)f_{,xx} + \omega^4\overline{\rho}_1^2f - \omega^2A_x\overline{\rho}_1f_{,xx}\right\}e^{i\omega t} = 0. \tag{7.65}$$

$$\left\{\left(B_x^2 - A_xD_x\right)g_{,xxxxxx} + \omega^2\left(D_x\overline{\rho}_1 - 2B_x\overline{\rho}_2 + A_x\overline{\rho}_3\right)g_{,xxxx} + \omega^4\left(\overline{\rho}_1\overline{\rho}_3 - \overline{\rho}_2^2\right)g_{,xx} + \omega^4\overline{\rho}_1^2g - \omega^2A_x\overline{\rho}_1g_{,xx}\right\}e^{i\omega t} = 0 \tag{7.66}$$

With transverse shear:

$$D_xW_{o,xxxx} + \left(\overline{\rho}_1\ddot{W}_o - q\right) - \overline{\rho}_3\ddot{W}_{o,xx} - \frac{D_x}{KA_{55}}\left(\overline{\rho}_1\ddot{W}_o - q\right)_{,xx} + \frac{\overline{\rho}_3}{KA_{55}}\left(\overline{\rho}_1\ddot{W}_o - q\right)_{,tt} = 0 \tag{7.67}$$

B.5 Laminated composite plate analysis

Equations of motion:

$$A_{11}U_{o,xx} + 2A_{16}U_{o,xy} + A_{66}U_{o,yy} + A_{16}V_{o,xx} + (A_{12} + A_{66})V_{o,xy} + A_{26}V_{o,yy} - B_{11}W_{o,xxx} - 3B_{16}W_{o,xxy} - (B_{12} + 2B_{66})W_{o,xyy} - B_{26}W_{o,yyy} - N^{\mathrm{T}}_{x,x} - N^{\mathrm{T}}_{xy,y} - N^{\mathrm{H}}_{x,x} - N^{\mathrm{H}}_{xy,y} = \rho\ddot{U}_o \tag{8.9}$$

$$A_{16}U_{o,xx} + (A_{12} + A_{66})U_{o,xy} + A_{26}U_{o,yy} + A_{66}V_{o,xx} + 2A_{26}V_{o,xy} + A_{22}V_{o,yy} - B_{16}W_{o,xxx} - (B_{12} + 2B_{66})W_{o,xxy} - 3B_{26}W_{o,xyy} - B_{22}W_{o,yyy} - N^{\mathrm{T}}_{xy,x} - N^{\mathrm{T}}_{y,y} - N^{\mathrm{H}}_{xy,x} - N^{\mathrm{H}}_{y,y} = \rho\ddot{V}_o \tag{8.10}$$

$$D_{11}W_{o,xxxx} + 4D_{16}W_{o,xxxy} + 2(D_{12} + 2D_{66})W_{o,xxyy} + 4D_{26}W_{o,xyyy} + D_{22}W_{o,yyyy} - B_{11}U_{o,xxx} - 3B_{16}U_{o,xxy} - (B_{12} + 2B_{66})U_{o,xyy} - B_{26}U_{o,yyy} - B_{16}V_{o,xxx} - (B_{12} + 2B_{66})V_{o,xxy} - 3B_{26}V_{o,xyy} - B_{22}V_{o,yyy} + M^{\mathrm{T}}_{x,xx} + M^{\mathrm{H}}_{x,xx} + 2M^{\mathrm{T}}_{xy,xy} + 2M^{\mathrm{H}}_{xy,xy} + M^{\mathrm{T}}_{y,yy} + M^{\mathrm{H}}_{y,yy} + \rho\ddot{W}_o = q \tag{8.11}$$

Plate bending:

$$D_{11}W_{o,xxxx} + 2(D_{12} + 2D_{66})W_{o,xxyy} + D_{22}W_{o,yyyy} = q(x, y) \tag{8.12}$$

$$q_{mn} = \frac{4}{ab}\int_0^b\int_0^a q(x,y)\sin\frac{m\pi x}{a}\sin\frac{n\pi y}{b}\mathrm{d}x\mathrm{d}y \tag{8.13}$$

$$D_{mn} = D_{11}m^4 + 2(D_{12}+2D_{66})(mnR)^2 + D_{22}(nR)^4 \tag{8.14}$$

$$A_{mn} = \frac{a^4}{\pi^4}\frac{q_{mn}}{D_{mn}} \tag{8.15}$$

$$W_o = \left(\frac{a}{\pi}\right)^4\sum_{m=1}^{\infty}\sum_{n=1}^{\infty}\frac{q_{mn}}{D_{mn}}\sin\frac{m\pi x}{a}\sin\frac{n\pi y}{b} \tag{8.16}$$

$$M_x = \left(\frac{a}{\pi}\right)^2\sum_{m=1}^{\infty}\sum_{n=1}^{\infty}\frac{q_{mn}}{D_{mn}}(m^2D_{11}+n^2R^2D_{12})\sin\frac{m\pi x}{a}\sin\frac{n\pi y}{b}$$

$$M_y = \left(\frac{a}{\pi}\right)^2\sum_{m=1}^{\infty}\sum_{n=1}^{\infty}\frac{q_{mn}}{D_{mn}}(m^2D_{12}+n^2R^2D_{22})\sin\frac{m\pi x}{a}\sin\frac{n\pi y}{b} \tag{8.17}$$

$$M_{xy} = -2RD_{66}\left(\frac{a}{\pi}\right)^2\sum_{m=1}^{\infty}\sum_{n=1}^{\infty}\frac{q_{mn}}{D_{mn}}mn\cos\frac{m\pi x}{a}\cos\frac{n\pi y}{b}$$

$$(\sigma_x)_k = \left(\frac{a}{\pi}\right)^2 z\sum_{m=1}^{\infty}\sum_{n=1}^{\infty}\frac{q_{mn}}{D_{mn}}\left(m^2(\bar{Q}_{11})_k + n^2R^2(\bar{Q}_{12})_k\right)\sin\frac{m\pi x}{a}\sin\frac{n\pi y}{b}$$

$$(\sigma_y)_k = \left(\frac{a}{\pi}\right)^2 z\sum_{m=1}^{\infty}\sum_{n=1}^{\infty}\frac{q_{mn}}{D_{mn}}\left(m^2(\bar{Q}_{12})_k + n^2R^2(\bar{Q}_{22})_k\right)\sin\frac{m\pi x}{a}\sin\frac{n\pi y}{b} \tag{8.18}$$

$$(\tau_{xy})_k = -2R(\bar{Q}_{66})_k\left(\frac{a}{\pi}\right)^2 z\sum_{m=1}^{\infty}\sum_{n=1}^{\infty}\frac{q_{mn}}{D_{mn}}mn\cos\frac{m\pi x}{a}\cos\frac{n\pi y}{b}$$

$$(\tau_{xz})_k = -\frac{a}{\pi}\sum_{m=1}^{\infty}\sum_{n=1}^{\infty}\int_{-h/2}^{z_k}\frac{mq_{mn}}{D_{mn}}\left(m^2(\bar{Q}_{11})_k + n^2R^2\left[(\bar{Q}_{12})_k + 2(\bar{Q}_{66})_k\right]\right)\cos\frac{m\pi x}{a}\sin\frac{n\pi y}{b}z\mathrm{d}z$$

$$(\tau_{yz})_k = \frac{a}{\pi}\sum_{m=1}^{\infty}\sum_{n=1}^{\infty}\int_{-h/2}^{z_k}\frac{nq_{mn}}{D_{mn}}\left(m^2\left[(\bar{Q}_{12})_k + 2(\bar{Q}_{66})_k\right] + n^2R^2(\bar{Q}_{22})_k\right)\sin\frac{m\pi x}{a}\cos\frac{n\pi y}{b}z\mathrm{d}z \tag{8.19}$$

Effects of shear deformation:

$$\begin{aligned}&A_{11}U_{o,xx} + 2A_{16}U_{o,xy} + A_{66}U_{o,yy} + A_{16}V_{o,xx} + (A_{12}+A_{66})V_{o,xy} + A_{26}V_{o,yy}\\&+B_{11}\Phi_{,xx} + 2B_{16}\Phi_{,xy} + B_{66}\Phi_{,yy} + B_{16}\Psi_{,xx} + (B_{12}+B_{66})\Psi_{,xy} + B_{26}\Psi_{,yy} = \bar{\rho}_1\ddot{U}_o + \bar{\rho}_2\ddot{\Phi}\end{aligned} \tag{8.21}$$

$$\begin{aligned}&A_{16}U_{o,xx}+(A_{12}+A_{66})U_{o,xy}+A_{26}U_{o,yy}+A_{66}V_{o,xx}+2A_{26}V_{o,xy}+A_{22}V_{o,yy}\\&+B_{16}\Phi_{,xx}+(B_{12}+B_{66})\Phi_{,xy}+B_{26}\Phi_{,xy}+B_{66}\Psi_{,xx}+2B_{26}\Psi_{,xy}+B_{22}\Psi_{,yy}=\overline{\rho}_1\ddot{V}_o+\overline{\rho}_2\ddot{\Psi}\end{aligned} \tag{8.22}$$

$$\begin{aligned}&B_{11}U_{o,xx}+2B_{16}U_{o,xy}+B_{66}U_{o,yy}+B_{16}V_{o,xx}+(B_{12}+B_{66})V_{o,xy}+B_{26}V_{o,yy}\\&+D_{11}\Phi_{,xx}+2D_{16}\Phi_{,xy}+D_{66}\Phi_{,yy}+D_{16}\Psi_{,xx}+(D_{12}+D_{66})\Psi_{,xy}+D_{26}\Psi_{,yy}\\&-K\left[A_{55}\left(\Phi+W_{o,x}\right)+A_{45}\left(\Psi+W_{o,y}\right)\right]=\overline{\rho}_2\ddot{U}_o+\overline{\rho}_3\ddot{\Phi}\end{aligned} \tag{8.23}$$

$$\begin{aligned}&B_{16}U_{o,xx}+(B_{12}+B_{66})U_{o,xy}+B_{26}U_{o,yy}+B_{66}V_{o,xx}+2B_{26}V_{o,xy}+B_{22}V_{o,yy}\\&+D_{16}\Phi_{,xx}+(D_{12}+D_{66})\Phi_{,xy}+D_{26}\Phi_{,yy}+D_{66}\Psi_{,xx}+2D_{26}\Psi_{,xy}+D_{22}\Psi_{,yy}\\&-K\left[A_{45}\left(\Phi+W_{o,x}\right)+A_{55}\left(\Psi+W_{o,y}\right)\right]=\overline{\rho}_2\ddot{V}_o+\overline{\rho}_3\ddot{\Psi}\end{aligned} \tag{8.24}$$

$$\begin{aligned}&K\left[A_{55}\left(\Phi_{,x}+W_{o,xx}\right)+A_{45}\left(\Phi_{,x}+\Psi_{,y}+2W_{o,xx}\right)+A_{44}\left(\Psi_{,y}+W_{o,yy}\right)\right]+q\\&+N_xW_{o,xx}+2N_{xy}W_{o,xy}+N_yW_{o,yy}=\overline{\rho}_1\ddot{W}_o\end{aligned} \tag{8.25}$$

Stability:

$$D_{11}\frac{\partial^4 W_o}{\partial x^4}+2(D_{12}+2D_{66})\frac{\partial^4 W_o}{\partial x^2\partial y^2}+D_{22}\frac{\partial^4 W_o}{\partial y^4}=N_x\frac{\partial^2 W_o}{\partial x^2}+N_y\frac{\partial^2 W_o}{\partial y^2} \tag{8.26}$$

$$N_o=\frac{\pi^2[D_{11}m^4+2(D_{12}+2D_{66})m^2n^2R^2+D_{22}n^4R^4]}{a^2(m^2+kn^2R^2)} \tag{8.27}$$

B.6 Laminated composite shell analysis

$$\begin{aligned}U(x,\theta,t)&=U_o(x,\theta,t)+z\beta_x(x,\theta,t)\\V(x,\theta,t)&=V_o(x,\theta,t)+z\beta_\theta(x,\theta,t)\\W(x,\theta,t)&=W_o(x,\theta,t)\end{aligned} \tag{9.1}$$

$$\begin{aligned}&\varepsilon_x=U_{o,x}+z\beta_{x,x}\quad \varepsilon_\theta=\frac{1}{1+z/R}\left[\frac{1}{R}(W_o+V_{o,\theta})+\frac{z}{R}\beta_{\theta,\theta}\right]\\&\gamma_{x\theta}=\frac{1}{1+z/R}\left(\frac{1}{R}\right)U_{o,\theta}+V_{o,x}+z\left(\frac{1}{R+z}\left(\frac{1}{R}\right)\beta_{x,\theta}+\beta_{\theta,x}\right)\\&\gamma_{xz}=\beta_x+W_{o,x}\quad \gamma_{\theta z}=\frac{1}{1+z/R}\left\{\beta_\theta+\frac{1}{R}(W_{o,\theta}-V_o)\right\}\end{aligned} \tag{9.3}$$

$$N_{x,x}+\frac{N_{\theta x,\theta}}{R}=-q_x+\overline{\rho}_1U_{o,tt}+\overline{\rho}_2\beta_{x,tt} \tag{9.13}$$

$$\frac{N_{\theta,\theta}}{R}+N_{x\theta,x}+\frac{Q_\theta}{R}=-q_\theta+\overline{\rho}_1 V_{o,tt}+\overline{\rho}_2\beta_{\theta,tt} \tag{9.14}$$

$$Q_{x,x}+\frac{Q_{\theta,\theta}}{R}-\frac{N_\theta}{R}=-p+\overline{\rho}_1 W_{o,tt} \tag{9.15}$$

$$M_{x\theta,x}+\frac{M_{\theta,\theta}}{R}-Q_\theta=-m_\theta+\overline{\rho}_2 U_{o,tt}+\overline{\rho}_3\beta_{x,tt} \tag{9.16}$$

$$\frac{M_{\theta x,\theta}}{R}+M_{x,x}-Q_x=-m_x+\overline{\rho}_2 V_{o,tt}+\overline{\rho}_3\beta_{\theta,tt} \tag{9.17}$$

Unidirectional shells:

$$N_x=K_x\left(U_{o,x}+\nu_{\theta x}\frac{W_o}{R}\right)-N_x^{\mathrm{T}}-N_x^{\mathrm{H}}\quad N_\theta=K_\theta\left(\frac{W_o}{R}+\nu_{x\theta}U_{o,x}\right)-N_\theta^{\mathrm{T}}-N_\theta^{\mathrm{H}} \tag{9.19}$$

$$M_x=-D_x W_{o,xx}-M_x^{\mathrm{T}}-M_x^{\mathrm{H}}\quad M_\theta=-\nu_{\theta x}D_x W_{o,xx}-M_\theta^{\mathrm{T}}-M_\theta^{\mathrm{H}} \tag{9.20}$$

$$Q_x=M_{x,x}=-D_x W_{o,xxx}-M_{x,x}^{\mathrm{T}}-M_{x,x}^{\mathrm{H}} \tag{9.21}$$

$$K_x\left(U_{o,xx}+\frac{\nu_{\theta x}}{R}W_{o,x}\right)=-q_x+N_{x,x}^{\mathrm{T}}+N_{x,x}^{\mathrm{H}} \tag{9.22}$$

$$-D_x W_{o,xxxx}-\frac{K_\theta}{R}\left(\nu_{x\theta}U_{o,x}+\frac{W_o}{R}\right)=-p-\frac{N_\theta^{\mathrm{T}}}{R}-\frac{N_\theta^{\mathrm{H}}}{R}+M_{x,xx}^{\mathrm{T}}+M_{x,xx}^{\mathrm{H}} \tag{9.23}$$

$$W_{o,xxxx}+4K^4W_o=\frac{1}{D_x}\left\{p+\frac{N_\theta^{\mathrm{T}}}{R}-M_{x,xx}^{\mathrm{T}}-M_{x,xx}^{\mathrm{H}}-\frac{\nu_{x\theta}K_\theta}{RK_x}\left[\int_0^x q_x dx-N_x^{\mathrm{T}}-N_x^{\mathrm{H}}-N_x(0)\right]\right\} \tag{9.24}$$

$$K^4=\frac{3E_\theta(1-\nu_{x\theta}\nu_{\theta x})}{R^2E_xh^2} \tag{9.25}$$

$$\begin{aligned}F_1(Kx)&=\mathrm{e}^{-Kx}(\cos Kx+\sin Kx)\\F_2(Kx)&=\mathrm{e}^{-Kx}(\cos Kx-\sin Kx)\\F_3(Kx)&=\mathrm{e}^{-Kx}\cos Kx=(F_1+F_2)/2\\F_4(Kx)&=\mathrm{e}^{-Kx}\sin Kx=(F_1-F_2)/2\end{aligned} \tag{9.27}$$

$$\begin{Bmatrix}W_o\\W_{o,x}\\W_{o,xx}\\W_{o,xxx}\end{Bmatrix}=\frac{1}{D_x}\begin{bmatrix}0&-M_o/2K^2&-Q_o/2K^3&0\\Q_o/2K^2&0&M_o/K&0\\-M_o&0&0&-Q_o/K\\0&-Q_o&0&2KM_o\end{bmatrix}\begin{Bmatrix}F_1(Kx)\\F_2(Kx)\\F_3(Kx)\\F_4(Kx)\end{Bmatrix} \tag{9.28}$$

$$\begin{bmatrix} 0 & -1 & 0 & 1 \\ K & K & -K & K \\ F_6(KL) & -F_5(KL) & -F_8(KL) & F_7(KL) \\ K\begin{bmatrix} F_5(KL) \\ -F_6(KL) \end{bmatrix} & K\begin{bmatrix} F_5(KL) \\ +F_6(KL) \end{bmatrix} & -K\begin{bmatrix} F_7(KL) \\ +F_8(KL) \end{bmatrix} & K\begin{bmatrix} F_7(KL) \\ -F_8(KL) \end{bmatrix} \end{bmatrix} \begin{Bmatrix} C_1 \\ C_2 \\ C_3 \\ C_4 \end{Bmatrix} = -\frac{1}{2K^2 D_x} \begin{Bmatrix} M_{\mathrm{A}} \\ Q_{\mathrm{A}} \\ M_{\mathrm{B}} \\ Q_{\mathrm{B}} \end{Bmatrix} \tag{9.29}$$

$$\begin{Bmatrix} W_o \\ W_{o,x} \\ W_{o,xx} \\ W_{o,xxx} \end{Bmatrix} = \begin{bmatrix} C_1 & C_2 & C_3 & C_4 \\ K(C_2-C_1) & -K(C_1+C_2) & K(C_3+C_4) & K(C_4-C_3) \\ -2K^2C_2 & 2K^2C_1 & 2K^2C_4 & -2K^2C_3 \\ 2K^3(C_1+C_2) & 2K^3(C_2-C_1) & 2K^3(C_4-C_3) & 2K^3(C_3+C_4) \end{bmatrix} \begin{Bmatrix} F_5(KL) \\ F_6(KL) \\ F_7(KL) \\ F_8(KL) \end{Bmatrix} \tag{9.30}$$

$$\begin{aligned} & N_{x,x} + \frac{1}{R} N_{x\theta,\theta} + q_x = 0 \\ & N_{x\theta,x} + \frac{1}{R} N_{\theta,\theta} + \frac{1}{R} M_{x\theta,x} + \frac{1}{R^2} M_{\theta,\theta} + q_\theta = 0 \\ & M_{x,xx} + \frac{1}{R^2} M_{\theta,\theta\theta} + \frac{z}{R} M_{x\theta,x\theta} - \frac{1}{R} N_\theta + p = 0 \end{aligned} \tag{9.31}$$

$$\begin{aligned} N_x &= \frac{E_x h}{1-\nu_{x\theta}\nu_{\theta x}} \left[U_{o,x} + \frac{\nu_{\theta x}}{R}\left(V_{o,\theta} + W_o\right) \right] - N_x^{\mathrm{T}} - N_x^{\mathrm{H}} \\ N_\theta &= \frac{E_\theta h}{1-\nu_{x\theta}\nu_{\theta x}} \left[\nu_{x\theta} U_{o,x} + \frac{1}{R}\left(V_{o,\theta} + W_o\right) \right] - N_\theta^{\mathrm{T}} - N_\theta^{\mathrm{H}} \\ N_{x\theta} &= G_{x\theta} h \left[V_{o,x} + \frac{1}{R} U_{o,\theta} \right] \\ M_x &= -\frac{E_x h^3}{12(1-\nu_{x\theta}\nu_{\theta x})} \left[W_{o,xx} + \frac{\nu_{\theta x}}{R^2}\left(W_{o,\theta\theta} - V_{o,\theta}\right) \right] - M_x^{\mathrm{T}} - M_x^{\mathrm{H}} \\ M_\theta &= -\frac{E_\theta h^3}{12(1-\nu_{x\theta}\nu_{\theta x})} \left[\nu_{x\theta} W_{o,xx} + \frac{1}{R^2}\left(W_{o,\theta\theta} - V_{o,\theta}\right) \right] - M_x^{\mathrm{T}} - M_x^{\mathrm{H}} \\ M_{x\theta} &= \frac{G_{x\theta} h^3}{12} \left[\frac{V_{o,x}}{R} - \frac{W_{o,x\theta}}{R} \right] \end{aligned} \tag{9.32}$$

$$\begin{aligned} & U_{o,xx} + \frac{G_{x\theta}(1-\nu_{x\theta}\nu_{\theta x})}{E_x R^2} U_{o,\theta\theta} + \frac{\nu_{x\theta} E_\theta + G_{x\theta}(1-\nu_{x\theta}\nu_{\theta x})}{E_x R} V_{o,x\theta} + \frac{\nu_{x\theta} E_\theta}{E_x R} W_{o,x} \\ & \quad = \frac{1-\nu_{x\theta}\nu_{\theta x}}{E_x h}\left(-q_x + N_{x,x}^{\mathrm{T}} + N_{x,x}^{\mathrm{H}} \right) + \frac{\rho(1-\nu_{x\theta}\nu_{\theta x})}{E_x} U_{o,tt} \end{aligned} \tag{9.33}$$

$$\frac{G_{x\theta}(1-\nu_{x\theta}\nu_{\theta x})}{E_\theta}V_{o,xx}+\frac{1}{R^2}V_{o,\theta\theta}+\frac{\nu_{x\theta}E_\theta+G_{x\theta}(1-\nu_{x\theta}\nu_{\theta x})}{E_\theta R}U_{o,x\theta}+\frac{1}{R^2}W_{o,\theta}$$
$$+\frac{h^2}{12R^2}\left\{\frac{G_{x\theta}(1-\nu_{x\theta}\nu_{\theta x})}{E_\theta}V_{o,xx}+\frac{1}{R^2}V_{o,\theta\theta}-\frac{\nu_{x\theta}E_\theta+G_{x\theta}(1-\nu_{x\theta}\nu_{\theta x})}{E_\theta}W_{o,xx\theta}-\frac{1}{R^2}W_{o,\theta\theta\theta}\right\}$$
$$=\left(\frac{1-\nu_{x\theta}\nu_{\theta x}}{E_\theta h}\right)\left(-q_\theta+\frac{N^{\mathrm{T}}_{\theta,\theta}+N^{\mathrm{H}}_{\theta,\theta}}{R}+\frac{M^{\mathrm{T}}_{\theta,\theta}+M^{\mathrm{H}}_{\theta,\theta}}{R^2}\right)+\frac{\rho(1-\nu_{x\theta}\nu_{\theta x})}{E_\theta}V_{o,tt} \tag{9.34}$$

$$W_{o,xxxx}+2\left[\frac{\nu_{\theta x}E_x+G_{x\theta}(1-\nu_{x\theta}\nu_{\theta x})}{E_xR^2}\right]W_{o,xx\theta\theta}+\frac{E_\theta}{R^4E_x}W_{o,\theta\theta\theta\theta}-\frac{\nu_{\theta x}E_x+2G_{x\theta}(1-\nu_{x\theta}\nu_{\theta x})}{E_xR^2}V_{o,xx\theta}$$
$$-\frac{E_x}{R^4E_\theta}V_{o,\theta\theta\theta}+\frac{12E_\theta}{E_xh^2}\left\{\frac{\nu_{x\theta}}{R}U_{o,x}+\frac{1}{R^2}\left(V_{o,\theta}+W_o\right)\right\}$$
$$=\frac{12(1-\nu_{x\theta}\nu_{\theta x})}{E_xh^3}\left(p+\frac{N^{\mathrm{T}}_\theta+N^{\mathrm{H}}_\theta}{R}-\frac{M^{\mathrm{T}}_{\theta,\theta\theta}+M^{\mathrm{H}}_{\theta,\theta\theta}}{R^2}-\left(M^{\mathrm{T}}_{x,xx}+M^{\mathrm{H}}_{x,xx}\right)\right)+\frac{12\rho(1-\nu_{x\theta}\nu_{\theta x})}{E_xh^2}W_{o,tt} \tag{9.35}$$

Anisotropic shells:

$$A_{11}U_{o,xx}+2\frac{A_{16}}{R}U_{o,\theta x}+\frac{A_{66}}{R^2}U_{o,\theta\theta}+A_{16}V_{o,xx}+\frac{(A_{12}+A_{66})}{R}V_{o,\theta x}+\frac{A_{26}}{R^2}V_{o,\theta\theta}$$
$$+B_{11}\beta_{x,xx}+2\frac{B_{16}}{R}\beta_{x,\theta x}+\frac{B_{66}}{R^2}\beta_{x,\theta\theta}+B_{16}\beta_{\theta,xx}+\frac{(B_{12}+B_{66})}{R}\beta_{\theta,\theta x}+\frac{B_{26}}{R^2}\beta_{\theta,\theta\theta}$$
$$+\frac{A_{12}}{R}W_{o,x}+\frac{A_{26}}{R^2}W_{o,\theta}+q_x=\overline{\rho}_1\ddot{U}_o+\overline{\rho}_2\ddot{\beta}_x$$

$$A_{16}U_{o,xx}+\frac{(A_{12}+A_{66})}{R}U_{o,\theta x}+\frac{A_{26}}{R^2}U_{o,\theta\theta}+A_{66}V_{o,xx}+2\frac{A_{26}}{R}V_{o,\theta x}+\frac{A_{22}}{R^2}V_{o,\theta\theta}$$
$$+B_{16}\beta_{x,xx}+\frac{(B_{12}+B_{66})}{R}\beta_{x,\theta x}+\frac{B_{26}}{R^2}\beta_{x,\theta\theta}+\frac{A_{45}}{R}\beta_x+B_{66}\beta_{\theta,xx}+2\frac{B_{26}}{R}\beta_{\theta,\theta x}$$
$$+\frac{B_{22}}{R^2}\beta_{\theta,\theta\theta}+\frac{A_{44}}{R}\beta_\theta+\frac{(A_{12}+A_{55})}{R}W_{o,x}+\frac{(A_{26}+A_{45})}{R^2}W_{o,\theta}+q_\theta=\overline{\rho}_1\ddot{V}_o+\overline{\rho}_2\ddot{\beta}_\theta$$

$$B_{11}U_{o,xx}+2\frac{B_{16}}{R}U_{o,\theta x}+\frac{B_{66}}{R^2}U_{o,\theta\theta}+B_{16}V_{o,xx}+\frac{(B_{12}+B_{66})}{R}V_{o,\theta x}+\frac{B_{26}}{R^2}V_{o,\theta\theta}$$
$$+D_{11}\beta_{x,xx}+2\frac{D_{16}}{R}\beta_{x,\theta x}+\frac{D_{66}}{R^2}\beta_{x,\theta\theta}-A_{55}\beta_x+D_{16}\beta_{\theta,xx}+\frac{(D_{12}+D_{66})}{R}\beta_{\theta,\theta x}$$
$$+\frac{D_{26}}{R^2}\beta_{\theta,\theta\theta}-A_{45}\beta_\theta+\left(\frac{B_{12}}{R}-A_{55}\right)W_{o,x}+\left(\frac{B_{26}}{R}-A_{45}\right)W_{o,\theta}+m_x=\overline{\rho}_2\ddot{U}_o+\overline{\rho}_3\ddot{\beta}_x$$

$$B_{16}U_{o,xx}+\frac{(B_{12}+B_{66})}{R}U_{o,\theta x}+\frac{B_{26}}{R^2}U_{o,\theta\theta}+B_{66}V_{o,xx}+2\frac{B_{26}}{R}V_{o,\theta x}$$
$$+\frac{B_{22}}{R^2}V_{o,\theta\theta}+\frac{A_{44}}{R}V_o+D_{16}\beta_{x,xx}+\frac{(D_{12}+D_{66})}{R}\beta_{x,\theta x}+\frac{D_{26}}{R^2}\beta_{x,\theta\theta}-A_{45}\beta_x$$
$$+D_{66}\beta_{\theta,xx}+2\frac{D_{26}}{R}\beta_{\theta,\theta x}+\frac{D_{22}}{R^2}\beta_{\theta,\theta\theta}-A_{44}\beta_\theta+\left(\frac{B_{22}}{R}-A_{45}\right)W_{o,x}$$
$$+\left(\frac{B_{22}}{R}-A_{44}\right)W_{o,\theta}+m_\theta=\overline{\rho}_2\ddot{V}_o+\overline{\rho}_3\ddot{\beta}_\theta$$

$$-\frac{A_{12}}{R}U_{o,x}-\frac{A_{26}}{R^2}U_{o,\theta}-\left(\frac{A_{26}+A_{45}}{R}\right)V_{o,x}+\left(\frac{A_{22}+A_{44}}{R^2}\right)V_{o,\theta}+\left(A_{55}-\frac{B_{12}}{R}\right)\beta_{x,x}$$
$$+\left(A_{45}-\frac{B_{26}}{R}\right)\left(\frac{1}{R}\right)\beta_{x,\theta}+\left(A_{45}-\frac{B_{26}}{R}\right)\beta_{\theta,x}+\left(A_{44}-\frac{B_{22}}{R}\right)\left(\frac{1}{R}\right)\beta_{\theta,\theta}$$
$$+A_{55}W_{o,xx}+2\frac{A_{45}}{R}W_{o,\theta x}-\frac{A_{22}}{R^2}W_{o,\theta\theta}+p=\overline{\rho}_1\ddot{W}_o$$

Glossary

The following terms are intended to aid those unfamiliar with the area of composite materials. This list does not exhaust the possibilities, but does contain many that a novice in the area of composites may find useful.

A

Advanced composites Generally considered to be composite materials with structural properties superior to those of aluminum. Composite material systems such as boron/epoxy and carbon/epoxy are included in this category.

Angle-ply laminate A laminate formed as result of orienting individual lamina at $+\theta$ and $-\theta$ with respect to a selected reference axis. The total number of lamina (plies) does not matter, and an angle-ply laminate is often referred to as a bidirectional laminate.

Anisotropy A material response in which the material properties vary with the orientation or direction of a set of reference axes.

Autoclave A special type of pressure vessel that can maintain specified temperatures and pressures for designated periods of time. Autoclaves are often used to cure organic matrix composites.

B

BFRA An acronym for boron-fiber-reinforced aluminum.

BFRP An acronym for boron-fiber-reinforced plastic.

Balanced laminate A laminate in which the total number of lamina (plies) oriented at an arbitrary angle of $+\theta$ are balanced by an equal number of lamina oriented at $-\theta$.

Balanced symmetric laminate A balanced laminate that is also symmetric.

Bending-extension Coupling The coupling between bending and extension that results from the existence of the $[B]$ matrix for a laminate.

Bending-twisting coupling The coupling between extension and shear terms D_{16} and D_{26} in the $[D]$ matrix. For a case of pure flexure, it is analogous to the shear-extension coupling that is present in off-axis unidirectional lamina.

Bidirectional laminate An angle-ply laminate in which the fibers are oriented in two distinct directions only.

Bleeder cloth A nonstructural cloth (usually made from fiberglass). It is typically placed around a composite component during curing to absorb excess resin, and is removed after curing.

Boron filament A manufactured filament that consists of B_4C vapor deposited onto a tungsten core.

Breather A porous material generally placed within a vacuum bag to aid in the removal of air, moisture, and volatiles during cure.

Bundle strength The strength resulting from a mechanical test of parallel filaments, with or without an organic matrix. The results of this test are generally used to replace those from tests of a single fiber.

C

Carbon fiber The general name of a wide range of fibers, all of which are made from carbon.

CCRP An acronym for carbon (or graphite)-cloth-reinforced plastic.

CFRP An acronym for carbon (or graphite)-fiber-reinforced plastic.

CLT An acronym for classical lamination theory.

Compliance A measurement of the softness of a material, as opposed to its stiffness. It is the inverse of the stiffness matrix.

Constituent material An individual material used to produce a composite material. Both the fiber and matrix are constituent materials.

Coupling The interaction of different individual effects into a combined effect. For a composite lamina this refers to the appearance of shear under the application of normal loads, and for a laminate it refers to the existence of curvature with application of normal loads.

Crack density The number of distinctive cracks (generally appearing in the matrix) per unit volume of composite.

Crazing The formation of matrix cracks, which may be confined to the matrix or located at the interface between matrix and fiber.

Cross-ply laminate A special case of an angle-ply laminate in which the individual lamina are oriented at either 0° or 90° to a reference axis. This laminate is bidirectional and can have an arbitrary number of lamina.

Cure The term typically reserved for the changing of properties of a thermosetting resin in order to process a composite. The chemical changes within the resin are irreversible.

Curvature A geometric measure of the bending and/or twisting of a plate, beam, or rod.

D

Dam An absorbent ridge surrounding a laminate during the cure process. The dam prevents resin from running out during the process.

Debond An area of separation within or between the individual plies of a laminate that generally results from contamination during the cure process, improper adhesion during cure, or interlaminar stresses.

Degradation The loss of material property characteristics (strength, stiffness, etc.) typically resulting from aging, corrosion, or fatigue.

Delamination The debonding of individual lamina, which primarily results from interlaminar stresses. Delamination can be controlled by proper design considerations.

E

Epoxy A thermosetting resin made from a polymerized epoxide. Epoxy is commonly used as a matrix.

Expansion coefficient A material-dependent measurement of the expansion (swelling) of a composite material due to temperature changes of moisture absorption.

F

Fiber A single filament, either rolled or formed in one direction, and used as the primary reinforcement for woven or nonwoven composite material systems.

Fiber volume fraction The percentage of fiber contained in a representative volume of a composite material system.

Fick's law A diffusion relationship used to describe moisture migration in a material.

Filament A continuous fiber with high stiffness and strength, used as the primary constituent in continuous fiber lamina.
Filament winding A manufacturing technique by which filaments (and resin) are placed on a mandrel in a specific manner. Its primary use is in constructing pressure vessels, pipes, or other axisymmetric structures.
First-ply failure load The load that causes the initial failure of a ply within a laminate.
Free expansion Thermal or hygral expansion without external stresses.

G

GFRP An acronym for glass-fiber-reinforced plastic.

H

Hybrid A composite material system composed of more than two constituents, such as a glass/graphite/epoxy composite. Intralaminar hybrids have individual plies made from two or more distinct fibers and matrix. Interlaminar hybrids have individual plies made from different fiber/matrix combinations.
Hygrothermal effect The change in properties resulting from moisture absorption and temperature changes.

I

Interface A boundary (or transition) region between constituents (fiber and matrix) or between individual lamina within a laminate.
Interlaminar stresses Stress components associated with the thickness direction of a plate.
Invariant Constant, regardless of the orientation of the coordinate system.

K

KFPR An acronym for Kevlar-fiber-reinforced plastic.
Kirchhoff–Love assumptions The basic assumptions from which classical lamination theory is established.

L

Lamina A single layer (ply) of unidirectional (or woven) composite material.
Laminate A collection of unidirectional lamina, stacked and arranged in a specific manner.
Laminated plate theory Sometimes referred to as classical lamination theory (CLT), it is the most commonly used method for initial analysis and design of composite laminates.

M

Macromechanics Term commonly used to describe the structural behavior of composites on the macroscopic level.
Mandrel A male mold generally used for filament winding.
Matrix The material that binds, separates, protects, and redistributes loads into the fibers of a composite.
Micromechanics Term commonly given to the analysis of a composite material's response based on a model of the constituent materials and their interaction with applied loads.
Midplane The geometric middle surface of a laminate, used as a reference position for determining laminate response characteristics. It is generally defined by $z=0$.
Moisture absorption The increase in moisture content resulting in swelling of the material.

Mold A cavity in which a composite part is placed so that it can be formed into the shape of the cavity.

Mold release agent A lubricant applied to the mold surface so that the part can be easily removed after curing.

Multidirectional laminate A laminate having multiple ply orientations through its thickness.

N

Neutral plane A plane that experiences no stretching, and subsequently no stress.

O

Off-axis Not coincident with the principal material directions.

On-axis Coincident with the principal material directions.

Orthotropic A material having three mutually perpendicular planes of symmetry. In an on-axis configuration, no extension-shear coupling exists.

P

Peel ply A fabric applied to a laminate prior to curing. It protects the laminate from dirt, etc., and is peeled off before curing.

Phenolic A thermosetting resin generally used for elevated temperatures.

Ply drop The reduction of the number of plies in a specific area of a laminate, thus decreasing its thickness.

Postcure An additional exposure to elevated temperatures after the initial cure process.

Preform A layup made on a mandrel or mockup that is eventually transferred to a curing tool or mold.

Prepreg A woven or unidirectional ply or roving impregnated with resin, and ready for layup or winding. Prepreg is a shortened term of the word 'preimpregnated'.

Q

Quasi-isotropic laminate A laminate that has an $[A]$ matrix similar to that of an isotropic material, with $A_{11}=A_{22}$, and $A_{13}=A_{23}=0$.

R

Residual stress In a composite, it is the stress generally resulting from cooldown after curing and/or moisture content.

Resin An organic material with a high molecular weight. Typically, it is insoluble in water and has no definite melting point and no tendency to crystallize.

Resin content The percentage of resin within a composite material.

Resin-rich area An area within a composite where the resin content is higher than the average content throughout the laminate. It generally results from improper compaction during cure.

Resin-starved area An area within a composite where the resin content is lower than the average content throughout the laminate. It has a dry appearance, and filaments or fabric do not appear to have been completely wetted during cure. It is probably a more severe condition that a resin-rich area in terms of structural integrity.

Roving A loose assembly of filaments that can be impregnated for use in filament winding, braiding, and unidirectional tapes.

Rule of mixtures A linear relationship between volume fractions and constituent material properties used for predicting macromechanical material behavior.

S

Scrim A reinforcing fabric woven into an open mesh and used in the processing of tape and other materials for handling purposes.

SDT An acronym for shear deformation theory.

Shear coupling The presence of a shear strain (or stress) under loading conditions generally associated with normal deformations only.

Sheet molding compound A short-fiber-reinforced composite generally designated by the acronym SMC.

Symmetric laminate A laminate that has both material and geometric symmetry with respect to the geometric central plane (midplane) of the laminate.

T

Tack A handling property characteristic generally associated with the stickiness of prepreg tape.

Thermal loads Laminate loads associated with hygrothermal effects resulting from the difference in operating and curing temperatures of the laminate and ply orientations throughout the laminate.

Thermoplastic An organic material characterized by a high strain capacity and a noncross-linked polymer chain. A thermoplastic can be easily reformed with the application of high temperatures.

Thermosetting plastic An organic material that has cross-linked polymer chains. A thermosetting plastic cannot be reformed after it is initially cured.

Tow A bundle of loose, untwisted filaments.

U

Unsymmetric laminate A laminate that does not have material and geometric symmetry with respect to its geometric central plane (midplane).

V

Vacuum bag An outer covering for a composite material-curing assembly. The vacuum bag can be sealed and evacuated to provide a uniform compaction pressure. It is most often made from a flexible nylon, Mylar, or other elastic film.

Volume fraction The fraction of either constituent (fiber of matrix) contained within a volume of composite material.

Void content The volumetric percentage of a composite that contains voids. For most curing procedures, the void content is generally less than 1%.

Index

Note: Page numbers followed by *f* indicate figures, *t* indicate tables and *ge* indicate glossary terms.

M

N

O

P

Q

R

S

Printed in the United States
By Bookmasters